Sustainable Water Treatment and Management

Sustainable Water Treatment and Management covers broad water and environmental engineering aspects relevant to water resources management as well as the treatment of storm water and wastewater. It provides a descriptive overview of complex 'black box' systems and related design issues and comprehensively discusses the design, operation, maintenance, as well as water quality monitoring and modelling of traditional and novel wetland systems. Further, it provides an analysis of asset performance, the modelling of treatment processes and the performance of existing infrastructure in both developed and developing countries as well as the sustainability and economic issues involved. The book serves as a useful reference for all concerned with the built environment, including town planners, developers, engineering technicians, water and agricultural engineers and public health workers.

Features:

- Presents the latest research findings in wastewater treatment.
- Includes international case studies and multi-disciplinary research projects.
- Explains treatment options that are applicable to any and all climatic regions.

Sustainable Water Treatment and Management

Miklas Scholz

CRC Press is an imprint of the
Taylor & Francis Group, an **informa** business

Designed cover image: Miklas Scholz

First edition published 2025
by CRC Press
2385 NW Executive Center Drive, Suite 320, Boca Raton FL 33431

and by CRC Press
4 Park Square, Milton Park, Abingdon, Oxon, OX14 4RN

CRC Press is an imprint of Taylor & Francis Group, LLC

First edition published by Willan 2008
Sixth edition published by Routledge 2009

Library of Congress Cataloging-in-Publication Data
Names: Scholz, Miklas, author.
Title: Sustainable water treatment and management / Miklas Scholz.
Description: First edition. | Boca Raton, FL : CRC Press, 2025. |
Includes bibliographical references and index. |
Identifiers: LCCN 2024027436 | ISBN 9781032834399 (hardback) |
ISBN 9781032834436 (paperback) | ISBN 9781003509370 (ebook)
Subjects: LCSH: Water–Purification–Technological innovations. |
Sewage–Purification–Technological innovations. | Sustainable engineering.
Classification: LCC TD430 .S363 2024 | DDC 628.1/62–dc23/eng/20241009
LC record available at https://lccn.loc.gov/2024027436

ISBN: 978-1-032-83439-9 (hbk)
ISBN: 978-1-032-83443-6 (pbk)
ISBN: 978-1-003-50937-0 (ebk)

DOI: 10.1201/9781003509370

Typeset in Times
by codeMantra

I dedicate this book to my wider family and friends, who supported me during my studies and career.

I am highly grateful to my partner Ramilya Galimova; children Philippa Scholz, Jolena Scholz, Felix Hedmark and Jamie Hedmark; twin sister Ricarda Lorey and mother Gudrun Spieshöfer.

Contents

About the Author

DProf. Miklas Scholz, cand ing, BEng (equiv), PgC, MSc, PhD, DSc, CWEM, CEnv, CSci, CEng, FHEA, FIEMA, FCIWEM, FICE, Fellow of IWA, Fellow of IETI, serves as a Distinguished Professor at Johannesburg University, South Africa. He is a Senior Researcher at the South Ural State University, The Russian Federation.

Miklas is also the Head of the Department for Urban Drainage at the Construction and Service Oberursel, Germany, a Technical Specialist for Nexus by Sweden and a Hydraulic Engineer at Kunststoff-Technik Adams, Germany.

He has published 9 books and 326 journal articles in 130 different journals. Prof. Scholz has total citations of 16,447 (10,497 citations since 2019), resulting in an h-index of 62 and an i10-index of 243. He belongs to the top 2% academics regarding the i10-index in the past five years. In addition, Miklas featured in the list of the World's Top 2% Scientists, published by Stanford University.

A bibliometric analysis of all constructed-wetland–related publications and corresponding authors with a minimum number of 20 publications and 100 citations indicates that Miklas ranked fifth in the world among 70 authors (including those who have sadly passed away).

In 2019, DProf. Scholz was awarded EURO 7M for the EU H2020 REA project – Water Retention and Nutrient Recycling in Soils and Streams for Improved Agricultural Production (WATERAGRI). He received EURO 1.52M for the JPI Water 2018 project – Research-based Assessment of Integrated approaches to Nature-based SOLUTIONS (RAINSOLUTIONS).

Foreword

MY PERSONAL CONNECTION WITH MIKLAS SCHOLZ

I met Miklas for the first time in my life on a sunny Sunday just before the start of the first WETPOL 2005 conference in Ghent, Belgium. A few years later, Miklas adopted me as his mentor and senior colleague in wetland science. I showed him how important wetlands are for the sustainable treatment and management of water in Ireland where I am based. *Sustainable Water Treatment and Management* is his tenth book on water management. I am pleased to have been an inspiring mentor!

MY CAREER AS A WETLAND SCIENTIST

When I met Miklas, I was still working for the Irish Government Department of Environment, Culture and Local Government. Together with my son Caolon and daughter Aila, Miklas and I undertook applied research on blue-green treatment solutions. After my retirement, I worked as a Senior Resident Engineer for the County Waterford and then as a Senior Scientist for Vesi Environmental Limited. We worked as partners in Miklas' WATER JPI project – RainSolutions.

SUSTAINABLE WATER TREATMENT AND MANAGEMENT IS MORE IMPORTANT THAN EVER BEFORE

The book provides evidence with the support of real applied research case studies covering the cutting-edge topics across various geographical regions, water types and innovative technologies. This makes the current work worth reading. Particularly, wetland systems take a deserving central role in the suite of innovations, which reflects their increasing importance as reliable sustainable blue-green infrastructure.

EXCELLENT INTEGRATION OF MULTI-FUNCTIONAL AND SUSTAINABLE TREATMENT SYSTEMS INTO THE LANDSCAPE

This is what the reader can expect from the proposed sustainable water systems such as wetlands covered in this book. For this purpose, we developed the well-presented integrated constructed wetland system concept that started in Ireland and has become the best practice across the world (see examples in Chapters 4 and 5). It has now become an integral part of the overall sustainable flood retention basin concept addressing the needs of multi-stakeholders for complex water resources challenges (Chapter 3).

Dr Rory Harrington

Dunhill, 25 May 2024

Senior Scientist, VESI Environmental, Unit B, Dunhill EcoPark, Dunhill, County Waterford, Ireland

Preface

WHAT IS THIS BOOK ABOUT?

The book entitled *Sustainable Water Treatment and Management* has a broad focus and attracts a wide range of audiences of academics and practitioners. The book covers all-inclusive water and environmental engineering aspects relevant to water resources management as well as the treatment of storm water and wastewater, providing a descriptive overview of complex 'black box' systems and general design issues involved. Fundamental science and engineering principles are explained to address the student and professional markets. Standard and novel design recommendations account for the interests of professional engineers and environmental scientists. The latest research findings in wastewater treatment are discussed to attract academics and senior consultants who could recommend the book to the final-year and postgraduate students as well as graduate engineers.

The book deals comprehensively with the design, operation, maintenance as well as water quality monitoring and modelling of traditional and novel wetland systems. It also provides an analysis of asset performance, the modelling of treatment processes and the performance of existing infrastructure in both developed and developing countries as well as the sustainability and economic issues involved.

The explained underlying scientific principles will also be of interest to all concerned with the built environment, including town planners, developers, engineering technicians, agricultural engineers and public health workers. The book has been written for a wide readership, but sufficient hot research topics have been addressed to guarantee a long shelf life of the book. Therefore, case study topics are diverse and research projects are multi-disciplinary, holistic, experimental and modelling-oriented.

WHAT IS THE TARGET AUDIENCE?

The book is essential for undergraduate and postgraduate students, lecturers and researchers in the civil and environmental engineering, environmental science, agriculture and ecological fields of sustainable water management. It is a standard reference for the design, operation and management of wetlands by engineers and scientists working for the water industry, local authorities, non-governmental organizations and governmental bodies. Moreover, consulting engineers will be able to apply practical design recommendations and refer to a large variety of practical international case studies, including large-scale field studies.

WHAT ARE THE KEY SELLING FEATURES?

This book has a broad focus on all applied aspects of sustainability research. The sustainable treatment and management of all water and wastewater types is covered. Applied research case studies independent from countries and climatic regions illustrate the application of innovative technologies and methodologies. Both urban and rural case studies with applied and academic aspects are addressed.

The book is split into seven inter-related chapters and sub-chapters to increase its readability. Each chapter answers topical overarching questions as outlined below.

HOW SHOULD URBAN WATER BE MANAGED?

Chapter 1 is concerned with urban water management on different scales. Section 1.1 looks closer at storm water quality associated with silt traps discharging into watercourses using a case study from Scotland. Values of suspended solids for treated storm water were often too high compared to international secondary wastewater treatment standards. Pollutants including heavy metals accumulated in the silt trap. However, high outflow velocities during heavy rainfall events did not result in clearly defined sediment layers due to sediment re-suspension. Metals did not accumulate in the receiving watercourse.

Finally, Section 1.2 is concerned with vertical subsurface flow constructed wetlands treating river water. The effects of intermittent artificial aeration and the use of polyhedron hollow polypropylene balls (PHPB) as part of the wetland substrate on the nutrient removal potential were also evaluated. A significantly positive contribution of PHPB to nutrient removal was obtained. The combination of artificial aeration and PHPB resulted in the augmentation of the first-order mean removal constants by 0.29, 3.12, 1.15, 0.65 and 0.54 m/d for chemical oxygen demand, ammonia-nitrogen, total nitrogen, soluble reactive phosphorus and total phosphorus, respectively. Findings from a brief cost-benefit analysis suggest that both artificial aeration and the presence of PHPB would result in enhanced nutrient removal that is cost-efficient for future projects, particularly if electricity costs are low.

WHAT ARE SUSTAINABLE DRAINAGE SYSTEMS?

Chapter 2 introduces the concept and application of sustainable (urban) drainage systems (SUDS). Section 2.1 discusses their management. The methodologies for an SUDS Option Decision Support Key and a corresponding SUDS Option Decision Support Matrix that are adaptable to different cities and even countries have been outlined. Development and management of storm water resources are introduced in Section 2.2. A combination of infiltration trenches or swales with ponds or underground storage was the most likely SUDS option for the majority of the areas studied in the example city Glasgow. Section 2.3 outlines an ecosystem services assessment system for retrofitting sustainable drainage systems. Permeable pavements, filter strips, swales, ponds, constructed wetlands and below-ground storage tanks are generally less preferred than infiltration trenches, soakaways and infiltration basins. The introduction of ornamental fish such as Goldfish into sustainable drainage ponds to increase their public acceptance is assessed in Section 2.4.

Section 2.5 introduces geothermal heat pumps integrated with permeable pavements for runoff treatment and reuse. Despite the relatively high temperatures in the indirectly heated sub-base of the pavement, potentially pathogenic organisms were not detected. Moreover, very high mean removal rates of biochemical oxygen demand, ammonia-nitrogen and orthophosphate-phosphates were recorded. Finally, the modelling of energy balances within geothermal paving systems is introduced in Section 2.6 using the same case study given in Section 2.5. The Runge-Kutta technique is proven to be an effective and reliable predictive tool.

WHAT ARE SUSTAINABLE FLOOD RETENTION BASINS?

Chapter 3 focuses on multi-functional sustainable flood retention basin (SFRB) use and management. Sustainable multi-functional land and urban management is covered in Section 3.1 discussing large European projects that have a holistic and multi-disciplinary assessment approach towards the total environment. Many projects propose decision-making tools partly supported by numerical models. Successful projects were identified as applying the criteria such as a communicative and holistic approach involving economic, environmental and social sciences throughout the project; sufficient geographic coverage; engagement of stakeholders from a wide variety of sectors and plans for dissemination of project outcomes and active knowledge sharing.

Furthermore, the ecological effects of water retention are evaluated in Section 3.2 summarizing the ecological effects of the use of floodplains and flood retention basins to control river flow in the River Rhine valley. Early river regulation strategies including channel straightening are assessed. The subsequent disappearance of alluvial hardwood forests has been highlighted as the major disadvantage. The response of trees to more recent strategies such as ecological flooding is also assessed. Water quality and habitat improvements due to these ecological control techniques are identified as effective. The development of flood retention basin classification methodologies and floodplain management decision support systems is recommended.

Finally, Section 3.3 is concerned with explaining the use of the classification of SFRBs to control runoff in a temperate climate. The most important classification variables were Rainfall, Dam Height, Flood Water Volume, Elevation, Dam Length, Flotsam, Floodplain Elevation, Forest and Animal Passage.

HOW ARE WATER AND WASTEWATER TREATMENT UNITS BE MANAGED SUSTAINABLY?

Chapter 4 talks about water and wastewater treatment technology and provides the data basis for design optimization through modelling. Sustainable treatment technologies are nature-based and blue-green; therefore, this book chapter puts semi-natural and constructed wetlands at its centre. Section 4.1 outlines a brief review of constructed wetlands and highlights a representative case study application. Wetlands are suitable for metal and nutrient removal under certain circumstances. Metals within watercourses are a considerable challenge in many countries. As an example, the removal of arsenic(V) in wetland filters treating drinking water with different substrates and plants is, therefore, outlined in Section 4.2.

Moreover, Section 4.3 deals with the reduction of ochre in groundwater wells. Ochre is a major challenge for groundwater abstraction, water distribution systems and occasionally also watercourses in mining areas. The recycling of less polluted waters such as greywater and industrial water has become interesting as freshwater resources such as groundwater become more scarce and polluted. However, standard wastewater types do not exist. Therefore, it is a challenge to compare different wastewaters with each other. Section 4.4 addresses this challenge by outlining the chemical simulation of a standard greywater type. Finally, the removal of nitrogen within an integrated constructed wetland treating domestic wastewater is evaluated in Section 4.5. This contribution is particularly relevant for semi-rural regions and developing countries where land costs are relatively low.

HOW ARE INDUSTRIAL WASTEWATER TREATMENT PROCESSES MODELLED?

Chapter 5 focuses on industrial wastewater treatment and modelling. Removal of nutrients and hydrocarbons within constructed wetlands is covered in Section 5.1. A research study was conducted to assess the potential of vertical-flow constructed wetlands to treat nutrients and to examine the effect of benzene concentration, presence of *Phragmites australis* (common reed) and temperature control on nutrient removal. Only the combination of the variables benzene and temperature impacted significantly on biochemical oxygen demand removal. The effluent biochemical oxygen demand concentrations in temperature-controlled benzene treatment wetlands were much lower than those located in the natural environment.

Section 5.2 is concerned with meso-scale integrated constructed wetland system operations to examine key operations including hydraulic loading rates, nutrient loading rates and nutrient recycling modes. Furthermore, Section 5.3 introduces hydrodynamic modelling of constructed wetlands. Simulation results showed that increasing the aspect ratio has a direct influence on the enhancement of the hydraulic efficiency λ in all cases. However, the aspect ratio should be at least 9 to achieve an appropriate rate for λ in rectangular constructed wetlands. Modified rounded rectangular constructed wetlands improved λ by up to 23%, which allowed for the selection of a reduced aspect ratio. Simulation results showed that constructed wetlands with low aspect ratios benefited from obstructions and optimized inlet/outlet configurations in terms of improved hydraulic retention time. Finally, Section 5.4 summarizes soft computing approaches in modelling aeration processes, highlighting that ANFIS-GA had the best performance.

HOW IS SLUDGE BEING TESTED FOR DEWATERABILITY?

The assessment and further development of sludge dewatering tests is covered by Chapter 6. Section 6.1 focuses on the assessment of the capillary suction time (CST) test, which is a commonly used method to measure the filterability and the easiness of removing moisture from slurry and sludge in numerous environmental and industrial applications. Furthermore, Section 6.2 deals with mixer impacts on sludge dewaterability characteristics. The CST test apparatus was used as a rapid measure to assess sludge dewaterability. Findings indicate that the use of magnetic stirrers leads to the lowest sludge dewaterability properties tested using the CST. The magnetic stirrer produced greater vortex and turbulence compared with other types of mixers, so rapid contact between the coagulant and the water occurred.

Finally, the relatively new dewaterability estimation test is outlined in Section 6.3. The DET is almost as simple as the CST, but considerably more reliable, faster, flexible and informative in terms of the wealth of visual measurement data collected with modern image analysis software. The standard deviations associated with repeated measurements for the same sludge are lower for the DET than for the CST test. In contrast to the CST device, capillary suction in the DET test is linear and not radial, allowing for a straightforward interpretation of findings. The new DET device may replace the CST test in the sludge-producing industries in the future.

WHAT IS THE CONNECTION BETWEEN CLIMATE CHANGE AS WELL AS WATER AND FOOD AVAILABILITY?

Chapter 7 is concerned with climate change, water availability and food. Digital filtering algorithms simulating climate variability on river flow are evaluated in Section 7.1. A baseflow separation methodology combining the outcomes of the flow duration curve and the digital filtering algorithms to cope with the restrictions of the traditional procedures has been assessed. Using this methodology as well as the monitored and simulated hydro-climatologic data, the baseflow annual variations due to climate change and human-induced activities were determined. The potential for wastewater reuse in irrigation is assessed in Section 7.2 using a case study in Yemen. Wastewater treatment with waste stabilization ponds is a very efficient, low-cost and low-maintenance operation. The treated wastewater from ponds should be considered as a valuable resource for reuse by water resources managers. A comparison with international guidelines reveals that it is possible to utilize the final effluent for only limited use in irrigation.

Finally, Section 7.3 critically assesses the impact of anthropogenic river regulation on water availability in watersheds. The adverse impacts of upstream anthropogenic regulation of a transboundary river watershed on the natural flow regime of the downstream country, by focusing on the Diyala (Sīrvān) river watershed shared between Iraq and Iran, were assessed. Transboundary watershed management difficulties in a three-level system entitled the transboundary three-scalar framework, which helps to sustainably manage water resources, were assessed.

Acknowledgements

The book is predominantly based on 26 previously published papers. Special credits go to all authors associated with the articles listed below according to the seven book chapters. All support received has been acknowledged in the individual original articles.

CHAPTER 1: URBAN WATER

Scholz, M., 2004. Storm water quality associated with a silt trap (empty and full) discharging into an urban watercourse in Scotland, *International Journal of Environmental Studies*, **61**(4), 471–483.

Tang, X., Huang, S., Scholz, M. and Li, J. 2011. Nutrient removal in vertical subsurface flow constructed wetlands treating eutrophic river water, *International Journal of Environmental Analytical Chemistry*, **91** (7–8), 727–739.

CHAPTER 2: SUSTAINABLE DRAINAGE SYSTEMS

Scholz, M., 2006. Best management practice: A sustainable urban drainage system management case study, *Water International*, **31**(3), 310–319.

Scholz, M. 2014. Rapid assessment system based on ecosystem services for retrofitting of sustainable drainage systems, *Environmental Technology*, **35**(9–12), 1286–1295.

Scholz, M. and Kazemi-Yazdi, S., 2005. How goldfish could save cities from flooding, *International Journal of Environmental Studies*, **62**(4), 367–374.

Scholz, M., Morgan, R. and Picher, A., 2005. Storm water resources development and management in Glasgow: Two case studies, *International Journal of Environmental Studies*, **62**(3), 263–282.

Tota-Maharaj, K., Grabowiecki, P. and Scholz, M., 2009. Energy and temperature performance analysis of geothermal (ground source) heat pumps integrated with permeable pavement systems for urban run-off reuse, *International Journal of Sustainable Engineering*, **2**(3), 201–213.

Tota-Maharaj, K. and Scholz, M., 2010. Permeable (pervious) pavements and geothermal heat pumps: Addressing sustainable urban storm water management and renewable energy, *International Journal of Green Economics*, **3**(3–4), 447–461.

Tota-Maharaj, K., Scholz, M., Ahmed, T., French, C. and Pagaling, E., 2010. The synergy of permeable pavements and geothermal heat pumps for storm water treatment and reuse, *Environmental Technology*, **31**(14), 1517–1531.

Tota-Maharaj, K., Scholz, M. and Coupe, S., 2011. Modelling temperature and energy balances within geothermal paving systems, *Road Materials and Pavement Design*, **12**(2), 315–344.

CHAPTER 3: MULTI-FUNCTIONAL SUSTAINABLE FLOOD RETENTION BASINS

Scholz, M., 2007. Ecological effects of water retention in the River Rhine Valley: A review assisting future retention basin classification, *International Journal of Environmental Studies*, **64**(2), 171–187.

Scholz, M., 2007. Expert system outline for the classification of sustainable flood retention basins (SFRBs), *Civil Engineering and Environmental Systems*, **24**(3), 193–209.

Scholz, M., Hedmark, Å. and Hartley, W., 2012. Recent advances in sustainable multifunctional land and urban management in Europe: A review, *Journal of Environmental Planning and Management*, **55** (7), 833–854.

CHAPTER 4: WATER AND WASTEWATER TREATMENT TECHNOLOGY AND MODELLING

Abed, S.N. and Scholz, M., 2016. Chemical simulation of greywater, *Environmental Technology*, **37**(13), 1631–1646.

Dzakpasu, M., Hofmann, O., Scholz, M., Harrington, R., Jordan, S.N. and McCarthy, V., 2011. Nitrogen removal in an integrated constructed wetland treating domestic wastewater, *Journal of Environmental Science and Health, Part A: Toxic/Hazardous Substances and Environmental Engineering*, **7**(7), 742–750.

Scholz, M. and Lee, B.-H., 2005. Constructed wetlands: A review, *International Journal of Environmental Studies*, **62**(4), 421–447.

Wu, M., Li, Q., Tang, X., Huang, Z., Lin, L. and Scholz, M., 2014. Arsenic(V) removal in wetland filters treating drinking water with different substrates and plants, *International Journal of Environmental Analytical Chemistry*, **94**(6), 618–638.

CHAPTER 5: INDUSTRIAL WASTEWATER TREATMENT AND MODELLING

Harrington, C., Scholz, M., Culleton, N. and Lawlor, P. D., 2011. Meso-scale systems used for the examination of different integrated constructed wetland operations, *Journal of Environmental Science and Health, Part A: Toxic/Hazardous Substances and Environmental Engineering*, **46**(7), 783–788.

Mahdavi-Meymand, A., Scholz, M. and Zounemat-Kermani, M., 2019. Challenging soft computing approaches in modelling complex hydraulic phenomenon of aeration process, *ISH Journal of Hydraulic Engineering*, **27**, 58–69.

Tang, X., Scholz, M., Eke, P.E. and Huang, S., 2010. Nutrient removal as a function of benzene supply within vertical-flow constructed wetlands, *Environmental Technology*, **31**(6), 681–691.

Zounemat-Kermani, M., Scholz, M. and Tondar, M.-M., 2015. Hydrodynamic modelling of free water-surface constructed storm water wetlands using a finite volume technique, *Environmental Technology*, **36**(20), 2532–2547.

CHAPTER 6: SLUDGE DEWATERING TESTS

Fitria, D., Swift, G.M. and Scholz, M., 2013. Impact of different shapes and types of mixers on sludge dewaterability, *Environmental Technology*, **34**(7), 931–936.

Sawalha, O. and Scholz, M., 2008. Assessment of Capillary Suction Time (CST) test methodologies, *Environmental Technology*, **28**(12), 1377–1386.

Scholz, M., Almuktar, S., Clausner, C. and Antonacopoulos, A., 2019. Highlights of the novel Dewaterability Estimation Test (DET) device, *Environmental Technology*, **41**(20), 2594–2602.

CHAPTER 7: CLIMATE CHANGE, WATER AVAILABILITY AND FOOD

Al-Faraj, F.A.M. and Scholz, M., 2015. Impact of upstream anthropogenic river regulation on downstream water availability in transboundary river watersheds, *International Journal of Water Resources Development*, **31**(1), 28–49.

Almas, A.A.M. and Scholz, M., 2006. Potential for wastewater reuse in irrigation: Case study from Aden (Yemen), *International Journal of Environmental Studies*, **63**(2), 131–142.

Mohammed, R. and Scholz, M., 2018. Flow duration curve integration into digital filtering algorithms simulating climate variability on river baseflow, *Hydrological Sciences Journal*, **63**(10), 1558–1573.

1 Urban Water

1.1 STORM WATER QUALITY ASSOCIATED WITH SILT TRAPS DISCHARGING INTO WATERCOURSES

1.1.1 Introduction

Stormwater runoff is usually transferred within a combined sewer (rainwater and sewage) or separate stormwater pipeline. Depending on the degree of pollution, it may require further conventional treatment at the wastewater treatment plant. Alternatively, stormwater runoff may be treated by more sustainable technology including stormwater ponds or simply by disposing it via ground infiltration or drainage into a local watercourse (Butler and Davies, 2000).

Common preliminary treatment steps for stormwater, which is to be disposed into a local watercourse, are extended storage within the stormwater pipeline and transfer through a silt trap (also referred to as a sediment trap). The silt trap is often the only active preliminary treatment of surface water runoff even though stormwater is frequently contaminated with heavy metals and hydrocarbons (Alloway, 1995; Ciszewski, 1998). Heavy metals within road runoff are associated with fuel additives, car body corrosion and tire and brake wear (Butler and Davies, 2000).

The dimensioning of silt traps is predominantly based upon the settlement properties of the particles and the maximum flow-through velocity. As a rule, the outflow should be located on the same axis as the inflow, so that continuous flow ensures particle settlement over the full length of the silt trap (Schmitt et al., 1999). The structure should be small and simple to keep capital costs low.

Various international and British guidelines (e.g., British Standards, British Research Establishment and Construction Industry Research and Information Association) exist for the design of outflow structures (Abwassertechnische Vereinigung e.V., 1994 and 1996; Pontier et al., 2001; Schmitt et al., 1999). In practice, designs vary considerably. However, silt traps are usually longer than wider by a factor of up to five. Silt trap depths vary considerably but are usually about 40 cm (Abwassertechnische Vereinigung e.V., 1994 and 1996; Pontier et al., 2001; Salehi et al., 1997; Schmitt et al., 1999).

However, it is often the lack of maintenance concerning outflow structures like silt traps that leads to problems years after the structure has been commissioned. Accumulated sediment must be removed from silt traps to retain their original design performance and to avoid re-suspension of accumulated pollutants during storms, causing pollution in the receiving water bodies (Abwassertechnische Vereinigung e.V., 1994 and 1996; Pontier et al., 2001; Salehi et al., 1997).

Watercourses in urban (built-up) areas that receive surface water runoff may be subject to pollution. This is particularly the case if silt traps, for example, are subject to insufficient maintenance (e.g., low frequency of sediment removal).

DOI: 10.1201/9781003509370-1

FIGURE 1.1 Outlet structure of the silt trap at The King's Buildings (Science and Engineering Campus, The University of Edinburgh) in winter 2003. For design drawings; Figure 1.2.

Furthermore, the additional water load may contribute to local flooding (Butler and Davies, 2000; Pontier at al., 2001).

In the case study presented in Section 1.1, a semi-natural urban watercourse (Braid Burn; Figure 1.1), which receives polluted surface water runoff, will be studied. Surprisingly, the water quality of the Braid Burn is generally considered as 'good' by the Scottish Environmental Protection Agency in terms of its chemistry, biology, nutrient content and esthetical appearance. However, the definition of 'good' with respect to stream location and sampling time is subject to further clarification (Abwassertechnische Vereinigung e.V., 1994 and 1996; Bullen Consulting, 2002; Pontier et al., 2001; Salehi et al., 1997).

The Braid Burn is a small stream (discharge of usually less than 1 m^3/s) with a total catchment area of approximately 30 km^2. The upstream reaches (Howden Burn and Bonaly Burn) of the catchment area are steep and rural. It follows that the Braid Burn can be considered as a mountain stream (high flow velocity; high bed load transport) near its spring. Downstream of this rural upland area, the stream passes through suburban areas of Edinburgh, Hermitage and subsequently the golf course development near The King's Buildings. The stream then flows within a culvert beneath a shopping centre at Cameron Toll and through Edinburgh where it discharges into the Firth of Forth (North Sea) at Portobello (Bullen Consulting, 2002).

The Braid Burn has been canalized and constricted into a deep, narrow, fast flowing, straightened channel along much of its length. These stream engineering methods are often considered to be a reason for urban flooding. Another reason could be the restrictive effects of culverts and bridges through blockage, and the absence of flood defences on stream embankments. For example, the semi-natural Braid Burn was subject to severe flooding on 26 April 2000, when 112 mm of rain fell within a 48-hour period. During spring 2000, the flow of the Braid Burn was the highest ever-recorded (Bullen Consulting, 2002).

The overall aim of Section 1.1 is to analyse the treatment efficiency of a full silt trap. The objectives are to assess (a) the performance in removal of pollutants including suspended solids (SS), turbidity and heavy metals during dry and wet weather conditions; (b) the distribution of sediment according to size within the full silt trap; (c) the spatial distribution of metals within the full silt trap and the receiving urban watercourse as an indication of pollutant accumulation; (d) the water quality of the receiving watercourse; (e) the silt trap design and operation performance before and after emptying and (f) alternative design, operation, maintenance and water quality sampling management options. Section 1.1 is based on an article by Scholz (2004), which has been revisited and updated.

1.1.2 Materials and Methods

1.1.2.1 Site and Boundary Conditions

A stormwater outlet (Figures 1.1 and 1.2) is located at The King's Buildings (Science and Engineering Campus, The University of Edinburgh, Scotland). All collected surface water runoff at The King's Buildings flows through the silt trap and discharges into the Braid Burn (Figure 1.1). The system performance of a full and empty silt trap for dry and wet weather conditions in cold climate was assessed. The mean temperature (measured at 11 AM) between 15 January and 28 March 2003 was 5.8°C (standard deviation: 4.05°C).

The diameter of the partially submerged (by approximately 37 cm) stormwater pipe leading into the silt trap chamber was 54 cm. The dimensions of the rectangular silt trap chamber were as follows: length = 184 cm, width = 184 cm, chamber height = 114 cm and silt trap outlet height = 42 cm (Figure 1.2). The stormwater pipe inlet to outlet angle is 90°. Approximately 27 cm of the silt trap was filled with potentially mature sediment (up to 8-year old according to Estates and Buildings) that was removed on 12 March 2003.

The sample site 'before inlet' was located 13-m upstream and the sample site 'after inlet' was located 6-m downstream of the sample station 'at inlet' (Figure 1.1). To investigate the mixing properties of the stream, a further sample station located at the street 'Liberton Brae' was chosen 77-m downstream of the station 'at inlet'.

Sediment samples were dried at 105°C. Every sample was weighted and subsequently relieved from clay (and partially fine silt) by using wet sieving procedures. The remaining material was dried and weighted and the difference in weight was considered as a total estimation of the clay and silt fractions. With the remaining material a sediment size distribution curve (or grading curve) was established. The methodology was according to British Standards (1990).

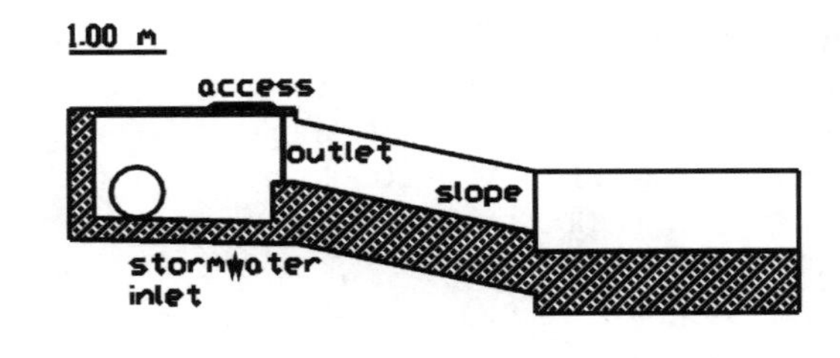

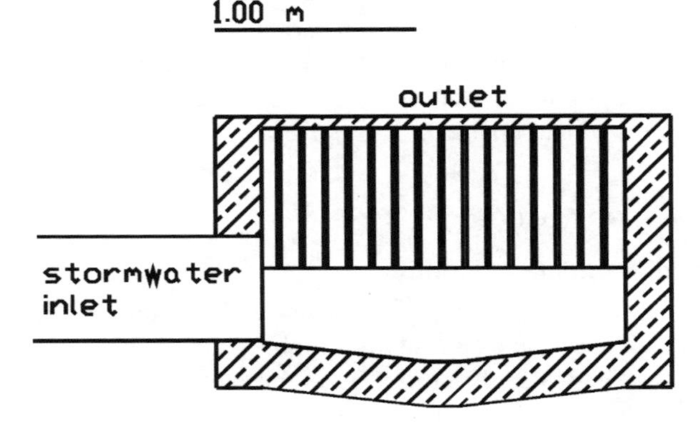

FIGURE 1.2 Silt trap design drawings (two side views). For picture; Figure 1.1.

1.1.2.2 Water Quality Analysis

Up to 2 L of composite sample water was taken at each sample station. The water quality analysis started not later than 20 minutes after sampling. The five-days @ 20°C ATU biochemical oxygen demand (BOD) was determined in all water samples with the OxiTop IS 12-6 system, a manometric measurement device, supplied by the Wissenschaftlich-Technische Werkstätten (WTW), Weilheim, Germany. The measurement principle is based on measuring pressure differences estimated by piezoresistive electronic pressure sensors. Nitrification was suppressed by using 0.05 mL of 5 g/L N-Allylthiourea (WTW Model No. NTH 600) solution per 50 mL of sample water.

A Whatman PHA 230 bench-top pH meter (for control only), a Hanna HI 9142 portable waterproof dissolved oxygen (DO) meter, a HACH 2100N turbidity meter and a Mettler Toledo MPC 227 conductivity and pH meter were used to determine pH, DO, turbidity and conductivity, respectively. Usually, one replicate measurement was taken. All other analytical procedures were performed according to the American standard methods (APHA, AWWA and WEF, 1998).

Composite samples were collected from the outflow of the silt trap, receiving watercourse (bulk water body), and occasionally from the sediment of both the silt trap and the receiving watercourse. Samples were stored frozen at −18°C until analysis.

Concerning the digestion methodology, 2 g of carefully sieved sediment samples (particles and sediment < 2 mm) were transferred to a 100 mL round bottom flask, and 21 mL of hydrochloric acid (strength of 37%) and 7 mL of nitric acid (strength of 69%) were added. The mixtures were heated on a Kjeldahl-Digestion shelve for 2 hours to support the chemical reaction. After cooling, the solutions were filled into separate flasks and filtered through Number 41 Ashless Whatmann filter papers. The filtered solutions were then filled up to 100 mL each with deionized water. This procedure was carried out for all sediment samples (plus three replicates each).

Water samples and the retrieved sediment extract (see above) were analysed for metals. An Inductively Coupled Plasma Optical Emission Spectrometer (ICP-OES) was used. Total concentrations of metals in filtered (Whatman 1.2-μm cellulose nitrate membrane filter) water samples were determined by ICP-OES using a TJA IRIS instrument (ThermoElemental, U.S.). Multi-element calibration standards with a wide range of concentrations were used and the emission intensity measured at appropriate wavelengths. For all elements, analytical precision (based on the relative standard deviation) was typically between 5% and 10% for three individual aliquots.

Oxidized aqueous nitrogen was determined in selected water samples as the sum of nitrate-N and nitrite-N (generally low concentrations). Nitrate was reduced to nitrite by cadmium and determined as an azo dye at 540 nm (using a Perstorp Analytical EnviroFlow 3,000 flow injection analyser) following diazotization with sulfanilamide and subsequent coupling with N-1-naphthylethylendiamine dihydrocloride.

Ammonia and phosphate were determined by automated colorimetry for selected water samples from reaction with hypochlorite and salicylate ions in solution in the presence of sodium nitrosopentacyanoferrate (nitroprusside), and reaction with acidic molybdate to form a phosphomolybdenum blue complex, respectively. The coloured complexes formed were measured spectrophotometrically at 655 and 882 nm, respectively, using a Bran and Luebbe autoanalyser (Model AAIII).

1.1.3 Results and Discussion

1.1.3.1 Water and Sediment Quality within the Silt Trap

Concerning the water quality, Table 1.1 summarizes the outflow water quality of the silt trap before and after emptying. The standard deviations for all variables are elevated due to the high natural variability of the data (Butler and Davies, 2000; Salehi et al., 1997; Schmitt et al., 1999). For example, the highly variable temporal distribution of SS is shown in Figure 1.3. The highlighted threshold value of 30 mg/L indicates where the water quality exceeds the internationally accepted standard for secondary treated wastewater (Tchobanoglous et al., 2003). Table 1.2 shows a comparison of the nutrient concentrations at the outlet of the silt trap.

SS and turbidity (as an indicator for the amount of predominantly small particles) are indicators for environmental water pollution (Salehi et al., 1997). For example, particulate and colloidal metal species are attached to SS. Furthermore, the discharge of SS leads to streambed sedimentation and subsequent deterioration of the ecosystem (Alloway, 1995; Ciszewski, 1998). The emptying of the silt trap had a

TABLE 1.1
Comparison of the Water Quality at the Outlet of the Silt Trap Before and After Emptying

Stream Location	Statistics	BOD[a] (mg/L)	SS[b] (mg/L)	Conductivity (μS)	pH (–)	Turbidity (NTU)	Total Solids (mg/L)
Before Emptying the Silt Trap (15/01–11/03/2003)							
Before	Count	33	33	33	33	33	24
Inlet	Mean	2.0	10.2	428	8.03	16.6	200.6
	Standard deviation	1.13	18.79	156.6	0.196	22.50	118.60
At	Count	33	33	33	33	33	25
Inlet	Mean	3.9	15.8	545	8.13	25.3	419.3
	Standard deviation	3.68	31.77	255.8	0.628	47.14	566.83
After	Count	33	33	33	33	33	24
Inlet	Mean	2.2	12.7	425	7.97	19.1	212.2
	Standard deviation	1.23	22.25	161.3	0.267	26.00	103.63
Liberton	Count	23	23	23	23	23	23
Brae[c]	Mean	2.5	5.5	476	8.11	14.2	218.4
	Standard deviation	1.66	10.14	169.5	0.111	26.90	89.44
After Emptying the Silt Trap (12/03–28/03/2003)							
Before	Count	5	5	5	5	5	5
inlet	Mean	1.4	2.7	346	8.01	4.1	170.4
	Standard deviation	0.41	2.29	55.4	0.048	1.93	49.03
At	Count	5	5	5	5	5	5
inlet	Mean	3.5	3.8	417	8.46	4.6	238.6
	Standard deviation	2.39	5.43	142.7	0.944	3.29	95.65
After	Count	5	5	5	5	5	5
inlet	Mean	1.3	3.6	353	7.98	2.76	190.4
	Standard deviation	0.44	3.02	46.3	0.037	0.89	42.53
Liberton	Count	5	5	5	5	5	4
Brae[c]	Mean	1.4	2.9	342	8.04	2.6	177.0
	Standard deviation	0.60	2.54	37.4	0.047	0.22	33.24

[a] BOD, five-days @ 20°C ATU biochemical oxygen demand.
[b] SS, suspended solids.
[c] Liberton Brae, sampling station located 77 m downstream of the station 'at inlet'.

positive effect on the water quality (Table 1.1) despite the unusually short length of the chamber. SS and turbidity values were reduced by 76% and 82%, respectively. However, the data set after emptying is small (4 or 5 values per variable) and does not include a storm event.

The recovered sediment was contaminated with zinc, copper and nickel (approximately 940, 420 and 60 mg/kg, respectively). Therefore, the sediment was not suitable for local recycling, for example, as manure on green fields located on the campus or as manure for the local agricultural industry (Alloway, 1995; Butler and Davies, 2000).

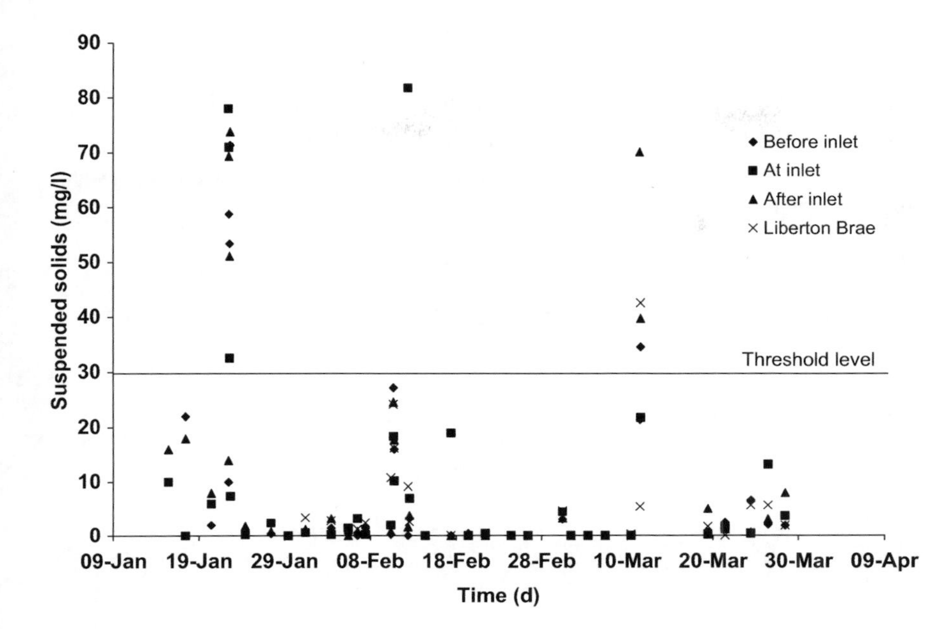

FIGURE 1.3 Distribution of suspended solids (SS) before, at and after the inlet where the silt trap discharges into the Braid Burn, and further downstream at Liberton Brae. On 11 March 2003, the SS concentration at the inlet was 141.6 mg/L. This value has been omitted to aid visual clarity.

TABLE 1.2
Comparison of the Nutrient Concentrations at the Outlet of the Silt Trap (06–26/02/03)

Stream Location	Statistics	Ammonia-N (mg/L)	Nitrate-N[a] (mg/L)	Phosphate-P (μS/cm)
Before	Count	6	6	6
Inlet	Mean	0.21	4.50	0.03
	Standard deviation	0.048	0.764	0.021
At	Count	13	13	13
Inlet	Mean	0.03	8.69	0.08
	Standard deviation	0.052	3.329	0.045
After	Count	6	6	6
Inlet	Mean	0.19	4.51	0.04
	Standard deviation	0.075	0.594	0.020
Liberton	Count	6	6	6
Brae[3]	Mean	0.18	4.59	0.03
	Standard deviation	0.036	0.816	0.007

[a] Nitrate-N, nitrate-N includes nitrite-N.

TABLE 1.3
Comparison of the Water Quality at the Outlet of the Silt Trap During Dry and Wet Weather Conditions between 15/01 and 28/03/03

Stream Location	Statistics	BOD[a] (mg/L)	SS[b] (mg/L)	Conductivi-ty (μS/cm)	pH (–)	Turbidity (NTU)	Total Solids (mg/L)
				During Dry Weather			
Before	Count	19	19	19	19	19	15
Inlet	Mean	2.0	2.0	367	8.05	6.4	193.8
	Standard deviation	1.34	5.35	86.8	0.182	8.35	119.15
At	Count	19	19	19	19	19	15
Inlet	Mean	2.8	2.2	562	8.10	4.7	300.9
	Standard deviation	1.39	4.81	124.0	0.278	2.79	182.66
After	Count	19	19	19	19	19	14
Inlet	Mean	1.9	2.5	372	8.01	6.8	171.0
	Standard deviation	1.23	5.45	102.7	0.216	9.15	58.55
Liberton	Count	13	13	13	13	13	13
Brae[c]	Mean	2.1	1.3	376	8.135	3.3	175.0
	Standard deviation	0.96	3.07	88.7	0.1211	1.31	50.32
				During Wet Weather			
Before	Count	14	14	14	14	14	9
Inlet	Mean	2.1	21.5	510	7.99	30.5	212.0
	Standard deviation	0.80	24.39	193.3	0.216	28.13	123.95
At	Count	14	14	14	14	14	10
Inlet	Mean	5.5	34.1	522	8.16	53.1	597.0
	Standard deviation	5.14	42.69	372.5	0.928	63.31	863.96
After	Count	14	14	14	14	14	10
Inlet	Mean	2.5	26.7	498	7.92	35.8	269.8
	Standard deviation	1.19	28.55	199.1	0.326	32.02	127.13
Liberton	Count	10	10	10	10	10	10
Brae[c]	Mean	3.0	10.9	605	8.08	28.2	274.8
	Standard deviation	2.22	13.44	164.4	0.096	37.11	99.62

[a] BOD, five-days @ 20°C ATU biochemical oxygen demand.
[b] SS, suspended solids.
[c] Liberton Brae, sampling station located 77 m downstream of the station 'at inlet'.

Table 1.3 indicates that the silt trap releases nitrate-N and phosphate-N into the Braid Burn. In contrast, the concentration of stormwater runoff in terms of its ammonia-N content is lower than the stream water. However, the associated data sets are small, and the variability is high. Nevertheless, official ammonia-N concentrations between 2001 and 2002 were approximately 0.28 mg/L (Ms. Fiona Logan, SEPA Southeast, personal communication) confirming the high stream concentrations recorded by the author.

Furthermore, Table 1.4 summarizes the water quality in terms of metals within the bulk water flow. Most metal concentrations (those not listed) were below the detection limit. The remaining metals (those listed) are associated with low concentrations.

TABLE 1.4
Element Concentrations within the Outflow Water of the Sedimentation trap (15/01–11/03/2003)

Stream Location	Statistics	Barium (mg/L)	Boron (mg/L)	Calcium (mg/L)	Magnesium (mg/L)	Manga-nese (mg/L)	Zinc (mg/L)
Dry and	Count	31	31	31	31	31	31
Wet	Mean	2.04	0.10	32.90	6.18	0.01	0.03
Weather	Standard deviation	1.388	0.048	11.991	2.165	0.006	0.025
Dry	Count	24	24	24	24	24	24
weather	Mean	2.08	0.09	30.20	5.81	0.01	0.03
	Standard deviation	1.209	0.039	10.825	2.073	0.006	0.019
Wet	Count	7	7	7	7	7	7
weather	Mean	1.91	0.15	42.17	7.44	0.01	0.05
	Standard deviation	2.000	0.051	11.842	2.142	0.006	0.037

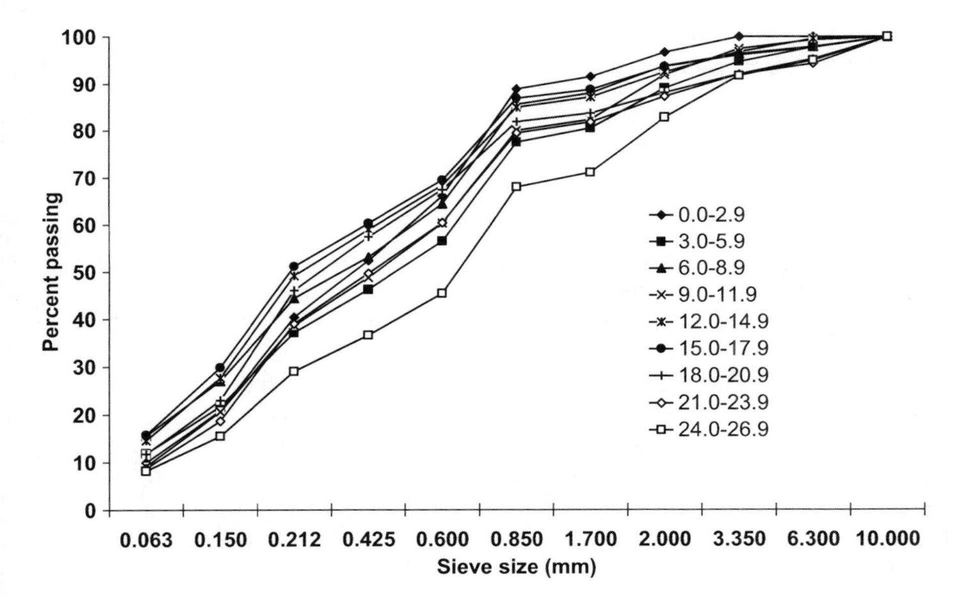

FIGURE 1.4 Sediment size distributions (or grading curves) within the silt trap on 12 March 2003.

Concerning the sediment quality, Figure 1.4 indicates the vertical sediment size distribution within the sediment of the full silt trap. The sediment is mixed in terms of its size distribution. However, the top layer of sediment (24.0–26.9 cm above the bottom) contains less fine sediment. This relative loss can be explained by re-suspension of relatively fine sediment during wet weather conditions (Pontier et al., 2001; Schmitt et al., 1999). Furthermore, re-suspension is enhanced by the unusual inflow to outflow angle of 90° (ideal angle: 180°).

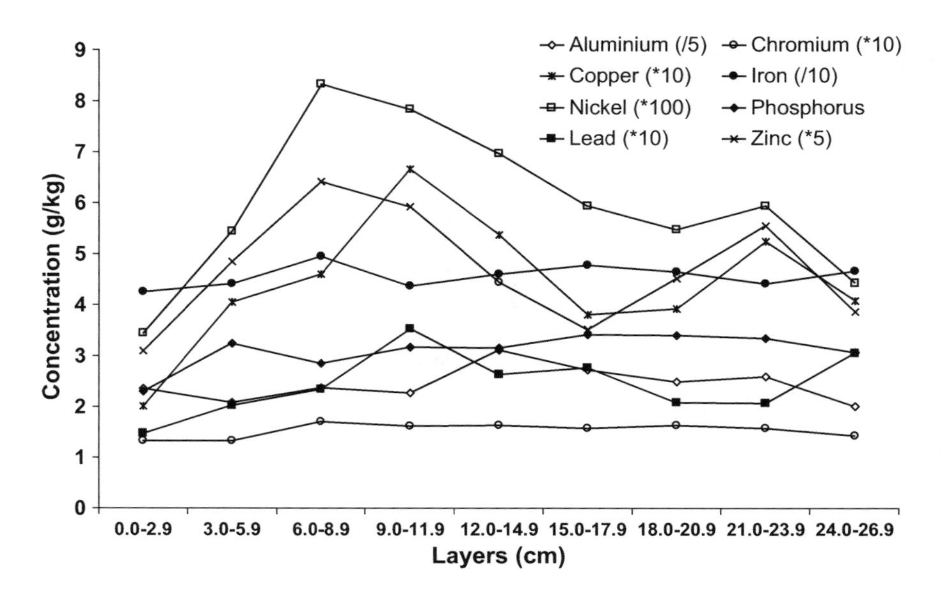

FIGURE 1.5 Metal concentrations within the different sediment layers of the silt trap on 12 March 2003.

In addition to Figure 1.3, Figure 1.5 indicates that sediment mixing during high flows leads to a random distribution of metals within the sediment. Previously accumulated material can be washed out into the Braid Burn. If the current length of the silt trap had been extended by up to a factor of five (relative to the width), more sediment would have been removed (Pontier et al., 2001; Schmitt et al., 1999). However, a more optimized design would have been expensive in terms of excavation costs due to the hilly terrain (Bullen Consulting, 2002; Butler and Davies, 2000).

The data set related to the duration before emptying the silt trap (top part of Table 1.1) was divided into two data sub-sets: Table 1.3 summarizes the water quality of the full silt trap during dry and wet weather conditions. As a result, BOD, SS, turbidity and total solids for the outflow water of the silt trap are higher during wet weather than during dry weather conditions. The outflow values for SS are frequently above the internationally accepted standard value of 30 mg/L for treated secondary wastewater (Tchobanoglous et al., 2003).

Furthermore, Table 1.4 summarizes the water quality in terms of metals within the bulk water flow. Most metal concentrations were below or close to the detection limit. There is no obvious difference between dry and wet weather conditions.

1.1.3.2 Water Quality of the Receiving Watercourse

Tables 1.1–1.3 summarize the water quality of the stormwater-receiving watercourse (Braid Burn; Figure 1.1) at one sampling station before and two stations after the inlet point. The water quality of the Braid Burn deteriorated after the discharge of virtually untreated stormwater (full silt trap). This is apparent for BOD, SS, turbidity and total solids particularly after rainfall events (Tables 1.1–1.3). However, the discharged pollutants (as indicated by SS and turbidity) mixed rapidly with the bulk water flow of the Braid Burn during wet weather conditions (Table 1.3).

Furthermore, high levels of conductivity were recorded due to road gritting (grit plus salt) after mid December 2002. Dissolved salt removal by silt traps and other water engineering structures is usually insignificant (Kjensmo, 1997). This led to additional salt loads being discharged into the Braid Burn and subsequently into the Firth of Forth (Tables 1.1 and 1.3).

Concerning metal pollution, Figure 1.6 indicates the spatial distribution of metals identified in the sediment of the Braid Burn (Figure 1.1). The metals do not accumulate along the channelled stream as previously indicated for other watercourses (Ciszewski, 1998; Pontier et al., 2001) but are likely to be washed out into the Firth of Forth during storm events. However, the last sampling station at Portobello is influenced by salty seawater. High conductivity values are often responsible for metal leaching leading to relatively low metal concentrations within the sediment of brackish water (Alloway, 1995). Therefore, the associated metal values are not directly comparable with the freshwater values recorded upstream.

Furthermore, copper concentrations within the sediment of the Braid Burn immediately downstream of the silt trap are approxmately 24 mg/kg. Figure 1.5 indicates that this relatively high concentration is likely to be linked to the surface water runoff from the Science and Engineering Campus at The King's Buildings.

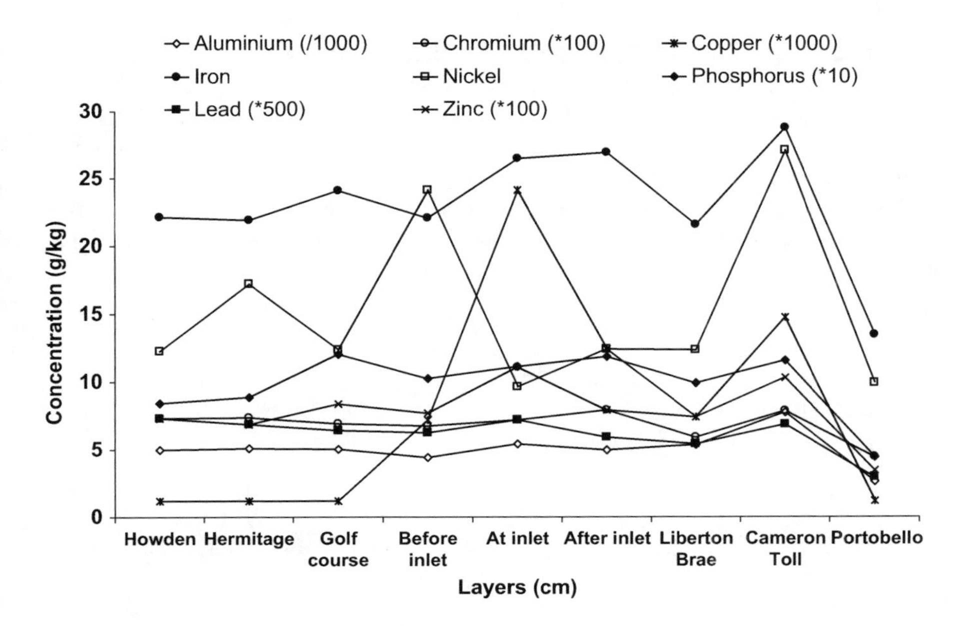

FIGURE 1.6 Metal concentrations within the sediment at different sampling locations of the Braid Burn (freshwater stream) on 13 February 2003 (after a prolonged period of rain). Howden is located near the spring and Portobello near the discharge location into the Firth of Forth (North Sea; salty water).

1.1.3.3 Future of the System

A new flood prevention scheme including a strategy to transform the Braid Burn into a more natural stream has been discussed (Pontier et al., 2001). This might lead not only to an improved physical and subsequently biological) environment but also to more sediment accumulation within the stream due to a reduction in flow velocity (Ciszewski, 1998; Jackman et al., 2001; Salehi et al., 1997).

The water quality of the Braid Burn is sensitive to surface water runoff including the discharge of the silt trap at The King's Buildings. To reduce the risk of stream bed pollution (Townsend and Riley, 1999; Zalewski et al., 1998), the author recommends the construction of a vertical flow (low footprint) constructed wetland after the silt trap for the removal of SS and particularly heavy metals (Scholz and Xu, 2001). Moreover, an oil separator should be located before the silt trap that requires at least annual emptying. A similar treatment chain is in operation for the Newbury Bypass in Southern England (Pontier et al., 2001).

Furthermore, the silt trap requires re-design: firstly, the length-to-width ratio should be increased and secondly, the incoming flow direction requires changing. The stormwater pipe inlet should be located opposite of the silt trap outlet (for current design: Figure 1.2) to reduce the risk of sediment re-suspension during storm events. The water quality should be monitored at least twice per year to assess the system performance.

1.1.4 Conclusions and Recommendations

The full silt trap was inefficient in holding back SS during wet weather conditions. Metals did not accumulate in the sediment of the full silt trap due to high flow-through velocities and an unconventional flow path design leading to re-suspension during periods of heavy rainfall. Metals did not accumulate along the channel of the semi-natural receiving watercourse.

The water quality of the receiving watercourse deteriorated due to the failing full silt trap. Emptying of the silt trap resulted in significant water quality improvements despite a length-to-width ratio of 1:1 and a stormwater inlet-to-outlet angle of 90°. Constructed wetlands for secondary treatment of stormwater runoff at source have been suggested.

1.2 VERTICAL SUBSURFACE FLOW CONSTRUCTED WETLANDS TREATING RIVER WATER

1.2.1 Introduction

Constructed wetlands are an emerging ecotechnology with optimized hydraulic control and management of vegetation (Lim et al., 2001; Scholz and Sadowski, 2009). Compared with conventional activated sludge and biofilm processes, low cost, easily operated and maintained constructed wetlands can be applied in developing countries with serious water pollution problems (Ciria et al., 2005; Li et al., 2008). Wetlands are most frequently used in river systems to manage flows (Chauvelon, 1998; Zalidis, 1998).

Only few studies report on the potential of constructed wetlands in treating eutrophic river water (Green et al., 1996; Jing et al., 2001). In China, constructed wetlands succeeded in treating eutrophic lake waters of Taihu (Li et al., 2008). However, there are currently only few full-scale constructed wetland applications for eutrophic river water treatment in China.

Constructed wetlands could be an effective ecotechnology to remove excess nutrients. Previous studies have demonstrated that the removal of BOD, chemical oxygen demand (COD) and SS in constructed wetlands can be satisfactory (Ansola et al., 2003; Scholz, 2006; Sun et al., 2005), although the removal of nitrogen and phosphorus tends to be variable and is frequently low (Brix et al., 2001; Vymazal, 2007). Therefore, constructed wetlands should be designed with some innovation to reach much higher nutrient removal rates. The role of oxygen availability in nutrient removal has frequently been discussed (Ouellet-Plamondon et al., 2006; Sun et al., 2005). Low oxygen content results in low aerobic organic matter decomposition (Brix et al., 2001). Moreover, nitrification may be the main limiting process for nitrogen removal in constructed wetlands, if the oxygen availability is low (Bezbaruah and Zhang, 2003).

Phosphorus removal is also indirectly affected by oxygen availability. Under aerobic conditions, oxidization of Fe^{2+} to Fe^{3+} (Fe, iron) may enhance the chemical precipitation of phosphorus (Kadlec and Knight, 1996). Furthermore, between 10% and 12% of phosphorus removal could be obtained by microbial assimilation with good aeration (Behrends et al., 2001). In planted constructed wetlands, the oxygen availability is enhanced by the presence of macrophytes through diffusion of oxygen in the sediment via the aerenchyma to the rhizomes (Brix, 1997). However, the contribution of plants to oxygen supply is still debated (Armstrong et al., 2000; Scholz, 2006).

In the past decade, some design advances have been proposed and applied to promote oxygen availability in constructed wetlands. For example, fluctuations of wetland water level, tidal flow or reciprocation systems are frequently used to enhance oxygen availability (Lee et al., 2006; Sun et al., 2005; Tanner et al., 1999). An alternative solution is the injection of compressed air into the bed matrix, which greatly increases the nutrient removal efficiency of constructed wetlands in cold climate (Nivala et al., 2007; Ouellet-Plamondon et al., 2006). Artificial aeration requires energy input at additional costs, but in some instances and in most developing countries, it may still be profitable. For eutrophic river water treatment in China, artificial aeration is a commonly used as an inexpensive option to increase the oxygen content within the water body to prevent the development of any odour.

Other ecotechnological research areas are, for example, focusing on the assessment of the potential of novel material such as constructed wetland substrate. Aggregates with large surface areas and high void spaces are prone to be rapidly colonized with biofilms. For example, polyhedron hollow polypropylene balls (PHPB) are useful in improving the nutrient removal rates in aerated biofilters (Liu et al., 2006; Yu et al., 2005). Similar plastic material is also used as solid substrate and biofilm carrier in recirculated aquaculture systems for enhanced nutrient removal (Boley et al., 2000; Pan et al., 2007). However, there are only few reports on the application of plastic biofilm carriers including PHPB in constructed wetlands (Chen et al., 2006). A potential drawback could be that PHPB are likely to increase the overall capital costs. However, the extent of cost increase would depend on the amount of PHPB used.

Before the start of most industrial-scale field projects, it is important to assess novel ecotechnologies in laboratories and pilot-scale trials. A sound experimental set-up is therefore required to statistically compare aerated and non-aerated constructed wetlands with and without PHPB. Eutrophic river water rather than synthetic wastewater should be used for pilot-scale experiments to simulate 'real' operational conditions. As constructed wetlands require a long period to reach maturity, any pilot-scale work can only deliver initial results to support decision-making.

The project aim was to assess different experimental systems to help decision-makers in selecting the best design option for subsequent industrial-scale applications. In light of the above rationale, the specific objectives were (a) to assess the potential of vertical subsurface flow (VSSF) wetlands to treat eutrophic Jinhe River water; (b) to examine the main and interactive effect of intermittent artificial aeration and PHPB application on nutrient removal within VSSF constructed wetlands; (c) to evaluate the contribution of plant biomass uptake to nutrient removal and (d) to conduct a preliminary cost-benefit analysis to determine the economic feasibility of introducing artificial aeration and PHPB in a future full-scale application. Section 1.2 is based on an updated original article by Tang et al. (2011).

1.2.2 Materials and Methods

1.2.2.1 Wetland System Design

Four pilot-scale VSSF-constructed wetland systems A, B, C and D were set up in the yard of the Tianjin Hydraulic Scientific Institute, Tianjin, China. Each VSSF wetland system consisted of a combined down-flow wetland and an up-flow wetland, which were both made of polyethylene columns (diameter, 0.5 m; height, 1.3 m and surface area, 0.196 m^2). Three different aggregates were used as substrate: coarse predominantly granite-based gravel (15.74 ± 2.97 mm in diameter; 48% porosity), shale dominated by quartz and feldspar (10.78 ± 1.47 mm in diameter; 46% porosity) and PHPB (25.00 mm in diameter; 81% porosity).

The packing order of the constructed wetlands A and D (without PHPB) was the same. Each down-flow wetland unit was filled with 0.3 m of gravel representing the bottom layer, and 0.6 m of shale as the main filter layer, followed by a gravel layer of 0.1 m thickness at the top to reduce the risk of clogging. The up-flow wetland was used for further secondary purification after the treatment of the influent within the down-flow wetland. The only difference in substrate packing between the down-flow and up-flow wetlands was the shale layer. The depth of the shale layer in the up-flow wetlands was 0.5 m. Thus, the water level of the up-flow wetland was 0.1 m lower than the one of the down-flow unit. This design allowed for the wastewater to flow naturally by gravity. Compared to wetlands without PHPB, regardless of the flow direction, a shale layer of 0.2-m thickness located above the bottom gravel layer was replaced by PHPB in the constructed wetlands B and C to examine the effect of PHPB on nutrient removal.

Artificial aeration of constructed wetlands A and B was performed to assess the effect of oxygen availability on nutrient removal. A perforated horizontal 0.3-m-diameter circular tube was installed at 0.05 m from the bottom of both the down-flow and up-flow wetlands. Wetlands C and D functioned as the non-aerated wetlands with no air diffuser present. All VSSF constructed wetlands were planted with cattail (*Typha latifolia* L.) at a density of eight rhizome cuttings per wetland (i.e. 16 per wetland system) on 1 May 2006. After 1 month, cattail was well established in each wetland with a mean individual height of approximately 0.4 m. New shoots were also detected. In August, mean plant heights achieved maximum values between 2.38 and 2.50 m.

1.2.2.2 Inflow Water and Wetland Operation

The Jinhe River, which was chosen as the source of the eutrophic influent, flows through downtown of Tianjin City (China). The approximate length of the landscaped river is 18.5 km. Relatively low heavy metal and toxic organic compound pollution has been recorded. However, the ammonia-nitrogen (NH_4^+-N), total nitrogen (TN) and total phosphorus (TP) concentrations were in the ranges between 1.15 and 1.85 mg/L, 4.11 and 6.18 mg/L and 0.04 and 0.22 mg/L, respectively. These previously measured (2004 and 2006) nutrient data are guide values but indicate that the Jinhe River is eutrophic (Ansola et al., 2003).

River water was directly pumped into the storage wells of the experimental rig, and then continuously discharged onto the wetland systems as influent. The inflow rates were adjusted manually and checked regularly to achieve a relatively high mean hydraulic loading rate of 0.8 m/d for each constructed wetland. The corresponding influent flow rate was 0.16 L/min. High hydraulic loading rates are preferable for eutrophic river or lake water treatment (Li et al., 2008).

In aerated constructed wetlands A and B, compressed air was slowly and continuously introduced via a perforated pipe into the wetland substrate for 8 hours between 8:30 and 16:30 at a corresponding ratio of air to water of 5:1. The purpose was to achieve full oxygen saturation of the water as far as practically possible. Thereafter, aeration was stopped for 16 hours until the next aeration cycle was started.

1.2.2.3 Water Sampling and Analysis

From June to November 2006, influent and effluent of the pilot-scale constructed wetlands were sampled once per week ($n=24$) under normal conditions to evaluate their treatment performances. All samples were analysed on the same day they were taken for the following parameters: COD, NH_4^+-N, nitrate-nitrogen (NO_3^--N), soluble reactive phosphorus (SRP), TP, DO, water temperature (T) and pH. Water quality parameters, including COD, NH_4^+-N, NO_3^--N, SRP and TP, were determined according to standard methods (APHA, AWWA and WEF, 1998), if not stated otherwise. A YSI 52 dissolved oxygen meter and a HANNA portable pH meter were used for DO, T and pH analysis, respectively.

1.2.2.4 Plant Harvesting and Analysis

All aboveground cattail biomass was harvested in November 2006 to estimate the contribution of plant harvesting to overall nutrient removal. After dividing the biomass into stems and leaves, they were subsequently oven-dried for approximately 48 hours at 80°C (Fraser et al., 2004). Dry weight of the harvested aboveground biomass was expressed in weight (g) per stems or leaves. Sub-samples of dried stems and leaves were powdered, wet-digested and analysed for TN and TP content according to a method by Huett et al. (1997). The results were expressed in weight (mg) of nutrient per weight (g) of dry stems or leaves.

1.2.2.5 Statistical Analysis and Modelling

All statistical tests were performed using the Statistical Package for the Social Sciences software package. Significances were defined as $p < 0.05$, if not stated otherwise. A one-way analysis of variance (ANOVA) and the Tukey's honestly significant difference multiple range tests (Fraser et al., 2004) were carried out to assess the differences between means of nutrient removal efficiency and effluent water quality variables in different constructed wetlands. For all ANOVA, the tested variables were normally distributed.

The removal rate constant k obtained from the first-order model (Equation 1.1) can be used to further test the contributions of wetland plants, substrate and novel operations including introduction of artificial aeration and the presence of PHPB on the nutrient removal efficiency in the experimental treatment wetlands. The background concentration C^* is often depleted (Kadlec and Knight, 1996), which simplifies Equation 1.1, leading to Equation 1.2.

$$C_e = \left(C_i - C^*\right)\exp\left(-\frac{k\text{HRT}}{h}\right) + C^* \tag{1.1}$$

where C_e and C_i are concentrations of nutrients (mg/L) in the effluent and influent, respectively, C^* is the background concentration (mg/L), k is the removal rate constant (m/d), HRT is the hydraulic residence time (d) and h is the effective depth of the wetland.

$$C_e = C_i \exp\left(-\frac{k\text{HRT}}{h}\right) \tag{1.2}$$

where C_e and C_i are concentrations of nutrients (mg/L) in the effluent and influent, respectively, k is the removal rate constant (m/d), HRT is the hydraulic residence time (d) and h is the effective depth of the wetland.

1.2.3 Results

1.2.3.1 Water Quality

During the entire monitoring period, most of the nitrogen (78%) in the influent occurred as NH_4^+-N and 77% of phosphorus as SRP (Table 1.1). SS (<10 mg/L) and BOD (<15 mg/L) concentrations were very low, and therefore not measured routinely. Effluent concentrations of COD, NH_4^+-N, NO_3^--N, TN, SRP and TP for wetlands A, B, C and D are summarized in Table 1.1. The concentrations in the effluent were lower than the corresponding ones in the influent. Effluent COD and NH_4^+-N concentrations were significantly lower in the aerated wetlands B than in the non-aerated wetlands D. However, effluent NO_3^--N concentrations were significantly higher in the aerated wetlands A than in the non-aerated wetlands D. For all wetlands, there were no significant differences in effluent SRP, TN and TP concentrations (Table 1.5).

Except for the above-mentioned nutrient variables, changes in values of the online measured parameters T, DO and pH were also presented in Table 1.1. The mean temperature recordings for the influent and effluent were not significantly different.

TABLE 1.5
Mean Concentrations± SD and Pollutant Removal Efficiencies for Chemical Oxygen Demand (COD), Ammonia-Nitrogen (NH_4^+-N), Nitrate-Nitrogen (NO_3^--N), Total Nitrogen (TN), Soluble Reactive Phosphorus (SRP) and Total Phosphorus (TP) and Other Variables Including Water Temperature (T), Dissolved Oxygen (DO) and pH in the Influent and Effluent Waters of Experimental Aerated Wetlands A (without PHPB) and B (with PHPB) and Non-aerated Wetlands C (with PHPB) and D (without PHPB)

		Effluent			
Variables	**Influent**	**Wetlands A**	**Wetlands B**	**Wetlands C**	**Wetlands D**
COD					
Concentrations (mg/L)	106.02 ± 13.70^{c}	$69.45 \pm 24.02^{a,b}$	63.07 ± 23.00^{a}	$69.23 \pm 23.08^{a,b}$	77.75 ± 24.03^{b}
Removal (%)		38 ± 10.2^{b}	43 ± 12.4^{b}	$37 \pm 12.0^{a,b}$	29 ± 10.6^{a}
NH_4^+-N					
Concentrations (mg/L)	5.73 ± 3.96^{c}	0.71 ± 0.84^{a}	0.66 ± 0.81^{a}	1.89 ± 1.38^{b}	2.34 ± 1.63^{b}
Removal (%)		89 ± 5.7^{c}	89 ± 6.9^{c}	67 ± 8.7^{b}	56 ± 14.9^{a}
NO_3^--N					
Concentrations (mg/L)	1.19 ± 0.23^{c}	0.76 ± 0.23^{b}	$0.63 \pm 0.20^{a,b}$	$0.44 \pm 0.11^{a,b}$	0.35 ± 0.10^{a}
Removal (%)		41 ± 26.7^{a}	50 ± 28.4^{a}	65 ± 9.9^{b}	72 ± 12.1^{b}
TN					
Concentrations (mg/L)	7.34 ± 3.61^{b}	1.98 ± 1.26^{a}	1.65 ± 1.15^{a}	2.43 ± 1.41^{a}	2.63 ± 1.38^{a}
Removal (%)		73 ± 9.0^{b}	78 ± 9.0^{c}	67 ± 8.7^{a}	63 ± 8.1^{a}
SRP					
Concentrations (mg/L)	0.40 ± 0.25^{b}	0.09 ± 0.05^{a}	0.09 ± 0.04^{a}	0.14 ± 0.10^{a}	0.17 ± 0.12^{a}
Removal (%)		74 ± 7.7^{b}	75 ± 10.9^{b}	65 ± 11.1^{a}	60 ± 14.6^{a}
TP					
Concentrations (mg/L)	0.52 ± 0.28^{b}	0.15 ± 0.07^{a}	0.15 ± 0.07^{a}	0.20 ± 0.12^{a}	0.22 ± 0.14^{a}
Removal (%)		68 ± 10.7^{b}	69 ± 9.9^{b}	60 ± 11.5^{a}	57 ± 14.9^{a}
T (°C)	24.56 ± 3.69	23.76 ± 3.65	23.70 ± 3.78	23.76 ± 3.65	23.70 ± 3.78
DO (mg/L)	3.42 ± 1.69^{b}	4.50 ± 1.79^{c}	4.53 ± 1.84^{c}	2.11 ± 0.55^{a}	1.85 ± 0.60^{a}
pH (−)	$7.73 \pm 0.39^{a,b}$	$7.91 \pm 0.34^{a,b}$	8.00 ± 0.45^{b}	$7.64 \pm 0.38^{a,b}$	7.57 ± 0.38^{a}

Note: Values with a different superscript letter (i.e. a, b and c) indicate significant difference at $p \leq 0.05$ based on Turkey's HSD.

In contrast, effluent pH values in wetlands with PHPB were considerably lower than in the influent. However, significantly higher effluent pH values in the aerated constructed wetlands compared to the influent were noted. Furthermore, DO concentrations in the effluent were significantly higher in the aerated wetlands than in the influent and effluent of the non-aerated systems. However, no significant differences in effluent DO concentrations were detected between wetlands with and without PHPB (Table 1.5).

1.2.3.2 Nutrient Removal

Significantly higher COD, NH_4^+-N, TN, SRP and TP removal efficiencies were recorded for the aerated compared to the non-aerated wetlands (Table 1.5). However, the NO_3^--N removal efficiency was significantly lower in the aerated wetlands than in the non-aerated wetlands. Wetlands C containing PHPB showed significantly higher COD and NH_4^+-N removal efficiencies in contrast to wetlands D without PHPB ($p < 0.05$). Without artificial aeration, however, no significant difference in NO_3^--N, TN, SRP and TP removal efficiency was observed between wetlands with and without PHPB (Table 1.5).

The first-order model (Equations 1.1 and 1.2) can estimate the actual k value, since the nutrient concentrations of the influent and effluent are known (in addition to the hydraulic retention time (HRT)). As shown in Table 1.6, the mean removal rate constant k (m/d) values for COD, NH_4^+-N, NO_3^--N, TN, SRP and TP removal in wetlands A, B, C and D were identified. According to the first-order model, removals of COD, NH_4^+-N, TN, SRP and TP were higher in the aerated wetlands A and B compared to the corresponding non-aerated wetlands D and C. Irrespectively of the presence or absence of aeration, COD, NH_4^+-N, TN, SRP and TP removal was higher

TABLE 1.6
Mean Removal Rate Constant k (m/d) Values for Chemical Oxygen Demand (COD), Ammonia-Nitrogen (NH_4^+-N), Nitrate-Nitrogen (NO_3^--N), Total Nitrogen (TN), Soluble Reactive Phosphorus (SRP) and Total Phosphorus (TP) Removal in Experimental Aerated Wetlands A (without PHPB) and B (with PHPB) and Non-aerated Wetlands C (with PHPB) and D (without PHPB)

	Wetlands			
Variable	A ($k_{a+}k_{p+}k_s$)	B ($k_{a+}k_{b+}k_{p+}k_s$)	C ($k_{b+}k_{p+}k_s$)	D ($k_{P+}k_s$)
COD	1.01 ± 0.44	1.02 ± 0.39	1.02 ± 0.47	0.73 ± 0.36
NH_4^+-N	4.75 ± 0.88	4.82 ± 1.54	2.37 ± 0.68	1.70 ± 0.77
NO_3^--N	0.96 ± 0.81	1.26 ± 1.03	2.38 ± 0.78	2.68 ± 1.13
TN	2.87 ± 0.76	3.10 ± 0.96	2.43 ± 0.73	1.95 ± 0.43
SRP	2.69 ± 0.55	2.60 ± 0.97	2.36 ± 0.76	1.95 ± 0.88
TP	2.26 ± 0.71	2.20 ± 0.86	2.02 ± 0.55	1.66 ± 0.67

k_a, k_p, k_b and k_s represents contributions of aeration, plant, biofilm carrier (PHPB) and traditional substrate (shale) to the mean removal constant values, respectively.

TABLE 1.7
Correlation Matrix for Chemical Oxygen Demand (COD), Ammonia Nitrogen (NH_4^+-N), Nitrite Nitrogen (NO_3^--N), Total Nitrogen (TN), Soluble Reactive Phosphorus (SRP), Total Phosphorus (TP), Dissolved Oxygen (DO) and Water Temperature (T) for the Representative Wetlands D

	COD	NH_4^+-N	NO_3^--N	TN	SRP	TP	DO	T
COD	1.000	(0.047)	(0.096)	(0.177)	(0.011)	(0.013)	(0.357)	(0.346)
NH_4^+-N	0.737	1.000	(0.454)	(0.007)	(0.002)	(0.001)	(0.187)	(0.076)
NO_3^--N	−0.616	−0.061	1.000	(0.245)	(0.251)	(0.264)	(0.060)	(0.333)
TN	0.465	0.904	0.354	1.000	(0.042)	(0.039)	(0.057)	(0.053)
SRP	0.876	0.951	−0.346	0.752	1.000	(<0.001)	(0.316)	(0.102)
TP	0.864	0.956	−0.326	0.762	0.997	1.000	(0.304)	(0.107)
DO	−0.193	0.447	0.703	0.710	0.251	0.268	1.000	(0.021)
T	0.209	0.662	0.227	0.721	0.604	0.594	0.828	1.000

The corresponding *p*-values are shown in parentheses.

in wetlands with PHPB than in wetlands without PHPB (Table 1.6). For NO_3^--N, however, the removal constants k were lower in the aerated wetlands than in the non-aerated ones. Moreover, among all the tested wetlands, the lowest NO_3^--N removal was noted for the aerated wetlands without PHPB (Table 1.6).

A linear regression analysis for wetlands D (representative example) was performed to test the relationships between each nutrient variable and other water quality parameters such as DO and T. The coefficients of the corresponding correlation matrix for all variables are shown in Table 1.7. The effluent COD concentrations were significantly and positively correlated with the effluent NH_4^+-N, SRP and TP concentrations ($p < 0.05$). Effluent NH_4^+-N concentrations were significantly and positively correlated with the effluent COD, TN, SRP and TP concentrations ($p < 0.05$). The effluent NO_3^--N concentrations were positively correlated with the DO concentrations and the effluent SRP. The TP concentrations were both significantly and positively correlated with effluent COD, NH_4^+-N and TN concentrations.

1.2.3.3 Plant Biomass Production and Nutrient Removal

Aboveground biomass (stems and leaves) harvesting contributed to TN and TP removal (Figure 1.7a and b). However, the contribution of <10% for total nitrogen removal was insignificant. In comparison, aboveground biomass harvesting was between 35% and 75% of the total phosphorus removal. Moreover, leaf harvesting removed more nitrogen and phosphorus than stem harvesting for all experimental wetlands (Figure 1.7a and b).

Nitrogen and phosphorus removal by aboveground biomass harvesting was much higher in wetlands with PHPB than in wetlands without PHPB. In addition, aerated wetlands showed higher nitrogen and phosphorus uptake and storage by aboveground biomass in contrast to non-aerated wetlands. The highest nutrient removal

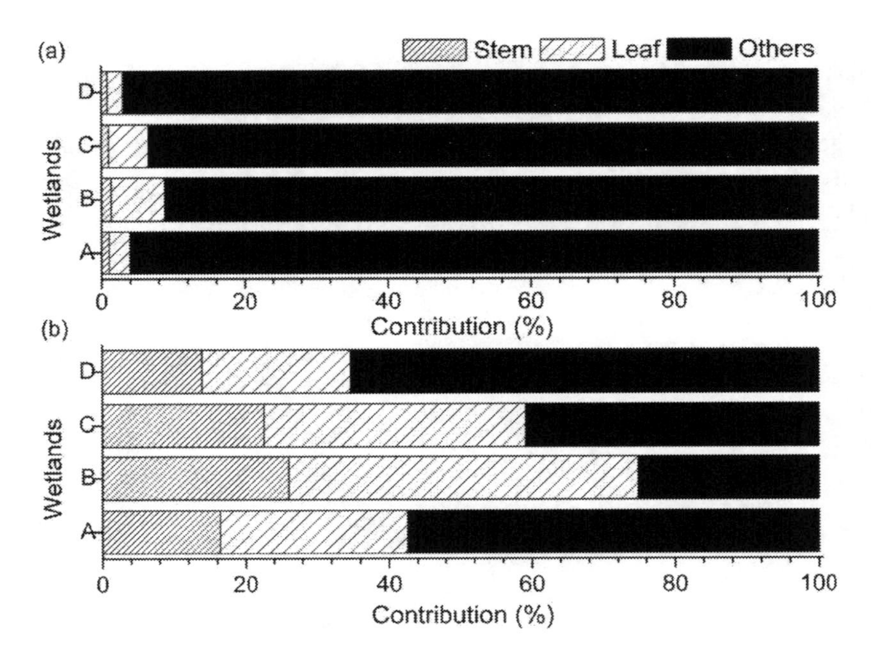

FIGURE 1.7 Contribution of the aboveground stems, leaves and other factors (e.g., substrate and microbes) to the (a) total nitrogen and (b) total phosphorus removal in wetlands A (aerated without polyhedron hollow polypropylene balls (PHPB)), B (aerated with PHPB), C (non-aerated with PHPB) and D (non-aerated without PHPB) in November 2006.

occurred in wetlands subjected to the interactive effect between intermittent artificial aeration and PHPB (Figure 1.7a and b).

1.2.3.4 Cost Analysis

The capital expenditure for items such as construction, substrates (gravel, shale and PHPB), plants, pumps, pipes, air condenser, diffuser and wetland units are listed in Table 1.4. Considering a total investment over a period of 20 years (i.e. estimated treatment plant lifetime) and a sustained treatment capacity of 0.16 L/min, the costs to treat eutrophic Jinhe River water are approximately 0.48, 0.50, 0.46 and 0.45 renminbi (RMB)/m^3 (RMB 1 = \$ 0.147 = £ 0.074) for wetland systems A, B, C and D, respectively (Table 1.8).

Concerning wetland D, artificial aeration and the presence of PHPB resulted in total cost increases of 8% and 3%, respectively. For sub-surface constructed wetlands, operational and maintenance costs, defined predominantly as electricity and labour wage, usually accounted for approximately 10% of the total capital costs in most developed countries (Rousseau et al., 2008). However, this figure might be as low as 5% for China, where energy and labour costs are still relatively low. It follows that the total costs (conservative estimate based on 10% for electricity and labour) are approximately 0.53, 0.54, 0.51 and 0.49 RMB/m^3 for wetland systems A, B, C and D, respectively.

TABLE 1.8
Expenditure for the Construction of Different Pilot-Scale Experimental Wetlands in 2006

	Cost for Tested Constructed Wetlands (RMB)			
Item	**A**	**B**	**C**	**D**
PHPB	–	50	50	–
Gravel	39	27	27	39
Shale	64	48	48	64
Plant	16	16	16	16
Pump, pipe and other facilities	150	150	150	150
Wetland apparatus fee	480	480	480	480
Air condenser and diffuser	60	60	–	–
Total	809	831	771	749

RMB, Ren Min Bi (currency of the People's Republic of China).

1.2.4 Discussion

The mean COD removal efficiency (<50%) in this study was lower than typically reported values between 80% and 99% (Ansola et al., 2003; Merlin et al., 2002; Sun et al., 2005). Low influent COD concentrations (106.02 ± 13.7 mg/L, Table 1.1) may limit the COD removal capacity, as COD concentrations below 50 mg/L are difficult to reduce any further (Korkusuz et al., 2005). Furthermore, the hydraulic loading rate applied in this research is considerably higher than those applied in similar systems, and the contact time and potential for biological degradation of organic matters may not be at an optimal level (Kadlec and Knight, 1996; Scholz, 2006; Scholz et al., 2002). Increased oxygen availability could improve organic matter mineralization and aerobic biodegradation (Vymazal, 1999). The findings confirmed that aerated wetlands performed better in COD removal compared to non-aerated wetlands (Nivala et al., 2007; Ouellet-Plamondon et al., 2006).

There are two reasons that wetlands with PHPB outperformed corresponding systems without PHPB. Firstly, due to the high porosity of PHPB (81%) compared to shale (46%), more organic compounds settled out and were retained in the wetland filter media for a long time. This allowed for improved hydrolysis of organic compounds and for rapid biodegradation (Lee et al., 2004). Secondly, the presence of PHPB supported for the accumulation of large numbers of attached bacteria colonization onto the substrate surface. This greatly improved biodegradation of COD (Chen et al., 2006; Lim et al., 2001). However, a further discussion on bacteria colonization is beyond the scope of Section 1.2 and would be speculation because microorganisms were not determined in this study.

Nitrogen removal efficiency differed significantly among the constructed wetlands (Table 1.1). Despite the high hydraulic loading rate in this study, considerable amount of NH_4^+-N was removed and the corresponding removal efficiency was reasonably comparable to the removal obtained in other wetlands (Nivala et al., 2007; Sun et al., 2005). Significantly high NH_4^+-N removal efficiency occurred in aerated wetlands confirming the positive effect of aeration on nitrifying bacteria (Ouellet-Plamondon et al., 2006). Artificial aeration allowed sufficient NH_4^+-N-substrate contact and enhanced the transfer of oxygen by drawing air from the atmosphere into the wetland bed media, which subsequently resulted in significantly higher effluent DO concentrations in aerated wetlands (Table 1.1). NH_4^+-N removal performed significantly better in wetlands with PHPB than in corresponding wetlands without PHPB can mean one of two things. One of them is that using PHPB as wetland substrate favoured biofilm attachment and thus enhanced bacteria nitrification (Liu et al., 2006). The second is that microbial assimilation removal of NH_4^+-N would be encouraged with the presence of PHPB.

There was more nitrate in the effluents of the aerated wetlands than the non-aerated wetlands (Table 1.1). This could be because injection of compressed air into the wetland substrate increased oxygen availability but simultaneously decreased the anaerobic conditions, which are necessary for the denitrification processes (Ciria et al., 2005). Furthermore, pH affected NO_3^--N removal (Li et al., 2008; Vymazal, 2007); however, the average values for measured pH of this study were within the range optimal (between 6.6 and 8.3) for denitrification (Šimek et al., 2002). Like supplemental artificial aeration, using PBHB as filter media also failed to improve NO_3^--N removal (Tables 1.1 and 1.2). This could be the case considering that microbes prefer the autotrophic uptake of NH_4^+-N over a corresponding uptake of NO_3^--N (Bigambo and Mayo, 2005).

The TN removal efficiency was between 63% and 67% for non-aerated wetlands, which are higher than 52% obtained in constructed wetlands treating eutrophic lake water in China (Li et al., 2008). Complete nitrification followed by denitrification was the most important approach for total nitrogen removal (Nivala et al., 2007; Vymazal, 2007). In this study, effluent TN concentrations in the tested wetlands were low, and the net accumulation in the concentration of nitrate was rarely observed (Table 1.1). Therefore, combined nitrification-denitrification performed not completely but well in TN removal. Furthermore, considering that the majority of TN occurred as NH_4^+-N, microbial assimilation of NH_4^+-N may also contribute greatly to the current TN removal as reported elsewhere (Bigambo and Mayo, 2005; Sun et al., 2005). In addition, PHPB lead to improved TN removal (Table 1.2), as the processes that facilitate TN removal in wetlands are sedimentation, nitrification-denitrification and uptake by plants and microbes (Bigambo and Mayo, 2005; Brix, 1997; Pan et al., 2007; Vymazal, 2007). It is reasonable to believe that these processes may be greatly enhanced by the presence of PHPB.

In relation to phosphorus removal, high average TP removal efficiencies achieved in this study were comparable to 60% and 64% reported previously (Li et al., 2008; Vymazal, 2007) for VSSF constructed wetlands, respectively. SRP and TP removal performed significantly better in aerated wetlands than in non-aerated ones (Table 1.2). Artificial aeration can reduce phosphorus loading at a level significantly lower than non-aerated wetlands (Nivala et al., 2007). Furthermore, low oxygen concentrations within constructed wetlands cause usually relatively low TP removal efficiencies (Chen et al., 2006). Aerobic conditions favour the chemical precipitation of phosphorus (Kadlec and Knight, 1996; Scholz, 2006). Alternatively, adsorption removal of phosphorus would also be encouraged in aerated wetlands with a much better degree of internal mixing than conventional non-aerated wetlands (Nivala et al., 2007).

Under intermittent artificial aeration conditions, wetlands with PHPB performed best in both SRP and TP removal (Table 1.1). Our findings verified that the presence of bioballs including PHPB favoured the microbial process responsible for phosphorus removal (Yu et al., 2005).

Wetland plants are known to take up nutrients, but the amount varies widely within and between constructed wetlands. In the present study, aboveground biomass nitrogen and phosphorus removal was between 21.5 and 79.9 g N/m^2 and between 14.8 and 41.6 g P/m^2 (Figure 1.7a and b), respectively. These ranges are comparable with those values (20–30 g N/m^2 and 3–8 g P/m^2) obtained from traditionally designed vertical-flow constructed wetlands treating eutrophic Taihu Lake (China) water (Li et al., 2008). Furthermore, the findings in this chapter compared well with previously reported values of 0.6–88.0 g N/m^2 (Johnston, 1991; Mitsch and Gosselink, 2000; Vymazal, 1995) and 0.1–45.0 g P/m^2 (Johnston, 1991; Lim et al., 2001; Vymazal, 1995).

If the uptake and storage of nitrogen and phosphorus by plants would have only occurred from the water column during the running period, a mass balance based on the nitrogen and phosphorus loading rate would have shown that aboveground biomass harvesting contributed to 4%, 9%, 7% and 3% N, and 43%, 75%, 59% and 35% P to the total nitrogen and total phosphorus removal in the tested wetlands A, B, C and D, respectively. Uptake and incorporation into plant tissues was a major factor responsible for the observed total nitrogen and phosphorus removal in VSSF wetlands treating eutrophic Jinhe River water. The results obtained were comparable to those of other eutrophic lake or pond water; for example, plant removal contributed 5%–26% and 41%–81% to total nitrogen and total phosphorus removal, respectively (Bigambo and Mayo, 2005; Li et al., 2008). Considering the above calculations, nutrient uptake into aboveground biomass could be greatly improved by artificial aeration and PHPB. The improvement of nutrient uptake contributed to the increase of total nitrogen and phosphorus removal.

The approximate treatment costs for wastewater in China are 1 RMB/m^3 in large cities (Chen et al., 2008). In comparison, the treatment costs for the wetland systems A, B, C and D were by approximately 47%, 46%, 49% and 51% lower, although artificial aeration and PHPB were introduced. Nevertheless, due to the relatively high construction costs of small wetland units, the purchase of only small quantities of PHPB and polyethylene for the wetland columns, and the lack of electrical appliance optimization, the total treatment costs were between 113% and 135% higher than those for the Chinese Longdao River constructed wetland, which is, however, very large in comparison; that is, >90,000 times bigger in area than the experimental wetlands studied in this paper (Chen et al., 2008).

Artificial aeration and the presence of PHPB, however, were only associated with less than 10% of the total construction costs (Table 1.4). Furthermore, the land occupied by traditional constructed wetlands is much greater than the land required for the novel compact wetland systems containing polyethylene columns. The current costs will be reduced with an increase in wetland size and user demand. Moreover, the provided cost-benefit analysis should be seen as preliminary, considering that the assessed wetland system is relatively small.

1.2.5 Conclusions and Recommendations

Both intermittent artificial aeration and using PHPB as part of the substrate enhanced the ability of nutrient removal in VSSF wetlands treating eutrophic river water. Findings indicate that plant uptake and storage played an important role in eutrophic river water treatment. Moreover, aboveground biomass nutrient uptake was significantly improved by artificial aeration as well as the presence of PHPB. This improvement accounted for the majority of the total enhancement in nitrogen and phosphorus removal.

Although a preliminary cost-benefit analysis has shown that the overall costs compare very well with traditional wastewater treatment plants for large cities, it is necessary to make a more detailed evaluation of investment costs for real-scale systems. Due to the scale effect and the short running period, the tested wetlands did not reach their full potential in nutrient removal.

Further research should target the assessment of the microbial development, biofilm attachment and root growth in the proposed wetland systems. Furthermore, different operating strategies including hydraulic loading rates should be evaluated, particularly to improve the COD removal. More information regarding the application of full-scale wetlands to purify eutrophic river water would also be desirable.

REFERENCES

Abwassertechnische Vereinigung e. V. (ATV), 1994 and 1996. *Richtlinien für die hydraulische Dimensioniering und Leistungsnachweis von Regenwasser-Entlastungsanlagen in Abwasserkanälen und –leitungen*, guidelines A 111 and 112 (Hennef: ATV), in German.

Alloway, B. J., 1995. *Heavy Metals in Soils*, 2nd edition (Suffolk: Chapman and Hall).

American Public Health Association (APHA), American Water Works Association (AWWA) and Water Environment Federation (WEF), 1998. *Standard Methods for the Examination of Water and Wastewater*, 20th edition (Baltimore, MD: United Book Press).

Ansola, G., Gonzalez, J.M., Cortijo, R. and de Luis, E., 2003. Experimental and full–scale pilot plant constructed wetlands for municipal wastewaters treatment, *Ecological Engineering*, **21**, 43–52.

Armstrong, W., Cousins, D., Armstrong, J., Turner, D.W. and Beckett, P.M., 2000. Oxygen distribution in wetland plant roots and permeability barriers to gas-exchange with the rhizosphere: a microelectrode and modelling study with Phragmites australis, *Annals of Botany*, **86**(3), 687–703.

Behrends, L., Houke, L., Bailey, E., Jansen, P. and Brown, D., 2001. Reciprocating constructed wetlands for treating industrial, municipal and agricultural wastewater, *Water Science and Technology*, **44**(11–12), 399–405.

Bezbaruah, A.N. and Zhang, T.C., 2003. Performance of a constructed wetland with a sulfur/limestone denitrification section for wastewater nitrogen removal, *Environmental Science and Technology*, **37**(8), 1690–1697.

Bigambo, T. and Mayo, A.W., 2005. Nitrogen transformation in horizontal subsurface flow constructed wetlands II: Effect of biofilm, *Physics and Chemistry of the Earth*, **30**, 668–673.

Boley, A., Müller, W.-R. and Haider, G., 2000. Biodegradable polymers as solid substrate and biofilm carrier for denitrification in recirculated aquaculture systems, *Aquacultural Engineering*, **22**(1–2), 75–85.

British Standards, 1990. Methods of Test for Soils for Civil Engineering Purposes, BS 1377 – 1 to 9 (London: British Standards).

Brix, H., 1997. Do macrophytes play a role in constructed treatment wetlands? *Water Science and Technology*, **35**(5), 11–17.

Brix, H., Arias, C.A. and Del Bubba, M., 2001. Media selection for sustainable phosphorus removal in subsurface flow constructed wetlands, *Water Science and Technology*, **44**(11–12), 47–54.

Bullen Consulting, 2002. Braid Burn Flood Prevention Scheme – Environmental Statement (Volume I). Reference 1018260 (Edinburgh: The City of Edinburgh Council).

Butler, D. and Davies, J.W., 2000. *Urban Drainage* (London: E & FN Spon).

Ciszewski, D. 1998. Channel processes as a factor controlling accumulation of heavy metals in river bottom sediments: Consequences for pollution monitoring (Upper Silesia, Poland), *Environmental Geology*, **36**(1–2), 45–54.

Chauvelon, P., 1998. A wetland managed for agriculture as an interface between the Rhone river and the Vaccares Lagoon (Camargue, France): Transfers of water and nutrients, *Hydrobiologia*, **373–374**, 181–191.

Chen, T.Y., Kao, C.M., Yeh, T.Y., Chien, H.Y. and Chao, A.C., 2006, Application of a constructed wetland for industrial wastewater treatment: A pilot-scale study, *Chemosphere*, **64**(3), 497–502.

Chen, Z.M., Chen, B., Zhou, J.B., Li, Z., Zhou, Y., Xi, X.R., Lin, C. and Chen, G.Q., 2008. A vertical subsurface-flow constructed wetland in Beijing, *Communications in Nonlinear Science and Numerical Simulation*, **13**(9), 1986–1997.

Ciria, M.P., Solano, M.L. and Soriano, P., 2005. Role of macrophyte Typha latifolia in a constructed wetland for wastewater treatment and assessment of its potential as a biomass fuel, *Biosystems Engineering*, **92**, 535–544.

Fraser, L.H., Carty, S.M. and Steer, D., 2004. A test of four plant species to reduce total nitrogen and total phosphorus from soil leachate in subsurface wetland microcosms, *Bioresources Technology*, **94**(2), 185–1992.

Green, M., Safray, I. and Agami, M., 1996. Constructed wetlands for river reclamation: Experimental design, start-up and preliminary results, *Bioresource Technology*, **55**, 157–162.

Huett, D.O., George, A.P., Slack, J.M. and Morris, S.C., 1997. Diagnostic leaf nutrient standards for low-chill peaches in subtropical Australia, *Australian Journal of Experimental Agriculture*, **37**(1), 119–126.

Jackman, A.P., Kennedy, V.C. and Bhatia, N., 2001. Interparticle migration of metal cations in stream sediments as a factor in toxics transport, *Journal of Hazardous Materials*, **82**(1), 27–41.

Jing, S.R., Lin, Y.F., Lee, D.Y. and Wang, T.W., 2001. Nutrient removal from polluted river water by using constructed wetlands, *Bioresource Technology*, **76**(2), 131–135.

Johnston, C.A., 1991. Sediment and nutrient retention by freshwater wetlands: Effects on surface water quality, *Critical Reviews in Environmental Control*, **21**(5–6), 491–565.

Kadlec, R. and Knight, R., 1996. *Treatment Wetlands* (Boca Raton, FL: Chemical Rubber Company Press).

Kjensmo, J., 1997. The influence of road salts on the salinity and meromictic stability of Lake Svinsjoen, Southeast Norway, *Hydrobiologia*, **347**, 151–158.

Korkusuz, E.A., Beklioğlu, M. and Demirer, N.G., 2005. Comparison of the treatment performances of blast furnace slag-based and gravel-based vertical flow wetlands operated identically for domestic wastewater treatment in Turkey, *Ecological Engineering*, **24**(3), 185–198.

Lee, B.-H., Scholz, M. and Horn, A., 2006. Constructed wetlands: Prediction of performance with case-based reasoning (Part B), *Environmental Engineering Science*, **23**(2), 203–211.

Lee, C.Y., Lee, C.C., Lee, F.Y., Tseng, S.K. and Liao, C.J., 2004. Performance of subsurface flow constructed wetland taking pretreated swine effluent under heavy loads, *Bioresources Technology*, **92**(2), 173–179.

Li, L.F., Li, Y.H., Biswas, D.K., Nian, Y.G. and Jiang, G.M., 2008. Potential of constructed wetlands in treating the eutrophic water: Evidence from Taihu Lake of China, *Bioresources Technology*, **99**(6), 1656–1663.

Lim, P.E., Wong, T.F. and Lim, D.V., 2001. Oxygen demand, nitrogen and copper removal by free-water-surface and subsurface-flow constructed wetlands under tropical conditions, *Environment International*, **26**, 425–431.

Liu, L., Xu, Z.H., Song, C.Y., Gu, Q.B., Sang, Y.M., Lu, G.L., Hu, H.L. and Li, F.S., 2006. Adsorption-filtration characteristics of melt-blown polypropylene fiber in purification of reclaimed water, *Desalination*, **201**(1–3), 198–206.

Merlin, G., Pajean, J.P. and Lissolo, T., 2002. Performances of constructed wetlands for municipal wastewater treatment in rural mountainous area, *Hydrobiologia*, **469**, 87–98.

Mitsch, W.J. and Gosselink, J.G., 2000. *Wetlands* (New York: Van Nostrand Reinhold Company).

Nivala, J., Hoos, M.B., Cross, C., Wallace, S. and Parkin, G., 2007. Treatment of landfill leachate using an aerated, horizontal subsurface-flow constructed wetland, *The Science of the Total Environment*, **380**(1–3), 19–27.

Ouellet-Plamondon, C., Chazarenc, F., Comeau, Y. and Brisson, J. 2006. Artificial aeration to increase pollutant removal efficiency of constructed wetlands in cold climate, *Ecological Engineering*, **27**(3), 258–264.

Pan, J., Sun, H., Nduwimana, A., Wang, Y., Zhou, G., Ying, Y. and Zhang, R., 2007. Hydroponic plate/fabric/grass system for treatment of aquacultural wastewater, *Aquacultural Engineering*, **37**(3), 266–273.

Pontier, H., Williams, J.B. and May, E., 2001. Metals in combined conventional and vegetated road run-off control systems, *Water Science and Technology*, **44**(11–12), 607–614.

Rousseau, D.P.L., Lesage, E., Story, A., Vanrolleghem, P.A. and De Pauw N., 2008. Constructed wetlands for water reclamation, *Desalination*, **218**(1–3), 181–189.

Salehi, F., Lagace, R. and Pesant, A.R., 1997. Construction of a year-round operating gauging station for sediment and water quality measurements of small watersheds, *Journal of Soil and Water Conservation*, **52**(6), 431–436.

Schmitt, F., Milisic, V., Bertrand-Krajewski, J.L., Laplace, D. and Chebbo, G., 1999. Numerical modelling of bed load sediment traps in sewer systems by density currents, *Water Science and Technology*, **39**(9), 153–160.

Scholz, M., 2004. Storm water quality associated with a silt trap (empty and full) discharging into an urban watercourse in Scotland, *International Journal of Environmental Studies*, **61**(4), 471–483.

Scholz, M., 2006. *Wetland Systems to Control Urban Runoff* (Amsterdam: Elsevier).

Scholz, M., Höhn, P. and Minall, R., 2002. Mature experimental constructed wetlands treating urban water receiving high metal loads, *Biotechnology Progress*, **18**(6), 1257–1264.

Scholz, M. and Sadowski, A.J., 2009. Conceptual classification model for sustainable flood retention basins, *Journal of Environmental Management*, **90**(1), 624–633.

Scholz, M. and Xu, J., 2001. Comparison of vertical-flow constructed wetlands for treatment of wastewater containing lead and copper, *Journal of the Chartered Institution of Water and Environmental Management*, **15**(4), 287–293.

Šimek, M., Jíšová, L. and Hopkins, D.W., 2002. What is the so-called optimum pH for denitrification in soil? *Soil Biology and Biochemistry*, **34**(9), 1227–1234.

Sun, G.Z., Zhao, Y.Q. and Allen, S., 2005. Enhanced removal of organic matter and ammoniacal-nitrogen in a column experiment of tidal flow constructed wetland system, *Journal of Biotechnology*, **115**(2), 189–197.

Tang, X., Huang, S., Scholz, M. and Li, J., 2011. Nutrient removal in vertical subsurface flow constructed wetlands treating eutrophic river water, *International Journal of Environmental Analytical Chemistry*, **91**(7–8), 727–739.

Tanner, C.C., D'Eugenio, J., McBride, G.B., Sukias, J.P.S. and Thompson, K., 1999. Effect of water level fluctuation on nitrogen removal from constructed wetland mesocosms, *Ecological Engineering*, **12**, 67–92.

Tchobanoglous, G., Burton, F.L. and Stensel, H.D., 2003. *Wastewater Engineering: Treatment and Reuse*, 4th edition (revised.), Metcalf & Eddy Inc. (New York: McGraw Hill Companies Inc.).

Townsend, C.R. and Riley, R.H., 1999. Assessment of river health: Accounting for perturbation pathways in physical and ecological space, *Freshwater Biology*, **41**(2), 393–405.

Vymazal, J., 1995. *Algae and Element Cycling in Wetlands* (Chelsea, MA: Lewis Publishers).

Vymazal, J., 1999. Removal of BOD5 in constructed wetlands with horizontal sub-surface flow: Czech experience, *Water Science and Technology*, **40**(3), 133–138.

Vymazal, J., 2007. Removal of nutrients in various types of constructed wetlands, *The Science of the Total Environment*, **380**(1–3), 48–65.

Yu, H.Y., Hu, M.X., Xu, Z.K., Wang, J.L. and Wang, S.Y., 2005. Surface modification of polypropylene microporous membranes to improve their antifouling property in MBR: NH3 plasma treatment, *Separation and Purification Technology*, **45**(1), 8–15.

Zalewski, M., Bis, B., Lapinska, M., Frankiewicz, P. and Puchalski, W., 1998. The importance of the riparian ecotone and river hydraulics for sustainable basin-scale restoration scenarios, *Aquatic Conservation-Marine and Freshwater Ecosystems*, **8**(2), 287–307.

Zalidis, G., 1998. Management of river water for irrigation to mitigate soil salinization on a coastal wetland, *Journal of Environmental Management*, **54**, 161–167.

2 Sustainable Drainage Systems

2.1 SUSTAINABLE URBAN DRAINAGE SYSTEM MANAGEMENT

2.1.1 Introduction

The European Union's (EU) Water Framework Directive (Council of European Communities, 2000), which came into force on 23 October 2000, requires all inland and coastal waters to reach a 'good status' by 2015. However, the directive fails to define the criteria for a 'good status'. This has led to much speculation and various interpretations.

The Directive sets a framework that should provide substantial benefits for the long-term water quality management of waters. The implementation of sustainable (urban) drainage systems (SUDS) based on current guideline (Jefferies et al., 1999; McKissock et al., 1999; CIRIA, 2000) in Glasgow can help prevent combined sewer overflows from spilling untreated sewage into receiving watercourses such as rivers and canals during storms (DEFRA, 2000).

Furthermore, SUDS can help to reduce the impact of diffuse pollution on urban watercourses by promoting passive treatment (D'Arcy and Frost, 2001). Example water quality variable concentrations and values for a typical urban watercourse and the corresponding storm water runoff are summarized elsewhere (Scholz, 2006b). Typical biochemical oxygen demand (BOD) and suspended solids concentration ranges for storm water runoff receiving watercourse are 2–25 and 10–150 mg/L, respectively. However, a detailed discussion is beyond the scope of Section 2.1.

In 1986, the enquiry into the housing of Glasgow identified the 20% most deprived areas in Glasgow based on the Grieve Plan (Grieve, 1986). Almost all of these areas such as Pollok (the selected representative case study of Section 2.1) and Drumchapel are located in the large peripheral estates. The aim was to regenerate these areas and to improve the status of many properties that were too badly damaged and ill maintained to be lettable. However, almost 20 years later, properties in the same areas are of similar status. Visible evidence in the form of hundreds of closed, boarded-up houses and shops exists in many of the demonstration areas such as the Pollok Centre area (typical example). Now the selected demonstration areas are the focus of regeneration efforts aimed at resolving the interlinked social, economic and environmental deficiencies of these former industrial areas.

The main urban drainage concern for Scottish Water (water utility) and Glasgow City Council is the lack of combined sewer system capacities for new surface water runoff. With future developments, this situation is likely to decline further, and the need for SUDS implementation becomes vital to Glasgow's continuing expansion.

DOI: 10.1201/9781003509370-2

Moreover, the government strongly encourages local authorities to implement SUDS for future development and regeneration sites. New guidelines incorporating a decision support tool that demonstrates which SUDS techniques are feasible for a particular site over a range of boundary and environmental conditions are therefore required to avoid annually reoccurring large-scale flooding events, for example, in the East End of Glasgow.

Furthermore, a previous project undertaken by Hyder Consulting for the East End of Glasgow did not show how SUDS implementation might contribute to a significant reduction of potential combined sewer overflow spills and flooding. However, this desk study failed to incorporate actual site conditions such as the percentage of build-up areas or the local topography of the landscape including watercourses and hills (Grieve, 1986; Greeman, 2004; Hyder Consulting, 2004).

As some SUDS techniques do not comply with a legal definition of a drain, water companies are often not required to maintain them. Consequently, property developers such as the Retail Property Holdings Ltd. regenerating the Pollok Centre area (the selected case study of this section) are often reluctant to apply them and prefer to implement traditional drainage methods despite relatively low water tables (less than 5 m below ground level) for most potential SUDS sites.

In April 2000, the Commission of European Communities established a Community Initiative concerning trans-European co-operation, known as INTERREG IIIB. The INTERREG IIIB initiative relates to the whole of the European Union. One of the projects funded by this initiative is entitled Transformation of Rural and Urban Spatial Structure (TRUST). This project aimed to develop new approaches to both spatial planning and land use to meet the challenges of continuing urbanization, along with reducing economic loss and reduction in biodiversity through the development of integral management methods.

Glasgow City Council's contribution to the TRUST project is known as the 'Glasgow Surface Water Management Project'. The deliverables of this are to come up with urban drainage recommendations (see below) to commission and construct at least one SUDS demonstration area, and to look into further expansion of the scheme in the wider area. The first of these deliverables incorporates 'The Glasgow Sustainable Urban Drainage System Management Project', led by the author and funded by The Royal Academy of Engineering and Glasgow City Council.

The Pollok Centre area is in Pollok that is located approximately 8 km to the Southwest of Glasgow City Centre (Figure 2.1). The area is owned by Retail Property Holdings Ltd. The Brock Burn is adjacent to the southern and western boundaries of the area. The general surface topography of the area is approximately 17 m above sea level, and almost rectangular.

The feasibility of implementing different SUDS throughout Glasgow and their likely contribution to the overall catchment dynamics is studied. This project aims to come up with SUDS demonstration areas that are representative of both different sustainable drainage techniques and different types of areas available for development and regeneration. The objectives were to (a) identify characteristics that determine the suitability of a site for the implementation of SUDS; (b) identify suitable SUDS sites within Glasgow; (c) outline an SUDS Option Decision Support Key to classify qualitatively and quantitatively potential SUDS sites suitable for different

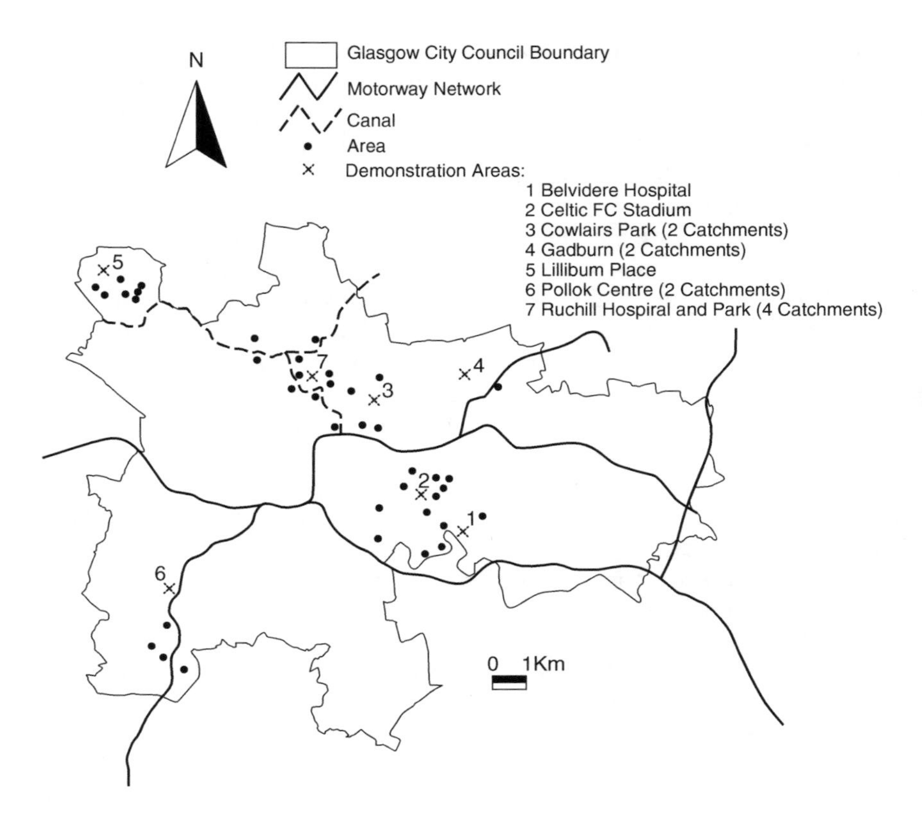

FIGURE 2.1 Indication of 46 potential areas comprising 57 sites for the implementation of sustainable (urban) drainage systems (SUDS).

SUDS technologies; (d) outline an SUDS Option Decision Support Matrix to determine relationships between different site characteristics and SUDS techniques; (e) identify representative SUDS technologies for representative sites that could be used for demonstration purposes; and (f) provide detailed design and management guidelines for a representative SUDS demonstration site (case study) and representative SUDS technologies for information and education purposes. Section 2.1 is based on an updated version of Scholz (2006a), which has been revisited.

2.1.2 Methodology

2.1.2.1 Site Classification with the SUDS Decision Support Key

Figure 2.1 shows a map of Glasgow highlighting the spatial distribution of 46 areas (associated with 57 sites) that were identified during an initial desk study as potentially suitable for the implementation of SUDS. Eight areas had the potential for more than one SUDS system and were therefore sub-divided into sub-areas or sites.

SUDS: The acronym for Sustainable (Urban) Drainage System (British concept) or also known as Best Management Practice (American concept). An individual or series of management structures and associated processes designed to drain surface water runoff in a sustainable approach to predominantly alleviate capacities in existing conventional drainage systems (predominantly combined sewers in Glasgow) in an urban environment (SEPA, 1999; Butler and Davis, 2000; CIRIA, 2000).

Infiltration trench: A linear drain (also known as a French Drain) consisting of a trench filled with a permeable material, potentially with a perforated pipe in the base, a number of perforated pipes at designated depths or drainage cells (e.g., Atlantis Water Management Ltd. system) to promote infiltration of surface water runoff to the ground. An infiltration trench may also convey water if the trench gradient is sufficiently steep.

Underground or belowground storm water storage: A sub-surface structure designed to accumulate surface water runoff, and where water is released from as may be required to increase the flow hydrograph. The structure may contain aggregates or drainage cells (e.g., Atlantis Water Management Ltd. system) and may function as a water recycler or infiltration device.

Fifty-seven sites were hierarchically classified (nine levels) according to their public acceptability, land costs, water supply, drainage issues, site dimensions, slope, groundwater table depth, fragmentation of ownership and ecological value (Figure 2.2). The classification was based on expert water-engineering understanding, rather than on statistical evaluation, and accounted for flexibility in selecting (numerical) thresholds (e.g., estimated land cost).

Sustainable urban drainage should not cause any public health problems, avoid pollution of the natural environment, minimize the use of resources, operate in the long term and be adaptable to change in requirements (Butler and Parkinson, 1997). Based on this definition, a site potentially suitable for SUDS was defined in Section 2.1 as an area:

1. That is acceptable for the council, developers and the wider public (e.g., Greenfield and Brownfield areas);
2. That is associated with a potential source of water (e.g., car park runoff) that results in sufficient runoff (to be defined on a case-by-case basis);
3. That is associated with an area where water can drain to (e.g., river or canal);
4. With sufficiently large dimensions (e.g., width ≥150 m and length ≥300);
5. Associated with no major contamination issues (e.g., total lead <450 mg/kg soil);
6. That is associated with not too high land costs (e.g., <£200/m^2) before the development or regeneration;
7. That is owned preferably only by a few individuals or organizations (e.g., <5 parties);
8. That is preferably associated with a low groundwater table (e.g., less than 2 m);
9. With a sufficient slope (e.g., ≥1 in 50 m) for conveyance structures, but not too steep for three-dimensional SUDS features;
10. Of potentially high ecological impact in the future but not during the planning phase (e.g., not a Site of Special Scientific Interest (SSSI)); and
11. With sufficiently high soil infiltration (e.g., total drainage within 48 hours), filtration is considered to be desirable (see also issue 8 above).

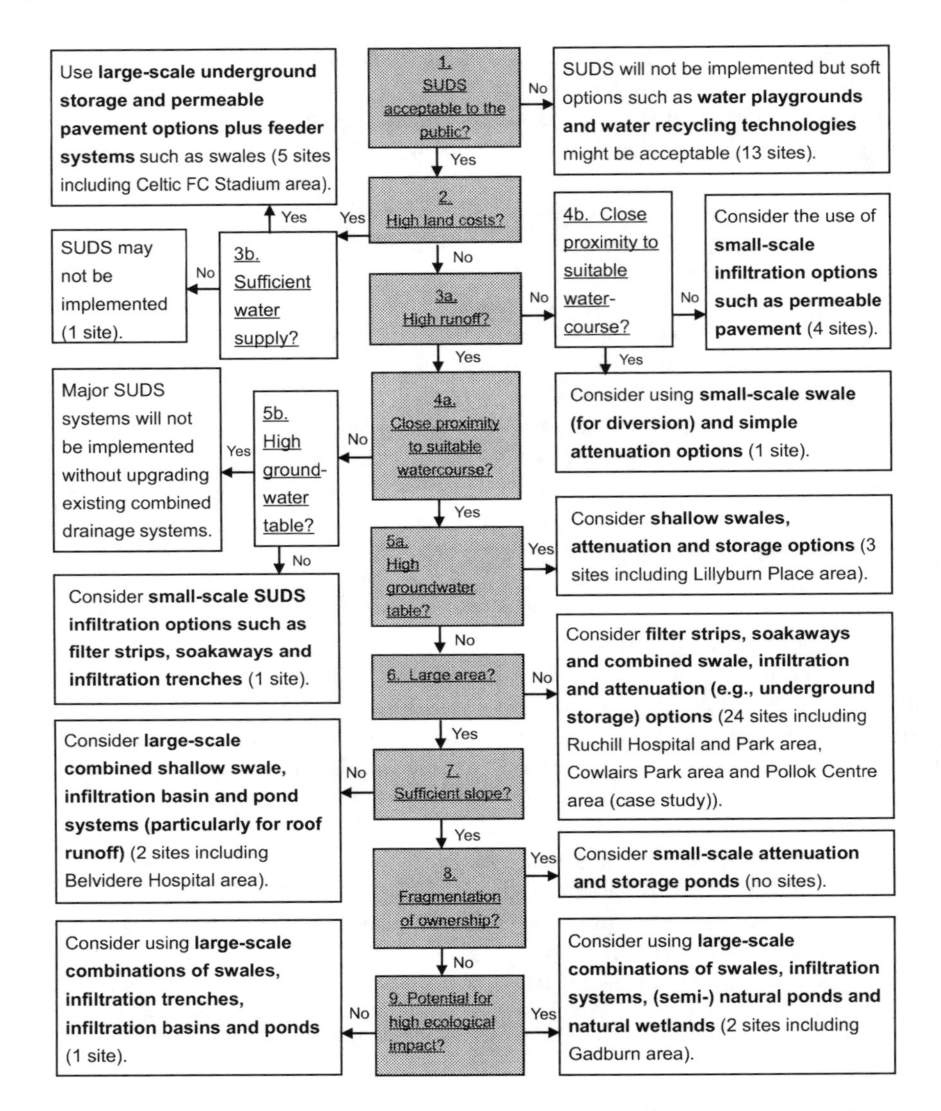

FIGURE 2.2 The sustainable (urban) drainage system (SUDS) Option Decision Support Key and the corresponding classification of 57 sites located in 46 areas available for regeneration and development.

2.1.2.2 SUDS Option Decision Support Matrix and Case Study

With the help of the findings based on the SUDS Option Decision Support Key, which is suitable for a quick assessment in the field, the SUDS Option Decision Support Matrix can subsequently be used for a detailed site assessment in the office. The criteria (most of them to be defined by national guidelines) for defining the best SUDS technique or combination of techniques as a function of a combination of site characteristics were defined as follows (Table 2.1):

TABLE 2.1
Sustainable Urban Drainage System (SUDS) Option Decision Support Matrix

	Runoff	Catchment (m^2)	Area for SUDS Feature	Serious Contamination	Land Value	Land Fragmented	High Ground-water	Sufficient Slope	High Ecological Impact	Soil Infiltration
Wetlands	High	>50,000	>5,000	No	<2	No	N/A	N/A	Yes	N/A
Ponds	High	>15,000	>50	No	<3	Yes	N/A	N/A	N/A	N/A
Lined ponds	High	>15,000	>50	Yes	<3	Yes	N/A	N/A	N/A	N/A
Infiltration basin	High	>15,000	>50	No	<3	Yes	No	N/A	N/A	High
Swale	High	N/A	>200	No	<3	No	No	Yes	N/A	N/A
Shallow swale	High	N/A	>200	No	<3	No	Yes	Yes	N/A	N/A
Filter strip	High	>15,000	>600	No	<3	Yes	No	Yes	N/A	High
Soakaway	Low	>3,000	>200	No	<3	Yes	No	Yes	N/A	High
Infiltration trench	Low	>3,000	>50	No	<3	No	No	Yes	N/A	High
Permeable pavement	Low/ High	N/A	N/A	No	N/A	Yes	N/A	N/A	N/A	High
Underground storage	Low/ High	N/A	>40	Yes	N/A	Yes	N/A	N/A	N/A	N/A
Water playground	Low	>200	>10	No	N/A	Yes	Yes	N/A	N/A	N/A

Land value: 1, low (<£100/m^2); 2, medium (≥£100/m^2 and ≤ £200/m^2); 3, high (>£200/m^2); N/A, not applicable.

1. Catchment size (specified for individual SUDS options) and associated hydrograph (Guo, 2000);
2. Area suitable for SUDS (specified for individual SUDS options);
3. Serious contamination as defined by national and international guidelines (yes or no);
4. Land value (low, medium, high or not applicable);
5. Fragmentation of ownership (yes or no);
6. High groundwater level (yes, no or not applicable);
7. Sufficient channel slope (yes, no or not applicable);
8. Potential of high ecological impact (yes, no or not applicable); and
9. Soil infiltration (low, high or not applicable).

The Pollok Centre Area is located at Longitude 4°20′ West and Latitude 55°49' North and lies approximately between 17 and 18 m above sea level. The area comprises a large shopping centre and associated car park, derelict land (Figure 2.3) and a meadow. The location of the area is at the intersection of the motorway M77 and Barrhead Road (Figure 2.4). It is approximately 8 km from Glasgow's city centre and forms part of the gateway to the city from the South (Figure 2.1).

Barrhead Road is a busy route that limits access from the North and is located near the motorway interchange. It is the major access road to the wider Pollok area. The Brock Burn marks a barrier in the West and South. The M77 is a barrier in the East. Cowglen Road is the only major access route in the Northwest. In the South, a footbridge and associated path create access to the area.

FIGURE 2.3 Pollok Centre area: site photograph taken on 20 May 2004.

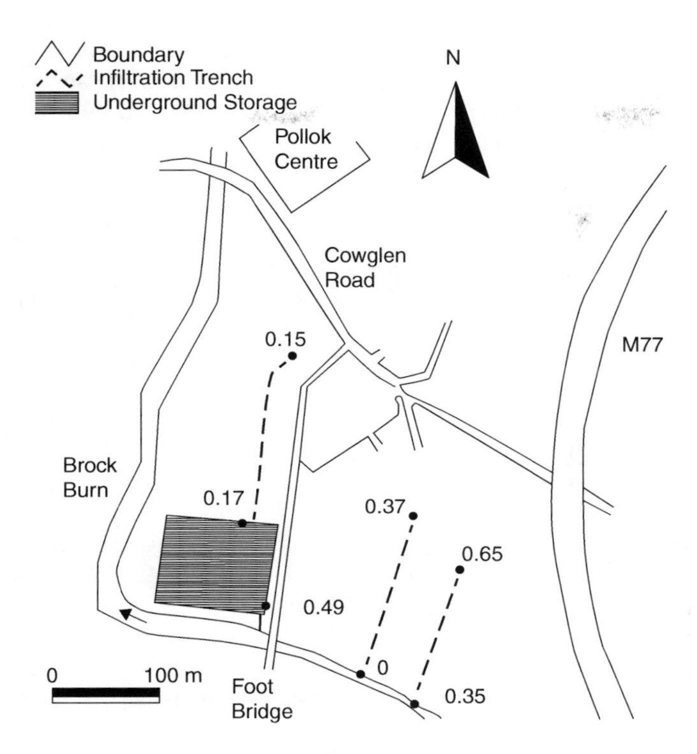

FIGURE 2.4 Pollok Centre area: proposed sustainable (urban) drainage system design drawing.

According to a study by Ove Arup & Partners Scotland (1996), the area is relatively flat, close to a stream and has a groundwater depth of only 1 m (Figures 2.3 and 2.4). The footpath separates the site proposed for redevelopment in the West and another site allocated for regeneration in the East. In the West, there is the site of the recently demolished Bellarmine School. This part of the site is protected against flooding by a large, vegetated flood defence bund designed to contain a design flood of 1 in 100 years return period (Ove Arup & Partners Scotland, 1996). In the East, a large meadow is located. This green space may be subject to flooding by the Brock Burn (part of the White Card River catchment) and functions therefore as a flood plain.

The strata underlying the area belong to the limestone coal group of the carboniferous series (sandstone, siltstone and mudstone with seams of coal and ironstone) which dips generally to the Southeast. The naturally occurring superficial deposits comprise glacial till (bolder clay; 10 m in thickness; in the Northeast only) with freshwater alluvial deposits (silt, clay, peat, sand and gravel; 5.0–12.5 m in thickness).

2.1.3 Results and Discussion

2.1.3.1 SUDS Option Decision Support Tools

The methodology outlines novel SUDS Option Decision Support Tools including a key and matrix that can be used to classify potential SUDS sites and to identify suitable SUDS techniques for potential SUDS sites. Figure 2.2 shows the outline of the SUDS Option Decision Support Key and the classification of all sites visited during the exploratory stage of this project. The key should be used in combination with Table 2.2 outlining an SUDS Option Decision Support Matrix.

The matrix has been tested with the exploratory data set collected during the site visits (Figure 2.1). Table 2.2 summarizes the outcomes of the application of this tool. The qualitative assessment (blank or cross(es)); Table 2.2), which led to this recommendation is purely based on the criteria summarized in this table. However, the quantitative assessment (one, two or three crosses; Table 2.2) that defines if particular SUDS techniques are applicable for specific sites was made by the site assessor based on his or her expert engineering judgement after using the key and matrix.

A weighting system for each criterion (matrix entry) in Table 2.2 could be defined to replace the bias introduced by the assessor. The nature of such a weighting system would obviously be influenced by national guidelines.

2.1.3.2 Case Study Application

The implementation of SUDS (wherever possible) is becoming mandatory for sites available for development or regeneration. This relieves the developer from undertaking a difficult and time-consuming cost-benefit analysis.

The proposed SUDS design for the Pollok Centre area is shown in Figure 2.5. Figure 2.4 shows a photograph of the site for which a major SUDS feature (underground storage) is planned. A planimeter investigation has shown that the horizontal catchment areas (West and East) of the Pollok Centre area available for integration of SUDS techniques would be 28,000 and 26,500 m^2, respectively.

A desk study was undertaken in the Pollock Centre area dominated by a shopping centre complex. From the current and historical site information, no significant levels of contamination were found to be present. However, it is known that dumping of untreated domestic refuse was carried out in the former recreation ground immediately to the east of the former Bellarmine School and South of the Pollock Shopping Centre car park (Figure 2.5). Two trenches of 3-m depth were dug and filled with refuse probably in the late 1960s and early 1970s during the strike action of the refuse workers (Ove Arup & Partners Scotland, 1996).

The sites available for regeneration in the West and development in the East will be an extension of the commercial retail park area of the Pollock Shopping Centre. The drainage system on this site will be divided into two independent parts (Figure 2.5). An option for the western site is to introduce underground storage in the South below a potential storage area, employee car park or customer overflow car park for the new shopping development. It is recommended that an infiltration trench runs from the existing Pollok Centre car park in the North, parallel to a new road, to the underground storage in the South (Figure 2.5).

The eastern area is to remain in its current capacity as a floodplain despite the presence of a car park that should be appropriately landscaped with tree planting at the end of bays, mixed internal planting and beech hedging. It is recommended that two semi-independent infiltration trenches should be constructed to drain all surface water runoff from a potential overflow car park in the Southeast (Figure 2.4).

TABLE 2.2
Sustainable Urban Drainage System (SUDS) Options Based on the SUDS Option Decision Support Matrix (Table 2.1)

Area	Catchment	Wetland	Pond	Infiltration Basin	Swale	Infiltration Trench	Soakaway	Filter Strip	Perm. Pavement	Underground Storage	Water Playground
Belvi Dere Hospital	Entire Area		XXX	X	XX	X			XX	X	XX
Celtic FC Stadium	Entire Area					X	X		XX	XXX	
Cowlairs	North		XXX	X	XX	X	X	X	XX	X	XX
Park	South		XXX		XX		X	XX	XX	X	XX
Gadburn	North	X	XXX		XX	X			XX	X	XX
	South	XXX	XX		XX				XX	X	XX
Lillyburn place	Entire area		X			XXX	X		X		X
Pollok	West				XX	X			XX	XXX	
Centre	East		X		XXX	XX	X		XX	X	X
Ruchill Hospital	North-east				XX				XX	XXX	
Park	South-east				XX				XX	XXX	
	South		XXX		XX				XX	X	XX
	West		XXX		XX	X	X	X	XX	X	XX

X, possible option; XX, recommended option; XXX, predominant SUDS design feature.

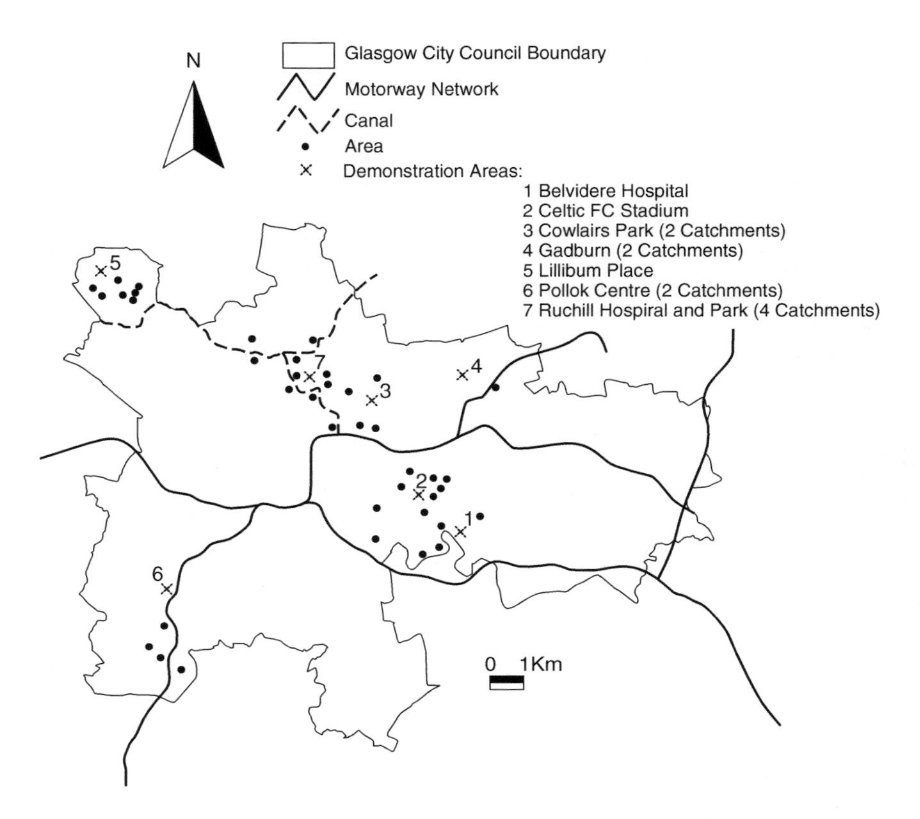

FIGURE 2.5 Indication of 46 potential areas comprising 57 sites for the implementation of sustainable urban drainage systems (SUDS). The SUDS demonstration areas have been highlighted.

The sub-catchment to the East of the trenches should remain a natural wetland (Scholz and Trepel, 2004), and there is the opportunity to create small ponds (Scholz, 2003) to increase the amenity and habitat value as well (SEPA, 1999; CIRIA, 2000). The infiltration trenches and the wetland should act as a passive flood attenuation measure. The wetland should be directly linked to the Brock Burn, a significant green corridor with potential for landscape enhancement: the stream is historically known as a 'trolley graveyard'.

2.1.4 Conclusions and Recommendations

A survey of 57 sites within 46 areas of Glasgow shows that it is feasible to implement different SUDS techniques throughout Glasgow. Characteristics that determine the suitability of a site for the implementation of SUDS in general have been identified. Representative areas and sites that are suitable for different representative SUDS techniques have been determined qualitatively and quantitatively.

The concept of using SUDS Option Decision Support Tools has been introduced. The methodologies for an SUDS Option Decision Support Key and a corresponding

SUDS Option Decision Support Matrix that are adaptable to different cities and even countries have been outlined. The key should be used during a preliminary site investigation to classify potential SUDS sites. Corresponding findings have to be compared with the criteria listed in the matrix, which will inform the planner about potentially suitable SUDS techniques for a particular site.

Seven entirely different SUDS demonstration areas that are representative of both different SUDS techniques and different types of areas available for development and regeneration within Glasgow have been identified. Design and management guidelines for a demonstration site, which should be constructed to inform and educate the public, developers and politicians have been formulated.

Underground storage tanks, ponds and wetlands linked with swales and infiltration trenches have been identified as the most useful sustainable drainage techniques for large sites within Glasgow. For example, the drainage of the Pollok Centre area (case study example representative for 24 sites) should be dominated by independent infiltration trenches draining either into an underground storage tank or directly into a nearby stream.

2.2 DEVELOPMENT AND MANAGEMENT OF STORM WATER RESOURCES

2.2.1 INTRODUCTION

The European Union's (EU) Water Framework Directive (Council of European Communities, 2000), which came into force on 23 October 2000, requires all inland and coastal waters to reach 'good status' by 2015. The Directive sets a framework that should provide substantial benefits for both long-term water resource development and water quality management. The implementation of SUDS based on current guidelines (CIRIA, 2000; Jefferies et al., 1999; McKissock et al., 1999) in Glasgow can help in preventing flooding from watercourses and sewer systems, and combined sewer overflows to spill untreated sewage into receiving watercourses such as rivers and canals during storms (DEFRA, 2000; Scholz, 2004a). Furthermore, SUDS can help to reduce the impact of diffuse pollution on urban watercourses by promoting passive treatment (D'Arcy and Frost, 2001).

The theme of TRUST is based upon multi-functional water resource development, water storage, integral surface water management and public and stakeholder participation. Six different authorities and institutions throughout Europe contributed to this project: British Waterways (Watford, UK), Gewestelijke Ontwikkelingsmaat-schappij (Brugge, Belgium), Glasgow City Council (Glasgow, UK), Hoogheemraadschap van Scieland (Rotterdam, The Netherlands), Provincie Noord Holland (Haarlam, The Netherlands) and the University of Osnabrück (Osnabrück, Germany).

Glasgow City Council's contribution to the TRUST project is known as the 'Glasgow Surface Water Management Project'. The project published a study of the inner-city canal corridor proposing innovative urban drainage recommendations. It is intended also to commission and construct at least one SUDS demonstration area (see below), and to investigate further expansion of the scheme in the wider region. The first study output is the 'The Glasgow Sustainable Urban Drainage System Management Project', led by the author of this book, and funded by The Royal Academy of Engineering and Glasgow City Council. The main focus is on the Ruchill Hospital and Park area of the city where the adjacent canal corridor provides an opportunity to devise an innovative surface water management plan based on SUDS. The project specifically looks at optimal methods and techniques to separate new surface runoff from the existing combined sewer system, redirecting it to the canal, and using it as a managed surface water collector system.

Lillyburn Place is a part of Drumchapel that is located in the Northwest of Glasgow (Figure 2.5). The area is predominantly owned by Glasgow City Council. Drumchapel is a suburb developed in the 1950s and 1960s, located on a valley edge amidst drumlins and the Kilpatrick Hills. The Lillyburn Place is part of the 'New Neighbourhood Initiative' managed by Glasgow City Council. The aim is to regenerate this area to create an attractive and sustainable residential suburb of Glasgow. The initiative plans to complement the existing housing by introducing new family houses and smaller homes (Glasgow City Council, 2004a).

In comparison, The Ruchill Hospital and Park area is located in the North of Glasgow (Figure 2.5), and incorporates the former Ruchill Infectious Diseases Hospital, and Ruchill Park. The former Ruchill Hospital is owned by Scottish Enterprise and Ruchill Park by Glasgow City Council. The surrounding area has traditionally been dominated by council housing, much of which, over the last 30 years, has fallen into a state of disrepair and become undesirable. This area is one of the 'key areas' in Glasgow's New Neighbourhood initiative and is, therefore, also an attractive demonstration site for the TRUST project due to its location near the Forth and Clyde Canal.

The feasibility to implement different SUDS throughout Glasgow and their potential contribution to the general water resource development and overall catchment dynamics has been studied. The data should help in understanding the challenges of holistic catchment management, diffuse pollution and the 'linking scales' in water resource development and catchment management.

This project aims to come up with SUDS demonstration areas that are representative of both different sustainable drainage techniques and different types of areas available for regeneration. The objectives are to:

i. Identify variables that determine the suitability of a site for the implementation of SUDS;
ii. Identify suitable SUDS sites within Glasgow;
iii. Classify qualitatively and quantitatively sites suitable for different SUDS technologies;
iv. Outline both a general SUDS implementation key and feasibility matrix;
v. Identify representative SUDS technologies for representative sites that could be used for demonstration purposes;
vi. Provide detailed design and storm water resources management guidelines for representative sites and representative SUDS technologies for information and education purposes.

Section 2.2 is based on an updated version of an original article by Scholz et al. (2005).

2.2.2 Methodology

2.2.2.1 SUDS Option Decision Support Key

Figure 2.6 shows a map of Glasgow highlighting the spatial distribution of 46 areas (associated with 57 sites) that were identified as potentially suitable for the implementation of SUDS. Eight areas had the potential for more than one SUDS system

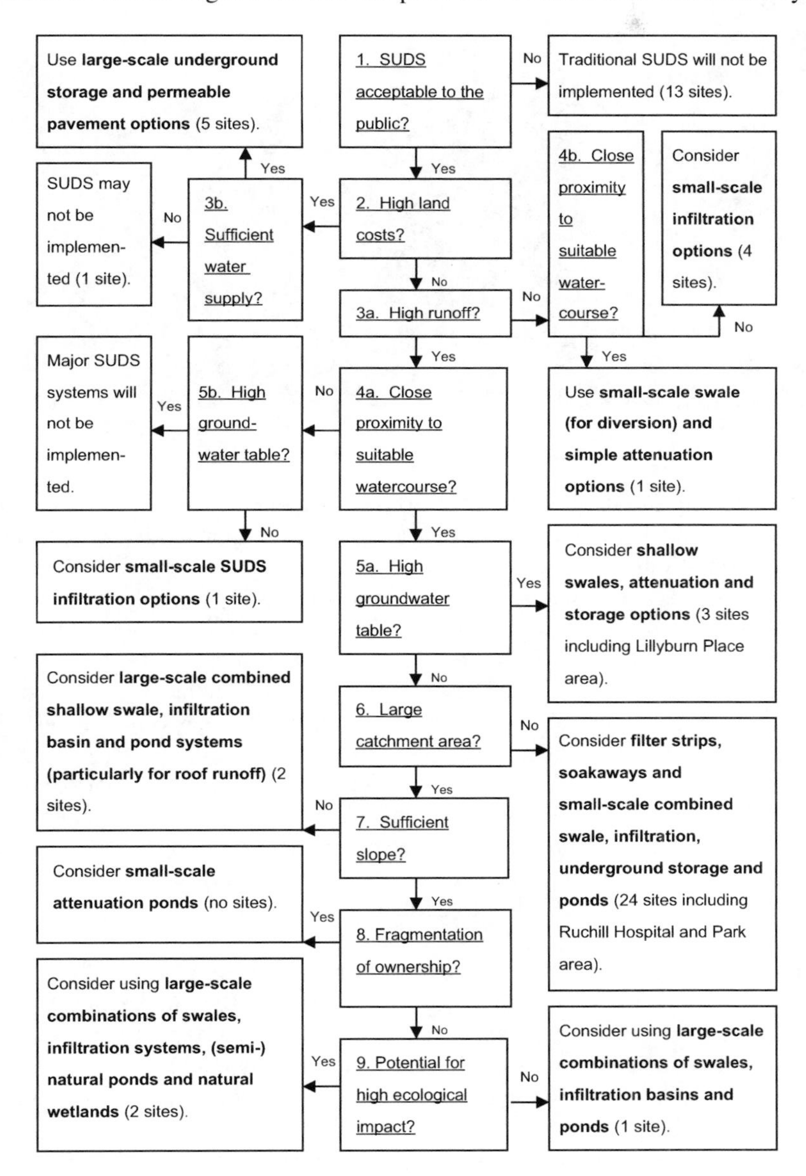

FIGURE 2.6 The sustainable urban drainage system (SUDS) feasibility key (showing nine levels of assessment) and classification of 57 sites located in 46 areas available for regeneration and development.

FIGURE 2.7 Lillyburn Place area: site photograph taken on 11 May 2004.

and were therefore sub-divided. Every effort has been made to also investigate areas currently represented only sparsely by discussing SUDS opportunities with planners employed by the Council.

Fifty-seven sites were hierarchically classified (nine levels) according to their public acceptability, land costs, water supply, drainage issues, site dimensions, slope, groundwater table depth, fragmentation of ownership and ecological value (Figure 2.7). The classification was based on expert water-engineering understanding, rather than on statistical evaluation, and accounts for flexibility in selecting (numerical) thresholds (e.g., estimated land cost).

Sustainable urban drainage should not cause any public health problems, avoid pollution of the natural environment, minimize the use of resources, operate in the long-term and be adaptable to change in water resources development and management requirements (Butler and Parkinson, 1997). Based on this definition, a site potentially suitable for SUDS has been defined as an area:

1. That is acceptable for development or regeneration to Glasgow City Council, developers and the wider public (e.g., Greenfield and Brownfield areas);
2. That is associated with a potential source of water (e.g., car park runoff) that results in sufficient runoff (to be defined on a case-by-case basis);
3. That is associated with a separate area to which water can drain (e.g., canal or river);
4. With sufficiently large dimensions (e.g., width $\geq$150 m and length $\geq$300 m);
5. Associated with no major soil contamination challenges;
6. That is associated with not too high land costs (e.g., <£200/m^2) before development or regeneration;

7. That is owned preferably only by a few individuals or organizations (e.g., < 5 parties);
8. That is preferably associated with a low groundwater table (e.g., <2 m);
9. With a sufficient slope (e.g., ≥1 in 50 m) for conveyance structures, but not too steep for three-dimensional SUDS features;
10. Of potentially high ecological impact in the future, but not during the planning phase (e.g., not an SSSI site); and
11. With sufficiently high soil infiltration, filtration is desirable.

The criteria for defining SUDS options were as follows (Table 2.3):

1. Runoff (low or high);
2. Catchment size (specified for individual SUDS options);

TABLE 2.3
Sustainable Urban Drainage System (SUDS) Option Feasibility Decision-Making Tool

SUDS Option	Runoff	Catchment Size (m^2)	Area of for SUDS Feature	Contamination	Land Value
Wetlands	High	>50,000	>5,000	No	Low
Ponds	High	>15,000	>50	No	Medium
Lined ponds	High	>15,000	>50	Yes	Medium
Infiltration basin	High	>15,000	>50	No	Medium
Swale	High	N/A	>200	No	Medium
Shallow swale	High	N/A	>200	No	Medium
Filter strip	High	>15,000	>600	No	Medium
Soakaway	Low	>3,000	>200	No	Medium
Infiltration trench	Low	>3,000	>50	No	Medium
Permeable pavement	Low/High	N/A	N/A	No	N/A
Underground storage	Low/High	N/A	>40	Yes	N/A
Water playground	Low	>200	>10	No	N/A

SUDS Option	Ownership Fragmented	High Groundwater Level	Sufficient Channel Slope	Ecological Impact	Infiltration
Wetlands	No	N/A	N/A	Yes	N/A
Ponds	Yes	N/A	N/A	N/A	N/A
Lined ponds	Yes	N/A	N/A	N/A	N/A
Infiltration basin	Yes	No	N/A	N/A	High
Swale	No	No	Yes	N/A	N/A
Shallow swale	No	Yes	Yes	N/A	N/A
Filter strip	Yes	No	Yes	N/A	High
Soakaway	Yes	No	Yes	N/A	High
Infiltration trench	No	No	Yes	N/A	High
Permeable pavement	Yes	N/A	N/A	N/A	High
Underground storage	Yes	N/A	N/A	N/A	N/A
Water playground	Yes	Yes	N/A	N/A	N/A

3. Area suitable for SUDS (specified for individual SUDS options);
4. Serious soil contamination (yes or no);
5. Land value (low, medium, high or not applicable);
6. Fragmentation of ownership (yes or no);
7. High groundwater level (yes, no or not applicable);
8. Sufficient channel slope (yes, no or not applicable);
9. Potential of high ecological impact (yes, no or not applicable); and
10. Soil infiltration (low, high or not applicable).

Two representative demonstration areas (see below) have been selected based on this SUDS Option Decision Support Key and the SUDS definition (see above). Moreover, an attempt has been made to select areas that are associated with as many SUDS suitability groups as possible. It follows that only seven areas suitable for the implementation of SUDS are not represented by the selected demonstration areas (Figure 2.7). However, it must be emphasized that the selection is rather qualitative than quantitative considering that most selection criteria do not require a numerical assessment. It follows that the SUDS classification is like an expert system and not a statistically unbiased assessment that would, however, not be suitable in this case anyway because of the lack of numerical information such as land value (e.g., recognizing also the future potential).

2.2.2.2 Analytical Work

The soil recording and pre-treatment before analysis were carried out in agreement with British Standards (British Standard Institute, 1999a, 2002). The determination of the particle size distribution was also carried out according to British Standards (British Standard Institute, 1999b).

Composite samples were collected and stored at −10°C prior to analysis. After thawing, approximately 2.5 g of each soil sample was weighed into a 100-mL digestion flask to which 21 mL of hydrochloric acid (strength of 37%, v/v) and 7 mL of nitric acid (strength of 69%, v/v) were added. The mixtures were then heated on a Kjeldahl digestion apparatus (Fisons, UK) for at least 2 hours. After cooling, all solutions were filtered through a Whatman Number 541 hardened ashless filter paper into 100-mL volumetric flasks. After rinsing the filter papers, solutions were made up to the mark with deionized water. The method was adapted from the section 'Nitric Acid-Hydrochloric Acid Digestion' of the American Public Health Association (1995).

An Inductively Coupled Plasma Optical Emission Spectrometer (ICP-OES) was used for the analysis of metals and other heavy elements. Total concentrations of elements in filtered (Whatman 1.2 µm cellulose nitrate membrane filter) samples were determined by ICP-OES using a TJA IRIS instrument (ThermoElemental, U.S.). Multi-element calibration standards with a wide range of concentrations were used, and the emission intensity was measured at appropriate wavelengths.

2.2.2.3 Demonstration Case Studies

The Lillyburn Place area (Figure 2.7) is located approximately Longitude 4°23′ West and Latitude 55°55′ North. Lillyburn Place is in the Cleddans part of Drumchapel, about 15 km Northwest of Glasgow City Centre (Figure 2.1). The approximately rectangular catchment area is about 25,500 m^2. The Lillyburn Place area can be divided into two sites, separated by the Achamore Road. However, the whole area is still one catchment in terms of its hydrology. The site in the West is currently undeveloped (Brownfield site), while the site in the East is predominantly dominated by social housing and Drumchapel Baptist Church in the Northeast of the area. The northern border is adjacent to a small stream (partly culverted in the East), and the southern border is demarcated by Fasque Place having a bus terminus. The Lillyburn Place area benefits from access to the A82 Great Western Road dual carriageway.

According to Glasgow City Council plans, two-storey family housing and associated landscaping, public open spaces, visitor parking, children's play areas and integrated SUDS should be constructed. However, two existing electrical substations within the area (North and Southwest) must be retained (Glasgow City Council, 2004a).

The topography of the Lillyburn Place area is relatively flat (Figure 2.8). A nearby high point with a water tower is located south-westerly of the area.

In comparison, the Ruchill Hospital and Park area is located at Longitude 4°16′ West and Latitude 55°53′ North, approximately 3 km from Glasgow City Centre (Figure 2.5) incorporating the former Ruchill Infectious Diseases Hospital (Figure 2.9) and Ruchill Park.

In 1891, the boundaries of Glasgow were extended to include Ruchill, and the Corporation of the City of Glasgow purchased land there for a new hospital for infectious diseases. A desk study into the wider Ruchill area revealed that until 1892,

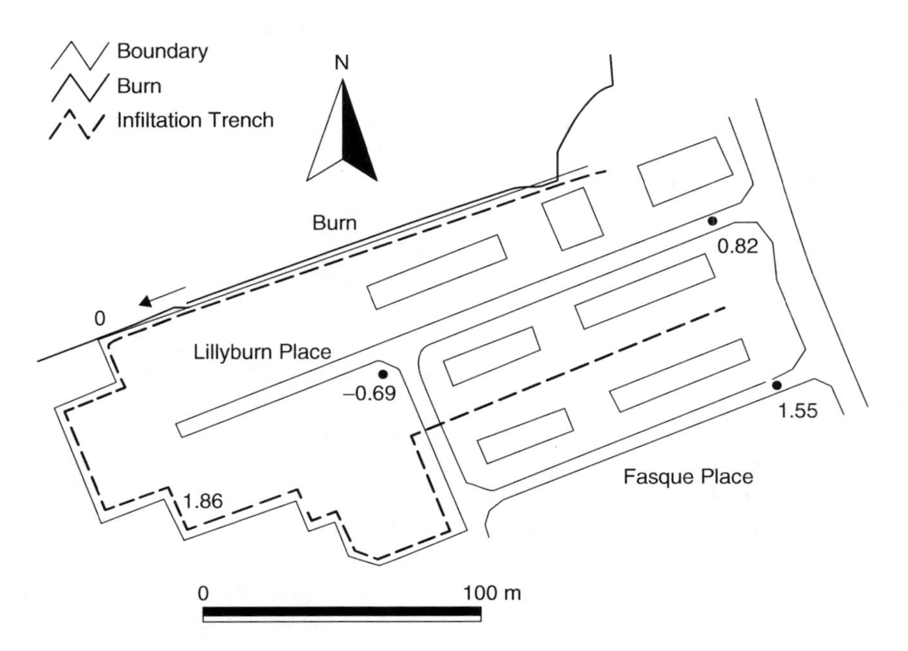

FIGURE 2.8 Lillyburn Place area: proposed sustainable urban drainage system design drawing and indication of relative heights to the infiltration trench overflow structure.

FIGURE 2.9 Ruchill Hospital area: site photograph taken on 20 May 2004.

when the construction of Ruchill Hospital began, the entire site was part of Ruchill Park (Glasgow City Council, 2003). Ruchill Hospital was designed by the famous City Engineer Mr. A. B. McDonald.

Ruchill Park lies adjacent to the West of the hospital and was opened in 1892. The poor quality of the soil and its high exposed location were not ideal for a public park, but under the direction of Park's Superintendent Mr. James Whitton, the area was transformed. The park's best-known feature is the panoramic view of Glasgow and its surroundings, which can be obtained from the top of the hill. This is topped by an artificial mound (with a flagpole) created from 24,000 cartloads of material excavated during the building of the hospital. It is known locally as 'Ben Whitton' after the Director of the Park at that time.

Most of the sub-areas of Ruchill Hospital have been cleared for new housing, with a few listed historical buildings remaining. However, due to these remaining structures, it is unlikely that the general layout of the land will change substantially (Glasgow City Council, 2004b). The overall topography is relatively uneven, with a hill in the centre and sharp drops on all sides of the hospital catchments (Figure 2.10).

The western end of the site borders Ruchill Park (Figure 2.11), a landscaped recreational sub-area that has low points in the North, West and South (Figure 2.10). The park is a significant green space area with a limited potential to store and convey surface water runoff from Ruchill Hospital. Moreover, since the Northeast corner of Ruchill Park is at a much lower elevation (Figure 2.10) than the Ruchill Hospital and the remaining areas of the park, it is therefore a potential location for a major SUDS feature.

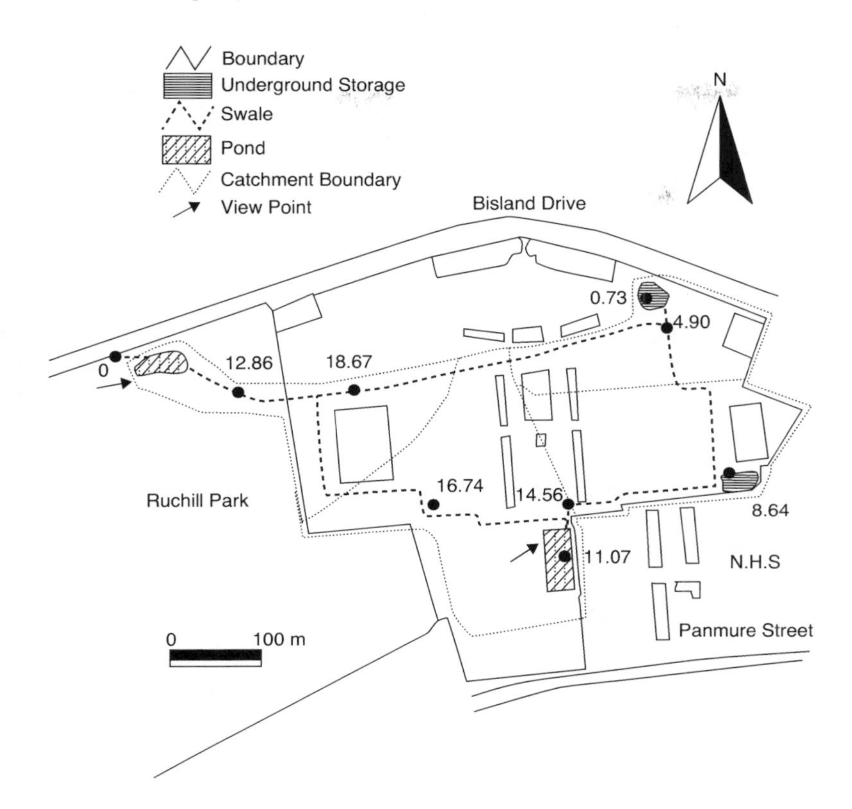

FIGURE 2.10 Ruchill Hospital area: proposed sustainable urban drainage system design drawing and relative heights to the lowest SUDS feature.

FIGURE 2.11 Ruchill Park area: site photograph taken on 7 May 2004.

No detailed geotechnical records exist for the Ruchill Hospital and Park area, as no substantial work has been done on the site for over 100 years. However, hospital-related contamination, such as remains of cleansing detergents, is expected to be located in the top soil. This is not considered a major problem, requiring little, if any, clean up.

2.2.3 Results

Figure 2.7 shows the outline of the SUDS feasibility key and the classification of all sites visited during the exploratory stage of this project. The key may be used in combination with Table 2.2 outlining an SUDS option feasibility decision-making tool. This tool has been tested with the exploratory data set collected during the site visits, and Table 2.4 summarizes the outcome of the application of this tool.

The sizes of the demonstration areas that should be occupied by major SUDS features have been summarized in Table 2.5. The planimeter investigation has shown that the horizontal area of the Lillyburn Place area available for the integration of SUDS techniques would be 25,500 m^2. In comparison, the horizontal catchment areas (Northeast, Southeast, South and West) of the Ruchill Hospital and Ruchill

TABLE 2.4
Sustainable Urban Drainage System (SUDS) Options Based on the SUDS Options Feasibility Decision-Making Tool (Table 2.2)

Area	Catchment	Wetland	Pond	Infiltration Basin	Swale	Infiltration Trench
Lillyburn Place	Entire area		X			XXX
Ruchill	Northeast				XX	
Hospital	Southeast				XX	
And	South		XXX		XX	
Park	West		XXX		XX	X
Area	**Catchment**	**Soak-away**	**Filter Strip**	**Permeable Pavement**	**Underground Storage**	**Water Playground**
Lillyburn Place	Entire area	X		X		X
Ruchill	Northeast			XX	XXX	
Hospital	Southeast			XX	XXX	
And	South			XX	X	XX
Park	West	X	X	XX	X	XX

X, possible option; XX, recommended option; XXX, predominant SUDS design feature.

TABLE 2.5
Catchment and Proposed Sustainable Urban Drainage System (SUDS) Areas Measured with a Planimeter

Area Name	Catchment or SUDS Feature name	Size (m^2)
Lillyburn Place	Catchment	25,500
	Impermeable area (approx.)	19,000
	Infiltration trenches	1,000
Ruchill Hospital and Park	Catchment in the Northeast	26,000
	Impermeable area (approx.)	19,500
	Swales in the Northeast	1,800
	Underground storage in the Northeast	830
	Catchment in the Southeast	25,000
	Impermeable area (approx.)	18,000
	Swales in the Southeast	1,300
	Underground storage in the Southeast	870
	Catchment in the South	38,400
	Impermeable area (approx.)	23,000
	Swales in the South	1,500
	Pond in the South	2,500
	Catchment in the West	16,500
	Impermeable area (approx.)	9,000
	Swales in the West	1,700
	Pond in the West	1,750

Assumed widths of small swales and infiltration trenches are 5.0 and 1.5 m, respectively.

Park areas available for the integration of SUDS techniques would be 26,000, 25,000, 38,400 and 16,500 m^2, respectively.

Figure 2.7 shows a photograph of the Lillyburn Place area (West) for which a major SUDS feature (infiltration trench with overflow structure) is planned. The proposed SUDS design for the Lillyburn Place area is shown in Figure 2.8. In comparison, Figures 2.9 and 2.7 show photographs of the Ruchill Hospital (South) and Park (Northwest) area for which storm water ponds are planned. The proposed SUDS design (ponds) for these catchments is illustrated in Figure 2.11. Figures 2.12 and 2.13 indicate the corresponding artist impressions for these sites after regeneration.

Furthermore, Table 2.6 summarizes the soil quality at 10-cm depth within both demonstration areas. Figure 2.14 shows particle size curves for a site close to the proposed storm water pond in the South of Ruchill Hospital.

FIGURE 2.12 Artist impression of the Ruchill Hospital site (catchment in the South): (a) pencil drawing and (b) computer animation.

FIGURE 2.13 Artist impression of the Ruchill Park site: (a) pencil drawing and (b) computer animation.

TABLE 2.6
Major Heavy Metals (mg/kg dry weight): Comparison of Soil Quality at 10-cm Depth during the Exploratory Investigation of 57 Sites

Area	Site	Al	Cr	Cu	Fe	Mn	Ni	Pb	Zn
Lillyburn Place	West	4,839	66	14	24,883	374	19	51	45
Ruchill	North-East	3,125	13	30	22,065	416	27	170	123
Hospital	East	2,765	7	22	15,809	483	17	374	217
And	Centre-East	3,653	15	41	21,629	283	23	451	231
Park	South	9,131	77	34	24,688	594	25	1,307	434
	North-West	11,507	78	37	23,606	311	30	194	158
	Park	4,515	21	33	33,096	504	35	298	135
All sites for all areas		12,538	96	72	27,375	485	34	198	180

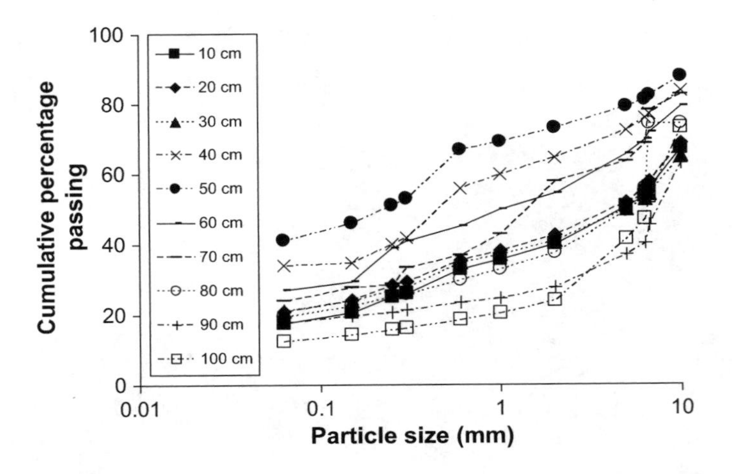

FIGURE 2.14 Ruchill Hospital area: particle size distribution curves for samples taken at 10 cm intervals within a ditch of 1.05-m depth.

2.2.4 Discussion

2.2.4.1 Management of Drainage Systems

For this study, the following definitions of SUDS and its associated features hold true throughout:

1. SUDS: This is an acronym for Sustainable (Urban) Drainage System (British concept), also known as Best Management Practice (American concept). A singular or series of management structures and associated processes designed to drain surface water runoff in a sustainable approach to predominantly alleviate capacities in existing conventional drainage systems (predominantly combined sewers in Glasgow) in an urban environment

(Butler and Davies, 2000; CIRIA, 2000; SEPA, 1999). Moreover, storm water runoff is seen as a potentially valuable water resource (and not as a waste product) that requires sustainable development and management. This philosophy contrasts the traditional management of storm water further downstream in the catchment, for example, construction of dams within river valleys for the purpose of flood control as discussed elsewhere (Takahasi, 2004).

2. Pond: A depression structure that increases the duration of the flow hydrograph with a consequent reduction in peak flow, with the depression always having a minimum depth of water present and an overflow outlet to an existing watercourse or sewer. Ponds can be used for attenuation, detention, retention, storage, infiltration and recreational purposes (Guo, 2001; Scholz, 2003). The outlet of the storm water pipe leading into the pond should have a silt trap constructed according to the best management practice (Scholz, 2004b). Moreover, storm water can be used as a valuable water resource and therefore recycled for irrigation purposes (Scholz, 2003).
3. Swale: A grass-lined conveyance structure of varying width designed to infiltrate but predominantly transport water from a site, while controlling the flow and quality of the surface water. A swale may convey water to a watercourse, SUDS feature or sewer.
4. Infiltration trench: A linear drain (also known as a French Drain) consisting of a trench filled with a permeable material, potentially with a perforated pipe in the base or several perforated pipes at designated depths to promote infiltration of surface water runoff to the ground. An infiltration trench may also convey water, if the trench gradient is sufficiently steep.
5. Underground storm water storage: A sub-surface structure designed to accumulate surface water runoff, and where water is released from as may be required to increase the flow hydrograph. The structure may contain aggregates or plastic boxes (e.g., Atlantis Water Management Ltd. system) and act also as a water recycler or infiltration device.

Optimizing the maintenance of SUDS structures is one of the greatest management challenges. Mowing grass and removing litter and debris are the most time-consuming and therefore costly maintenance tasks (Jefferies et al., 1999; McKissock et al., 1999; Scholz, 2003).

Maintenance of all public SUDS structures above ground is usually the responsibility of the local authority (The Stationery Office, 1998). Aboveground SUDS structures are defined as swales, ponds, basins and any other ground depression features. In contrast, the maintenance of underground SUDS structures is usually the responsibility of the local water authority. Underground SUDS structures include infiltration trenches, filter strips, underground storage and culverts (Butler and Davis, 2000; CIRIA, 2000; Nuttall et al., 1998).

2.2.4.2 Soil Contamination

The soil contaminants summarized in Table 2.6 should be seen in context with soil contamination guidelines. The guidelines specify thresholds for heavy metals such as chromium, copper, manganese, nickel, lead and zinc. However, only lead was identified as the major contaminant: The threshold for residential properties (with and without plant uptake) is 450 mg/kg dry weight. In contrast, the threshold for commercial and industrial land is 750 mg/kg dry weight (Environment Agency, 2002). In comparison, the Dutch intervention concentration is 530 mg/kg dry weight (Ministry of Housing, Spatial Planning and Environment, 2000), while the Kelly Indices Guidelines for Contaminated Soils (specifically developed for gasworks sites in London) is 1,000 mg/kg dry weight (Society of the Chemical Industry, 1979).

2.2.4.3 Classification and Demonstration Areas

The classification is particularly useful for practitioners without a deep understanding of SUDS. Moreover, it can be adapted in a revised form for water resources developers and managers of urban areas in industrialized and developing countries. It is of practical use also for policymakers (level 1), sociologists (levels 1 and 8), economists (level 2), hydrologists (level 3), geographers (levels 4, 6 and 7), geologists (level 5) and limnologists (level 9); see Figure 2.7 for a brief description of the levels.

Future building design plans for Lillyburn Place have been defined in the Glasgow City Plan (Glasgow City Council, 2004a). It has been proposed to build different types of houses: 9 detached, 34 semi-detached and 18 terraced houses. All properties should have private gardens.

According to the characteristics, size, topography and geology, the Lillyburn Place area comprises only one small catchment area that will be built up completely. Therefore, only one SUDS technique, an infiltration trench with an emergency overflow to the burn in this case, has been proposed (Figure 2.8)

The infiltration trench will follow the boundaries of the undeveloped part of the area in the West and crosses private gardens in the East. The drain will allow the infiltration of surface water runoff in this area. The outlet of the existing burn is in the Northwest. The recommended location for the infiltration trench is shown in Figure 2.8.

Overall, the concentration of organic compounds found was low (estimated to be less than 0.5 mg/kg). Pollutants found included polycyclic aromatic hydrocarbons (PAH). For example, pyrene, fluoranthene, chrysene, benz(a)anthracene, benzo(k) fluoranthene, benz(a)pyrene and benzo(g,h,i)perylene were found at very low levels in the Lillyburn Place area (near the proposed overflow structure). Other compounds found included aliphatic hydrocarbons such as tetracosane, eicosane, heptadecane, heptacosane and nonadecane, which are commonly constituents of diesel-type fuels. Phenol derivatives and carbolic-acid-related compounds were also identified. These types of compounds were often used as cleaning agents and disinfectants.

In comparison, future regeneration plans for the area combining the Ruchill Hospital and Ruchill Park are unclear. However, it is believed that low-to-medium-density residential properties will be in the majority of future housing developments. This is the fundamental assumption on which the following recommendations have been based upon.

Due to the large size of the area (Table 2.5), there appears to be a capacity for several different SUDS techniques (Table 2.4 and Figure 2.11). The area is split into four sub-catchments relevant to the proposed SUDS based on the current topography, with the area to the North being regarded as unfeasible for SUDS implementation due to the steep slope and extensive vegetation, which is expected to remain. Although minor changes of the area topography can be expected due to levelling, the remaining listed buildings will limit future housing developments. Each sub-catchment supplies surface water runoff to the different SUDS features (either ponds or underground storage) via a series of interconnected swales that surround the entire site, as can be seen in Figure 2.11.

At the Northeast and Southeast of Ruchill Hospital, two areas of underground storage are proposed. This decision was made on the limited space available on sloping areas, and the flexibility that underground storage offers. For example, a car park or recreational green space can be placed on top. Although the water from these sub-catchments will ultimately end up in the combined sewer system, due to its physical distance to any other outlet, the main function of the underground storage will be flood alleviation. However, if in the future a separate sewer system is retrofitted into the wider Ruchill area, the stored runoff water may be integrated into the separate drainage system with ease. A connection between the pond and canal via a separate culvert may be envisaged in the future.

The proposed solution for the south side of the Ruchill Hospital area is a retention pond, as there is sufficient space and the pond can be incorporated into an existing green space area (Figure 2.11). The steep slope evident on this side of the site would make major infiltration unfeasible because of the risk of flooding properties located further South and East below the pond level. Moreover, soil analysis has revealed the presence of bolder clay (below a depth of about 0.7 m) at the location of the proposed pond in the South of Ruchill Hospital (Figure 2.14). It is therefore likely that infiltration properties are low.

Considering the pond's proximity to the Forth and Clyde Canal, it is likely that it could overflow into the Firhill Basin via a culvert, instead of draining directly into the combined sewer system. It is also envisaged that stored water could be used for irrigation purpose by the Ruchill Mental Health Hospital for its greenhouse and garden scheme, which will be located just across the site boundary to the East of the proposed retention pond in the South (Figures 2.9, 2.10 and 2.12).

In the Northwest of the Ruchill Hospital sub-area, there is the potential to divert runoff to the Northeast of the Ruchill Park sub-area, creating another retention pond. In addition to being a feature for the park, this pond will be an important system for the potential expansion of SUDS in the wider Ruchill area. For example, roof runoff from the proposed indoor recreational area can be collected, and due to the site's low elevation, nearby road (Bisland Drive) and roof runoff can also be collected. It is then suggested that the water detained can be released into the Forth and Clyde Canal via a culvert.

In the future, there is also potential for expansion into the Hugo Street area, as this is another area on which regeneration is expected. However, it is recognized that serious contamination including cyanide is an issue here and only a lined SUDS system could be used without undertaking major excavation work.

The ponds and underground storage facilities will allow runoff to be attenuated (and partly recycled) before it is diverted to the canal. This will directly address the requirement of the TRUST project (see above).

The top soil is contaminated with lead (Table 2.6). However, lead is likely to be present in particulate matter and is therefore unlikely to cause any problems with the water quality associated with future SUDS effluent (Scholz et al., 2002).

Overall, the concentration of organic compounds found was low (estimated to be less than 0.5 mg/kg) for Ruchill Park and very low for Ruchill Hospital – near the proposed pond in the South (estimated to be less than 0.1 mg/kg). Contaminants found included PAH. For example, pyrene, fluoranthene, chrysene, benz(a)anthracene, benzo(k)fluoranthene, benz(a)pyrene and benzo(g,h,i)perylene were identified at very low levels at Ruchill Park (proposed pond location). Other compounds found included aliphatic hydrocarbons such as tetracosane, eicosane, heptadecane, heptacosane and nonadecane, which are common constituents of diesel-type fuels.

2.2.5 Conclusions and Recommendations

A survey of 57 sites within 46 areas of Glasgow shows that it is feasible to implement various SUDS techniques throughout Glasgow. The likely contribution of future SUDS to the development of water resources and the overall catchment dynamic of two representative demonstration areas has been assessed. The preliminary designs help to understand the challenges of holistic water resource development and catchment management, diffuse pollution and the linking scales in catchment management.

Characteristics that determine the suitability of a site for the implementation of SUDS in general have been proposed. Representative areas and sites that are suitable for different representative SUDS techniques have been identified qualitatively and quantitatively. A general SUDS implementation key and feasibility matrix that is adaptable to different cities has been outlined.

Design and management guidelines for demonstration sites that should be constructed to inform and educate the public, developers and politicians have been proposed: (a) the drainage of the Lillyburn Place area should be characterised by infiltration trenches draining into a nearby burn during extreme rainfall events and (b) the drainage of the Ruchill Hospital and Park area should be dominated by a single network of swales conveying water either into underground storage tanks or retention ponds. The runoff will ultimately drain into a sewer or canal, but some water could be recycled for irrigation as well.

2.3 ECOSYSTEM SERVICES ASSESSMENT SYSTEM FOR RETROFITTING OF SUSTAINABLE DRAINAGE SYSTEMS

2.3.1 Introduction

2.3.1.1 Rethinking the Philosophy of Drainage

Traditionally, combined sewer systems are used to deal with wastewater and storm water runoff. These sewerage systems operate on the philosophy of preventing local flooding by conveying surface runoff away as quickly as possible. Combined sewers function by carrying both wastewater and storm water in a single pipeline to a wastewater treatment plant, where it is treated and discharged into a suitable natural watercourse such as a river (Scholz, 2006b). During periods of medium or heavy rainfall, when sewers are incapable of carrying an increased flow, a structure called the combined sewer overflow discharges untreated wastewater directly into natural watercourses to relieve combined sewers from high runoff loads (Butler and Davies, 2004; Scholz, 2006b, 2010).

Separate sewer systems are nowadays being designed to reduce the pressure caused by medium and heavy rainfall, by carrying surface runoff and wastewater in separate pipes. Surface runoff is conveyed in a dedicated pipe and discharged straight into a watercourse without being treated (Butler and Davies, 2004). This more modern sewerage system is advantageous over the combined sewer system, as it does not discharge wastewater directly into a receiving watercourse. However, the untreated surface runoff still contains some unwanted contaminants (CIRIA, 2007; Scholz, 2006b, 2010).

Traditional drainage often creates flooding and pollution problems in the lower catchment. The implementation of SUDS can help to achieve these goals at similar or reduced construction costs (Scholz, 2006b, 2010). The philosophy of SUDS is to mimic the natural drainage into the ground, as closely as possible, prior to its development (CIRIA, 2007). Most SUDS techniques are able to do this in a number of ways such as attenuation of runoff before entering the watercourse, storage of water in natural contours, infiltration of partially treated runoff into the ground and evapotranspiration of surface water by vegetation (CIRIA, 2010; Scholz, 2010).

The main objective of SUDS is to reduce the negative impact of urbanization on the quantity and quality of surface runoff, while simultaneously increasing amenity and biodiversity opportunities, where possible. SUDS are capable of managing and controlling surface runoff through techniques such as infiltration, detention/attenuation, conveyance and/or rain harvesting (CIRIA, 2007; Scholz et al., 2006). In general, they make use of physical, chemical and/or biodegradation processes to improve the quality of surface runoff by minimizing the amount of storm water-based pollutants washed into nearby watercourses (Eriksson et al., 2007; Scholz, 2010). However, potential improvement opportunities in terms of ecosystem services including amenity and biodiversity by introducing SUDS are often neglected by engineers and planners in practice (Scholz, 2010).

2.3.1.2 Sustainable Drainage System Techniques

This section provides a brief and generic overview of the key SUDS techniques assessed and tested in this study. For further information on these techniques and related ones, the reader may wish to refer to other publications (Butler and Davies, 2004; CIRIA, 2004, 2007, 2010; Scholz, 2006b, 2010).

Permeable pavements: These systems allow surface runoff to infiltrate through their surface and underlying construction layers, as opposed to flowing over it. They are mainly used for car parks and roads where traffic intensity is relatively low. The infiltrated rainwater is usually treated and subsequently stored before it infiltrates into the ground, reused or released to a drainage system or surface watercourse (CIRIA, 2004; Scholz and Grabowiecki, 2007).

Filter strips: These techniques are a form of passive treatment, which are designed to treat runoff from adjacent impermeable areas (CIRIA, 2004). A typical filter strip is a wide area of grass, or other dense vegetation, that is characterized by its gentle slope. Filter strips are usually located between surface water bodies, small car parks and at the side of roads. High groundwater levels and steep gradients can generally be overcome by filter strips (Ellis et al., 2004).

Swales: These structures are a form of permeable conveyance system. A typical swale is a broad and shallow channel, which is lined with suitable vegetation such as grass. As in the case of filter strips, the vegetation that covers the swale slows down the rate of surface runoff, thus reducing peak flows, as well as filtering the particulate pollutants contained within it (CIRIA, 2004).

Green roofs: These roofs are covered with vegetation and are ideal for a range of flat or gently sloping roofs and are well suited for urban areas where space is limited. These roofs are capable of removing pollutants from rainwater by filtering, adsorption onto the substrate and retention by plants (CIRIA, 2004).

Ponds: These water bodies act as a form of passive treatment. They are usually cost-effective (due to a high-volume-to-area ratio) SUDS techniques making them popular to control storm water runoff. Ponds are able to provide enhanced wildlife and amenity benefits and should be designed to do so without compromising the primary function of it being part of a storm water management system. The degree of treatment achieved depends greatly on the residence time of the temporary storage water, which typically ranges between 24 and 48 hours (CIRIA, 2004; Scholz, 2004b).

Constructed wetlands: These structures contain water of varying depths across their area and consist of marsh or wetland vegetation. This is one of the most effective SUDS techniques for providing diverse wildlife habitat and pollutant removal. However, there are also long-held concerns over the dangers of using wetlands designed for pollution accumulation as wildlife habitat (Helfield and Diamond, 1997). Wetlands are able to eliminate pollutants by both plants and aggregates filtering and screening particles. Inlet and outlet sumps are recommended to deal with excessive sediment, which can quickly overpower the shallow ends of the wetland (Scholz and Lee, 2005).

Infiltration trenches: These trenches are shallow excavations lined with a geotextile material and backfilled with stones, creating a small belowground storage reservoir. Storm water runoff that flows into the trench slowly infiltrates into the subsoil. Infiltration trenches are capable of removing pollutants by adsorption, filtration and microbial decomposition in the soil underlying the trench (Scholz, 2006b).

Soakaways: These SUDS techniques are a form of source control, operated by dispersing surface runoff into the ground. Recent types of soakaways consist of open chambers (in contrast to holes in the ground filled with aggregates) to store large quantities of water (CIRIA, 2007; Scholz, 2006b).

Infiltration basins: These basins are open and uncovered areas of ground, and they are relatively shallow features, which can be constructed either by excavating depressions or embankments. If landscaped, they can be aesthetically pleasing and also add amenity value. Infiltration basins store storm water runoff, which gradually percolates through the soil of the basin. The soil's permeability and the water table depth are mainly responsible for the efficiency of an infiltration basin (Scholz, 2006b).

Storage tanks: These belowground (or underground) storage techniques are subsurface structures that entrap and store surface runoff. The stored water is released at a slow rate to reduce peak flows during medium or heavy rainfalls. If soil conditions are suitable and the water table is located at a significant depth below the chamber, the storage tanks can be designed to allow stored water to infiltrate into the ground thus encouraging groundwater recharge (Nanbakhsh et al., 2007).

Water playgrounds: These SUDS have little effect on managing the quantity and quality of surface runoff. Their main purpose is, however, to enhance amenity value through recreational benefits by providing a variety of water features that individuals (particularly children) can interact with (Scholz, 2006b; Scholz et al., 2006).

2.3.1.3 Ecosystem Services

Ecosystem services are often defined as the benefits human beings can obtain from the semi-natural (managed) environment, for example, wildlife, green space, open countryside, forest, farmland, river, stream, lake and sea (Busch et al., 2012; Millennium Ecosystem Assessment, 2005; Moore and Hunt, 2012). Furthermore, DEFRA (2013) characterizes ecosystem services as the benefits human gain from the products and services generated by the natural environment. The natural resources and functioning natural systems (Walsh et al., 2012) and a regulated climate are essential for humans (Millennium Ecosystem Assessment, 2005). A high biodiversity helps to sustain the natural environment and is thus an important factor for ecosystem service provision.

The Natural Environment White Paper (DEFRA, 2013), the UK National Ecosystem Assessment (2011) and the Economics of Ecosystems and Biodiversity Manual for Cities (TEEB, 2013) have identified the following four ecosystem services categories: supporting, regulating, provisioning and cultural services. All existing ecosystem services are strongly linked to one another and to other types of ecosystem services. Supporting services are strongly interrelated to one another by an extensive range of chemical, physical and biological interactions (UK National Ecosystem Assessment, 2011).

The goods obtained can be distinguished depending on the degree of human interference. Goods that have been yielded from nature with minimal interference from humans can be referred to as natural products, while goods that have had a higher level of human interference, such as the use of fertilizers and pesticides, can be referred to as joint products (Slootweg et al., 2010). The provisioning service of fresh water (FW) is particularly complex in the context of the urban water cycle and the interactions between potential water uses including drinking water supply, irrigation and maintenance of the water supply to urban watercourses (Walsh et al., 2012).

A list of ecosystem service variables relevant for SUDS and their respective categories (DEFRA, 2013; Moore and Hunt, 2012; TEEB, 2013; UK National Ecosystem Assessment, 2011; Walsh et al., 2012) used in Section 2.3 is provided in Table 2.7, which is based on the Economics of Ecosystems and Biodiversity Manual for Cities (TEEB, 2013) proposing a comprehensive list of ecosystem service variables of generic nature and Moore and Hunt (2012) who chose variables of relevance to wetlands. The variables in Table 2.7 also recognize the definitions of the 'Making Space for Nature' initiative (Slootweg et al., 2010) and the Water Framework Directive (European Union, 2013).

TABLE 2.7
Universal Ecosystem Service Categories and Variables for Sustainable (urban) Drainage Systems (SUDS)

Service Category	Variable
Supporting	1. Habitats for species (HS) including water, food and shelter
	2. Maintenance of genetic diversity (MGD), which is the diversity of genes within and between populations of species
Regulating	3. Local climate and air quality regulation (LCAR), particularly by green spaces planted with trees
	4. Carbon sequestration and storage (CSS) by ecosystems such as wetlands and urban forests
	5. Moderation of extreme events (MEE) such as storms, floods and landslides
	6. Storm runoff treatment (SRT) particularly by sustainable drainage techniques
	7. Erosion prevention and maintenance of soil fertility (EPMSF)
	8. Pollination (P)
	9. Biological control (BC)
Provisioning	10. Food (F), particularly from agro-ecosystems and freshwater resources
	11. Raw materials (RM) such as wood, biofuel and oil from plants
	12. Fresh water (FW) such as drinking water
	13. Medicinal resources (MR) provided by plants used for traditional medicine
Cultural	14. Recreation, and mental and physical health (RMPH) associated with activities close to the drainage system
	15. Tourism and area value (TAV) associated with an attractive ecosystem featuring a high biodiversity
	16. Aesthetic and educational appreciation and inspiration for culture, art and design (AEAICAD), which can be of high importance to individuals and groups
	17. Spiritual experience and sense of place (SESP) associated with specific parks, watercourses and woods

2.3.1.4 Case Study and Objectives

Manchester and Salford form the core of the Greater Manchester urban example case study region providing homes to approximately two million people in the North-west of England and comprising 100 tested sites for SUDS retrofitting. However, a few potential SUDS sites are also located in neighbouring municipalities (Bury, Oldham, Tameside and Trafford), which are less urbanized (White and Alacron, 2009). Figure 2.15 shows an overview of the assessed sites where SUDS could potentially be retrofitted.

Due to the interconnectivity between local authorities, flooding in one area of the conurbation will usually have a knock-on effect in the remaining local authorities (AGMA, 2008). It is through recognizing this that the ten local authorities joined together in 2011 to form the Greater Manchester Combined Authorities (GMCA) to tackle common problems such as flooding.

Storm water runoff from impermeable surfaces has been identified by strategic flood risk assessments undertaken by local authorities as one of the main flood

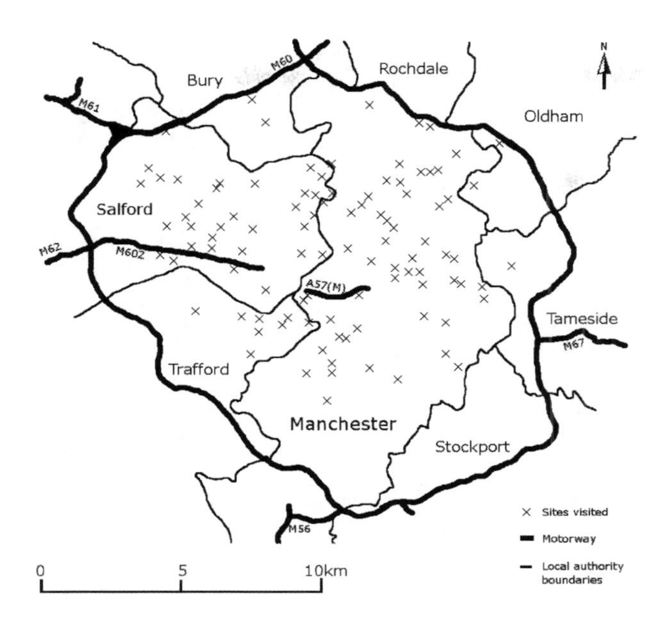

FIGURE 2.15 Overview of the assessed sites where sustainable (urban) drainage systems could be retrofitted.

sources in the conurbation. Concerns with this traditional method of dealing with storm water runoff only arose after a serious flood incident in 1998. With the turn of the century, new national policies such as the Planning Policy Guidance Note 25 on the Development and Flood Risk Management (DLTR, 2001) were released to address flooding issues. This guidance note (not in force anymore) formally introduced the use of SUDS to deal with storm water management (DLTR, 2001; White and Alarcon, 2009).

The aim is to develop a unique and rapid decision support tool based on ecosystem service variables for retrofitting of key SUDS techniques in urban areas leading to an improvement of urban water, soil and air quality. The main objectives to achieve this aim are: (a) to assess the suitability of potential SUDS sites within an example case study region based on traditional 'community and environment' variables; (b) to evaluate the suitability of these SUDS sites based on ecosystem service variables; (c) to gauge a combination of both approaches for example sites; and (d) to compare the above assessment outcomes with each other. Scholz (2014) provided the basis for Section 2.3.

2.3.2 Methodology

2.3.2.1 Case Study Evaluation

Figure 2.16 outlines the key seven steps of the proposed methodology of retrofitting SUDS based on estimated ecosystem service variables. A total of 100 potential SUDS sites were identified using the Ordnance Survey and Google maps of Greater Manchester. The purpose of focusing the study on this example region was to demonstrate that the implementation of SUDS even within densely built-up cities is possible.

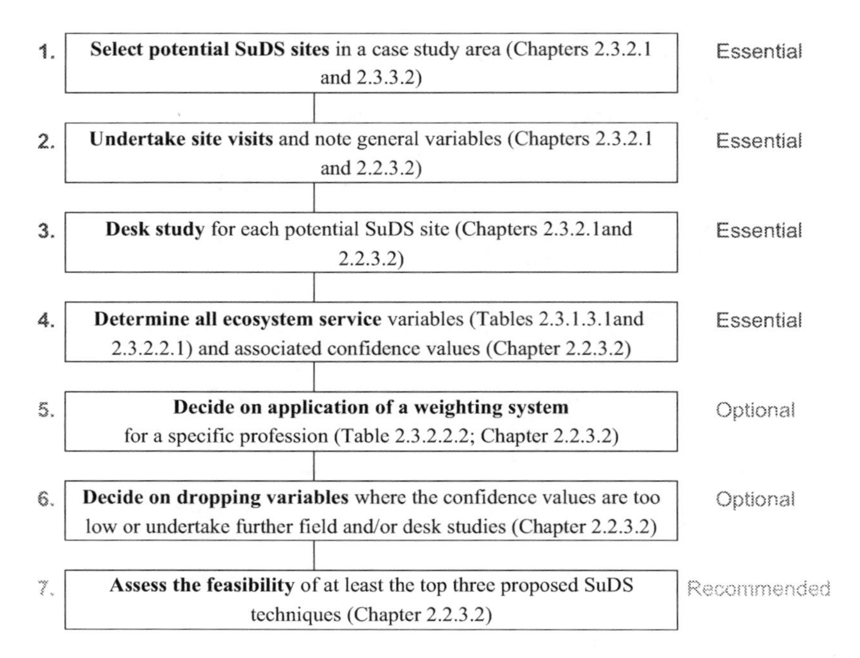

FIGURE 2.16 Outline of the proposed methodology of retrofitting sustainable (urban) drainage systems (SUDS) based on estimated ecosystem services variables.

The site assessment template was based on a combination of previously published frameworks (CIRIA, 2004, 2007; Scholz, 2006b; Scholz et al., 2006). Each potential SUDS site was assessed by two to five research team members (author and students of his research group) to reduce subjectivity (Munoz-Pedreros, 2004). A subsequent desk study supplemented the site evaluation. The reliability of the assessment was judged by the provision of a mark out of 100. Unreliable appraisals were double-checked. The following information was collected to support the assessment team in determining the variables required for the traditional and ecosystem services approaches:

- General site information and site acceptability for SUDS and presence of existing SUDS.
- Photos of the key site features were taken for each potential SUDS site and its catchment.
- Land ownership information and estimated site value (£).
- Proportions (%) of site development, regeneration, retrofitting and recreation.
- Surrounding area characteristics, total area of the catchment (m^2) and catchment shape.
- Location description and distance (m) to the nearest receiving watercourse, if located within a reasonable distance within or at the border of the catchment.

- Estimated current and future surface permeability (%) of the proposed SUDS site and its catchment.
- Estimated proportions (%) of current and future roof runoff.
- Estimated proportions (%) of current and future road runoff.
- For each sub-catchment, area (m^2) and gradient in the two main directions have an angle of 90° to each other in the horizontal plain.
- Hydro-geological information such as contaminated land (present or absent), soil infiltration (low, medium or high) and groundwater level (below or above 2-m depth).
- SUDS technology feasibility proportion (%) for the technologies permeable pavement, filter strip, swale, green roof, pond, constructed wetland, infiltration trench, soakaway, infiltration basin, belowground storage tank and water playground.

2.3.2.2 Assessment of Ecosystem Services

Table 2.8 shows the new 17 ecosystem service variables, which belong to the established four ecosystem service categories (Table 2.7). The quantitative and qualitative approaches to assessing ecosystem services (Busch et al., 2012) have been applied.

The ecosystem service variables for SUDS retrofitting assessment are described in Table 2.8 Characteristics for low and high estimations (out of 100 points) are provided. A measure of certainty (%) was given to each variable to indicate the reliability

TABLE 2.8
Estimation (Maximum of 100 Points) of New Ecosystem Service Variables to be Used for the Generic Assessment of Retrofitting Sustainable Drainage System Techniques

Ecosystem Service Variable	Characteristics for Rather Low Estimations	Characteristics for Rather High Estimations
1. Habitats for species (HS)	Wildlife benefits of the proposed SUDS area are low (e.g., little green spaces) due to a virtually unsuitable surrounding area, but mainly due to the impermeable surface coverage of the site	Wildlife benefits of the proposed SUDS area are high due to a suitable surrounding area (e.g., sufficient mature green space) and due to the very permeable surface coverage of the site
2. Maintenance of genetic diversity (MGD)	Site is very isolated from other habitats (at least 5 km away) and does not consist of a variety of ecosystems, thus can only maintain a limited number of species; SUDS techniques having a short lifespan (e.g., swale) will have no effect on providing a new habitat and thus creating wider diversity	Site is interconnected to neighbouring habitats (less than 1 km away) and consists of a large variety of ecosystems, thus maintaining a high number of species; SUDS techniques having a long lifespan (e.g., pond and wetland) will have a high impact on providing new habitats and thus creating even wider diversities

(*Continued*)

TABLE 2.8 (*Continued*)
Estimation (Maximum of 100 Points) of New Ecosystem Service Variables to be Used for the Generic Assessment of Retrofitting Sustainable Drainage System Techniques

Ecosystem Service Variable	Characteristics for Rather Low Estimations	Characteristics for Rather High Estimations
3. Local climate and air quality regulation (LCAR)	Areas of trees and surface water are scarce, if any at all	Site is almost entirely covered by dense trees contributing to a great improvement of the air quality for the benefit of human well-being; a mature surface water body is also present
4. Carbon sequestration and storage (CSS)	Small site comprising areas of a few trees	Large greenspace, which is entirely covered by dense trees; presence of a wetland
5. Moderation of extreme events (MEE)	In the case of events such as flooding and drought, the site is inadequate to moderate for the event (i.e. SUDS site becomes ineffective); direct harm to receiving watercourses	In the case of extreme events such as flooding, droughts and fire, the site will moderate these events well (i.e. SUDS site retains most of its functions and stays fit-for-purpose for the direct benefit of human well-being).
6. Storm runoff treatment (SRT)	Low potential to remove pollutants not even through physical processes such as straining; direct harm to receiving watercourses	High potential to remove pollutants through plenty of physical and chemical processes, and biodegradation
7. Erosion prevention and maintenance of soil fertility (EPMSF)	Low erosion prevention potential harming the urban landscape and receiving watercourses, and low likelihood of maintenance of soil fertility (e.g., unprotected soil; i.e. not even covered by grass or gravel)	High erosion prevention potential (e.g., reinforced structure) and high likelihood of maintenance of soil fertility (e.g., no wash-out of nutrients); rarely applicable variable for SUDS
8. Pollination (P)	Site has a low, if any, potential for the presence of animals (e.g., dense urban area with no green space)	Site has a high potential for the presence of animals such as insects to pollinate surrounding areas (e.g., rural area)
9. Biological control (BC)	Site has very little potential for the presence of predatorily animals and insects to regulate pests and diseases in the surrounding areas (e.g., virtual absence of any green spaces and mature water bodies)	Site has a high potential for the stable presence of predatorily animals and insects to regulate pests and diseases in the surrounding areas (e.g., rich terrestrial and aquatic habitat diversity); no pest nuisance benefiting human well-being
10. Food (F)	Small and contaminated site having no or little potential to produce food (e.g., small site within a dense urban area)	Large and fertile site having a high potential to produce food for the well-being of humans (e.g., large site fully integrated into an agricultural landscape)

(*Continued*)

TABLE 2.8 (*Continued*)
Estimation (Maximum of 100 Points) of New Ecosystem Service Variables to be Used for the Generic Assessment of Retrofitting Sustainable Drainage System Techniques

Ecosystem Service Variable	Characteristics for Rather Low Estimations	Characteristics for Rather High Estimations
11. Raw materials (RM)	Small site having no or very little potential to produce any raw materials (small site within a dense urban area)	Large site with great potential to increase raw material production (e.g., wood from tree production)
12. Fresh water (FW)	Low amount of surface runoff; high pollution harming receiving watercourses (e.g., heavily trafficked urban street runoff)	High amount of surface runoff; low pollution (e.g., large, roofed areas from retail parks)
13. Medicinal resources (MR)	Little potential for plants to be used for medicinal purposes; rarely applicable variable in developed countries	High potential for plants that can be used as medicinal resource; rarely applicable variable in developed countries
14. Recreation, and mental and physical health (RMPH)	Site can be considered as unsafe and provides virtually no recreational opportunities for anybody; SUDS site requires fencing-in	Site provides safe and recreational opportunities of relatively high quality for everybody, directly benefiting human well-being (e.g., bird watching, walking, fishing and group sports)
15. Tourism and area value (TAV)	Site does provide little value for tourism; property value around the site is likely to decrease; rundown estate (e.g., site is fenced-in and has drainage function only)	Site would attract much attention and a large number of visitors from far away; high increase of property value likely (e.g., site integrated within a mature park located within the city centre)
16. Aesthetic and educational appreciation and inspiration for culture, art and design (AEAICAD)	An SUDS would not increase the attraction of the area or provide additional inspiration (e.g., fenced-in site with pure drainage function)	An SUDS would create an area of outstanding semi-natural beauty providing much inspiration for people with diverse backgrounds; highly valuable education resource
17. Spiritual experience and sense of place (SESP)	Provides people with virtually no connection to the land (e.g., fenced-in site with predominantly drainage function)	The site makes people feel connected to the area and have a sense of strong belonging (e.g., site as part of a community and/or educational project, directly benefiting human well-being)

of the assessment; the higher the value given, the more certain was the group of assessors. Only values greater than 50% were considered to be acceptable to progress to the next estimation without conducting further studies.

Table 2.8 can be adapted by the reader for case studies of concern. More quantitative guidance can be introduced to cater for any specific situation. However, a detailed discussion on this matter is beyond the scope of Section 2.3.

Weightings recognizing differences in regions and stakeholders could be introduced. For example, variables of relatively low relevance for a drainage engineer such as medicinal resource (MR) in Greater Manchester could be assigned with a low weight of, for example, 1, while variables with a medium (e.g., RMPH) or high (e.g., MEE) relevance could be assigned with a medium (2) or high (3) weight, respectively. Such a proposed weighting system has not been introduced for the case study to keep the methodology simple and transparent. Nevertheless, Table 2.9 proposes potential weights from the viewpoint of a drainage engineer, ecologist, planner and social scientist to support the reader with additional guidance. The weights can be revised for any other case study area and weights for even more specific viewpoints (e.g., structural engineer, ornithologist and housing developer) may be proposed.

The site assessment was based on previous work (CIRIA, 2004; Scholz, 2006b; Scholz et al., 2006). The guideline C609 (CIRIA, 2004) bases the selection of an SUDS

TABLE 2.9
Proposed Weights as a Function of User Preference (Not Applied for the Greater Manchester Case Study Example to Avoid the Introduction of Bias)

		Weight			
Category	Variable	Drainage Engineer	Ecologist	Planner	Social Scientist
Supporting Services	1. Habitats for species (HS)	1	3	2	1
	2. Maintenance of genetic diversity (MGD)	1	3	1	1
Regulating Services	3. Local climate and air quality regulation (LCAR)	1	2	2	2
	4. Carbon sequestration and storage (CSS)	1	2	1	1
	5. Moderation of extreme events (MEE)	5	2	4	2
	6. Storm runoff treatment (SRT)	5	2	2	2
	7. Erosion prevention and maintenance of soil fertility (EPMSF)	2	2	2	2
	8. Pollination (P)	1	3	1	1
	9. Biological control (BC)	1	2	2	2
Provisioning Services	10. Food (F)	1	1	1	1
	11. Raw materials (RM)	1	1	1	1
	12. Fresh water (FW)	4	2	2	1
	13. Medicinal resources (MR)	1	1	1	1
Cultural Services	14. Recreation, and mental and physical health (RMPH)	2	1	2	3
	15. Tourism and area value (TAV)	1	1	2	3
	16. Aesthetic and educational appreciation and inspiration for culture, art and design (AEAICAD)	1	1	2	3
	17. Spiritual experience and sense of place (SESP)	1	1	2	3

type on assessments regarding hydrology, land use, physical site characteristics, community and environment, economics and maintenance. These criteria have been adapted from a previous report (Ellis et al., 2003). Scores for each criterion range from one to five, where one refers to a SUDS technique being very unsuitable, and five signifies an SUDS technique being highly appropriate for that particular criterion. The SUDS type obtaining the highest sum of scores is likely to be most suitable for a particular site. The minimum and maximum overall scores for all criteria were 0 and 25, respectively.

The traditional 'community and environment' approach comprises the conventional variables safety, pond premium, aesthetics, wildlife habitat and acceptance (CIRIA, 2004). Variables that are not relevant to ecosystem services were ignored for the purpose of this study. Each potential SUDS type was assessed for each site according to safety with respect to people and pets, water premium recognizing property value, aesthetics, wildlife habitat and public acceptance by the local community. In comparison, Tables 2.6 and 2.7 were used to estimate numerical values for the proposed ecosystem service variables. A numerical comparison between the traditional method and new approach recognized that the latter method comprises more variables and higher maximum values than the former.

A combination of the traditional and new approach was also tested. In the combined assessment, the traditional criteria aesthetic and wildlife habitat were replaced by the four ecosystem service categories shown in Table 2.7. Those SUDS techniques that were associated with the highest preferences for a site were recommended to landowners for subsequent implementation.

2.3.3 Results and Discussion

2.3.3.1 Ecosystem Service Variables

This research study combining new ecosystem service variable assessments for all key SUDS techniques with a simple assessment system applied for a large database of real case studies is unique. However, Danso-Amoako et al. (2010) assessed sustainable flood retention basins with respect to dam failure and a limited set of ecosystem variables (Scholz and Yang, 2010) in Greater Manchester as well. Moreover, Gill et al. (2007) as well as White and Alarcon (2009) were concerned with green infrastructure in the context of climate change, planning and drainage in the same study area. Nevertheless, ecosystem services were not assessed in a similar context. The specific ecosystem service variable assessment is outlined below:

- Habitats for species (HS): This assessment was influenced by the permeability of a potential SUDS site and the surrounding urban area. Green areas with highly permeable surfaces and plenty of vegetation provide wildlife benefits were rare in Greater Manchester.
- Maintenance of genetic diversity (MGD): The interconnectivity between sites providing habitats for a wide variety of ecosystems is often responsible for a relatively large number of species. The interconnectivity between and the quality of green spaces within Greater Manchester is relatively poor. The implementation of SUDS techniques such as wetlands and ponds, having a long-life span, will further enhance the site's ecosystem service potential.

- Local climate and air quality regulation (LCAR): Tree coverage rates and surface water numbers were rather low in the study area.
- Carbon sequestration and storage (CSS): Woodlands and wetlands were rare within the study area.
- Moderation of extreme events (MEE): The ability of a potential SUDS site to manage extreme events such as flooding and drought was relatively high in Greater Manchester.
- Storm runoff treatment (SRT): The likelihood of runoff treatment was high for most sites.
- Erosion prevention and maintenance of soil fertility (EPMSF): Only a few sites were covered by dense vegetation. Sites with bare soil and poor grass cover did not provide much erosion protection.
- Pollination (P): Green spaces such as parks, woodlands and fields, which act as a habitat for pollinators, were rare in city areas.
- Biological control (BC): Predatory animals capable of regulating pests and diseases were rare for most of the smaller case study sites.
- Food (F): The assessment was based on the potential of a SUDS site to provide food. The size of a site as well as its soil and associated contamination are important indirect evaluation parameters. A cultural change in the study area and a deepening of the current recession would be required to realize the potential of transforming parts of the potential SUDS sites into allotments and gardens used to grow food and rear small livestock.
- Raw materials (RM): This evaluation considered the potential of a site to provide a range of RM such as wood, grass and water. The active harvesting of RM is underutilized within most parts of the study area due to a lack of local policies promoting the multi-purpose use of green spaces.
- Fresh water (FW): The quantity and quality of surface runoff for most sites was sufficiently high.
- Medicinal resources (MR): Some plants covering a potential SUDS site may have medicinal benefits for people and animals. This variable is unlikely to be relevant for the UK in the medium-term future.
- Recreation, and mental and physical health (RMPH): There is an underutilized potential for the multi-purpose use of potential SUDS sites predominantly due to cultural and political reasons.
- Tourism and area value (TAV): There is a considerably underutilized potential for attracting visitors to score high in Greater Manchester, mainly due to the presence of a few large parks suffering from under-investment.
- Aesthetic and educational appreciation and inspiration for culture, art and design (AEAICAD): There is an underutilized potential for aesthetics to score high in Greater Manchester, mainly due to public under-investment in park infrastructure.
- Spiritual experience and sense of place (SESP): A potential SUDS site's ability to encourage people to feel connected to the area and their associated community, giving them a strong sense of belonging, was evaluated. Considering the high multi-cultural diversity in Greater Manchester, there is a potential for SESP to score high in some areas.

The strengths of the proposed methodology, particularly in comparison to the community and environment approach adopted by Ellis et al. (2003) and CIRIA (2004), are the generic retrofitting approach based truly on universal ecosystem service variables and not on conventional engineering understanding. The evaluation is also inexpensive and easy-to-understand. On the other side, weaknesses include methodological subjectivity, which was addressed by involving groups and using uncertainty values for all estimations (Danso-Amoako et al., 2012; Munoz-Pedreros, 2004; Scholz and Yang, 2010). Some ecosystem service variables are also rarely applicable in the developed world such as the UK. Finally, the possibility of multicollinearity among variables can be seen as a potential risk (McMinn et al., 2010).

2.3.3.2 Comparison of Assessment Approaches

Table 2.10 indicates a comparison of all assessment approaches, which follow similar methodological principles. However, the main difference lies in the selection of variables. The relative proportions for each SUDS technique have been expressed in percentage points for each column to allow for a direct comparison between approaches and preferences for the example case study area. High confidence values were only obtained for the first three preferences.

TABLE 2.10
Comparison of Assessment Approaches in Terms of Proposed Sustainable (Urban) Drainage System (SUDS) Techniques for all Selected Sites in Greater Manchester

	Proportion (%) of Sites at Which SUDS Techniques are Given First, Second or Third Order of Preference for the Ecosystem Service Approach			Proportion (%) of Sites at Which SUDS Techniques are Given First, Second or Third Order of Preference for the Community and Environment Approach			Proportion (%) of Sites at Which SUDS Techniques are Given First, Second or Third Order of Preference for the Combined Approach		
SUDS Technique	**First**	**Second**	**Third**	**First**	**Second**	**Third**	**First**	**Second**	**Third**
Permeable pavement	13	12	12	19	11	7	19	12	5
Filter strip	6	16	21	19	23	5	16	24	5
Swales	0	5	6	6	5	18	0	1	10
Green roof	0	0	0	0	1	0	0	0	0
Pond	17	7	5	30	6	0	31	6	0
Constructed wetland	3	1	0	6	0	0	6	0	0
Infiltration trench	14	16	22	0	13	22	4	16	28
Soakaway	30	24	4	0	19	21	2	29	31
Infiltration basin	1	0	3	0	6	7	0	2	2
Below-ground storage	16	10	17	20	14	13	21	9	9
Water playground	0	9	10	0	2	7	1	1	10

TABLE 2.11
Comparison of the Inter-Site Variability for a Given Sustainable Drainage Technique for Greater Manchester

	Standard Deviations (Based on Relative Percentage Points Awarded)		
SUDS Technique	Ecosystem Services Approach	Community and Environment Approach	Combined Approach
Permeable pavement	12.55	32.04	21.97
Filter strip	14.36	29.87	21.91
Swale	12.42	24.04	18.03
Green roof	3.76	6.19	5.01
Pond	35.10	39.63	35.75
Constructed wetland	20.35	25.04	22.23
Infiltration trench	8.45	26.77	17.46
Soakaway	5.67	19.06	12.00
Infiltration basin	9.70	19.65	14.73
Below-ground storage	11.83	30.46	20.67
Water playground	9.52	28.99	18.26

Table 2.11 shows a comparison of the inter-site variability expressed with the help of the standard deviation capturing the variance around the mean for a given sustainable drainage technique for Greater Manchester and helps to interpret the preference distributions in Table 2.10. The standard deviation is an appropriate statistic to explain data spread as a result of subjective assessments. The new ecosystem services and the traditional assessment approaches have the lowest and highest inter-site variability, respectively. The relatively high variability for most variables such as ponds and constructed wetlands cannot be explained by factors relating to specific planning policies for Greater Manchester (White and Alarcon, 2009). Ponds are associated with the greatest inter-site variability for all three approaches because of their potentially relatively small size and great popularity (Scholz, 2010; Scholz and Grabowiecki, 2007; Scholz et al., 2006), particularly with the traditional and combined approaches (Table 2.10). However, it would not be right to assume that the Greater Manchester case study findings apply necessarily to other areas as well. The key contribution to knowledge is the proposed generic methodology for SUDS retrofitting and not the example case study findings.

Findings based on Table 2.10 indicate for first preferences that permeable pavements, filter strips, swales, ponds, constructed wetlands and belowground storage tanks were not favoured using the assessment based on the ecosystem service variables compared to the traditional approach. In contrast, all high-rate infiltration techniques (infiltration trench, soakaway and infiltration basin) were favoured by the new approach compared to the old one. This can be explained by the fact that different sets of variables were applied.

Table 2.10 also indicates that the first preferences of the combined approach for the SUDS techniques filter strip, infiltration trench and soakaway are more similar numerically to the traditional than the new ecosystem services approach. Moreover, the percentage point for belowground storage is the highest for the combined approach. However, the opposite is the case for the second and third preference.

It may come as a surprise that permeable pavements and belowground storage techniques scored relatively highly on ecosystem service variables, which contradicts the common belief among some engineers that there has to be a strong bias towards natural and soft techniques when using ecosystem service assessment techniques (Butler and Davies, 2004; Scholz, 2010). However, permeable pavements (Imran et al., 2013; Tota-Maharaj et al., 2010) and belowground storage (Scholz, 2006b, 2010) are likely to attract high values for variables such as SRT and MEE, respectively, if properly designed and managed. Nevertheless, these specific findings relate to the Greater Manchester case study area and might therefore not apply to other regions.

The author also developed a rapid decision support tool based on more specific ecosystem service variables, particularly for retrofitting of permeable pavement systems in the presence of mature trees (Scholz and Uzomah, 2013). Findings indicate that permeable pavements score even higher on ecosystem services, if mature trees are present. This is the case because of the important role of trees in terms of water and air quality improvement and flood alleviation.

2.3.4 Conclusions and Recommendations

A rapid assessment methodology for retrofitting of SUDS was successfully introduced to reduce the currently high level of subjectivity in practice. Retrofitting of SUDS is possible for a high number of sites within a densely built-up area such as Greater Manchester. Generic ecosystem service variables suitable for SUDS were determined and their assessment indicated that most sites had a relatively low ecosystem service potential.

The suitability of sites for SUDS retrofitting was assessed based on traditional 'community and environment' variables and the ecosystem service variables. A comparison shows a slight bias of the old tool towards semi-natural SUDS techniques such as wetlands, ponds, filter strips and swales. However, belowground storage tanks received also high scores, because of their great impact in terms of flood control. In contrast, the new approach favours infiltration techniques.

A combination of the traditional and new approach shifts first preferences back towards the traditional community and environment approach. However, differences for some SUDS techniques are insignificant.

All sites were suitable for the retrofitting of SUDS when the traditional assessment based on 'community and environment' variables was carried out. In comparison, the ecosystem services approach shows that nearly half of the sites visited are valued as having a relatively low ecosystem services potential, making them of limited use for retrofitting of most SUDS techniques. This finding can be used to prioritize sites for SUDS retrofitting, which is particularly important during challenging financial times. The application of the new tool is therefore likely to change the drainage infrastructure promoting ecosystem services and reducing urban pollution. This will also lead to an improvement of urban water, soil and air quality.

More research is recommended to develop the ecosystem service assessment approach further. Additional urban but also rural case studies with a larger number of sites could be assessed to test the robustness of the new approach and to subsequently refine it.

Specific weighting systems for the ecosystem service variables as a function of individual SUDS techniques, the preference of the user and different climatic regions and cultures could be introduced to reduce the impact of what may be perceived as less relevant ecosystem service variables. However, this would introduce extra bias considering that, for example, an engineer would have a different weighting system than an ecologist.

2.4 GOLDFISH AND SUSTAINABLE DRAINAGE PONDS

2.4.1 Introduction

Conventional storm water systems are designed to dispose of surface water runoff as quickly as possible. This results in 'end of pipe' solutions that often involve the provision of large interceptor and relief sewers, huge storage tanks at downstream locations and centralized wastewater treatment facilities. These traditional civil engineering solutions often lead to flooding and environmental pollution due to combined sewer overflows during storm events (Butler and Davies, 2000; CIRIA, 2000).

In contrast, SUDS such as combined attenuation and infiltration systems can be applied as cost-effective local 'source control' drainage solutions; for example, delaying storm runoff leads to a reduction of the peak flow (Scholz, 2004b). It is often possible to divert all storm runoff for infiltration or storage and subsequent water reuse. As runoff from roads is a major contributor to the quantity of surface water requiring disposal, this is a particularly beneficial approach where suitable ground conditions prevail (Zheng et al., 2005). Furthermore, infiltration of storm runoff can reduce the concentration of diffuse pollutants such as dog faeces and leaves, thereby improving the water quality of surface water runoff (D'Arcy and Frost, 2001).

Cities have now found a new ally against flooding: *Carassius auratus* (discovered by Linnaeus in 1758), usually known as common goldfish (Edwards, 2004; Mckie, 2005; Zheng et al., 2005). Artificial ponds as part of SUDS should be used to hold storm runoff water within cities, and *C. auratus* can help to increase public acceptance, and to keep them clean. Goldfish could reduce unsightly and sometimes smelly mats of algae and help to keep estates aesthetically pleasing and good smelling. Every time a new housing estate is built, flood problems may arise due to a lack of existing sewer system capacity. One solution is to create ponds and small lakes that will attenuate storm runoff water and/or allow it to slowly filter back into the ground (CIRIA, 2000; Guo, 2001; Scholz et al., 2005).

The aim of Section 2.4 is to show if SUDS can be kept clean, healthy and pleasing to look at with the help of *C. auratus*. The impact of *C. auratus* on the water quality of ponds and the associated inflow water quality on *C. auratus* will be assessed. Section 2.4 is based on an original article by Scholz and Kazemi-Yazdi (2005).

2.4.2 Materials and Methods

Since 1 April 2003, both planted and unplanted runoff demonstration ponds as part of a SUDS at The King's Buildings campus of The University of Edinburgh are in operation (Figures 2.17 and 2.18). The dominant macrophyte of the constructed

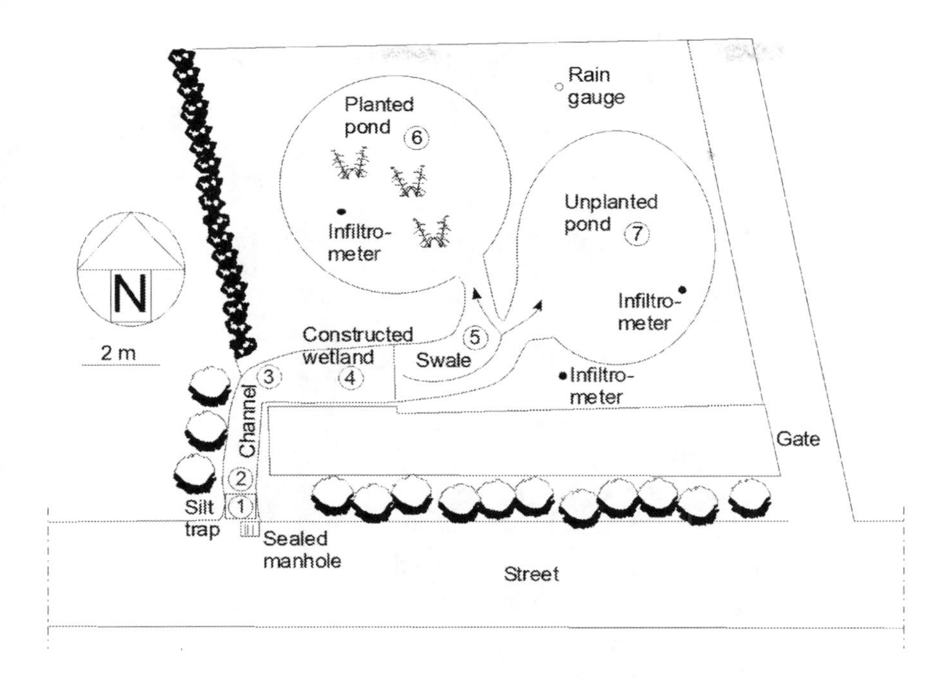

FIGURE 2.17 Runoff flows from the road (partly shown) into the silt trap (1), then via the gravel ditch (2) into the constructed wetland (3 and 4) and finally via the swale (5) into the infiltration ponds (6 and 7).

FIGURE 2.18 Miklas Scholz presenting a *C. auratus* (common goldfish) to The Observer (picture taken by M. MacLeod on 21 April 2005 (Mckie, 2005)).

FIGURE 2.19 Case study site (picture taken by M. Scholz on 12 April 2005).

wetland and planted pond were *Phragmites australis* (common reed) *and Typha latifolia* (broadleaf cattail), respectively.

Precipitation from a road of the campus was channelled to infiltration ponds, but (filamentous) green algae began to grow until *C. auratus* (Figure 2.18) was introduced on 1 April 2004. Twenty healthy *C. auratus* of approximately 180-g total weight were introduced into each pond. Both watercourses were covered with a plastic mesh (Figure 2.19) to prevent animals such as *Ardea cinerea* (grey heron) and *Felis catus* (cat) to prey on *C. auratus* (Zheng et al., 2005).

Since 1 April 2005, approximately 400 g/week of fresh dog excrements are currently added directly to the silt trap protecting the ponds predominantly from solid contaminants. A constructed wetland located between the silt trap and the ponds should prevent contamination from dissolved organic pollutants and potentially pathogenic organisms, but it is an open research question if *C. auratus* can cope with any additional nutrient (particularly nitrogen and phosphorus) load and total coliforms including *Escherichia coli* (pathogenic bacteria) build-up. All water quality determinations were undertaken according to standard methods (Clesceri et al., 1998).

2.4.3 Results and Discussion

2.4.3.1 Water Quality Management

The water qualities of the constructed wetland inflow, constructed wetland outflow, unplanted infiltration pond and planted infiltration pond (Figures 2.17 and 2.19) are shown in Tables 2.11–2.14, respectively. After two years of operation, the water

TABLE 2.12
Summary Statistics: Water Quality of the Inflow to the Constructed Wetland (Figure 2.17) before (1 April 2003 to 31 March 2004) and after (1 April 2004 to 31 March 2005) the Introduction of *Carassius auratus* (Common Goldfish)

Variable	Unit	Sampling Number		Mean		Standard Deviation	
Temperature	°C	54	94	9.9	12.1	3.15	3.92
BOD[a]	mg/L	36	43	10.6	17.7	14.59	14.20
Suspended solids	mg/L	47	93	152.4	1,174.0	245.45	1,458.11
Ammonia-N	mg/L	32	73	0.5	1.1	0.77	2.54
Nitrate-N	mg/L	28	70	1.8	1.7	3.16	7.55
Phosphate-P	mg/L	32	73	0.09	0.68	0.085	1.144
Conductivity	µS	56	93	246.2	209.7	200.54	149.44
Turbidity	NTU	56	94	105.8	695.4	167.58	839.42
Dissolved oxygen	mg/L	18	94	4.6	3.3	1.59	1.15
pH	–	56	94	7.0	7.1	0.76	0.21

[a] Five-day @ 20°C biochemical oxygen demand.

TABLE 2.13
Summary Statistics: Water Quality of the Outflow to the Constructed Wetland (Figure 2.17) before (1 April 2003 to 31 March 2004) and after (1 April 2004 to 31 March 2005) the Introduction of *Carassius auratus* (Common Goldfish)

Variable	Unit	Sampling Number		Mean		Standard Deviation	
Temperature	°C	56	94	9.7	12.0	3.14	4.01
BOD[a]	mg/L	36	43	6.0	14.2	8.01	17.17
Suspended solids	mg/L	42	93	100.7	366.9	117.19	582.08
Ammonia-N	mg/L	30	73	0.2	0.4	0.15	0.43
Nitrate-N	mg/L	26	70	2.1	2.6	1.33	5.46
Phosphate-P	mg/L	31	73	0.08	0.30	0.040	0.496
Conductivity	µS	58	93	171.5	124.4	98.05	58.15
Turbidity	NTU	58	94	68.5	184.9	126.90	268.66
Dissolved oxygen	mg/L	52	94	6.0	3.8	8.67	1.15
pH	–	58	94	7.0	7.0	0.69	0.34

[a] Five-day @ 20°C biochemical oxygen demand.

TABLE 2.14
Summary Statistics: Water Quality of the Planted Pond (Figures 2.17 and 2.19) Receiving the Outflow from the Constructed Wetland before (1 April 2003 to 31 March 2004) and after (1 April 2004 to 31 March 2005) the Introduction of *Carassius auratus* (Common Goldfish)

Variable	Unit	Sampling Number		Mean		Standard Deviation	
Temperature	°C	56	94	8.7	11.6	4.60	4.90
BOD[a]	mg/L	36	43	15.5	19.3	18.91	14.25
Suspended solids	mg/L	47	93	58.7	24.7	116.61	55.45
Ammonia-N	mg/L	34	71	0.3	0.1	0.58	0.21
Nitrate-N	mg/L	28	69	0.7	0.4	2.25	0.84
Phosphate-P	mg/L	33	72	0.18	0.25	0.149	0.238
Conductivity	μS	58	93	310.5	246.9	116.86	83.21
Turbidity	NTU	58	94	18.4	14.2	20.02	29.84
Dissolved oxygen	mg/L	52	94	6.1	3.5	7.01	1.54
pH	-	57	94	7.2	7.2	0.24	0.25

[a] Five-day @ 20°C biochemical oxygen demand.

quality of the unplanted infiltration pond (Tables 2.13 and 2.14) was acceptable for disposal and recycling according to discussions particularly on potential five-day @ 20°C BOD and suspended solids threshold concentrations (e.g., 20 and 30 mg/L, respectively) (Butler and Davies, 2000; Scholz, 2004b; Zheng et al., 2005). However, water quality monitoring is not required for closed drainage systems (zero discharge) in Scotland (Zheng et al., 2005).

Suspended solids and turbidity measurements are high in the constructed wetland inflow due to high loads of organic material such as decomposing leaves (Table 2.12). These concentrations are reduced due to treatment within the constructed wetland (Tables 2.12 and 2.13). Nevertheless, nutrient concentrations were sufficiently high to cause an algal bloom. Nitrate-nitrogen increased in the constructed wetland due to nitrification of ammonia-N. Mats of algae swimming partly on top of a watercourse are usually considered unpleasant in their appearance by the public (CIRIA, 2000; Richardson and Whoriskey, 1992; Zheng et al., 2005).

2.4.3.2 Control of Algae with Goldfish

Algae began to grow in the infiltration ponds until *C. auratus* (Figure 2.18) was introduced on 1 April 2004. The result was a pleasant and clean SUDS during the second year of operation despite fears of water quality deterioration voiced elsewhere (Richardson and Whoriskey, 1992) (Tables 2.14 and 2.15).

Carassius auratus (similar to *Cyprinus carpio* or also known as common carp) is classified as herbivores with wild specimens predominantly feeding on plants.

TABLE 2.15
Summary Statistics: Water Quality of the Unplanted Pond (Figures 2.17 and 2.19) Receiving the Outflow from the Constructed Wetland before (01/04/03–31/03/04) and after (01/04/04–31/03/05) the Introduction of *Carassius auratus* (Common Goldfish)

Variable	Unit	Sampling Number		Mean		Standard Deviation	
Temperature	°C	56	94	8.6	11.6	4.74	5.35
BOD[a]	mg/L	33	43	18.1	19.2	24.80	14.38
Suspended solids	mg/L	44	93	25.6	21.7	52.03	39.57
Ammonia-N	mg/L	33	73	0.6	0.8	1.49	2.25
Nitrate-N	mg/L	30	71	0.8	1.3	2.27	5.91
Phosphate-P	mg/L	35	73	0.24	0.35	0.458	0.551
Conductivity	μS	58	93	220.6	193.4	139.62	88.04
Turbidity	NTU	58	94	12.6	19.1	19.45	36.92
Dissolved oxygen	mg/L	52	94	6.4	4.2	10.80	1.75
pH	-	58	94	7.2	7.3	0.46	0.29

[a] Five-day @ 20°C biochemical oxygen demand.

This particularly applies to closed pond systems (Figures 2.17 and 2.19). Therefore, *C. auratus* could be used to control aquatic weeds and potentially algae in ponds (Gouveia and Rema, 2005; Mikheev, 2005; Richardson and Whoriskey, 1992; Zheng et al., 2005).

Concerning the field experiment, relatively high numbers of filamentous green algae (Chlorophyta) were counted in pond samples taken on 29 March 2004. The dominant alga present was *Oedogonium capillare* that is cosmopolitan in freshwater. *O. capillare* can form mats in small ponds and is often mistaken for the more common *Cladophora glomerata* (blanket weed) according to Zheng et al. (2005).

Carassius auratus was introduced to control predominantly filamentous green algae and to increase public acceptance of SUDS. Concerning samples of algae taken on 4 October 2004, both the unplanted and planted ponds were less dominated by *O. capillare* in comparison to estimations on 29 March 2004. Moreover, the unplanted pond developed a greater diversity of filamentous green algae if compared to the planted pond. This may be due to the absence of macrophytes that would compete with algae for nutrients (particularly phosphorus). Moreover, large macrophytes (located in the planted pond; Figure 2.17) provide shade leading to a reduction of sunlight penetrating the water, and subsequently reducing the growth of algae (CIRIA, 2000; Zheng et al., 2005).

Nevertheless, the estimated algal biomass was considerably higher (at least one order of magnitude) in the planted if compared to the unplanted pond on 28 April 2005. This can be explained with the obvious observation that algae are the dominant (virtually only) plant food source in the unplanted pond.

2.4.3.3 Urban Drainage and Hygiene Planning

Flood protection management and the recreational value of urban landscapes can be improved at the same time by integrating SUDS (in contrast to conventional below-ground drainage) into the urban planning and development processes. Recreational activities may include watching ornamental fish such as *C. auratus* (Figure 2.18) and birds, walking, fishing, boating, holding picnics and teaching children about aquatic ecology (Butler and Davies, 2000; Scholz et al., 2005).

The confidence of town planners towards SUDS and public acceptance of infiltration ponds can both be increased by correct dimensioning of sustainable systems (Zheng et al., 2005) to avoid flooding, enhance water pollution control by using a robust pre-treatment train (e.g., silt trap, constructed wetland and swale; Figure 2.17) (D'Arcy and Frost, 2001) and control algae by biological (e.g., *C. auratus*) and not chemical (e.g., copper sulphate) means (Scholz, 2004b; Zheng et al., 2005). Moreover, storm water can be reused for watering gardens and flushing toilets as part of an urban water resources protection program (Butler and Davies, 2000; Scholz, 2004b; Zheng et al., 2005).

The challenge of urban water hygiene requires consideration. Runoff water could sweep some animal faeces into SUDS. Particularly dog faeces being carried in by floodwaters are a problem in urban environments despite local government efforts to encourage dog owners to scoop up droppings (Mckie, 2005).

Can *C. auratus* cope with the additional nutrient load? Could there be a potentially dangerous build-up of *E. coli* from excrements? Further detailed research on the health of *C. auratus* is therefore in progress. Findings indicate that the additional nutrient load is very small in comparison to the background load (e.g., leaves and soil), and that no accumulation of bacteria in the system is detectable.

2.4.4 Conclusions and Outlook

The research has attracted wide national and even international public interest; for example, The Observer, sundayherald, THE SUNDAY POST, Daily Mail and Deutschlandfunk. The public relate well to their urban environment and common goldfish (*C. auratus*), which are often used as pets in aquariums and garden ponds. Moreover, the rather unappetizing character of experiments with dog droppings interests the public – usually a taboo for both the 'dog-loving' public (an obvious but very 'human' contradiction), and even the scientific and engineering community (first scientific study according to the Web of Knowledge and Scopus).

Findings show that the introduction of *C. auratus* to closed (zero discharge) systems such as infiltration ponds do improve most water quality variables (e.g., reductions of algae suspended solids, nitrate-N and turbidity) and should lead to an increase in public acceptance of SUDS. Moreover, the water qualities of the infiltration ponds were acceptable for water reuse (below likely future thresholds for BOD and suspended solids) after the set-up period of the SUDS and the introduction of *C. auratus*. A bloom of filamentous green algae dominated by *O. capillare* during the springs of 2003 and 2004 was observed. However, *C. auratus* decimated the algae particularly well in the unplanted pond, where no other (plant) food source was abundant.

2.5 GEOTHERMAL HEAT PUMPS INTEGRATED WITHIN PERMEABLE PAVEMENTS FOR RUNOFF TREATMENT AND REUSE

2.5.1 Introduction

Unsustainable construction practices worldwide are the major contributors to the depletion of natural resources, resulting in air and water pollution, solid waste and health hazards. Urbanization has detrimental effects on surface water systems (Scholz and Grabowiecki, 2007). Water harvesting and reuse as well as renewable energy solutions is the way forward towards greater sustainability of ecosystems for mankind and as a result receives great attention amongst academics, environmentalists, public health experts and economic development advocates.

Greater environmental awareness has led to innovative research such as The Hanson Ecohouse, which is a demonstration site located at the British Research Establishment's Innovation Park at Garston, Watford, England, UK. The Ecohouse demonstrates integrated smart home solutions, which are cost-effective and provide significant benefits for the management of energy and water usage incorporating permeable (pervious) pavements as well as ground-source heat pumps (GSHPs) (Formpave, 2008).

Permeable pavements are a type of SUDS, which facilitates attenuation and treatment of surface water. Storm water runoff sometimes requires treatment preventing pollutants and other harmful substances from entering surface waters and groundwater. This approach is beneficial as it considers water quality, quantity, environmental and amenity issues. Permeable paving effectively provides a structural pavement suitable for pedestrians and vehicular traffic while allowing water to infiltrate freely through the surface into the pavement sub-base, where it can be temporarily stored, treated and recycled (Grabowiecki et al., 2008). It is an alternative to conventional pavement surfaces whereby surface runoff percolates through the pavement into a sub-base. Concrete permeable pavements are made of fine filler fractions in a modular form to allow for free flow through of water and are mainly used where there is light traffic.

Permeable pavement systems (PPS) have many potential benefits such as prevention of pollution by suspended solids, increased BOD and elevated ammonia concentrations via processes including filtration, sedimentation, absorption, adsorption and biodegradation (Pratt et al., 1999; Scholz and Grabowiecki, 2007). When compared to traditional, impervious asphalt, permeable pavements can reduce runoff quantity, lower peak runoff rates and delay peak flows due to their high surface infiltration rates (Bean et al., 2007; Brattebo and Booth, 2003; Pratt et al., 1999). Aquifer recharge by infiltrated storm water and the reduction for urban drainage is also advantageous with PPS. Infiltration rates for newly laid PPS are commonly more than 1,000 mm/h, and maintenance is carried out by vacuum sweeping approximately every six months (Pratt et al., 1995, 1999).

Permeable pavements can be combined with water recycling technology. The purpose is to collect treated runoff in a tanked belowground collection system for subsequent recycling. Applications may include car washing, garden sprinkling and toilet flushing. More recently, PPS have been combined with GSHPs to either heat or cool nearby buildings (Scholz and Grabowiecki, 2007; Grabowiecki et al., 2008).

Geothermal energy is seen as renewable and sustainable, and geothermal heat pump (GHP) systems are established energy-saving tools, predominantly used in new and sustainable commercial and residential areas (Ozgener and Hepbasli, 2007). Geothermal space conditioning (combined heating and cooling) has expanded considerably, permitting the utilization of heat associated with low-temperature bodies such as the ground, shallow aquifers and ponds (Sanner et al., 2003).

GHPs use refrigerants to move thermal energy (heat) from buildings during summer or hotter months and transfer heat when required during winter. A GHP draws energy from relatively moderate and consistent temperatures found just below the Earth's surface. GSHPs use refrigerant to move unwanted energy (i.e. heat) out of buildings during summer and into them (if required) during winter according to the U.S. EPA (Bose, 2005). They use constant temperatures of surrounding grounds, which are lower than the corresponding air temperatures during warm seasons (heat sinks) and higher during cold seasons (heat sources). There is no need for burning fossil fuels to transfer energy either side; therefore, this is a sustainable technology, which also reduces carbon dioxide emissions in comparison to traditional techniques (Geothermal Heat Pump Consortium, 2009).

GSHP systems draw energy from the relatively moderate and consistent temperatures found just below the Earth's surface. Approximately 50% of solar energy that reaches the Earth is absorbed into the ground (Frederick et al., 1989). This allows the Earth to maintain warmer ground temperatures than the air temperatures in the winter, and cooler ground temperatures in the summer.

Although the heat transfer process may seem complicated, the GHP is like a refrigerator (Tota-Maharaj et al., 2009). It is simply a process of moving heat energy from one point to another using heat exchangers. During heating, the heat pump extracts heat from the earth loop (or permeable pavement's water) and, using a refrigeration process, the heat is intensified and transferred, which is then delivered into a residential or commercial building. During cooling, the process is reversed. Heat is removed from the building by the cool refrigerant and then transferred to the ground (Tota-Maharaj et al., 2009).

The aim is to assess the water quality and associated health risks related to GSHP operation within a permeable pavement sub-base. The corresponding objectives are to assess (a) the combined system (PPS and GHP) performance with respect to the general water quality in terms of oxygen demand and nutrient removal for a controlled indoor and uncontrolled outdoor experimental set-up; (b) the microbiological contaminant removal for various system designs; (c) the system operation during various natural environmental conditions such as seasonal changes and artificial temperature fluctuations during heating and cooling cycles and (d) the GSHP's efficiency by analysing the coefficient of performance (COP). Section 2.5. is based on the original articles by Tota-Maharaj and Scholz (2010) and Tota-Maharaj et al. (2009, 2010).

2.5.2 Methodology and Experimental Set-up

2.5.2.1 Design, Construction and Composition of Experimental Rigs

Two identical experimental PPS set-ups comprising six different rigs each were located at The University of Edinburgh's Institute of Infrastructure and Environment. The first set-up of six rigs was operated under controlled indoor (mean ambient temperature of 16°C) and a further identical set-up under uncontrolled outdoor (system located below ground level but subject to natural changes) environmental conditions. No further replicate set-up was operated. The main component of each rig was a bin containing the PPS. A heating and cooling system was integrated at the base of bins 1, 2, 4 and 5 for both PPS set-ups to simulate different seasons by adjusting temperatures. Figure 2.20 shows a general system overview for the combined PPS and GHP rig.

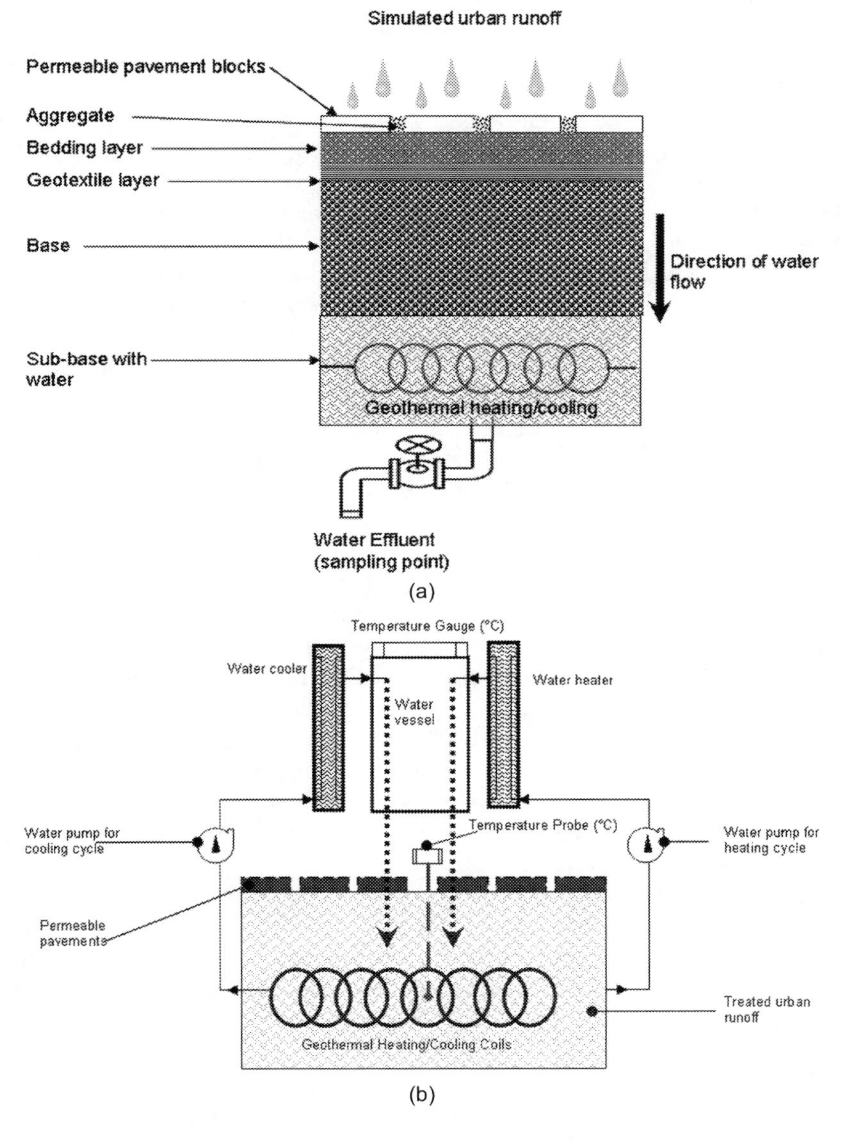

FIGURE 2.20 (a) Schematic outline of the main principles of the experimental set-up and (b) schematic of temperature probes and temperature gauges for geothermal heat pump simulation within permeable pavement systems.

Commercial 240-l wheelie-bins (height: 1,070 mm; maximum length: 730 mm; maximum width: 580 mm) predominantly filled with aggregates were used for each rig as basic construction devices, mimicking impermeable tanked systems and providing suitable conditions for water collection. Aggregate sizes were determined according to British Standards (British Standards Institution, 1992) and Highway Agency (2009) specifications.

The layers and corresponding layer thicknesses (mm) and aggregate diameters (mm, if applicable) within each bin were as follows: air (800; not applicable), clean stone (500; 50), geotextile (20; not applicable), upper sub-base (1,000; 50–200), upper part of the lower sub-base (2,500, 63–100) and lower part of the lower sub-base (2,500; 63–100). For research purposes, the lower sub-base within each bin was increased in depth from the recommended 350–500 mm, because GHP coils need to be submerged fully in water to operate appropriately.

Bins 1–3 and bins 4–6 of each experimental set-up contained Inbitex Composite and Inbitex Geotextile, respectively. The Inbitex geotextile (Terram Geosynthetics, 2009) made of polyethylene and polypropylene fibres was placed in the top part of the upper base of selected bins, as this is the area where microbial degradation of pollutants is likely to take place. Either Inbitex on its own or Inbitex together with an impermeable membrane (Terram Drainage Composite, Terram Geosynthetics, Gwent, UK) called a 'composite' were used in both the indoor and outdoor experiments.

On the top of the geotextile, an additional 5-mm layer of pea gravel was placed and covered with the pervious (permeable) paving blocks provided by Hansen Formpave. Clean single-sized 3-mm pea gravel was used as grout and aggregate to seal and fill the gaps between the individual permeable blocks. Temperature probes and gauges were placed at the bottom and within the top layers of the tanked PPS where heating and cooling took place periodically (Figure 2.19).

Each system received a mixture of ten parts of de-chlorinated tap water and one part of gully pot liquor once per week providing approximately 2.5 L of simulated urban runoff per rig. A gully pot is a chamber where the initial collection and treatment of urban runoff occurs. It is normally located at the edge of a pathway or road with a protective cast iron gating on top. The runoff is introduced into this chamber first and fills it. Thereafter, the overflow is usually directed into a traditional sewerage system. Gully pots allow for heavy particle sedimentation and contain organic matter such as floral debris, pesticides, insecticides, animal faeces and other storm water pollutants.

Freshly collected dog faeces (3.1 g) were added weekly to bins 1 and 4 of both the inside and outside experimental set-ups to simulate the worst-case scenario of urban runoff contaminated with potentially pathogenic organisms. Approximately 2.2 l of treated water was sampled from the bottom of each bin weekly between June 2006 and March 2009.

2.5.2.2 Physico-Chemical Analyses

Water samples were collected approximately twice per week from a thoroughly mixed 0.5–l sample obtained from the bottom of each bin with a hand pump. No replicate sampling was therefore undertaken. The samples were stored during collection in a clean and large glass beaker. The subsequent analyses were carried out immediately after sampling according to American standard methods (Clesceri et al., 1998) unless stated otherwise. The five-day BOD (stored at 20°C; N-Allylthiourea nitrification inhibitor used) was determined with the OxiTop system (Wissenschaftlich-Technische-Werkstätten, Weilheim, Germany). The amount of sample water used was usually 0.432 L for the outflows, and between 0.192 and 0.250 L for the inflows. After 0.5 hour of aeration with air pumps, a nutrient inhibitor was added to the sample, and bottles were incubated at a constant temperature of 20°C for five days. After this time, relevant values were displayed electronically.

Nitrate was reduced to nitrite by hydrazine in alkaline solution, and subsequently determined at 550 nm by a colorimetric method using a Bran Luebbe AA3 flow injection analyser (Bran and Luebbe, 1999) following ISO standard methods (Clesceri et al., 1998). Ortho-phosphate-phosphorus was determined using ammonium molybdate, ascorbic acid and sulphuric acid. Ammonium molybdate and sodium bicarbonate were also used in the analyses performed by automated colorimetry with an AA3 colorimeter (Clesceri et al., 1998). Blue-coloured compounds were measured at the same time using a Bran and Luebbe (1999) autoanalyser.

American standard methods (Clesceri et al., 1998) were also used for the examination of suspended solids, pH, conductivity and total dissolved solids. A Hanna HI 991 300 meter was applied for the examination of the latter three parameters. Dissolved oxygen and the redox potential were measured with a WTW Oxi 315i meter and the Redox 201 meter, respectively. All meters were calibrated weekly for quality control assurance purposes.

2.5.2.3 Microbiological Analyses

American standard methods (Clesceri et al., 1998) were applied for microbiological assessments unless stated otherwise. Microbiological determinations were conducted immediately after sampling. The same sample water was used for both standard water quality and microbiological parameters. Gram staining was performed where necessary (see below). Three replicates were used for all microbiological determinations.

Nutrient agar, a non-selective growth medium, allowed for the development of a wide variety of oxygen-tolerant genera such as *Escherichia coli*, which is one of the key microbial water quality indicators. MacConkey Agar allowed for the detection and isolation of *Salmonella* and *Shigella* species occurring frequently, for example, in pathological and food specimens. The Slanetz and Bartley Agar was designed to favour the growth of faecal enterococci (Tejedor et al., 2001).

Legionella GVPC agar was specially developed for the isolation of most *Legionella* species, notably *Legionella pneumophila*, which is the species most frequently involved in Pontiac fever. This selective medium also enables the enumeration of *Legionella* in water according to published standards (ISO International Organization for Standardization, 2008; Leoni and Legnani, 2001).

All culture media for growing total coliforms, faecal coliforms, faecal streptococci and *Legionella* spp. were ordered from Oxoid Ltd (Solar House, Mercers Row, Cambridge, UK). Petri plates, filter papers (MF 200 with diameter of 125 mm) and primers were supplied by Fisher Scientific, UK (Bishop Meadow Road, Loughborough, UK).

Viable cell counts were determined by the spread plate method (Cappuccino and Sherman, 1996). Nutrient Agar plates were incubated at 36°C ± 1°C for two days. Slanetz and Bartley agar plates were kept at 45°C for five days. The *Legionella* GVPC agar plates were incubated and inoculated in a moist chamber at 36°C ± 1°C for ten days and examined at intervals between one and two days during the incubation period (Cappuccino and Sherman, 1996; Leoni and Legnani, 2001).

Figure 2.21 provides a general overview of the selected methodology. Colonies from selective plates were sub-cultured on Nutrient Agar to test for purity. The organisms were then identified by sequencing of the 16S ribosomal ribonucleic acid (rRNA) gene. The 16S rRNA gene was obtained by polymerase chain reaction (PCR) amplification using 'universal' bacterial primers (Weisburgh et al., 1991). PCR was performed in a total reaction volume of 50 µL containing 34-µL water, 10-µL reaction buffer, 2-µL primer mixture fD1 + rD1 (10 pmol/µL each), 1-µL dNTP mixture (10 mM each), 2-µL cell suspension and 1-µL Taq polymerase.

Rather than purifying genomic DNA for use as a template, a suspension of cells in sterile water was added directly to the PCR reaction to act as a template. The

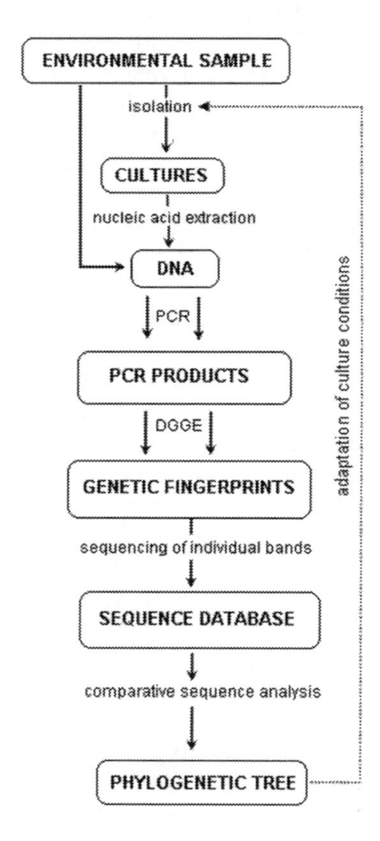

FIGURE 2.21 General overview of key molecular microbiological techniques. DNA, deoxyribonucleic acid; RNA, ribonucleic acid; PCR, polymerase chain reaction; DGGE, denaturing gradient gel electrophoresis.

release of DNA from cells lysed during the heating stages is sufficient to provide a template for the amplification, greatly reducing the amount of sample preparation required and opportunities for contamination with exogenous bacterial DNA. The primer sequences used were fD1: agagtttgatcctggctcag and rD1: aaggaggtgatccagcc. The cycling conditions were as follows: denaturation at 95°C for 2 minutes, followed by 30 cycles at 95°C for 0.5 minute, annealing at 58°C for 0.5 minute and extension at 72°C for 1.5 minutes and a final extension step of 72°C for 10 minutes. Samples were kept at 4°C until further analysis.

The PCR products were analysed by agarose gel electrophoresis, and if a PCR product of the correct size was observed, the DNA was purified using standard techniques (Sambrook et al., 2000). If no product was obtained, the PCR was repeated using a lower annealing temperature. After purification, the PCR products were submitted for sequencing by the Gene Pool, The University of Edinburgh Sequencing Service, using the same primers as for PCR. The analysis of sequence data and homology comparisons were performed using the Basic Local Alignment Search Tool (BLAST) according to Altschul et al. (1990). Phylogenetic analysis was done using the Molecular Evolutionary Genetics Analysis (MEGA) software version 4.0 (Tamura et al., 2007), using the Jukes and Cantor nucleotide substitution model for sequence alignment and the neighbour-joining method of tree inference. Confidence estimates of the branches in this tree were determined by bootstrap resampling analysis with 1,000 replicates.

Denaturing gradient gel electrophoresis (DGGE) was used to analyse community DNA. Total waterborne DNA was extracted from 1 mL of each water sample using an UltraClean™ soil and water DNA extraction kit (CamBio Ltd, UK) according to the manufacturer's protocol. The bacterial 16S rRNA genes were amplified from the community DNA or pure cultures as described above.

Nested PCR was then performed with universal bacterial primers to amplify the V3 variable region of the bacterial 16S rRNA gene (Fedorovich et al., 2009). The extension products were loaded onto DGGE gels containing a linear gradient of between 30% and 70% denaturant, electrophoresed at 60°C for 960 minutes at 75 V and silver-stained (Fedorovich et al., 2009). The stained gels were scanned at 600 dots per inch (DPI) using an Epson Perfection V700 scanner. The bands obtained from isolated bacteria were compared to community profiles to obtain an indication as to whether potentially pathogenic organisms were in fact common in the community.

2.5.3 Results and Discussion

2.5.3.1 Water Quality Comparisons

The inflow and outflow water quality of both the indoor and outdoor experimental systems were assessed. The variability in treatment can be judged by comparing inflow and outflow concentrations with each other (Table 2.16). High standard deviations were recorded for sample temperatures (°C), pH, conductivity (μS), total dissolved solids (mg/L), suspended solids (mg/L), dissolved oxygen (mg/L), redox potential (mV), nitrate-nitrogen (mg/L), ortho-phospahte-phosphorus (mg/L), ammonia-nitrogen (mg/L) and BOD (mg/L) for mean indoor and outdoor rigs between June 2006 and March 2009. This can be explained by the highly variable properties of gully pot liquor, which change subject to season, and rain events (dilution but also sediment washout). Moreover, the highest standard deviations were computed for the inside and outside bins 1 and 4, which received dog faeces (see above).

Mean temperature values were higher and more variable for the outside set-up in comparison to the inside one (Table 2.16). This is due to the less controlled operation of the outside system, which is subject to natural weather fluctuations. Nevertheless, the total dissolved solids, suspended solids, pH, conductivity, redox potential and dissolved oxygen values were relatively constant, indicating similar water quality characteristics.

The highest mean differences between inflow and outflow were recorded for the nitrogen fractions. Phosphorus concentrations are usually very low (i.e. 0.01–0.1 mg/L) in waters, because phosphates are relatively difficult to dissolve. Sources of phosphates are variable including agricultural (containing fertilizers), urban and rural runoff, failing septic systems (Spellman, 2008). The experimental system showed relatively high phosphorus removal rates (Figure 2.22a and b). The removal rates were high and very stable (standard deviations (SD): 0.01–0.02 mg/L) for the inside bins with (82%–94% removal) and without (60%–75% removal) additional dog faeces load. Concerning the outside rig, the corresponding rates were also high and relatively stable for the inside bins with (76%–100% removal; SD: 0.01–0.03 mg/L) and without (79%–81% removal; SD: 0.01–0.05 mg/L) additional contamination. The ortho-phosphate-phosphorus influent concentrations for In and In+Faeces were 2.1 and 3.8 mg/L, respectively.

The findings were statistically evaluated. A one-sample t-test at 99% confidence interval was carried out for removal efficiencies >99.5%, indicating that the removal of ortho-phosphate-phosphorus was highly statistically significant. A one-way analysis of variance was performed for indoor versus outdoor bins, indicating that there was no statistically significant difference in terms of removal efficiencies associated with temperature differences.

Ammonia-nitrogen removal rates were relatively high (97%–98%) and stable (Figure 2.22), indicating that the urban runoff was rich in nutrients and organic material, and easily biodegradable. A one sample t-test was applied for ammonia-nitrogen, indicating that the corresponding removal rates were statistically significant ($p < 0.05$). A one-way analysis of variance for the indoor versus the outdoor experimental set-up indicated that there was no significant difference.

Nitrate-nitrogen concentrations (mg/L) were the highest for the inside and outside bin 4 (Table 2.16), which contained faecal matter from dogs. This additional contamination contributed to the effluent's high organic matter content. If inflow and outflow values are compared with each other (Table 2.16), negative removal efficiencies for nitrate-nitrogen are apparent. This can be explained by the virtually complete transformation of ammonia-nitrogen to nitrate-nitrogen (nitrification). It follows that all bins acted as sources of nitrate-nitrogen. According to the Waste Water Framework Directive (EU Waste Water Treatment, 1991), the concentration of total nitrogen into the receiving water should be less than 10 mg/L, which is higher than the combined nitrogen fractions measured in this study.

TABLE 2.16
Inflow and Outflow Water Quality Parameter (Mean/Standard Deviation (SD)) for the Indoor and Outdoor Experimental Set-ups (June 2006–March 2009)

Water Parameters	Unit	Inflow	Inflow with Faeces	Outflow of the Indoor Set-up						Outflow of the Outdoor Set-up					
				1	2	3	4	5	6	1	2	3	4	5	6
Biochemical oxygen demand (mean/SD)	(mg/L)	40.5/15.3	90.8/32.7	0.8/0.1	0.4/0.2	0.3/0.1	0.6/0.2	0.9/0.3	0.7/0.2	0.5/0.1	1.5/0.1	1.2/0.2	0.7/0.2	0.8/0.3	0.6/0.4
Ammonia-nitrogen (mean/SD)	(mg/L)	0.56/0.00	1.8/0.05	0.03/0.01	0.02/0.01	0.02/0.00	0.01/0.00	0.02/0.01	0.03/0.00	0.03/0.00	0.06/0.00	0.06/0.00	0.08/0.00	0.09/0.00	0.04/0.00
Nitrate-nitrogen (mean/SD)	(mg/L)	1.4/0.08	1.6/0.07	2.15/1.32	2.86/1.1.4	4.37/2.37	5.13/2.25	2.64/1.22	4.05/2.31	4.82/2.76	3.84/1.78	4.32/1.84	6.17/2.67	4.85/2.33	4.74/2.47
Ortho-phosphate-phosphorus (mean/SD)	(mg/L)	2.1/1.4	3.8/1.6	0.68/0.01	0.64/0.02	0.24/0.01	0.44/0.02	0.52/0.01	0.83/0.02	0.90/0.03	0.14/0.02	0.84/0.01	0.28/0.02	0.39/0.01	0.45/0.05
Total dissolved solids (mean/SD)	(mg/L)	58/35	88/39	212/21	167/34	156/27	187/31	165/26	153/28	183/31	180/37	152/23	198/27	198/31	167/28
Suspended solids (mean/SD)	(mg/L)	179/12	254/13	30/14	38/12	50/12	77/13	88/15	76/16	54/16	39/13	110/12	86/12	70/10	104/11

(Continued)

TABLE 2.16 (*Continued*)
Inflow and Outflow Water Quality Parameter (Mean/Standard Deviation (SD)) for the Indoor and Outdoor Experimental Set-ups (June 2006–March 2009)

Water Parameters	Unit	Inflow	Inflow with Faeces	Outflow of the Indoor Set-up						Outflow of the Outdoor Set-up					
				1	2	3	4	5	6	1	2	3	4	5	6
Sample temperature (mean/SD)	(°C)	16.6/1.7	16.2/2.3	15.5/2.1	16.3/3.2	14.9/1.4	15.1/1.7	15.7/1.4	16.1/1.8	14.9/2.2	15.8/2.7	17.3/1.9	17.5/2.3	16.1/2.2	17.1/1.4
pH (mean/SD)	(–)	6.8/0.1	7.1/0.2	7.0/0.2	6.8/0.3	6.7/0.3	7.1/0.1	7.2/0.2	6.7/0.3	7.2/0.3	7.4/0.4	7.3/0.3	7.2/0.3	7.1/0.2	6.9/0.1
Conductivity (mean/SD)	(μS)	127/33	189/38	440/31	320/26	315/42	384/53	345/41	350/37	378/39	356/40	380/36	365/27	328/52	334/34
Dissolved oxygen (mean/SD)	(mg/L)	8.8/0.1	8.6/0.3	4.3/0.3	4.9/0.4	5.6/0.2	5.8/0.6	6.2/0.3	5.3/0.2	6.5/0.1	4.3/0.7	5.7/0.5	4.7/0.9	5.7/0.8	3.2/0.8
Redox potential (mean/SD)	(mV)	129/43	196/59	175/51	165/55	160/54	171/52	172/48	178/50	166/46	159/57	165/46	177/52	172/55	136/58

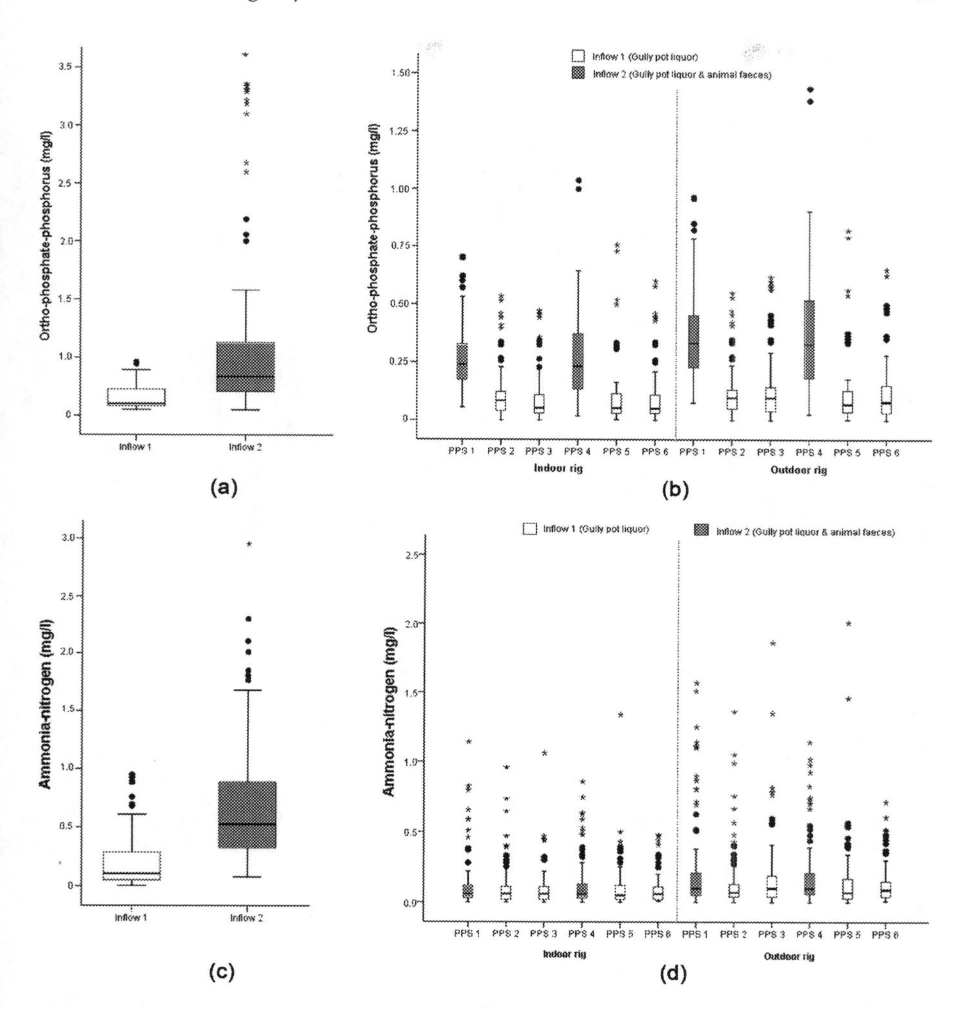

FIGURE 2.22 (a) Ortho-phosphate-phosphorus (mg/L) within Inflow 1 (gully pot liquor) and Inflow 2 (gully pot liquor and dog faeces); (b) ortho-phosphate-phosphorus (mg/L) outflow for the indoor and outdoor bins; (c) ammonia-nitrogen (mg/L) within the Inflow 1 and Inflow 2 and (d) ammonia-nitrogen (mg/L) outflow for indoor and outdoor bins. The plots represent the 25th percentile, median and the 75th percentile. The whiskers represent the 10th and 90th percentiles (June 2006 to March 2009, $n = 190$). Solid circles and stars indicate outliers and extreme outliers, respectively.

The presence of suspended solids contributes to turbid waters, which may be objectionable for aesthetic reasons. Moreover, suspended particles may interfere with disinfection processes. Solids provide shelter for microorganisms, thus inhibiting the ability of a disinfectant to destroy potentially pathogenic organisms. The mean suspended solids removal rates were between 70% and 100%, and between 70% and 90% for the indoor and outdoor rigs, respectively, indicating how effective permeable pavements are in suspended solids and associated turbidity removal (Figure 2.23).

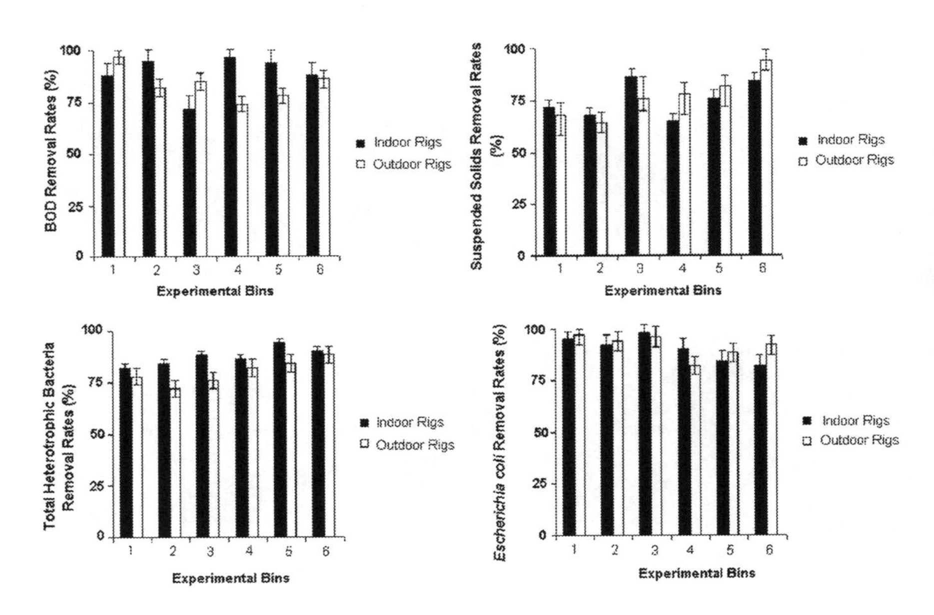

FIGURE 2.23 Mean removal rates for the biochemical oxygen demand, suspended solids, total heterotrophic bacteria and *Escherichia coli* (June 2006 to March 2009).

High suspended solids concentrations are often associated with high BOD values. The BOD inflow concentrations strongly depended on the gully pot liquor characteristics; 82 and 106 mg/L were recorded for the inflow and the inflow spiked with dog excrements, respectively. The BOD reductions due to water treatment within the sub-base of the pavement were relatively high; that is, between 60% and 90%, and between 75% and 100% for both the indoor and outdoor set-up, respectively (Figure 2.23).

2.5.3.2 Microbiological Assessment

Waterborne diseases such as cholera, typhoid, typhus, giardiasis, legionellosis and gastroenteritis have historically all been linked to microbiological contamination (Weiner, 2008). Identification and elimination of the corresponding microorganisms is essential to assess the efficiency of microbiological removal within PPS. There are several routes of infection that these biological contaminants can take such as ingestion, respiration or inhalation regarding water reuse such as toilet flushing and garden sprinkling (Salvato et al., 2003).

Counts for all microbiological indicators were more stable for the inside compared to the outside bins due to the control of temperature. The temperature range in which most potentially pathogenic waterborne bacteria grow and survive is between 40°C and 60°C at a pH range between approximately 3.5 and 8.7 (Salvato et al., 2003). Bins with higher temperatures are therefore more likely to support the growth of potentially pathogenic bacteria.

Bacterial counts were higher for the raw runoff samples in comparison to the treated water as indicated in Figure 2.23. The mean total heterotrophic bacteria

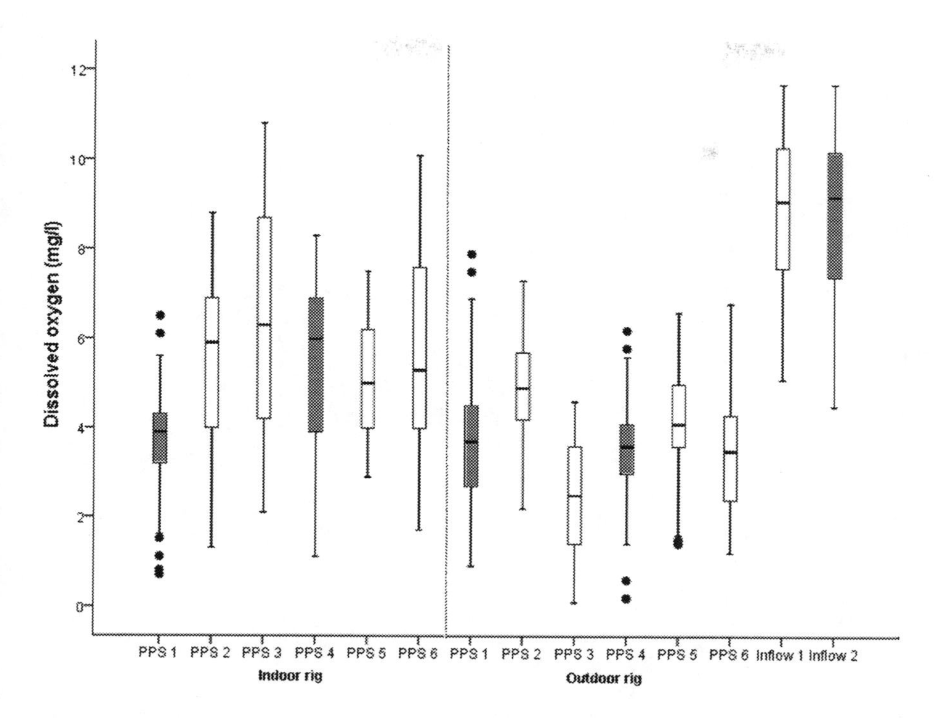

FIGURE 2.24 Comparison of the dissolved oxygen (mg/L) for the indoor and outdoor bins (June 2006–March 2009). Grey shade is indicative of water contaminated with dog faeces. Solid circles indicate outliers.

count for the influents without and with dog faeces were 4.13×10^7 colony forming unit (CFU)/100mL and 3.8×10^6 CFU/100mL, respectively. Outflows of the indoor rig reached maximum numbers of 3.7×10^5 CFU/100mL for influents without and 4.5×10^5 CFU/100mL for bins with dog faeces. In comparison, the lowest counts were recorded for the indoor bin 3 (2.5×10^4 CFU/100mL) and the outdoor bin 5 (1.74×10^4 CFU/100mL). Mean *E. coli* counts were 18.8×10^3 and 36.7×10^5 CFU/100mL for influents without and with dog excrements, respectively. All bins performed well regarding the removal of *E. coli* (Figure 2.23). Figure 2.24 indicates reductions in dissolved oxygen. The consumption of oxygen is a result of the biogeochemical processes whereby biological material is being broken down aerobically reducing the counts of total heterotrophic bacteria.

Traditionally, the classification or taxonomy of microbiological organisms has been based on their morphology (i.e. their visible structure and form). Gram staining (Cappuccino and Sherman, 1996) was performed to assess the morphology of the bacteria. Both Gram-positive and Gram-negative bacterial species were found in the system. The colony morphology of each isolated bacterium was studied on nutrient agar. An indication of the health significance of selected waterborne organisms is shown in Table 2.17.

Different colonies were isolated and identified by molecular microbiological techniques (see above). Members of different phylogenetic groups were isolated from the

TABLE 2.17
Qualitative Deoxyribonucleic Acid Sequencing Results for Samples (Two Replicates) Collected from Different Indoor and Outdoor Bins

Seq	Outdoor Rig (Bin 1–6)						Indoor Rig (Bin 1–6)						Inflow	Inflow + Dog Faeces	Accession No./Closest Match	SimilArity (%)	Closest Matched Organism	Risk Group
T1 F	x	x					X	±	x				X	X	AM262151	99	*Aeromonas hydrophila* subsp. *ranae* type strain CIP 107985T	*2*
T2 F	±	±									x		X	X	AB021192	100	*Bacillus mycoides*	*1*
T4 F			X			x			x			±	X	X	FM955876	100	*Brevundimonas vesicularis* strain Asd M7-3	*1*
T5 F	x				x	x			x	X	x		X	X	AM062693	99	*Enterobacter amnigenus* isolate p412	*2*
T6 F											x		X	X	FM955853	100	*Stenotrophomonas rhizophila* strain Asd M1–7	*1*
T7 F	x	x				x	X		x			x	X	X	AY987751	100	*Aeromonas salmonicida* subsp. *salmonicida* strain CECT 894T	*2*
T16 F		x				x			x				±	X	Y17658	99	*Klebsiella (Raoultella) terrigena* strain ATCC33257T	*1*
T22F		x		x	x	x					x		x	X	EF204209	99	*Brevundimonas nasdae* strain G124	*1*
TA1 F	x	x					X			x	x		x	X	FJ485825	100	*Bacillus subtilis* strain S52	*1*
M2 F		x	X				X				x		x	X	EU003535	98	*Pseudomonas* sp. 122 (not *P. aeruginosa*)	*1*

Blank, negative; ±, weekly positive; x, positive; X, strongly positive.
The risk group classification is based on guidance provided by the UK Advisory Committee on Dangerous Pathogens.
The closest matches are based on BLAST database searches performed in February 2009.

experimental systems. Potentially pathogenic species such as *Aeromonas salmonicida* were found in some samples regardless of contamination with animal faecal matter (Table 2.17). Gram negative organisms were dominant among the isolates identified. Most frequently recovered bacterial isolates were of the genera *Bacillus*, *Brevundimonas*, *Pseudomonas* and *Aeromonas*.

No growth of faecal streptococci, *E. coli*, *Salmonella*, *Shigella* and faecal enterococci was observed in any effluent samples, indicating the potential suitability of the treated runoff for recycling purposes. The inflow waters contaminated with dog faeces did, however, contain *E. coli* and faecal streptococci. Moreover, no growth of *Legionella* spp. was observed in the sub-base of the pavements. In contrast, *Pseudomonas* spp., *Aeromonas* spp., *Bacillus* spp. *Brevundimonas* spp. and *Enterobacter* spp. were identified in the inside and outside rigs. In addition, *Klebsiella* spp. and *Stenotrophomonas* spp. were only isolated and identified from indoor and outdoor bins, respectively.

Molecular microbiological methods (16S rRNA gene sequencing and DGGE fingerprinting) were applied to study the bacterial community and to compare the microbial diversity of the inflow and outflow samples. Table 2.17 indicates the presence of different organisms in different bins. Figure 2.25 shows a neighbour-joining phylogenetic tree of the bacteria that were present in the saturated zone of the PPS, based on the 16S rRNA gene sequences of the isolates. These heterotrophic bacteria were closely related to seven known bacterial species: *Brevundimonas* spp., *Pseudomonas* spp., *Aeromonas* spp., *Klebsiella* spp., *Enterobacter* spp., *Stenotrophomonas* spp. and *Bacillis* spp. (98%–100% similarity; see Table 2.17). Furthermore, DGGE

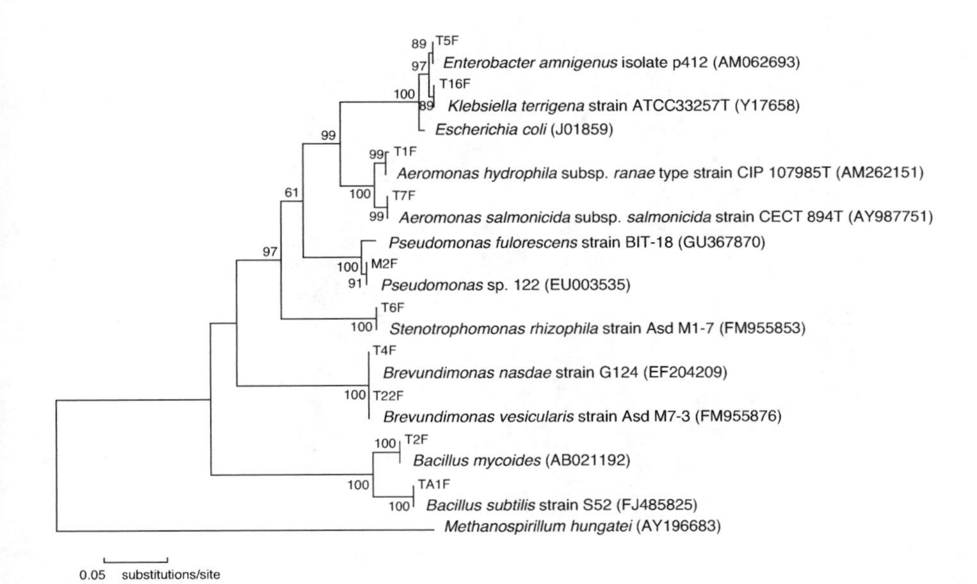

FIGURE 2.25 Phylogenetic tree illustrating the bacterial isolates present in the saturated zone of the experimental permeable pavement system and the closest related organisms (Table 2.18). *Methanospirillum hungatei* was added as the outlier organism.

fingerprinting of the bacterial communities in the 12 test bins and the 2 inflow samples (Figure 2.25) shows that the bacterial communities of the inflow with and without dog faeces were like each other, but despite this similarity, the outflow from the replicate bins showed very different outflow communities. There does not appear to be any similarity between the indoor and outdoor bins. The isolates were identified in the inflow and outflow and are represented in Figure 2.26 by open ellipses. This showed that they were not necessarily the most abundant organisms in the communities. Bacterial communities are similarly diverse in both the inflow and outflow, showing 20–30 possible bacterial species in all communities.

Molecular analysis has shown that the microbial diversity in the inflow and outflow is much higher than plating suggests. Further studies are required to determine the microbial community functionality and how this contributes to nutrient removal and bioremediation in permeable pavements. Information on persistence of microorganisms from sample to pavement is important since information on the presence or absence of species in the system helps to make an informed assessment on potential health risks.

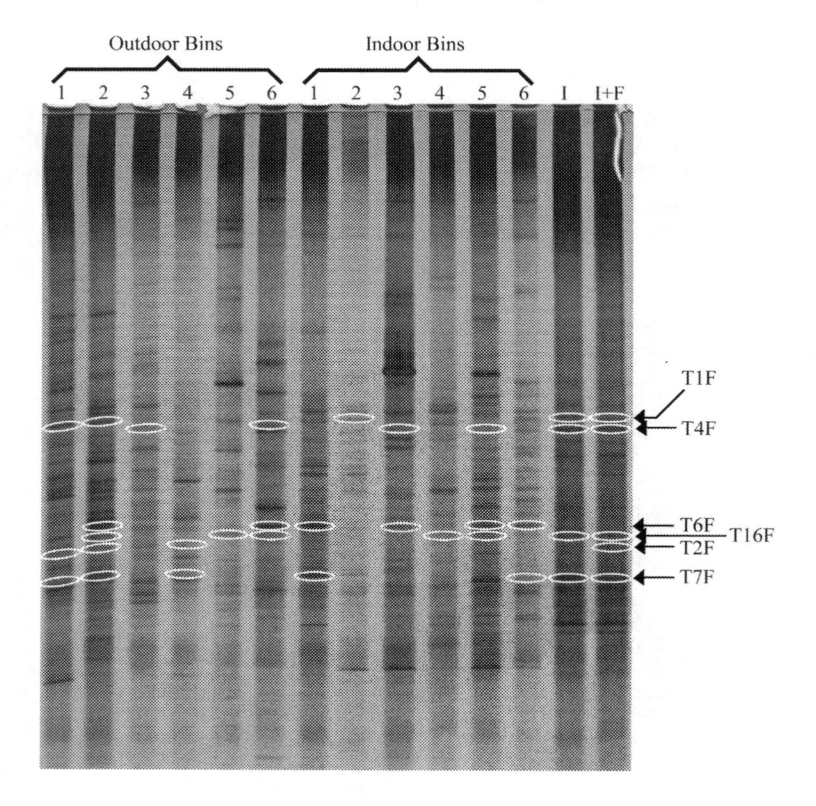

FIGURE 2.26 Denaturing gradient gel electrophoresis showing the bacterial communities in the inflow and outflow of the experimental permeable pavement system. I and I+F represent bacterial communities in the inflow without and with dog faeces, respectively. The positions to which the bands from the known isolates have migrated in the gel are indicated by arrows. These isolates are indicated in the inflow and outflow communities, represented by open ellipses.

2.5.3.3 Performance of the Geothermal Heat Pumps

GHP energy efficiency ratings differ depending on the individual application. However, the heating COP usually ranges between 3.0 and 4.0, although the closed-loop applications have COP values between 2.5 and 4.0 (Omer, 2008).

The performances of the simulated GHP systems were determined by the COP ratio, which measures the steady-state cooling and heating efficiency of the heat pumps. A high COP is a good indicator of an efficient unit. As the COP increases, either less heat is supplied by the GSHP or more heat is rejected by the system for the same amount of mechanical and electrical energy consumed (Cengel and Boles, 2007).

The heating and cooling modes of the simulated GSHP system were applied for different seasons to assess the long-term system performance. The COP can be calculated with the help of temperature measurements for the systems during heating and cooling cycles. For an 'ideal' GHP operating between corresponding low and high temperature reservoirs, the maximum COP during the heating cycle is obtained by Equation 2.1. The 'real' performance in terms of efficiency for a GHP system can be expressed by the second law efficiency, which is the ratio of the maximum COP divided by the actual COP during a heating cycle (Equation 2.2). This is the best method to assess the energy efficiency of a system (Healy and Ugursal, 1997; Hepbasli, 2005).

$$\mathrm{COP}_{\mathrm{Heating}} = \frac{T_H}{T_H - T_L} \tag{2.1}$$

where $\mathrm{COP}_{\mathrm{Heating}}$ = maximum coefficient of performance during the heating cycle; T_H = high temperature reservoir and T_L = low temperature reservoir (Cengel and Boles, 2007; Healy and Ugursal, 1997; Hepbasli, 2005).

$$\eta_{II} = \mathrm{COP}_{\mathrm{Heating}} / \mathrm{COP}_{\mathrm{Actual}} \tag{2.2}$$

where η_{II} = second law GHP efficiency; $\mathrm{COP}_{\mathrm{Heating}}$ = maximum COP during the heating cycle and $\mathrm{COP}_{\mathrm{Actual}}$ = actual COP during the heating cycle (Cengel and Boles, 2007; Healy and Ugursal, 1997; Hepbasli, 2005).

The temperature probes placed throughout selected bins (Figure 2.20b) recorded values for the heating and cooling cycles. They provided mean daily readings during the heating and cooling cycles. The mean efficiencies of the GHPs were calculated and summarized in Table 2.18, utilizing data obtained from the temperature analysis and the calculation of actual geothermal heating COP (Figure 2.20b). Table 2.18 indicates how effectively the simulated pumps operated for a heating cycle when compared to residential GHPs. The findings indicate a good correlation and show how two hybrid sustainable systems function together in harmony.

2.5.4 Conclusions

The performance of 12 PPS integrating simulated GSHPs was good with respect to water quality treatment and energy use performance indicators. Microbial count

TABLE 2.18
Mean Geothermal Heat Pump Thermodynamic Efficiency η (%) for Selected Indoor and Outdoor Bins (June 2006–March 2009)

Year	Location	Bin 1	Bin 2	Bin 3	Bin 4
2006	Indoor	75.4	76.6	77.3	72.5
	Outdoor	70.2	72.3	70.4	74.5
2007	Indoor	83.6	84.2	84.8	81.5
	Outdoor	72.7	74.3	73.1	73.7
2008	Indoor	78.3	79.4	86.7	84.3
	Outdoor	68.4	71.8	70.6	71.5
2009	Indoor	78.2	79.4	77.6	73.9
	Outdoor	71.8	77.5	72.4	70.3

reductions were relatively high during the treatment process despite the simulation of a worst-case pollution scenario involving the introduction of dog faeces to the simulated runoff. The effluent could be used for recycling (e.g., garden watering and toilet flushing), considering that it does not pose an elevated risk to human health. The additional heat provided by the GSHPs did not result in a deterioration of microbial pollution due to regrowth during various operational modes. The efficiency of the GSHP compares well with other systems used for residential developments.

2.6 MODELLING OF ENERGY BALANCES WITHIN GEOTHERMAL PAVING SYSTEMS

2.6.1 INTRODUCTION

Geothermal paving is the combination of permeable or pervious pavements with GSHPs. This combined pavement system is structured around engineered heating, ventilation and air conditioning technology (Hanson Formpave, 2009). The pavement-based system can reduce a building's reliance on gas or electricity for heating and cooling. The permeable pavement system collects water from all impermeable surfaces, pathways, roofs and driveways, while the GHP moves the heat-treated storm water through underfloor heating, enlarged radiators or fan coils. In the reverse cycle, the chilled water from the ground is introduced (Omer, 2008). GSHPs are receiving increasing interest because of their potential to reduce primary energy consumption, emissions of greenhouse gases and thus the effects of climate change (Ozgener and Hepbasli, 2007). Permeable pavements are a sustainable urban drainage system whereby water from urban runoff can be treated by filtration and sedimentation for recycling, harvesting or reuse purposes (Scholz and Grabowiecki, 2007). This hybrid paving technology is already established in the UK and has been known to subsist in two separate systems (GSHPs and permeable pavements). Section 2.6 provides a detailed thermodynamic analysis and simulation for the combined systems and details how they perform incorporated with each other.

GHPs, also referred to as GHSPs, or geoexchange systems, are recognized globally as energy-saving devices. Most GHPs are regarded as a sustainable technology, as they reclaim and recycle thermal energy from the earth (Omer, 2008; Phetteplace, 2007). In climates with a near balance in the annual heating and cooling loads, GHPs function essentially as a seasonal energy storage scheme. They are mainly used in new and sustainable commercial and residential areas. In general, GHPs are also a very interesting and growing research area (Ozgener and Hepbasli, 2007).

The earth's energy is utilized as a heat source when operating in heating mode, with a fluid (water or a water-antifreeze mixture) as the medium that transfers the thermal fluxes from the earth to the evaporator of the heat pump, thus utilizing geothermal energy. In cooling mode, the earth, lakes or groundwater sources are used as a heat sink where cooler temperatures are taken from the ground and transferred to a nearby building. One of the many advantages of GHPs is that they can offer both heating and cooling at virtually any location, with great flexibility to meet various demands (Banks, 2008).

Geothermal paving in an established achievement and concept of sustainability and has a sound design. Geothermal space conditioning (heating and cooling) has expanded considerably since the 1980s, following on the introduction and widespread use of heat pumps (Phetteplace, 2007; Banks, 2008). The various systems of heat pumps available permit to economically extract and utilize the heat content of low-temperature bodies, such as the ground and shallow aquifers and ponds (Sanner et al., 2003).

The GSHPs use refrigerants to move unwanted energy (i.e. heat) out of buildings during summer and into them (if required) during winter according to the *United States Environmental Protection Agency* (U.S. EPA) (Bose, 2003). They use constant temperatures of surrounding grounds, which are lower than the corresponding air temperatures during warm seasons (heat sinks) and higher during cold seasons (heat sources) (Bose, 2003). There is no need for burning fossil fuels to transfer energy either side; therefore, this is a sustainable technology, which also reduces carbon dioxide (CO_2) emissions (Geothermal Heat Pump Consortium, 2009).

The GHP energy performance and efficiency are based on the COP in a heating cycle and the Energy Efficiency Ratio (EER) in a cooling cycle. For domestic systems, COP ranges from 3.5 to 6.0 in the heating mode and 9.5–20.0 in the cooling applications (Omer, 2008). The performance of the heat pumps is directly linked to the temperature analysis. Tota-Maharaj et al. (2009) showed when the GHP is combined with permeable pavements, the sub-base zone (simulated aquifer) is ideal for the heat transfer process. There is a requirement for power to run the heat pump usually relying on the main electricity supply system.

Permeable or pervious pavement systems (PPS) are a type of SUDS, which facilitate the attenuation and treatment of urban runoff. The use of permeable surfaces can reduce flooding and other adverse impacts associated with increased rainfall and runoff in urban areas (Balades et al., 1995). Permeable or pervious pavements allow for water to soak into the ground beneath or provide underground storage.

Storm water runoff sometimes requires treatment to prevent pollutants and other harmful substances from re-entering surface waters or groundwater. The large surface area in the pores of pervious pavements provides the setting for treatment of

FIGURE 2.27 Hanson Formpave installed Aquaflow-Thermapave systems into the car parking areas of a new office development at the Hanson Stewartby offices (Bedford, UK) in August 2009. The car park of 6,500 m^2 provides heating, cooling and drainage functions.

water by attached microbial growth (Ferguson, 2005). This approach is beneficial as it considers water quality and quantity as well as environmental and amenity issues.

Permeable paving can effectively provide a structural pavement suitable for pedestrians and vehicular traffic, in addition allowing surface water to infiltrate freely through the surfaces into the pavement construction for temporary storage, storm attenuation and dispersal (Figure 2.27). Pervious pavements are also used to store and support ground water recharge and to treat and remove waterborne bacteria and pollutants (Scholz and Grabowiecki, 2007). Most PPS are particularly suited for providing a hard surface within a SUDS framework as an opportunity to reuse water and save potable water supplies (Pratt et al., 1999). Permeable pavement systems are regarded as an effective tool for managing storm water (Pratt et al., 1999; Scholz and Grabowiecki, 2007). Permeable pavements can be used in confined urban designs that admit surface waters over the area, reducing the need for deep excavations in drainage resulting in a lower cost when compared to conventional surfacing and drainage solutions (CIRIA, 2007).

Harmful contaminants such as hydrocarbons and heavy metals that have the potential to endanger watercourses and groundwater resources have been infiltrated and removed by permeable pavements (Brattebo and Booth, 2003). In comparison to other permeable paving materials such as asphalt, concentrations of zinc, copper and lead were significantly lower from the effluent of permeable structures (Brattebo and Booth, 2003). Pratt et al. (1999) stated that PPS give a high reduction in suspended solids, BOD and chemical oxygen demand (COD) for urban runoff consisting of mineral oil deposition. Permeable pavements can function as an effective hydrocarbon trap and a powerful in situ bioreactor (Coupe et al., 2003). Biodegradation within permeable pavements is enhanced by bacteria and fungi. When inoculated with microorganism, the protozoan population diversity increases at a faster rate when compared to non-inoculated systems (Coupe et al., 2003).

Recent studies by Scholz and Grabowiecki (2009) have found that permeable pavements integrated within ground-source heating and cooling systems not only reduce runoff and save energy costs, but also improve water quality. Tota-Maharaj and Scholz (2010) have shown that the combined hybrid PPS system has many environmental benefits as a pollution prevention device, reducing suspended solids, BOD, COD, turbidity, ammonia and phosphate levels in addition to microbiological contaminant removal in comparison to alternative highway gullies or other urban drainage techniques. When compared to traditional impervious asphalt, permeable pavements can reduce runoff quantity, lower peak runoff rates and delay peak flows due to their high surface infiltration rates (Newman et al., 2002; Brattebo and Booth, 2003).

The warmest pavements tend to be impermeable and dark in colour with solar reflectance values under 25%. These pavements can heat to 65°C (150°F) or hotter in tropical climates (Gartland, 2008). Two ways of making pavements cooler are by (a) increasing the solar reflectance by making the pavements lighter in colour, using lighter coloured ingredients in the pavement mix or by applying lighter coatings over the pavement surface and/or (b) increasing its ability to store and evaporate water, by making the pavements permeable (Gartland, 2008). Permeable pavements not only allow water to drain through during rainfall events but to evaporate back out during hot, dry and sunny weather. The evaporating process of the water draws heat from the pavements, keeping the pavement cooler in the sun making permeable pavement an ideal option for reducing urban temperature (Gartland, 2008).

Water temperature is a critical parameter in urban runoff reuse and recycling. Survival of microbiological organisms and growth rates or degradation kinetics of bacterial parameters are linked to the environmental stresses that occur within water at varying temperatures which can affect the environmental impacts and reduce the sustainability of urban runoff reuse. Furthermore, pathogens may thrive within a given temperature range (Maier et al., 2008).

The water temperature of permeable pavement systems can also affect management practices. Saturated water temperatures for the heating cycle in the aquifer zone of the pavement occur between 21°C and 26°C. This mode of temperature range has a moderately faster bacterial decomposition rate when compared to the range between 4°C and 7°C for the cooling mode (Coupe et al., 2009). Moreover, there is enough dissolved oxygen to support the decomposition process (Pond, 2005).

There are several reasons why temperature plays a pivotal role in the development of geothermal paving systems (GPS) regarding urban runoff reuse. From commercial perspectives, applications of GHPs for seasonal cycles (winter and summer) justify the control and monitoring of the treated storm water's temperature at the sub-base of the pavement system. As a result of the risk of legionellosis, caused by the proliferation of *Legionella* bacteria in warm water systems (20°C–45°C), it is important to model the temperature and geothermal heat fluxes of the system (Pond, 2005; Maier et al., 2008; Coupe et al., 2009).

An energy balance for GPS was never conducted before. Such an energy balance would identify dominant energy transfer mechanisms and allow engineers to properly design geothermal heating and cooling systems for commercial and industrial applications. The same energy balance could be applied as a management tool in maintaining the desired permeable pavement aquifers' temperatures. Buildings and institutions in the UK are generally heating-dominated, with only some cooling applied. Therefore, a review of the theory pertaining to energy transfer mechanism will allow for the development of a differential equation describing the energy fluxes and temperature changes in integrated permeable pavements and GSHPs. Note that Section 2.6 is based on an updated version of the original article by Tota-Maharaj et al. (2011).

2.6.2 Methodologies, Modelling and Equations

2.6.2.1 Experiments

The GPS was constructed as a laboratory-tanked system, and data collection has been conducted by the Institute of Infrastructure and Environment, School of Engineering, The University of Edinburgh. Two GPS were constructed, one consisting of geotextile and the other without as seen in Figure 2.28. The pilot-scale geothermal pavement rigs are rectangular with a plan area of 0.45×0.76 m. Heat is absorbed/rejected to each artificial aquifer by circulating heated or cooled water through a slinky coil heat exchanger (a pipe coiled in a circular design such that each loop overlaps the adjacent loop) installed at the base of the permeable pavements (constructed aquifer). Fibreglass material was placed externally to the bins to prevent heat transfer through the tank walls.

The rigs were constructed from materials such as aggregates, gravel, clean stones and geotextiles provided by Hanson Formpave. The rig layout consisted of

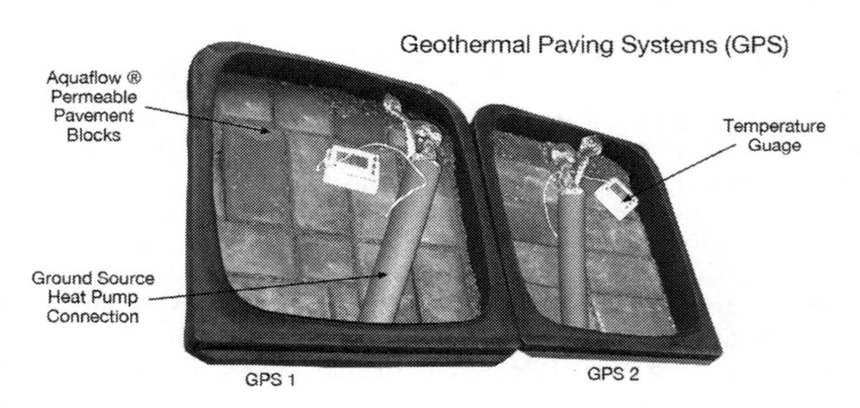

FIGURE 2.28 Top view of experimental geothermal pavement rigs showing the temperature gauges, permeable pavement and connections to the ground-source heat pumps.

the Aqauflow® permeable pavement blocks (80 mm in depth), the bedding layer consisting of pea gravel and clean stone (250 mm in depth), for the GPS 1 the geotextile layer (5 mm in depth), followed by the base made up of gravel (500 mm in depth) and finally the constructed aquifer zone, which comprised the filtered storm water. The ground-source heating and cooling coils were fitted in the aquifer zone of the permeable pavement system. All pavement materials met the British Standards (British Standards Institution BS 882: 1992) and Highways Agency specifications (Highways Agency, 2009). The aquifer zone depth depends on the individual application of the permeable pavements; water is to be stored temporarily, dispersed back into the watercourse or used as the energy transfer mechanism for the GHP. A recommended depth of 350 mm is required for fitting the slinky coils into the system.

The Inbitex geotextile layer is made of polyethylene and polypropylene fibres and is placed at the top part of the upper base to aid in further reduction of microbial and other pollutants that exists from urban runoff. The addition of geotextiles results in a higher removal efficiency for many water quality parameters and provides a higher stand of water quality from the permeable pavement structure (Scholz and Grabowiecki, 2009; Tota-Maharaj and Scholz, 2009).

Digital LCD aquarium thermometers with a measurement range of (−50°C to 70°C or −58°F to 158°F) were placed at the top surface of the pavement blocks and bottom in the tanked saturated water zone for the permeable pavement systems to measure the actual temperature of the aquifer for both the heating and cooling mode. Weather data used in the model runs were taken from The University of Edinburgh's Institute of Atmospheric and Environmental Science, School of Geosciences Weather Station, which has a monitoring site at The King's Building campus, The University of Edinburgh with daily data available online (http://www.geos.ed.ac.uk/abe/Weathercam/station). Weather parameters recorded included air temperature, atmospheric pressure, solar radiation, wind speed and relative humidity.

2.6.2.2 Modelling

The aquifer at the base of the permeable paving system was in the form of dynamic hydrological and thermal equilibrium with the environment. In assessing the significance of heat exchange and water flow from the permeable pavement medium, Darcy's Law indicates that flow is dependent on both the hydraulic gradient and the hydraulic conductivity of materials (Banks, 2008) and is adopted to model the filtration and thermal processes. Darcy's equations are linked through suitably chosen conditions that describe the fluid across the surface of the porous media. It provides the simplest linear relation between velocity and pressure within the pavement media under the physically reasonable assumption that the water flow rates are usually very slow, and all the inertial (non-linear) terms may be neglected.

Heat transfer is dependent on the flow velocity and thermal properties of the pavement rig (Banks, 2008; Charbeneau, 2000). The porosity of the medium within permeable pavement systems is important in controlling the influence of hydraulic conductivity. Materials with higher porosity on average have a higher conductivity (Charbeneau, 2000).

2.6.2.3 Governing Equations and Parameterization

Solar radiation and subsequent heat transfer reach and warm the pavement's aquifer during the summer. Convection processes are likely to occur, ensuring that water is mixed homogenously throughout. As temperature drops during winter, cold water will sink to the base and the geothermal pavement system subsequently utilizes this change in temperature with the heat exchanger (Chiasson, 1999).

The solar radiation and thermal heat transfer fluxes are essential to the overall mechanisms with regard to the paving system energy it receives from the earth's surface (Chiasson, 1999). Shortwave solar radiation that reaches the surface after atmospheric mixing may be rejected or partially heats the ground. Thermal infrared radiation is emitted from both the surface and the atmosphere, with a net loss to the surface since the ground is usually warmer than the atmosphere above. The net radiative energy is used to evaporate water (latent heat), to directly heat the atmosphere and to heat the deeper layers of the soil. Below the pavement surface through the various layers, surface heat diffusion takes place (Hermansson, 2001, 2004).

Figure 2.29 indicates the applied heat balance and Figure 2.30 illustrates the main processes and components of the energy and water exchange at the surface. The latent heat flux couples the surface energy with the water balance. If rain reaches the ground, water may be stored on the surface, infiltrate into the soil or be transported horizontally as surface runoff (Hermansson, 2001, 2004). These processes mainly depend on soil type and the total amount of water already stored in the soil column. Below the surface, the water may reach greater depths through gravitational drainage and diffusion processes and may form interflow or groundwater runoff. When computing the thermal convective and conductive heat transfers of the pavement rigs, hydraulic properties (e.g., hydraulic conductivity, porosity and velocity) are required in addition to thermal ones including thermal conductivity and volumetric heat capacity (Table 2.19).

The conventional methodology of performing an energy balance requires the definition of a control volume. Therefore, for this research, all the filtered storm water within the constructed aquifer zone is defined as the control volume and the thermal energy inertia contained by the GPS $E_{\text{Geothermal Permeable Pavement}}$ is the total energy within

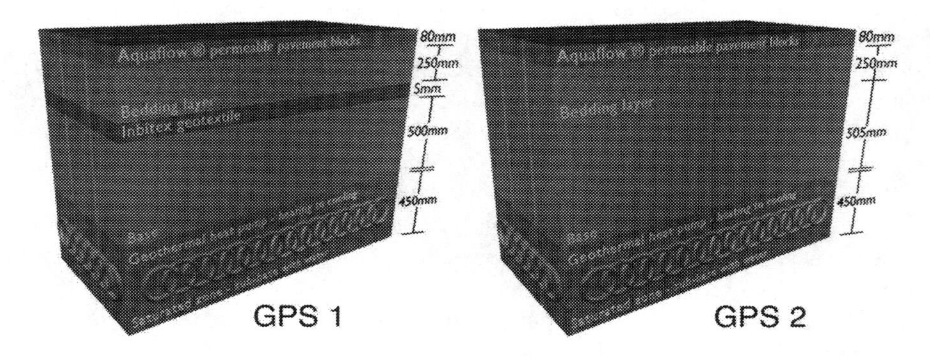

FIGURE 2.29 Cross-section of geothermal paving systems with geotextile layers (GPS 1) and without geotextile layer (GPS 2).

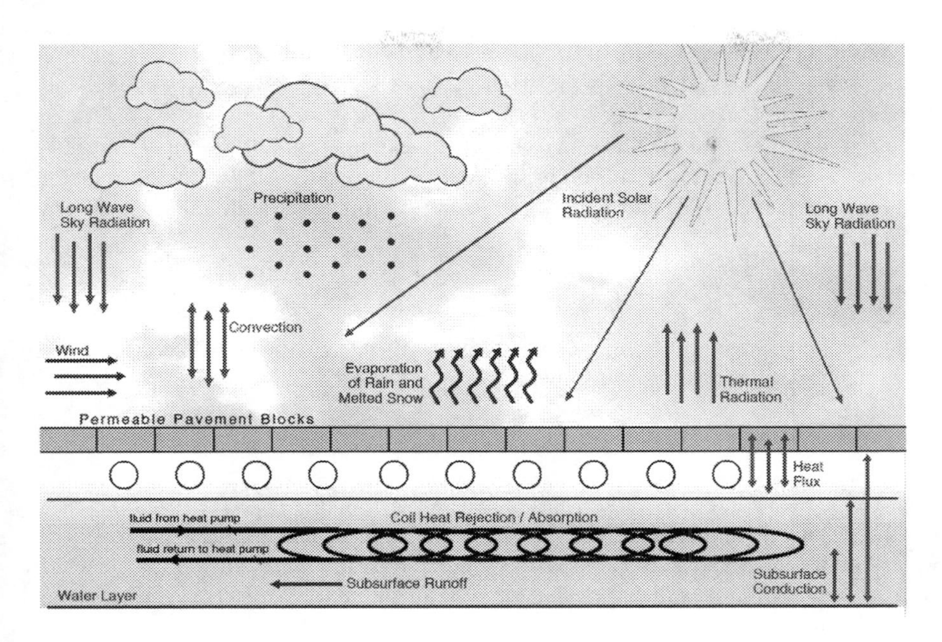

FIGURE 2.30 Schematic of water and energy balance for the geothermal paving system energy and temperature model.

TABLE 2.19
Typical Values of Hydraulic and Thermal Properties of Permeable Pavement Systems

	Hydraulic Properties			Thermal Properties	
Permeable Pavement Medium	**Hydraulic Conductivity *K* (m/s)**	**Porosity¹n**	**Velocity *v* (m/year)**	**Thermal Conductivity *k* (W/m °C)**	**Volumetric Heat Capacity $\rho_s c_s$ (J/m³°C)**
Permeable pavement	1.0×10^{-6}	0.42	18.95×10^{4}	1.23	1.05×10^{6}
Geotextile	3.5×10^{-7}	0.52	3.21×10^{7}	0.98	2.23×10^{6}
Clean stone/Pea gravel	3.0×10^{-3}	0.38	1.26×10^{7}	0.8	1.40×10^{6}
Gravel (base)	9.8×10^{-2}	0.31	3.05×10^{3}	0.8	1.40×10^{6}

Source: Adapted from Freeze and Cherry (1979), Ling et al. (1993), Charbeneau (2000), Hermansson (2001, 2004), Banks (2008), Herb et al. (2008).

the boundaries of the control volume (Equation 2.3). The amount of internal energy or thermal inertia inside the system at any time can be calculated (Lamoureux et al., 2006).

$$E_{\text{Geothermal Paving System}} = \rho V c_p \Delta T \tag{2.3}$$

where ρ is the density of water at the sub-base of the pavement (kg/m^3), V is the volume of water in the system, c_p is the specific heat capacity of water (KJ/kg °C) and ΔT is the temperature fluctuation (°C).

It is assumed that the GPS is set up such that the temperature throughout the aquifer is the same for heating and cooling. The total energy in the GPS is the summation of the top surface and the sub-base aquifer zone (Lamoureux et al., 2006) as shown in Equation 2.4.

$$E_{\text{GPS}} = E_{\text{surface}} + E_{\text{aquifer}} = \left(\rho V c_p \Delta T\right)_{\text{surface}} + \left(\rho V c_p \Delta T\right)_{\text{aquifer}} \tag{2.4}$$

The governing equation of the model is an overall energy balance of the GPS using the lumped parameter approach shown by Equation 2.5.

$$q_{\text{in}} - q_{\text{out}} = \rho V c_p \frac{dT}{dt} \tag{2.5}$$

where q_{in} is the heat transfer to the permeable paving system, q_{out} is the heat flux leaving the pavement structure, *dT/dt* is the rate of change of temperature for the storm water runoff in the aquifer zone. If the temperature fluctuates in the constructed aquifer zone of the permeable pavement system, the amount of energy within the control volume also changes and is described by Equation 2.6.

$$\left(\frac{dE}{dt}\right)_{\text{GPS}} = \left(\frac{d\rho c_p VT}{dt}\right)_{\text{GPS}} = \rho c_p \left[T\frac{dV}{dt} + V\frac{dT}{dt}\right]_{\text{GPS}} \tag{2.6}$$

These fluctuations of temperature and thermal inertia are caused by absorption of solar radiation, heat exchange within the layers and structure of the permeable pavement system, heat fluxes with air because of convection, evaporation and black radiation, in addition to the bulk movement of water and transportation of energy across the control boundary system (Charbeneau, 2000; Herb et al., 2008).

The heat movement vector quantities can be quantified and balanced with the rate at which energy within the system is changing (Mengelkamp et al., 1999). The mathematical expression of the summation of heat transfer is schematically represented in Figure 2.30 and follows the expression shown in Equation 2.7.

$$\left(\frac{dE}{dt}\right)_{\text{GPS}} = q_{\text{solar}} - q_{\text{themal}} + q_{\text{sky}} - q_{\text{evap}} + q_{\text{snow/rain}} \pm q_{\text{conv}} \pm q_{\text{pps}} - q_{\text{seepage}} \pm q_{\text{coils}} \pm q_{\text{resp}} \tag{2.7}$$

where E is the total energy at any given time (t) within the GPS. It is assumed that energy entering the system is positive (heat gain), while energy exiting the system is negative (heat loss). Rearranging Equation 2.7 as a function of temperature T, it becomes the expression shown in Equation 2.8.

$$\left(\frac{dT}{dt}\right)_{\text{GPS}} = \frac{q_{\text{solar}} - q_{\text{themal}} + q_{\text{sky}} - q_{\text{evap}} + q_{\text{snow/rain}} \pm q_{\text{conv}} \pm q_{\text{pps}} - q_{\text{seepage}} \pm q_{\text{coils}} \pm q_{\text{resp}}}{V\rho c_p} \tag{2.8}$$

where q_{solar} is the solar radiation heat gain to the GPS, $q_{thermal}$ is the thermal radiation heat transfer, q_{sky} is the radiative long-wave emissions from the sky, q_{evap} is the heat and mass transfer as a result of evaporation throughout the pavement system as a result of the latent heat of fusion and vaporization, $q_{snow/rain}$ is the sensible and latent heat fluxes resulting from the mass transfer of snow and rain, q_{conv} is the convection heat transfer at the pavement surface, q_{pps} is conduction heat flux across the permeable pavement medium in addition to the heat transfer due to inflow and outflow of urban runoff, $q_{seepage}$ is the internal energy of the liquid movement from seepage, q_{coils} are the heat exchange from the slinky coils at the base of the GPS in the aquifer zone for the heating and cooling cycles respectively and q_{resp} is the metabolic heat rate from microbial respiration at the base of the permeable pavement structure

The following assumptions were used to simplify the solution of Equation 2.8:

- The water density and specific heat capacity remained constant, regardless of changes in water temperature.
- The pavement aquifer volume was constant.
- The sky was cloudless when calculating the emitted atmospheric long-wave radiation. Note that long-wave radiation from a cloudless sky is the worst-case scenario to ensure that the model did not under-predict the thermal fluxes.
- The hydraulic and thermal properties of the permeable pavement medium were uniformly distributed.
- No evaporation occurred when the relative humidity of air is at or above 100%.
- The concrete permeable pavement blocks had a solar reflectance value of 40%.
- The heat flux for seepage through the pavement rig was ignored as it is a tanked system. However, in practical GPS, seepage of urban runoff will occur.

Solar radiation heat gain is the net solar radiation absorbed by the paving system. It is assumed that all solar radiation absorbed by the GPS becomes a heat gain and is computed according to Duffie and Beckman (2006) as shown in Equation 2.9.

$$q_{solar} = \alpha I \tag{2.9}$$

where I is the solar radiation incident on the pavement surface and α is the absorptivity coefficient of the permeable pavement material. The absorptivity coefficient is computed for the solar incidence angle (θ) using an empirical correlation (Duffie and Beckman, 2006). The model accepts solar radiation in the form of direct and diffused components in which case I is computed from Equation 2.10. The angle of incidence (θ) of the sun's rays is calculated at each time step from correlations given by Duffie and Beckman (2006) and ASHRAE (1997).

$$I = I_{direct}\cos\theta + I_{diffuse} \tag{2.10}$$

Thermal radiation can be viewed as the propagation of energy in the form of electromagnetic waves or discrete photons. No medium is required in its propagation form (source). Thermal radiation with wavelengths ranging from λ =0.2 to 1,000 μm will be considered (Duffie and Beckman 2006; Dieter and Stephan, 2006). The thermal heat transfer mechanisms are spilt into two major components: (a) the heat transfer from the aquifer surface throughout the system and (b) the heat fluxes at the pavements top surface. For both heat transfers, the model uses a linearized radiation coefficient h_r defined in Eqs. 2.11 and 2.12 (Dieter and Stephan, 2006).

$$h_{r(\text{aquifier})} = 4\varepsilon_w\sigma\left[\frac{T_{\text{aquifer}} + T_{\text{sky}}}{2}\right]^3 \text{ for the permeable pavement aquifer} \quad (2.11)$$

$$h_{r(\text{pavement})} = 4\varepsilon_P\sigma\left[\frac{T_{\text{pavement}} + T_{\text{sky}}}{2}\right]^3 \text{ for the permeable pavement surface} \quad (2.12)$$

where ε_w and ε_p are the emissivity coefficients for the water in the constructed aquifer zone and the permeable pavement, respectively; σ is the Stefan-Boltzman constant; T_{aquifer} is the water temperature at the base of the pavement rig in absolute units; T_{pavement} is the pavements surface temperature in absolute units and T_{sky} (Equation 2.13) is computed from the relationship, which relates sky temperature to the dew point temperature (T_{dp}) and the dry bulb temperature (T_{db}) according to Van Buren et al. (2000).

$$T_{\text{Sky}} = T_{\text{db}}\left[0.8 + \frac{T_{\text{dp}}}{240}\right]^{\frac{1}{4}} \quad (2.13)$$

All temperatures are in absolute units. The thermal radiation heat transfer (q_{thermal}) is then computed by Equation 2.14 (Van Buren et al., 2000).

$$q_{\text{thermal}} = h_{r(\text{aquifier})}A_{\text{aquifer}}\left(T_{\text{sky}} - T_{\text{aquifer}}\right) + h_{r(\text{pavement})}A_{\text{pavement}}\left(T_{\text{sky}} - T_{\text{pavement}}\right) \quad (2.14)$$

For a cloudless sky, the emissivity can be approximated with Equation 2.15 (Bliss, 1961; Diefenderfer et al., 2006).

$$\varepsilon_{\text{sky}} = \frac{1}{c_1 - c_2T_{\text{dew}} + c_3T_{\text{dew}}^2} \quad (2.15)$$

where $c_1 = 1.2488219$, $c_2 = -0.0060896701$, $c_3 = 4.8502935 \times 10^{-5}$ and T_{dew} is the dew temperature (°C). The dew temperature is calculated from Equation 2.16 (Bliss, 1961).

$$T_{\text{dew}} = \frac{1}{\left(\frac{1}{T_{\text{air}} + 273}\right) - (1.846 \times 10^{-4})\ln(rh)} - 273 \quad (2.16)$$

where T_{air} is the air temperature (°C), rh is the relative humidity. Using the apparent emissivity, the long-wave sky radiation is expressed by Equation 2.17 (Bliss, 1961; Dieter and Stephan, 2006).

$$q_{\text{sky}} = \varepsilon_{\text{sky}} A_{\text{pavement}} \sigma \left(T_{\text{air}} + 273\right)^4 \tag{2.17}$$

where T_{air} is the air temperature recorded in °C.

In practice, the properties of the GPS can vary tremendously because of the local climate, soil composition and temperature. Two properties of interest are the thermal conductivity of the material (k) and the volumetric specific heat (C_v). The rate at which heat is exchanged across the permeable pavement system can be described by Fourier's Law of heat conduction as shown by Equation 2.18 (Dieter and Stephan, 2006; Incropera et al., 2007; Diefenderfer et al., 2006).

$$q_{\text{pps}} = -k_n A \left.\frac{\partial T}{\partial z}\right|_{z=0} \tag{2.18}$$

where k is the thermal conductivity of materials within the GPS (Table 2.20), A is the area of the aquifer (m^2), T is the temperature fluxes of the materials, z is the depth (m) and $\left.\frac{\partial T}{\partial z}\right|_{z=0}$ is the temperature gradient across the GPS.

For the scenario where the water temperature and the flow characteristics along the permeable pavement medium remain constant, Equation 2.19 may be applied (Dieter and Stephan, 2006; Incropera et al., 2007).

$$-k_n A \left.\frac{\partial T}{\partial z}\right|_{z=0} = -hA\left(T_{\text{pavement medium}} - T_{\text{Aquifer}}\right) \tag{2.19}$$

where h is the heat transfer coefficient (W/m^2/K). Setting the initial and final boundary conditions, it was assumed that the temperature throughout the pavement medium was initially the same at all depths for both heating and cooling cycles (see also Eqs. 2.20 and 2.21). The solution to the heat diffusion equation with the boundary conditions is given by Equation 2.22.

$$T_{\text{pavement medium}}(z,\ t) \approx T_{\text{initial}} \tag{2.20}$$

TABLE 2.20
The Use of Different Empirical Statistical Equations to Measure the Accuracy of the Fourth-Order Runge-Kutta Method in Predicting the Temperatures

		Statistical Validation		
Geothermal Paving System	**Thermodynamic Cycle**	**Bias μ (°C)**	**S.D. (°C)**	**Pearson's Correlation Coefficient *r***
1	Cooling	−1.91	1.2	0.947
	Heating	0.64	1.9	0.871
2	Cooling	−0.75	1.5	0.825
	Heating	1.26	1.3	0.881

$$T(z=0,t) \approx T_{\text{Aquifer}} \tag{2.21}$$

$$\frac{T(z,t) - T_{\text{aquifer}}}{T_{\text{initial}} - T_{\text{aquifer}}} = \text{erf}\,\frac{z}{2\sqrt{\alpha t}} \tag{2.22}$$

where erf is the Gauss error function and α is the thermal diffusivity, which can be calculated from the pavement medium parameters using Equation 2.23.

$$\alpha = \frac{k_p}{C_v} \tag{2.23}$$

where k_p is the thermal conductivity (W/m/K) of the pavement medium and C_v is the volumetric specific heat (J/m^3/K).

This mechanism accounts for heat transfer at the geothermal pavement top and bottom (aquifer) surfaces due to free and forced convection. The convection coefficient (h_c) is a function of the Nusselt Number (Nu). For a pavement surface, correlations for a horizontal flat plate would be the most applicable, similarly for that of the aquifers surface. In free convection heat transfer mechanism, the Nusselt number (Nu) is correlated to the Rayleigh Number (Ra) expressed by Equation 2.24 (Dieter and Stephan, 2006; Incropera et al., 2007):

$$\text{Ra} = \frac{g\beta(\Delta T)L^3}{v\zeta} \tag{2.24}$$

where g is the acceleration due to gravity, ζ is the thermal diffusivity of air, β is the volumetric thermal expansion coefficient of air, v is the kinematic viscosity of air, ΔT is the temperature difference between the GPS and the air and L is the length of the pavement system. In external free convection flows over a horizontal flat plate, the critical Rayleigh Number (Ra) is approximately 10^7 (Incropera et al., 2007). The empirical Equations 2.25 and 2.26 for the Nu are used in the model for free convection for the upper surface of a heated plate and/or the lower surface of a cooled plate (Dieter and Stephan, 2006; Incropera et al., 2007).

$$\text{Nu} = 0.54(\text{Ra})^{[1/4]} \quad \left(10^4 < \text{Ra} < 10^7\right) \text{ for laminar flow} \tag{2.25}$$

$$\text{Nu} = 0.15(\text{Ra})^{[1/3]} \quad \left(10^4 < \text{Ra} < 10^7\right) \text{ for turbulent flow} \tag{2.26}$$

The convection coefficient (h_c) for free convection can then be determined by Equation 2.27.

$$h_c = \frac{(\text{Nu})k}{L} \tag{2.27}$$

where k is the thermal conductivity of air and L is the length.

With regard to forced convection heat transfer, the Nusselt Number (Nu) is a function of the Reynolds (Re) and Prandtl (Pr) Numbers. The Reynolds Number is described by Equation 2.28 and the Prandtl Number is defined by Equation 2.29.

$$\text{Re} = \frac{\upsilon L}{v} \tag{2.28}$$

where v is the wind speed, υ is the kinematic viscosity of air and L is the length.

$$\text{Pr} = \frac{c_p \mu}{k} \tag{2.29}$$

where c_p is the specific heat capacity of air, μ is the dynamic viscosity of air and k is the thermal conductivity of air.

For forced convection over a flat plate (pavement surface), the critical Reynolds Number is approximately 10^5 (Incropera et al., 2007). Again, two empirical relations (Eqs. 2.30 and 2.31) for the Nusselt Number (Nu) are used in the model as described by Incropera et al. (2007):

$$Nu = 0.664\,\text{Re}^{[1/2]}\,\text{Pr}^{[1/3]} \quad \text{for laminar flow regime} \tag{2.30}$$

$$Nu = 0.037\,\text{Re}^{[4/5]}\,\text{Pr}^{[1/3]} \quad \text{for mixed and turbulent flow} \tag{2.31}$$

The convection coefficient (h_c) is determined by Equation 2.27. The convection heat transfer across the GPS is computed by Equation 2.32.

$$q_{\text{convection}} = h_{c(\text{aquifier})} A_{\text{aquifer}} \left(T_{\text{air}} - T_{\text{aquifer}} \right) + h_{c(\text{pavement})} A_{\text{pavement}} \left(T_{\text{pavement}} - T_{\text{pavement}} \right) \tag{2.32}$$

where T_{air} is the ambient air temperature and h_c is the maximum free and forced convection coefficient.

This heat transfer mechanism contributes to the pavement cooling. The mass transfer of evaporating water ($m'w$) at the pavement surface is expressed by Equation 2.33 (Freeze and Cherry, 1979).

$$m'_w = h_d \left(w_{\text{air}} - w_{\text{pavement-surf}} \right) \tag{2.33}$$

where h_d is the mass transfer coefficient, w_{air} is the humidity ratio of the ambient air, $w_{\text{pavement-surf}}$ is the humidity ratio of saturated air at the pavement surface and $w_{\text{pavement-surf}}$ is computed from psychometric charts and tables (not shown).

The mass transfer coefficient (h_d) is defined using the Chilton-Colburn analogy (Tosun, 2007) as indicated in Equation 2.34.

$$h_d = \frac{h_c}{c_p \text{Le}^{\frac{2}{3}}} \tag{2.34}$$

where h_c is the convection coefficient defined previously in Equation 2.27, c_p is the specific heat capacity of air and Le is the Lewis Number computed as shown in Equation 2.35 (Tosun, 2007).

$$Le = \frac{\alpha}{D} \tag{2.35}$$

where α is the thermal diffusivity of air and D represents the diffusion coefficient (Mills, 1995) computed as shown by Equation 2.36 (Tosun, 2007).

$$D = \frac{1.87 \times 10^{-10} T^{2.072}}{P_{\text{air}}} \quad \left(280\,\text{K} < T < 450\,\text{K}\right) \tag{2.36}$$

where P_{air} is the atmospheric pressure in atmospheres. The heat transfer due to evaporation (q_{evap}) is then computed by Equation 2.37.

$$q_{\text{evap}} = h_{fg} A_{\text{pavement}} \dot{m}_w \tag{2.37}$$

where h_{fg} is the latent heat of vaporization. The heat transfer mechanism includes both latent and sensible heat fluxes. The model uses a simple energy and mass balance of water at the pavement surface to account for the heat and mass transfer. The sensible heat flux as a result of snow or rainfall onto the pavement surface $\left(q'_{\text{rain,snow}}\right)$ is given by Equation 2.38.

$$q'_{\text{rain,snow}} = \dot{m}_{\text{rain,snow}} c_p \left(T_{\text{air}} - T_{\text{pavement}}\right) \tag{2.38}$$

where $\dot{m}_{\text{rain,snow}}$ is the snowfall or rainfall (expressed in water equivalent mass per unit time per area of the pavement calculated from a modified version of Equation 2.33) and c_p is the specific heat capacity of water at air temperature.

The urban runoff water entering or leaving the GPS has an internal energy, as movement of (liquid) water across the system boundary represents energy gains or losses. The rate of bulk energy moved across the system by water seepage can be calculated with Equation 2.39 (Freeze and Cherry, 1979).

$$q_{\text{seepage}} = \dot{m}_{\text{water-seepage}} c_p T_{\text{aquifer}} \tag{2.39}$$

where $\dot{m}_{\text{water-seepage}}$ is the mass flow rate of water from seepage, c_p is the specific heat capacity of water and T_{aquifer} is the water temperature in the aquifer zone. $\dot{m}_{\text{water-seepage}}$ can be calculated from Darcy's Equation expressed by Equation 2.40.

$$\dot{m}_{\text{water-seepage}} = k_{\text{hyd}} i A \tag{2.40}$$

where k_{hyd} is the hydraulic conductivity (m/s), which can be assumed to be in the range of 10^{-11} to 10^{-9} cm/s, i is the hydraulic gradient (dimensionless) ≈ 0.001, and A is the area (m^2) of the aquifer. As a result of the experiment being based on a tanked system, seepage losses were ignored. However, considering permeable pavement systems in practice, this must be taken into account for both water movement and energy losses.

The heat transfer from the heating and cooling coils represents the heat flux occurring for both cycles. The heat flux from the coils can be computed from Equation 2.41.

$$q_{\text{coils}} = U_{\text{coils}}\left(T_{\text{coils}} - T_{\text{aquifer}}\right) \tag{2.41}$$

where U_{coils} is the overall heat transfer coefficient of the slinky coils and can be expressed by Equation 2.42 (Dieter and Stephan, 2006; Incropera et al., 2007).

$$U_{\text{coils}} = \frac{1}{\dfrac{1}{h_{\text{coils}}} + \dfrac{l}{k_{\text{coils}}}}, \tag{2.42}$$

where h_{coils} is the convection coefficient, k_{coils} is the thermal conductivity of the slinky coil material and l_{coils} is the wall thickness of the coils. For the slinky coils, the convection coefficient h_{coils} can be determined using the Nusselt Number as laminar flow through a horizontal cylinder and is approximately 4.36 (Dieter and Stephan, 2006; Incropera et al., 2007). Equation 2.43 should be used for its computation.

$$h_{\text{coils}} = \frac{4.36k}{D} \tag{2.43}$$

where D is the inner diameter of the slinky coils (m).

Decomposition of organic matter is a source of energy within GPS. The energy released within the urban runoff is a by-product of decomposer respiration (Datta, 2002). Chemically, the aerobic respiration of glucose can be described by Equation 2.44 (Datta, 2002).

$$C_6H_{12}O_6 + 6O_2 \rightarrow 6CO_2 + 6H_2O + \Delta H_c \tag{2.44}$$

where ΔHc is the heat of combustion for glucose = 16 KJ/mol of glucose.

Permeable pavement systems without geotextiles consume 1–2g O_2/m^2/day, while those with geotextiles consume 3–4 g O_2/m^2/day (Tota-Maharaj and Scholz, 2009). If most of the generated energy comes from the decomposition of organic matter, the total energy produced by the anaerobic decomposers would be in the range of 82–163 J/m^2/day without geotextiles and approximately 325 J/m^2/day at the presence of geotextiles. Factors that may cause alterations in the rate at which the aquifer respiration occurs include temperature, oxygen accessibility, pH and nutrient availability (Datta, 2002).

2.6.2.4 Solving the Energy Balance Equation

Runge-Kutta integration is an extension methodology that allows for a substantially improved accuracy for solving complex differential equations, without imposing a severe computational burden (Chapra, 2005). The idea is to step into the interval and evaluate derivatives, providing a more accurate solution with a smaller increase in computation derivation. Second-order Runge-Kutta integration is associated with an error that is proportional to the time step cubed for an integration step and proportional to the time step squared from the whole simulation (Chapra, 2005). Fourth-order Runge-Kutta integration has an error that is proportional to a time step to the fifth power for an integration step and proportional to a time step to the fourth power for the whole simulation, thus improving the computational efficiency and reducing the error linked to the Runge-Kutta techniques (Chapra, 2005). The differential equation describing the overall energy balance of the GPS is rearranged in the form of Equation 2.45.

$$\left(\frac{dT}{dt}\right)_{\text{GPS}} = \frac{\Sigma q}{\rho V_{\text{pps}} c_p} = f(t,T) \tag{2.45}$$

where T is the temperature, t is the time, and q is the energy vectors. The linear first-order differential equation is solved at each time step. To solve this differential equation, the fourth-order Runge-Kutta numerical technique was used in Matlab 7.0 (The MathWorks Inc, Natick, Massachusetts, U.S.). The fourth-order Runge-Kutta numerical technique is recognized as an accurate method in evaluating and solving differential equations (Chapra, 2005). With the use of initial conditions, the Runge-Kutta technique evaluates the GPS aquifer's temperature with Equations 2.46–2.50.

$$T(t+\Delta t) = T(t) + \frac{\xi_1 + 2\xi_2 + 2\xi_3 + \xi_4}{6} \tag{2.46}$$

$$\xi_1 = (\Delta t) f(t,T) \tag{2.47}$$

$$\xi_2 = (\Delta t) f\left(t + \frac{\Delta t}{2}, T + \frac{\xi_1}{2}\right) \tag{2.48}$$

$$\xi_3 = (\Delta t) f\left(t + \frac{\Delta t}{2}, T + \frac{\xi_2}{2}\right) \tag{2.49}$$

$$\xi_4 = (\Delta t) f\left(t + \Delta t, T + \xi_3\right) \tag{2.50}$$

where T (Δt) is the average GPS temperature at the previous time step and ξ is the mean GPS temperature at the new time step (Chapra, 2005). The model was run and completed, and the output was analysed with the measured data. A convergence criterion for the permeable pavement aquifer's temperature of 1×10^{-5}°C (1.8×10^{-5}°F) was used.

2.6.2.5 Validation of Fourth-Order Runge-Kutta Numerical Analysis

To compare the predicted and measured data, three statistical parameters were applied: (a) the average bias; (b) standard deviation of the average bias; and (c) the *Pearson's linear correlation* coefficient '*r*'. These techniques are an effective methodology in testing the precision of a simulation (Lamoureux et al., 2006).

The average bias μ is the bias measurement of how accurate the model is when estimating the actual GPS temperature (i.e. the predicted temperature minus the measured temperature). The average bias is the mean of all the biases at every time step, for n time steps and is calculated by Equation 2.51.

$$\mu_{\text{bias}} = \frac{\sum \left(T_{\text{Predicted}} - T_{\text{Measured}}\right)}{n} \tag{2.51}$$

The standard deviation of the average bias measured the variations in the bias and is an indicator of the model's consistency (Equation 2.52).

$$S.D = \sqrt{\frac{\sum T_{\text{Predicted}} - T_{\text{Measured}} - \mu_{\text{bias}}}{n-1}} \tag{2.52}$$

The *Pearson's linear correlation* coefficient (*r*) between the measured and predicted temperatures was also calculated to view the model's data set in comparison to measured variables. When $r \approx 1$, the model predictions are near perfect, on the other hand, when $r \approx 0$, the model has failed to predict the temperature.

2.6.3 Results and Discussion

The validation and verification of the model process were compared by scatter plots of the modelled (predicted) data set versus the measured temperatures for both GPS in a heating and cooling mode. This comparison allowed the data to be checked for the accuracy of the simulation, which included environmental heat fluxes and energy mechanisms across the pavement. Simulated and actual temperatures are shown in Figure 2.31a for GPS 1 in a cooling cycle for approximately six months. The Pearsons' correlation coefficient $r = 0.871$ is statistically a very good link between the simulated points and measured data set (Figure 2.31b). The equation $y = 0.65x + 3.2$ represents a linear relationship between the predicted and measured temperatures. However, when compared to the unity line $y = x$, the modelled values are offset several times, showing the level of error in prediction.

For the heating cycle concerning GPS 1, the temperatures successfully mimicked the measured data set illustrating that the fourth-order Runge-Kutta method supported the simulation with a relatively good accuracy and high Pearsons' correlation coefficient $r = 0.947$. Figure 2.32a shows the calculated and observed temporal variations of the treated urban runoff's temperature for GPS 2 from the experimental rig during the period April to September 2008. An assessment of the plots in Figure 2.32a illustrates that the model cumulative heat rejected compares well to the measured heating element and GHP input. At the end of the six-month test period, the percent difference between the cumulative simulated heat rejected compared to the actual data is negligible. Nonetheless, Figure 2.32b depicts a slightly less efficient comparison between the calculated and observed diurnal variations of the temperature during the cooling mode for GPS 2. Since the water temperatures in the surface layer fluctuate relatively more than those of GPS 1, because of the geotextile layer, the numerical model gives values that can differ from the observed surface temperatures.

This model is featured in Figure 2.33 and is particularly good at simulating water temperature. The model gave comparable results with the measurements for both data sets, suggesting that the model can simulate the GPS well. However, there were some inaccuracies (variations) for the cooling cycle (Figure 2.33b).

A review of the plots in Figure 2.34a and b shows also that the model temperatures compare favourably with the actual temperatures within a certain error range

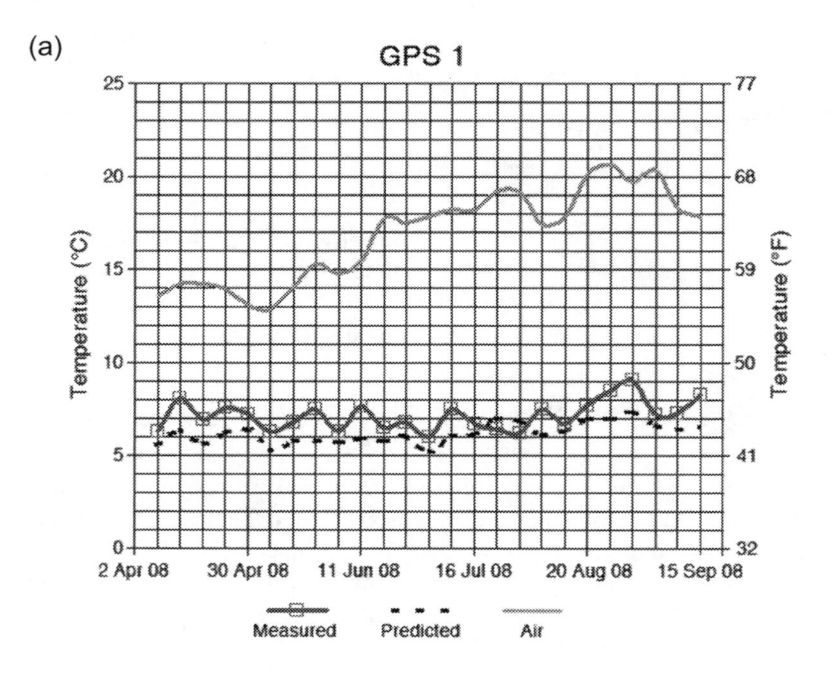

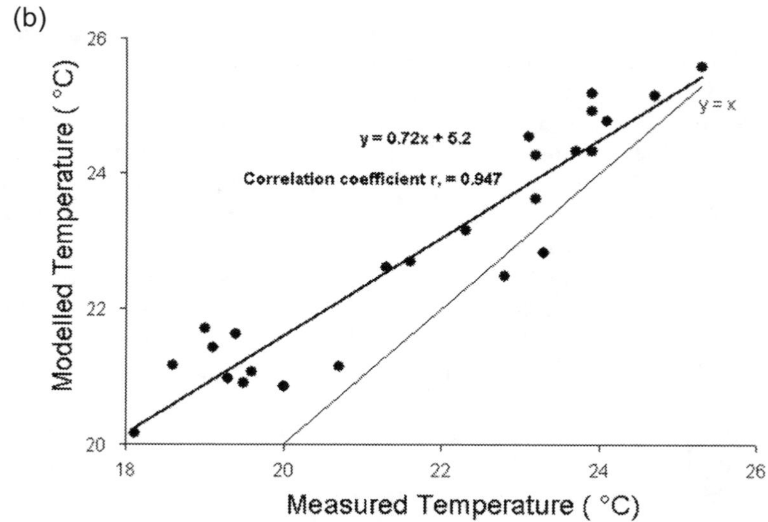

FIGURE 2.31 (a) The modelled, predicted and environmental temperatures for the geothermal paving system GPS 1 in a cooling cycle plotted against time and (b) the scatter plot for the best-fit line and the unity line $y=x$ for the cooling cycle measurement of the model's accuracy. For GPS 1, the model was run for the period from 2 April 2009 to 11 September 2009 with $n=40$ sample points.

(a)

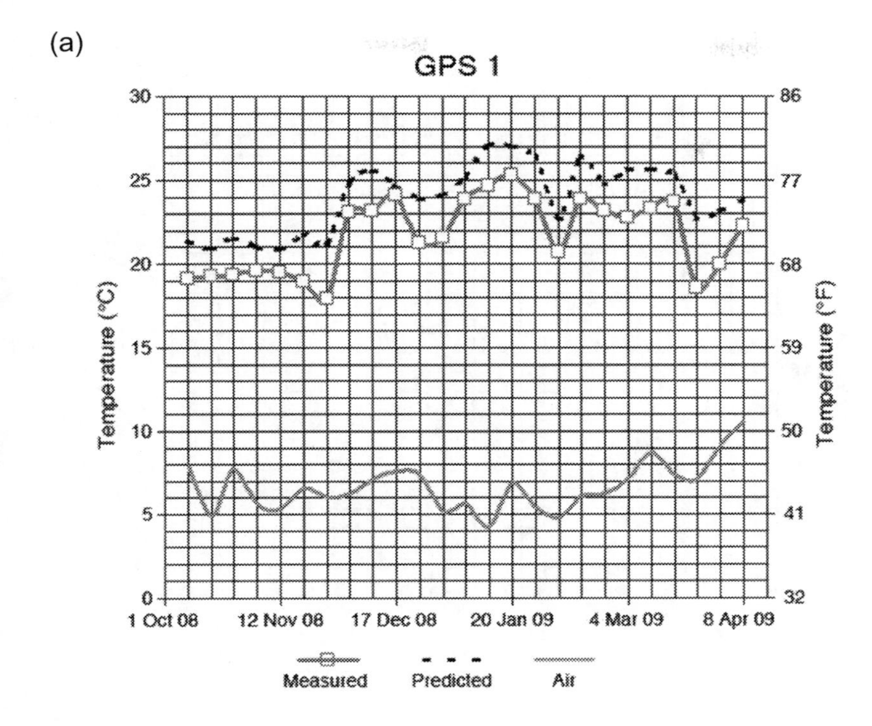

(b)

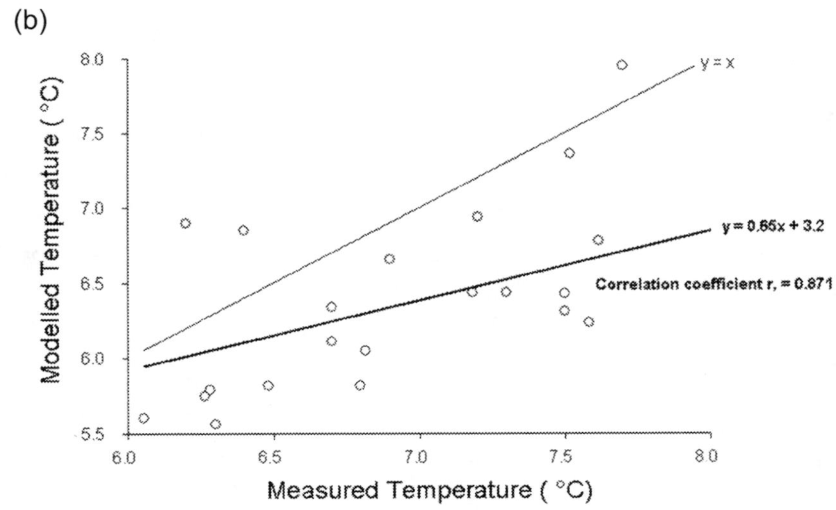

FIGURE 2.32 (a) The modelled, predicted and environmental temperatures for the geothermal paving system GPS 1 in a heating cycle plotted against time and (b) the scatter plot for the best-fit line and the unity line $y=x$ for the heating cycle measurement of the model's accuracy. For GPS 1, the model was run for the period from 1 October 2008 to 8 April 2009 with $n=40$ sample points.

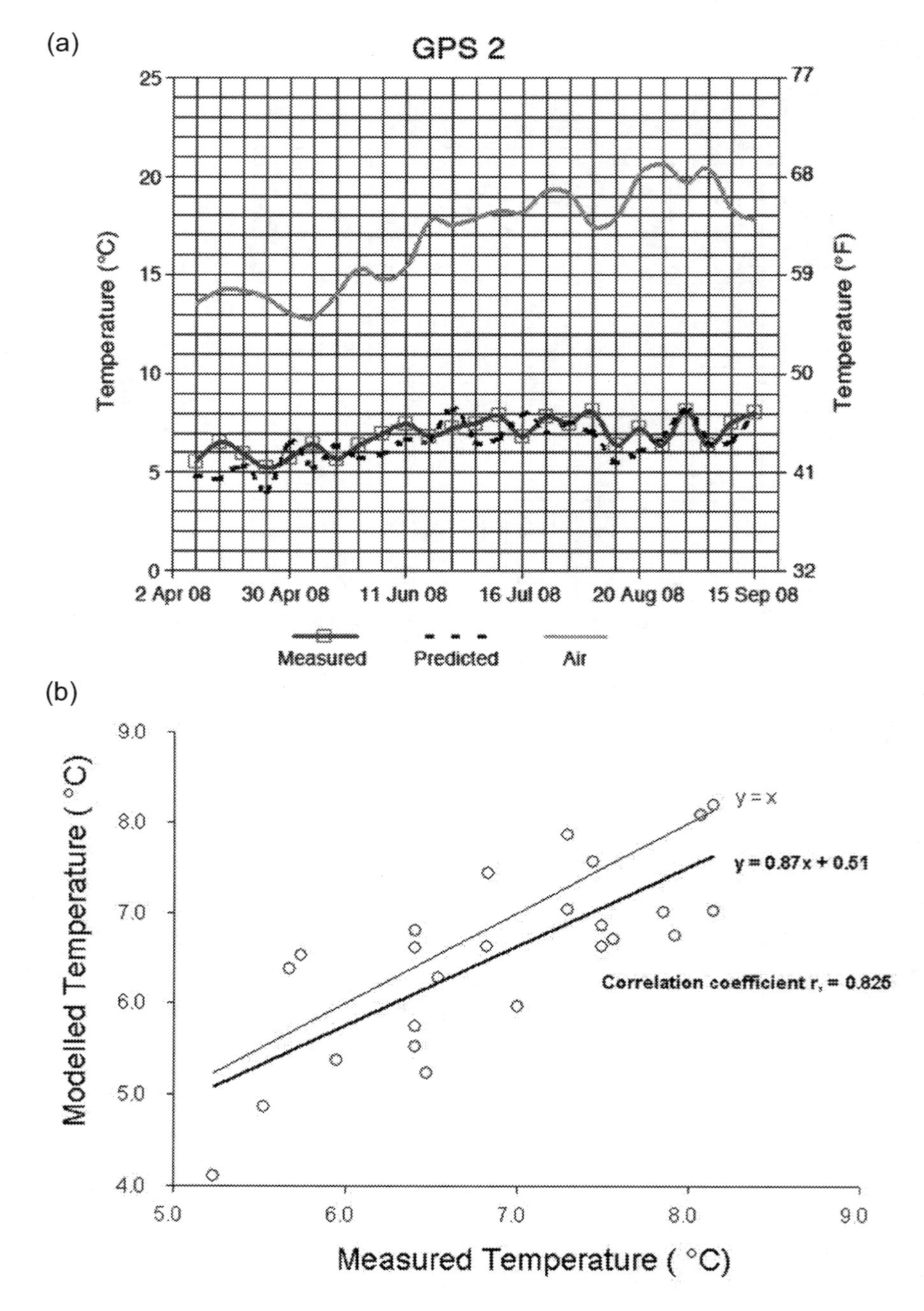

FIGURE 2.33 (a) The modelled, predicted and environmental temperatures for the geothermal paving system GPS 2 in a cooling cycle plotted against time and (b) the scatter plot for the best-fit line and the unity line $y = x$ for the cooling cycle for measurement of the model's accuracy. For GPS 2, the model was run for the period from 2 April 2008 to 11 September 2009 with $n = 40$ sample points.

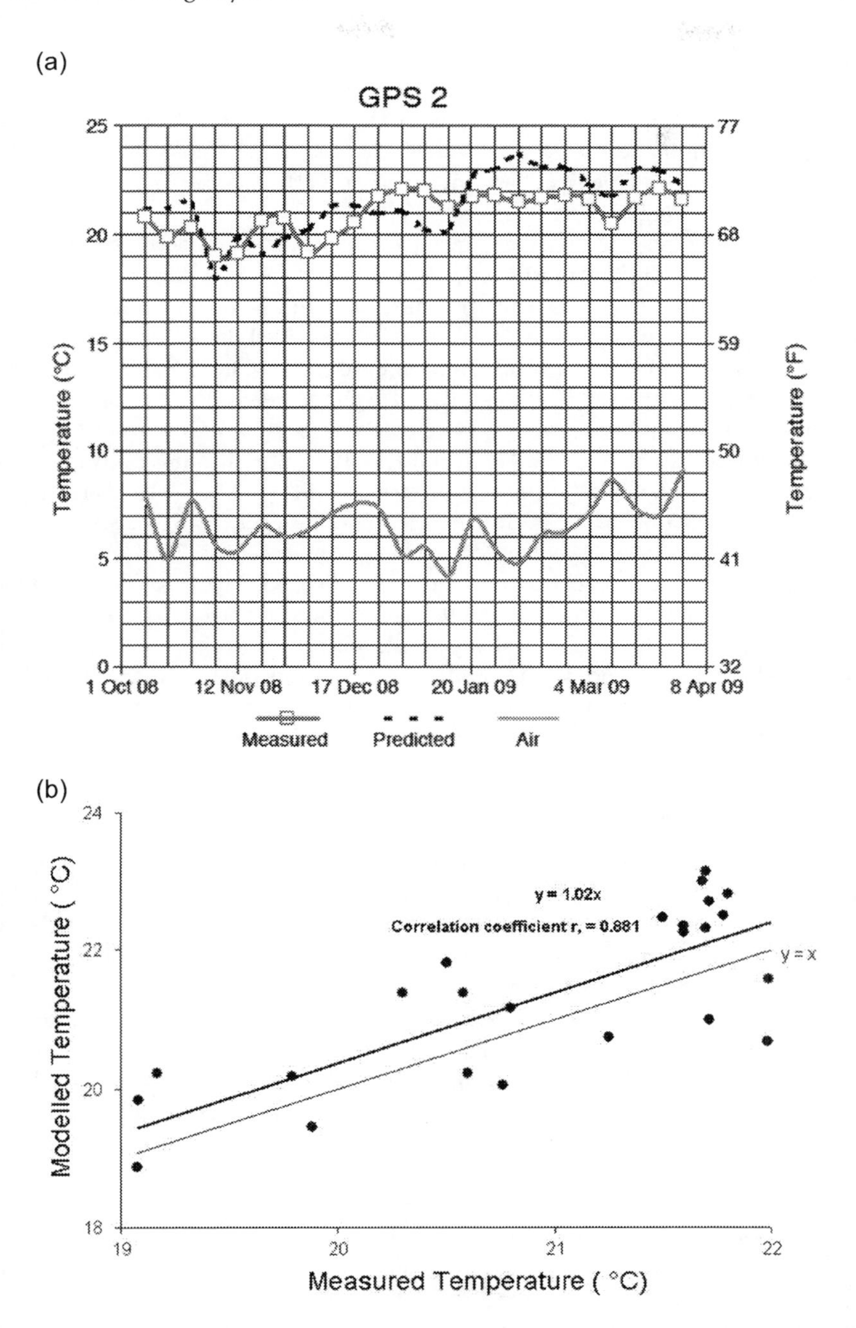

FIGURE 2.34 (a) The modelled, predicted and environmental temperatures for the geothermal paving system GPS 2 in a heating cycle plotted against time; and (b) the scatter plot for the best-fit line and the unity line $y=x$ for the cooling cycle for measuring the model's accuracy. For GPS 2, the model was run for the period from 1 October 2008 to 8 April 2009 with $n=40$ sample points.

of approximately 2°C. The Runge-Kutta analysis simulated the temperatures within 1.46°C (34.63°F) throughout the testing period. The correlation coefficient was 0.881, which shows a near linear relationship between the predicted and measured temperatures for GPS 2 in the heating mode. On the other hand, there is an offset between the unity line ($y=x$) and the linear best-fit equation $y=0.65\,x+3.2$, which illustrates the margin of error that occurred during simulation.

Table 2.20 presents the statistical results indicating the model accuracy. The numerical method from the fourth-order Runge-Kutta method gave a good accuracy and showed stability in the prediction of temperatures. All correlations showed strong linear relationships with the measured and modelled data. The average bias, standard deviation and Pearson's correlation coefficient for GPS 1 and 2 in a heating and cooling cycle were all statically positive. The model tends to under-estimate the temperatures of GPS 1 for the cooling cycle by 1.9°C (SD = 1.3°C).

The model over-estimated the GPS temperature for the heating cycle from 11 October 2008 to 24 April 2009 (bias $\mu=0.64$°C and SD = 1.8°C). The average bias for GPS 2 was less than that for GPS 1 in the cooling cycle, whereas the average bias was very close to that of the heating cycle as seen in Table 2.20. The variations and discrepancies in the findings can be a result of weather fluctuations between the summer and winter period for 2008–2009. Underestimations and/or overestimations of energy vectors to the paving system include influences such as solar radiation, long-wave sky radiation and the evaporation rate. For practical paving systems, the importance of air convection and conduction through the pavement medium would account for variations in the thermal inertia of the system.

2.6.4 Conclusion and Recommendations

GPS are a sustainable pavement technology, hybridizing two unique eco-friendly systems (permeable pavements and GSHPs). The study modelled and experimentally validated thermogeological energy balances. An energy and temperature balance model for two geothermal paving rigs was designed, developed, tested and validated based on the temperature in two constructed lab-based GPS.

The model estimated the energy surpluses and deficits, which were required to balance for temperature control within the system. The numerical tool developed can be coupled to practical GPS for modelling short-time steps (hourly or even minutely) as part of a system analyses. The models were validated by comparing simulation results to experimental data.

Major vectors of energy transfer for unheated ponds were found to be radiation heat transfer mechanisms. For heated ponds, bulk energy flow rates were also important. The model accounts for several environmental heat transfer mechanisms. The heat transfer processes, which were simulated by the Runge-Kutta numerical method, included solar radiation heat gain, heat and mass transfer as a result of evaporation, convective heat transfer to the atmosphere, thermal and long-wave solar radiation heat transfer, conduction heat transfer to the surroundings from the pavement medium and the pavement's aquifer's zone.

The model's output is variable and can be influenced by several environmental conditions including solar radiation, convection and flow of water into and out of the system. Model runs showed that for the cooling cycle, the transport of energy through radiation dominated all energy transfer mechanisms. The energy balance on the GPS initially developed a system whereby temperatures can be assessed or controlled at the aquifer zones of permeable pavements. It is hoped that this data and the model presented can be developed and applied appropriately when monitoring built-in GPS and for future sizing and design systems.

The potential for this model can be applied in the optimization and design for this hybrid pavement system in terms of reduction in the size of the slinky coils heat exchangers, the sizing of the aquifer zone as well as the energy efficiency of the system. The additional support offered by the numerical performance of the methods confirms that it is promising and deserves further research.

REFERENCES

AGMA, 2008. *Strategic flood risk assessment for Greater Manchester* (Manchester: Association of Greater Manchester Authorities (AGMA)).

Altschul, S.F., Gish, W., Miller, W., Myers, E.W. and Lipman, D.J., 1990. Basic local alignment tool, *Journal of Molecular Biology*, **215**(3), 403–410.

American Public Health Association, 1995. *Standard Methods for the Examination of Water and Wastewater*, 19th edn. (Washington, DC: American Public Health Association, American Waterworks Association and Water and Environment Federation).

ASHRAE, 1997. *ASHRAE Handbook, Fundamentals* (Atlanta: American Society of Heating, Refrigeration and Air-Conditoning Engineering, Inc.).

Balades, J.D., Legret, M. and Madiec, H., 1995. Permeable pavements: Pollution management tools, *Water Science and Technology*, **32**(1), 49–56.

Banks, D., 2008. *An Introduction to Thermogeology: Ground Source Heating and Cooling* (Oxford: Blackwell Publishing Ltd.).

Bean, E. Z., Hunt, W.F. and Bidelspach, D.A., 2007. Field survey of permeable pavement surface infiltration rates, *Journal of Irrigation and Drainage Engineering – ASJE*, **133**, 247–255.

Bliss, R.W., 1961. Atmospheric radiation near the surface of the ground, *Solar Energy*, **5**(3), 103–120.

Bose, J.E., 2003. Ground source heat pump systems: State of the industry, *Water Well Journal*, **57**(8), 29–32.

Bose, J.E., 2005. Space conditioning: The next frontier. In: *Ground source heat pump (GSHP) Association Air Innovations Conference (24–26 August)*. US Environmental Protection Agency, Office of Air and Radiation, Chicago, IL.

Bran and Luebbe, 1999. *Bran Luebbe AA3 Autoanalyser Methods: G-109–93 and 94* (Norderstedt: Bran and Luebbe).

Brattebo, B.O. and Booth, D.B., 2003. Long-term stormwater quantity and quality performance of permeable pavement systems, *Water Research*, **37**(18), 4369–4376.

British Standards Institution, 1992. *BS 882 Specification for Aggregates from Natural Sources for Concrete* (London: British Standards Institution).

British Standard Institute, 1999a. *Soil Quality – Part 3: Chemical Methods* – Section 3.5*: Pre-treatment of Samples for Physico-chemical Analysis*, BS 7755-3.5:1995 and ISO 11464:1994 (London: British Standard Institute).

British Standard Institute, 1999b. *Soil Quality – Part 5: Physical Methods* – Section 5.4*: Determination of Particle Size Distribution in Mineral Soil Material – Method by Sieving and Sedimentation*, BS 7755-5.4:1998 and ISO 11277:1998 (London: British Standard Institute).

British Standard Institute, 2002. *Soil Quality – Format for recording Soil and Site Information*, BS ISO 15903:2002 (London: British Standard Institute).

Busch, M., Notte, A.L., Laporte, V. and Erhard, M., 2012. Potentials of quantitative and qualitative approaches to assessing ecosystem services, *Ecological Indicators*, **21**, 89–103.

Butler, D. and Davies, J.W., 2000. *Urban Drainage*, 1st ed. (London: Spon).

Butler, D. and Davies, J.W., 2004. *Urban Drainage*, 2nd ed. (London: Spon).

Butler, D. and Parkinson, J. 1997. Towards sustainable urban drainage, *Water Science and Technology*, **35**(9), 53–63.

Cappuccino, J.G. and Sherman, N., 1996. *Microbiology, a Laboratory Manual*, 4th ed. (Suffern, NY: Rockland Community College).

Cengel, Y.A. and Boles, M.A., 2007. *Thermodynamics: An Engineering Approach* (New York: McGraw-Hill Education).

Chapra, S.C., 2005. *Applied Numerical Methods with MATLAB for Engineers and Scientists* (New York: McGraw-Hill).

Charbeneau, R.J., 2000. Groundwater Hydraulics and Pollutant Transport (New Jersey, Upper Saddle River: Prentice-Hall, Inc.).

Chiasson, A.D., 1999. *Advances in Modeling of Ground-Source Heat Pump Systems*, MSc thesis (Oklahoma: Oklahoma State University).

CIRIA, 2000. *Sustainable Urban Drainage Systems: Design Manual for Scotland and Northern Ireland*, Construction Industry Research and Information Association (CIRIA) Report C521 (London: Cromwell Press).

CIRIA, 2004. *Sustainable Drainage Systems. Hydraulic, structural and water quality advice*, Construction Industry Research and Information Association (CIRIA) Report C609 (London: CIRIA).

CIRIA, 2007. *The SuDS Manual*, Construction Industry Research and Information Association (CIRIA) (London: CIRIA).

CIRIA, 2010. *Planning for SuDS – Making It Happen*, Construction Industry Research and Information Association (CIRIA) (London: CIRIA).

Clesceri, L.S., Eaton, A.D., Greenberg, A.E. and Franson, M.A.H., 1998. *Standard Methods for the Examination of Water and Wastewater*, 20th ed. (Washington, DC; American Public Health Association, American Water Works Association, and Water Environment).

Council of European Communities, 2000. Directive of 23 October 2000 establishing a framework for community action in the field of water policy (2000/60/EC), *Official Journal*, **43**(L327), 1–73.

Coupe, S.J., Charlesworth, S. and Faraj, A. (2009). Combining permeable paving with renewable energy devices: Installation, performance and future prospects. In: *9th. International Conference on Concrete Block Paving 2009. Small Element Paving Technologists*, Buenos Aires, Argentina, 18–21 October, 2009, Argentinean Concrete Block Association, pp. 1–10.

Coupe, S.J., Smith, H.G., Newman, A.P. and Puehmeier, T., 2003. Biodegradation and microbial diversity within permeable pavements, *European Journal of Protistology*, **39**(4), 495–498.

D'Arcy, B. and Frost, A., 2001. The role of best management practices in alleviating water quality problems associated with diffuse pollution, *The Science of the Total Environment*, **265**(1), 359–367.

Danso-Amoako, E., Kalimeris, N., Scholz, M., Yang, Q. and Shao, J., 2010. Predicting dam failure risk for sustainable flood retention basins: A generic case study for the wider Greater Manchester area, *Computers, Environment and Urban Systems*, **36**(5), 423–433.

Datta, A.K., 2002. *Biological and Bioenvironmental Heat and Mass Transfer* (New York: Marcel Dekker, Inc.).

DEFRA, 2000. *Second Consultation Paper on the Implementation of the EC Water Framework Directive (2000/60/EC)* (London: Department for Environment, Food and Rural Affairs (DEFRA)).

DEFRA, 2013. *Overarching impact assessment for the Natural Environment White Paper* (London: Department of the Environment, Food and Rural Affairs).
Diefenderfer, B.K., Al-Qadi, I.L. and Diefenderfer, S.D., 2006. Model to predict pavement temperature profile: Development and validation, *Journal of Transportation Engineering*, **132**(2), 162–167.
Dieter, H. and Stephan, B.K., 2006. *Heat and Mass Transfer*, 2nd edn. (Heidelberg: Springer).
DLTR, 2001. *Planning Policy Guidance 25: Development and Flood Risk*, Department for Transport, Local Government and the Regions (London: Her Majesty's Stationery Office).
Duffie, J.A. and Beckman, W.A., 2006. *Solar Engineering of Thermal Processes*, 3rd edn. (New York: John Wiley and Sons).
Edwards, R., 2004, Goldfish will save us from flood, interview with Miklas Scholz plus picture, sundayherald, 29/02/04, news, 12.
Ellis, J.B., Deutsch, J.C., Mouchel, J.M., Scholes, L. and Revitt, M.D., 2004. Multicriteria decision approaches to support sustainable drainage options for the treatment of highway and urban runoff, *The Science of the Total Environment*, **334–335**, 251–260.
Ellis, J.B., Shutes, R.B.E. and Revitt, M.D., 2003. *Constructed wetlands and links with sustainable drainage systems*, Technical Report P2–159/TR1 (Bristol: Environment Agency).
Environment Agency, 2002. *Soil Guideline Values for Lead, Chromium, Nickel and Other Contaminants*, individual papers (Bristol: Department for Environment, Food and Rural Affairs, Environment Agency).
Eriksson, E., Baun, A., Scholes, L., Ledin, A., Ahlman, S., Revitt, M. and Noutsopoulos, C. and Mikkelsen, P.S., 2007. Selected stormwater priority pollutants – a European perspective, *The Science of the Total Environment*, **383**, 41–51.
EU Waste Water Treatment, 1991. EU Directive 91/271/EEC (Brussels: European Communities).
European Union, 2013. *Directive 2000/60/EC of the European Parliament and of the Council establishing a framework for the community action in the field of water policy.* https://eur-lex.europa.eu/LexUriServ/LexUriServ.do?uri=CELEX:32000L0060:EN:NOT (accessed on 25 February 2024).
Fedorovich, V., Knighton, M.C., Pagaling, E., Ward, F.B., Free, A. and Goryanin, I., 2009. Novel electrochemically active bacterium phylogenetically related to arcobacter butzleri, isolated from a microbial fuel cell, *Applied and Environmental Microbiology*, **75**(23), 7326–7334.
Ferguson, B.K., 2005. *Porous Pavements, Integrative studies in Water Management and Land Development* (Boca Raton, FL: CRC Press).
Frederick, J.E., Snell, H.E. and Haywood, E.K., 1989. Solar ultraviolet radiation at the Earth's surface, *Photochemistry and Photobiology*, **50**(4), 443–450.
Freeze, R.A. and Cherry, J.A., 1979. *Groundwater* (Englewood Cliffs, NJ: Prentice Hall, Inc.).
Gartland, L., 2008. *Heat Islands-Understanding and Mitigating Heat in Urban Areas* (London: Earthscan).
Geothermal Heat Pump Consortium, 2009. *Fact sheet GB-034.*
Gill, S.E., Handley, J.F., Ennos, A.R. and Pauleit, S., 2007. Adapting cities for climate change: The role of the green infrastructure, *Built Environment*, **33**(1), 115–133.
Glasgow City Council, 2003. *The Ruchill-Keppoch New Neighbourhood Initiative – Urban Design Framework* (Glasgow: Development and Regeneration Services, Glasgow City Council).
Glasgow City Council, 2004a. *Drumchapel New Neighbourhood Sites, Planning and Design Guidelines: Site specific Brief (Site H)*, consultation draft (Glasgow: Development and Regeneration Services, Glasgow City Council).
Glasgow City Council, 2004b. *Ruchill Hospital Planning Document – Scoping Report for Ruchill Hospital Development Brief* (Glasgow: Glasgow City Council, Development and Regeneration Services).

Gouveia, L. and Rema, P., 2005. Effect of microalgal biomass concentration and temperature on ornamental goldfish (*Carassius auratus*) skin pigmentation, *Aquaculture Nutrition*, **11**(1), 19–23.

Grabowiecki, P., Tota-Maharaj, T., Scholz, M. and Coupe, S., 2008. Combined permeable pavement and ground source heat pump system to treat urban runoff during storms and recycle energy. In: *10th British Hydrological Society National Hydrology Symposium. Sustainable Hydrology for the 21st Century* (15–17 September 2008). University of Exeter, Exeter.

Greeman, A. 2004. Glasgow drain brains, *New Civil Engineer, Section Water and Drainage*, 20, 31–32.

Grieve, R. 1986. *Inquiry into the Housing of Glasgow* (Glasgow: City Press of Glasgow Ltd.).

Guo, Y., 2001. Hydrologic design of urban flood control detention ponds, *ASCE Journal of Hydrologic Engineering*, **127**(6), 472–479.

Guo, Y.C.Y., 2000. Storm hydrographs from small urban catchments, *Water International*, **2**(3), 481–487.

Hanson Formpave, 2009, Stormwater source control system Aquaflow permeable paving leaflet, Coleford, UK.

Healy, P.F. and Ugursal, V.I., 1997. Performance and economic feasibility of ground source heat pumps in cold climate, *International Journal of Energy Research*, **21**, 857–870.

Helfield, J.M. and Diamond, M.L., 1997. Use of constructed wetlands for urban stream restoration: A critical analysis, *Environmental Management*, **21**, 329–341.

Hepbasli, A., 2005. Thermodynamic analysis of a ground source heat pump system for district heating, *International Journal of Energy Research*, **29**, 671–687.

Herb, W.R., Janke, B., Mohseni, O. and Stefan, H.G., 2008. Ground surface temperature simulation for different land covers, *Journal of Hydrology*, **356**(3–4), 327–343.

Hermansson, Å., 2001. Mathematical model for calculation of pavement temperatures: Comparison of calculated and measured temperatures, *Transportation Research Record*, 1764(1), 180–188.

Hermansson, Å., 2004. Mathematical model for paved surface summer and winter temperature: Comparison of calculated and measured temperatures, *Cold Regions Science and Technology*, **40**(1–2), 1–17.

Highways Agency, 2009. *Road pavements, unbound cement and other hydraulically bound mixtures*. Manual of Contract Documents for Highway Works Series **800**. Vol. **1**, Specification for Highway Works.

Hyder Consulting, 2004. *Glasgow Strategic Drainage Plan – East End/ Dalmarnock SUDS Solutions*. Report Number NE02310/D10/1, Scottish Water (client).

Imran, H.M., Akib, S. and Karim, M.R., 2013. Permeable pavement and stormwater management systems: A review, *Environmental Technology*, **34**, 2649–2656.

Incropera, F.P., DeWitt, D.P., Bergman, T.L. and Lavine, A.S., 2007. *Fundamentals of Heat and Mass Transfer* (Hoboken, NJ: John Wiley & Sons Inc.).

ISO, 2008. 11731:1998 Water quality - Detection and enumeration of *Legionella*, International Standard Organisation (ISO), ISO now withdrawn and replaced. Available online at https://www.iso.org/iso/catalogue_detail.htm?csnumber=19653 (accessed on 20 March 2024).

Jefferies, C., Aitken, A., Mclean, N., MacDonald, K. and G. McKissock, 1999. Assessing the performance of urban BMPs in Scotland, *Water Science and Technology*, **39**(12), 123–131.

Lamoureux, J., Tiersch, T.R., and Hall, S.G., 2006. Pond heat and temperature regulation (PHATR): Modeling temperature and energy balances in earthen outdoor aquaculture ponds, *Aquacultural Engineering*, **34**(2), 103–116.

Leoni, E. and Legnani, P.P., 2001. Microbiology comparison of selective procedures for isolation and enumeration of Legionella species from hot water systems, *Journal of Applied Microbiology*, **90**(1), 27–33.
Ling, H., Tatsuoka, F., Wu, J.T.H. and Nishimura, J., 1993. Hydraulic conductivity of geotextiles under typical operational conditions, *Geotextiles and Geomembranes*, **12**(6), 509–542.
Maier, R.M., Pepper, I.L. and Gerba, P.C., 2008. *Environmental Microbiology*, 2nd edition. (Burlington, NJ: Academic Press).
Mckie, R., 2005. How Philippa the Goldfish's unappetising tastes could save Britain's cities from the danger of flooding, interview (including picture) with Miklas Scholz, The Observer, 24/04/05, news, 12.
McKissock, G., Jefferies, C. and D'Arcy, B.J., 1999. An assessment of drainage best management practices in Scotland, *Journal of the Chartered Institution of Water and Environmental Management*, **13**(1), 47–51.
Mengelkamp, H.T., Warrach, K. and Rascke, E., 1999. SEWAB – a parameterization of the surface energy and water balance for atmospheric and hydrologic models, *Advances in Water Resources*, **23**(2), 165–175.
McMinn, W.R., Yang, Q. and Scholz, M., 2010. Classification and assessment of water bodies as adaptive structural measures for flood risk management planning, *Journal of Environmental Management*, **91**(9), 1855–1863.
Mikheev, V.N., 2005. Individual and cooperative food searching tactics in young fish, *Zoologichesky Zhurnal*, **84**(1), 70–79.
Millennium Ecosystem Assessment, 2005. *Ecosystems and Human Well-Being* (Washington, DC: United Nations Environment Programme).
Ministry of Housing, Spatial Planning and Environment, 2000. *Circular on Target Values and Intervention Values for Soil Remediation*, Reference No. DBO/1999226863 (The Hague: Ministry of Housing, Spatial Planning and Environment).
Moore, T.L. and Hunt, W.F., 2012. Ecosystem service provision by stormwater wetlands and ponds – A means for evaluation? *Water Research*, **46**(20), 6811–6823.
Munoz-Pedreros, A., 2004. Landscape evaluation: An environmental management, *Revista Chilena de Historia Natural*, **77**(1), 139–156.
Nanbakhsh, H., Kazemi-Yazdi, S. and Scholz, M., 2007. Design comparison of experimental storm water detention systems treating concentrated road runoff, *The Science of the Total Environment*, **380**, 220–228.
Newman, A.P., Pratt, C.J., Coupe, S.J. and Cresswell, N., 2002. Oil bio-degradation in permeable pavements by microbial communities, *Water Science Technology*, **45**(7), 51–56.
Nuttall, P.M., Boon, A.G. and Rowell, M.R., 1998. *Review of the design and Management of Constructed Wetlands*, Construction Industry Research and Information Association (CIRIA) Report 180 (London: CIRIA).
Omer, A.M., 2008. Ground-source heat pumps systems and applications, *Renewable and Sustainable Energy Reviews*, **12**(2), 344–371.
Ove Arup & Partners Scotland, 1996. Proposed Pollok Retail Park. Rev1/96 (Glasgow: WD Ltd.).
Ozgener, O. and Hepbasli, A., 2007. Modeling and performance evaluation of ground source (geothermal) heat pump systems, *Energy and Buildings*, **39**(1), 66–75.
Phetteplace, G., 2007. Geothermal heat pumps, *Journal of Energy Engineering*, **133**(1), 32–38.
Pond, K., 2005. *Water Recreation and Disease: Plausibility of Associated Infections: Acute Effects, Sequelae and Mortality* (London: IWA Publishing).
Pratt, C.J., Mantle, J.D.G and Schofield, P.A., 1995. UK research into the performance of permeable pavement reservoir structures in controlling storm water discharge quantity and quality, *Water Science and Technology*, **32**(1), 63–69.

Pratt, C.J., Newman, A.P. and Bond, P.C., 1999. Mineral oil bio-degradation within a permeable pavement: Long term observations, *Water Science and Technology*, **39**(2), 103–109.

Richardson, M.J. and Whoriskey, F.G., 1992. Factors influencing the production of turbidity by goldfish (*Carassius auratus*), *Canadian Journal of Zoology*, **70**(8), 1585–1589.

Salvato, A.J. Nemerow, L.N. and Agardy, F.J., 2003. *Environmental Engineering*, 5th edition. (Hoboken, NJ: Wiley).

Sambrook, J., Fritsch, E.F. and Maniatis, T., 2000, *Molecular Cloning: A Laboratory Manual* (Cold Spring Harbor, NY: Cold Spring Harbor Laboratory Press).

Sanner, B., Karytsas, C., Mendrinos, D. and Rybach, L., 2003. Current status of ground source heat pumps and underground thermal energy storage in Europe, *Geothermics*, **32**(4–6), 579–588.

Scholz, M., 2003. Sustainable operation of a combined flood-attenuation wetland and dry pond system, *Journal of the Chartered Institution for Water and Environmental Management* **17**(3), 171–175.

Scholz, M., 2004a. Stormwater quality associated with a silt trap (empty and full) discharging into an urban watercourse in Scotland, *International Journal of Environmental Studies*, **61**(4), 471–483.

Scholz, M., 2004b. Case study: Design, operation, maintenance and water quality management of sustainable stormwater ponds for roof run-off, *Bioresource Technology*, **95**(3), 269–279.

Scholz, M., 2006a. Best management practice: A sustainable urban drainage system management case study. *Water International*, **31**(3), 310–319.

Scholz, M., 2006b. *Wetland Systems to Control Urban Runoff* (Amsterdam: Elsevier).

Scholz, M., 2010. *Wetland Systems – Storm Water Management Control* (Berlin: Springer).

Scholz, M., 2014. Rapid assessment system based on ecosystem services for retrofitting of sustainable drainage systems, *Environmental Technology*, **35**(9–12), 1286–1295.

Scholz, M., Corrigan, N.L. and Yazdi, S.K, 2006. The Glasgow sustainable urban drainage system management project: Case studies (Belvidere Hospital and Celtic FC Stadium Areas), *Environmental Engineering Science*, **23**, 908–922.

Scholz, M. and Grabowiecki, P., 2007. Review of permeable pavement systems, *Building and Environment*, **42**, 830–836.

Scholz, M. and Grabowiecki, P., 2009. Combined permeable pavement and ground source heat pump systems to treat urban runoff, *Journal of Chemical Technology and Biotechnology*, **84**(3), 405–413.

Scholz, M., Höhn, P. and Minall, R., 2002. Mature experimental constructed wetlands treating urban water receiving high metal loads, *Biotechnology Progress*, **18**(6), 1257–1265.

Scholz, M. and Kazemi-Yazdi, S., 2005. How goldfish could save cities from flooding. *International Journal of Environmental Studies*, **62**(4), 367–374.

Scholz, M. and Lee, B.-H., 2005. Constructed wetlands: A review, *International Journal of Environmental Studies*, **62**(4), 421–447.

Scholz, M., Morgan, R. and Picher, A. 2005. Storm water resources development and management in Glasgow: Two case studies. *International Journal of Environmental Studies*, **62**(3), 263–282.

Scholz, M. and Trepel, M., 2004. Water quality characteristics of vegetated groundwater-fed ditches in a riparian peatland, *The Science of the Total Environment*, **332**(1–3), 109–122.

Scholz, M. and Uzomah, V., 2013. Rapid decision support tool based on novel ecosystem service variables for retrofitting of permeable pavement systems in the presence of trees, *The Science of the Total Environment*, **458–460**, 486–498.

Scholz, M. and Yang, Q., 2010. Guidance on variables characterizing water bodies including sustainable flood retention basins, *Landscape and Urban Planning*, **98**(3–4), 190–199.

SEPA, 1999. *Protecting the Quality of our Environment – Sustainable Urban Drainage: An Introduction* (London; Stationary Office).

Slootweg, R., Rajvanshi, A., Mathur, V.B. and Kolhoff, A., 2010. *Biodiversity in Environmental Assessment. Enhancing Ecosystem Services for Human Well Being* (Cambridge: Cambridge University Press).

Society of the Chemical Industry, 1979. Site investigation and materials problems. In: *Proceedings of the Conference on Reclamation of Contaminated Land* (Eastborne: Society of the Chemical Industry).

Spellman, F.R., 2008. *The Science of Water: Concepts and Applications*, 2nd ed. (Boca Raton: CRC Press).

Takahasi, Y., 2004. Public-private partnership as an example of flood control measures in Japan, *Journal of Water Resources Development*, **20**(1), 97–106.

Tamura, K., Dudley, J., Nei, M. and Kumar, S., 2007. MEGA4: Molecular Evolutionary Genetics Analysis (MEGA) software version 4.0, *Molecular Biology and Evolution*, **24**(8), 1596–1599.

TEEB, 2013. *The Economics of Ecosystems and Biodiversity (TEEB) Manual for Cities: Ecosystem Services in Urban Management*, Geneva, Switzerland.

Tejedor, J.M.T. Gonzalez, M.M., Pita, M.L., Lupiola, G.P. and Martin, B.J.L., 2001. Identification and antibiotic resistance of faecal enterococci isolated from water samples, *International Journal of Hygiene and Environmental Health*, **203**(4), 363–368.

Terram Geosynthetics, 2009. *Geotextiles for civil engineering*, Maldon, UK.

The Stationery Office, 1998. *Sewerage (Scotland) Act 1968*. The Stationery Office Books, Section 7, 35 (London: The Stationary Office).

Tosun, İ., 2007. *Modeling in Transport Phenomena: A Conceptual Approach* (Amsterdam: Elsevier).

Tota-Maharaj, K., Grabowiecki, P. and Scholz, M., 2009. Energy and temperature performance analysis of geothermal (ground source) heat pumps integrated with permeable pavement systems for urban run-off reuse. *International Journal of Sustainable Engineering*, **2**(3), 201–213.

Tota-Maharaj, K. and Scholz, M., 2010. Permeable (pervious) pavements and geothermal heat pumps: Addressing sustainable urban storm water management and renewable energy, *International Journal of Green Economics*, **3**(3–4), 447–461.

Tota-Maharaj, K., Scholz, M., Ahmed, T., French, C. and Pagaling, E., 2010. The synergy of permeable pavements and geothermal heat pumps for storm water treatment and reuse. *Environmental Technology*, **31**(14), 1517–1531.

Tota-Maharaj, K., Scholz, M. and Coupe, S., 2011. Modelling temperature and energy balances within geothermal paving systems, *Road Materials and Pavement Design*, **12**(2), 315–344.

UK National Ecosystem Assessment, 2011. *The UK national ecosystem assessment technical report* (Cambridge United Nations Environment Programme World Conservation Monitoring Centre).

Van Buren, M.A., Watt, W.E., Marsalek, J. and Anderson, B.C., 2000. Thermal enhancement of stormwater runoff by paved surfaces, *Water Resources*, **34**(4), 1359–1371.

Walsh, C.J., Fletcher, T.D. and Burns, M.J., 2012. Urban stormwater runoff: A new class of environmental flow problem, *PLoS ONE*, **7**(9), e45814.

Weiner, E.R., 2008. *Applications of Environmental Aquatic Chemistry: A Practical Guide*. 2nd ed. (Boca Raton, FL: CRC Press).

Weisburgh, W.G., Brans, S.M., Pelletier, D.A. and Lane, D.J., 1991. 16S Ribosomal DNA amplification for phylogenetic study, *Journal of Bacteriology*, **73**(2), 697–703.

White, I. and Alarcon, A., 2009. Planning policy, sustainable drainage and surface water management: A case study of Greater Manchester, *Built Environment*, **35**(4), 516–530.

Zheng, J., Nanbakhsh, H. and Scholz, M., 2005, Case study: Design and operation of sustainable urban infiltration ponds treating storm runoff, *Journal of Urban Planning and Development*, **132**(1), 36–41.

3 Multi-functional Sustainable Flood Retention Basins

3.1 SUSTAINABLE MULTI-FUNCTIONAL LAND AND URBAN MANAGEMENT

3.1.1 INTRODUCTION

The debate on sustainable land and urban development emphasizes the importance of integrating environmental policy into all other policy sectors. It is increasingly recognized that this integration is needed at both national and regional levels of governance. Integration in the context of land and urban development is important because of sustainability, as well as economic and social factors such as maintaining development towards a healthy population and controlling environmental impacts. Therefore, the European Union is funding large international research consortia to address integration challenges in a holistic manner.

Recent research into land and urban management has been concerned with promoting sustainable development. Land outside urban areas is often still subject to intense agricultural farming methods. These stresses may negatively impact on biodiversity and rural society. The trend in these areas, therefore, needs to be towards less intensive methods of food production, distribution and consumption, promoting sustainable agriculture methods (Elliot, 2006).

This chapter is based on original work that reviewed international trends in sustainable multi-functional land management that are relevant to environmental policymaking and is focused on large European projects with potential to make a significant impact on policy changes (Scholz et al., 2012). A database summary of the key research projects useful for policy uptake is shown in Table 3.1. The following criteria were used during the selection process to limit the review to research output of importance and relevance for European stakeholders:

- Selected studies must have wide geographical validity and thematic relevance. This excluded work that is only of relevance to a small region.
- Work undertaken at an overall 'systems' level received more attention than subject-specific projects that would have been less representative.
- Findings that required urgent changes to policies and strategies were highlighted. This excluded standard and routine work that confirmed established knowledge.
- Papers that have high innovation potential were favoured over documents that did not present any new or possible solution(s) to a problem.

DOI: 10.1201/9781003509370-3

TABLE 3.1
Sub-set of Representative European Union Projects

Reference	Summary	Programme
URBS PANDENS (2010)	Solved urban sprawl problems.	Framework 5
LUTR (2010)	Helped cities to optimize land use.	Framework 5
SPICOSA (2010)	Improved the management of coastal areas.	Framework 6
SEAMLESS (2010)	Assessed the impact of policy changes on agriculture.	Framework 6
SENSOR (2010)	Was concerned with sustainable impact assessments in general.	Framework 6
PLUREL (2010)	Focused on the interface between rural and urban areas.	Framework 6
Susta-Info (2010)	Provided access to sustainable development information for planners.	Framework 6
SWITCH (2010)	Highlighted sustainable urban water management.	Framework 6
DeSurvey (2010)	Helped to assess and monitor desertification.	Framework 6
KASSA (2010)	Accumulated knowledge on conservation agriculture.	Framework 6
RUFUS (2010)	Assessed how policy regimes can be combined to ensure more sustainable development.	Framework 7
SUME (2010)	Focused on the design of sustainable urban systems in general.	Framework 7
BRIDGE (2010)	Will develop sustainable urban planning strategies.	Framework 7
SAWA (2010)	Will support sustainable water management solutions.	Interreg

- Work adding substantially to the critical mass of knowledge and understanding and providing a significant step forward was selected.
- Research that addressed priorities of European institutions involved in finding solutions took preference over findings associated with no urgency or wider interest.
- The focus was on projects supporting the green economy and associated technologies as well as a sustainable environment. Published work that did not lead, for example, to significant enhancement of natural capital, ecosystem services (Daily and Matson, 2008), resource efficiency or sustainable job and wealth creation received no attention in this review.
- Predominantly relevant ISI Web of Knowledge-listed journal papers published between 1999 and 2012 have been reviewed. Table 3.2 provides an overview of how well some representative European Union projects achieved the key selection criteria for inclusion in the review.

The target audience of this review are policy decision-makers and representatives from business, industry, academia and non-governmental organizations. Section 3.1 is therefore deliberately written for generalists interested in a holistic and interdisciplinary overview of current and internationally relevant urban and land management problems linked to social and economic issues, and not for technical specialists. Figure 3.1 provides an overview of the key projects identified as important and relevant for sustainable land management.

TABLE 3.2
Overview of Key Selection Criteria for Representative European Union Projects

Project	A	B	C	D	E	F
URBS PANDENS	+	++	++	++	++	++
LUTR	+	++	++	++	++	++
SPICOSA	+	++	++	++	++	++
SEAMLESS	+	++	++	++	++	++
SENSOR	++	++	++	++	++	++
PLUREL	++	++	++	++	++	++
Susta-Info	+	N/A	N/A	–	N/A	N/A
SWITCH	+	++	++	++	++	++
DeSurvey	–	++	++	++	++	++
KASSA	+	++	++	++	–	++
RUFUS	+	+	+	+	–	++
SUME	+	+	+	+	–	+
BRIDGE	+	+	+	+	–	+
SAWA	++	+	+	+	+	+

A, wide geographical validity and thematic relevance. B, holistic work undertaken at an overall systems level. C, findings that require urgent changes to policies and strategies. D, high innovation potential. E, work adding substantially to the critical mass of knowledge. F, support of green technology, economy and sustainable environment. ++, criteria have been fully met. +, criteria have been partly met. –, criteria have not been met. N/A, not applicable.

European Projects Concerned with Sustainable and Multi-functional Land Management				
Urban	**Semi-urban**	**Rural**	**Water**	**Coastal**
URBS PANDENS	PLUREL	SEAMLESS	SWITCH	SPICOSA
LUTR	Susta-Info	SENSOR	DeSurvey	
SUME		KASSA	SAWA	
BRIDGE		RUFUS		

FIGURE 3.1 Overview of important European projects discussed in detail. URBS PANDENS, Urban Sprawl: European Patterns, Environmental Degradation and Sustainable Development; PLUREL, Peri-urban Land Use Relationships - Strategies and Sustainability Assessment Tools for Urban-Rural Linkages; SEAMLESS, System for Environmental and Agricultural Modelling; Linking European Science and Society; SWITCH, Managing Water for the City of the Future; SPICOSA, Science and Policy Integration for Coastal System Assessment; LUTR, Land Use and Transportation Research: Policies for the City of Tomorrow; Susta-Info, Global Knowledge for Local Sustainability; SENSOR, Sustainability Impact Assessment: Tools for Environmental Social and Economic Effects of Multifunctional Land Use in European Regions; DeSurvey, Surveillance System for Assessing and Monitoring Desertification; SUME, Sustainable Urban Metabolism for Europe; KASSA, Knowledge Assessment and Sharing on Sustainable Agriculture; SAWA, Strategic Alliance for Integrated Water Management Actions; BRIDGE, Sustainable Urban Planning Decision Support Accounting for Urban Metabolism and RUFUS, Rural Future Networks.

3.1.2 Urban Management Promoting Sustainable Cities

The environmental policy integration principle is receiving more attention from urban management and planning scholars. The principle is under-researched and, in many countries, its implementation, particularly sub-nationally as in urban planning, is hindered by organizational and administrative weaknesses in local authority planning departments (Lafferty and Hovden, 2003).

Simeonova and van der Valk (2009) discussed about organization theory and communicative planning in their literature review where they showed how a communicative approach can be used to improve environmental policy integration in the context of urban planning policy. Their review sheds light on the relevance of a communicative approach to environmental policy by comparing existing approaches. The authors conclude that a communicative approach (i.e. where different stakeholders meet to discuss potential issues and solutions throughout the planning process) is potentially helpful in changing organizational structures and altering the way in which individual actors interact in urban planning processes.

This section highlights key European projects that have made major contributions to advances in urban management. Projects such as Sustainable Urban Metabolism for Europe (SUME), Sustainable Urban Planning Decision Support Accounting for Urban Metabolism (BRIDGE), Urban Sprawl: European Patterns, Environmental Degradation and Sustainable Development (URBS PANDENS) and Land Use and Transportation Research: Policies for the City of Tomorrow (LUTR), cover urban metabolism, decision support systems, sprawl and mobility, respectively. Wherever relevant, communicative approaches used to improve environmental policy integration have been highlighted and assessed.

SUME, an FP7 collaborative research project costing a total of approximately €3,630,000, ran between 2010 and 2011. It engages ten partners from nine countries and two continents and is focusing on how urban buildings and spatial structures can be used to reduce energy and resource consumption. The concept of urban metabolism (Heynen et al., 2006) helps stakeholders to understand and analyse the way urban societies use environmental system resources such as energy and land for maintenance and reproduction (SUME, 2010).

Using the urban metabolism approach, the SUME (2010) consortium explores the flows of resources, energy and waste which maintain urban systems. The spatial qualities of built urban systems have an impact on the qualities and quantities of resources needed to subsequently maintain them. The SUME project analyses the impacts of existing urban forms on resource use and estimates the future potential to transform urban buildings and spatial structures to reduce resource and energy consumption.

A key outcome of SUME has been the improvement in communication between research communities. Dissemination and communication were deemed to be a very important component of this project, which addresses research questions such as how to increase awareness and understanding of built structures while assessing the current resource levels that maintain them with the intention of providing alternative development strategies (SUME, 2010). The project has developed a new planning tool (Metabolic Impact Analyses) for assessing resource efficiency in urban development projects. It has also ensured engagement of stakeholders from a wide variety of sectors together with active knowledge sharing, which is the project's main outcome, but it has also set out to improve communication between different communities. However, SUME's impact on policy and society remains to be seen, but the planning tool MIA already appears to be a successful decision support tool for land management issues.

BRIDGE is a three-year (2008–2011) FP7 collaborative research project costing a total of about €3,100,000. The project team comprises 14 European organizations from 11 EU countries. The consortium aims to incorporate sustainability aspects into urban planning processes. BRIDGE (2010) assists urban planners to present and evaluate planning alternatives with the aim of creating more sustainable towns and cities.

BRIDGE consortium members work on the assumption that urban communities are living systems that consume resources and then eliminate the waste products. The key objectives of BRIDGE (2010) are to define urban metabolism in terms of energy, water, carbon and air pollution fluxes at the local scale and to bridge the gap between bio-physical sciences and urban planners with the intention of developing a decision support system based on these indicators. This system can then be used to evaluate planning alternatives for representative case studies. Finally, the project devises sustainable planning strategies based on case study evaluations.

Urban sprawl is one of the most important types of land-use change in Europe. It is increasingly diminishing the quality of life in some parts of Europe, with major impacts on the environment via surface sealing, emissions through increased transport activity and ecosystem fragmentation. Urban sprawl also impacts on the social structure of an area via segregation, lifestyle changes or neglected urban centres. Also affected is the economy via distributed production, fluctuations in land prices and issues of scale (URBS PANDENS, 2010).

Between 2002 and 2005, the European Union project URBS PANDENS carried out an integrated impact assessment of social, economic and environmental issues related to urban sprawl in Europe, including regulations, incentives, economic instruments and infrastructure measures (Jaeger et al., 2010). Within the project, several policies imposed by European, national and regional agencies were investigated, and options for improvement were developed (URBS PANDENS, 2010).

The main outcome of this project was the creation of a policy guide to be used by urban planners based on seven case studies from different European countries where urban sprawl had become a challenge. Within the framework of the URBS PANDENS project, researchers focused on two declining European cities, Liverpool (UK) and Leipzig (Germany). The case study in Liverpool identified that green belt and urban regeneration policies had been effective at preventing urbanization on green field land and hence halted urban sprawl. With the urban regeneration policy in place, the focus was on redeveloping derelict and underused land (brownfield). The main conclusion in terms of policy was that with urban decline, problems of sprawl were reduced because there was not the same demand for housing (Couch et al., 2005). This project had sufficient geographical coverage (seven case studies) and appears to have adopted a communicative and holistic approach addressing economic, environmental and social issues throughout the project related to urban sprawl. Considering that there were and are on-going studies on urban sprawl in various countries, it is difficult to judge which of them have led to policy changes. However, it is likely that a combination of different studies has influenced change and helped stakeholders both to recognize patterns of urban sprawl and to make recommendations on general policies (URBS PANDENS, 2010).

The European Union project Land Use and Transportation Research (LUTR) cluster, which was funded under the Fifth Framework Programme, linked several different projects in sustainable urban mobility (Camagni et al., 2002), including land use, transportation and the environment. The key objectives were to develop strategic approaches and methodologies in urban planning, which would all contribute to the promotion of sustainable urban development. The cluster included 12 projects adopting a holistic approach and covering many different topics, including decision-making regarding the re-construction of arterial streets (ARTISTS), promoting non-motorized transport in cities (focusing on pedestrian traffic (PROMPT)) and developing settlement patterns with priority given to the requirements of sustainable transport (ECOCITY).

The projects addressed issues such as transportation demands and related land-use planning, the design and provision of efficient and innovative transportation services, such as alternative means of transportation and the minimization of negative environmental and socio-economic impacts (LUTR, 2010). From these projects, it is possible to make specific policy recommendations. Important project outcomes were that transport and integrated land use strategies were more successful than isolated individual policies, while the strategies themselves were only successful if car travel was made less attractive, possibly through higher costs, and with supportive alternatives put in place for suburban living.

Dissemination of the project outcomes and active knowledge sharing was considered vital. Social acceptance was seen as very important to the success of these policies, and therefore, it was considered that better communication and marketing was required to inform both the public and the ultimate decision-makers. Involvement of the public, politicians and the media in future research was important to achieve successful project outcomes. The proposed approaches and methodologies for strategic urban planning are currently being assessed by project member countries.

3.1.3 Urban and Rural Interface Management

A relatively small number of European consortia are concerned with the urban and rural interface (Bella and Irwin, 2005) with respect to all subject areas such as transportation, utilities and planning. For example, the projects Peri-urban Land Use Relationships – Strategies and Sustainability Assessment Tools for Urban-Rural Linkages (PLUREL) and Global Knowledge for Local Sustainability (Susta-Info) discuss urban-rural linkage issues and summarize urban and rural interface management problems, respectively.

The PLUREL project (2007–2010) was a European integrated research project within the European Commission's 6th Framework Programme. The consortium comprised 31 partners (14 European countries and China). This project supported the development of new strategies, and planning and forecasting tools for the interface between rural and urban regions (Bella and Irwin, 2005).

The project aimed to understand how sustainable relationships could be developed between urban and rural areas in terms of land use by evaluating six European case studies. The project made use of information from stakeholders to benchmark spatial planning and governance strategies for sustainable relationships between urban, peri-urban and rural areas, and to develop corresponding forecasting tools (PLUREL, 2010).

The descriptions and benchmarking of these strategies in relation to main trends such as globalization, climate change and demographic development were among the most important results of the PLUREL project. Other key findings were stakeholder-based participatory scenarios and web-based sustainable impact assessment tools for urban regions. The corresponding tools for the interface between rural and urban regions were tailored to the assessment of local, regional and European policies within the context of these regions.

The project team analysed impacts of land use policy options in environmental, social and economic terms (PLUREL, 2010). PLUREL appears to have successfully adhered to certain criteria (see above) by ensuring engagement of stakeholders from a wide variety of sectors. A published newsletter (PLUREL, 2010) provided views of the stakeholders involved. One of the most important outcomes of the project was to get academics to make their work understandable for policymakers. The outcome of PLUREL was the development of a toolbox that can be used by practitioners for planning strategies.

Susta-Info (2006–2008) is a global database of case studies and publications on sustainability, validated by research institutes, city associations and expert groups. Susta-Info is a European Union-supported project in the context of the Sixth Framework Programme. It prioritizes ecosystems all over the world that are changing because of pressures from urbanization and climate change (Schröter et al., 2005; Susta-Info, 2010).

Susta-Info supports local authorities and experts in attaining sustainable development by establishing a portal and web-based database and by collecting knowledge items on local sustainable development accessible to large groups of targeted users. Moreover, Susta-Info provides access to selected research projects funded under the Fifth Framework Programme, as well as to Sixth Framework projects on urban management, sustainable land use, water treatment and management and urban mobility. Urban research projects supported by United Nations-HABITAT (UN-HABITAT) and case studies from the UN-HABITAT best practices database can also be accessed (Susta-Info, 2010; UN-HABITAT, 2010).

Relevant and pre-selected high-impact articles are part of SUSTA-INFO; for example, the holistic and multi-disciplinary paper by Jansson and Colding (2007), who estimated the effects of urban development through the quantification of nitrogen leakage to the Baltic Sea, under two urban development scenarios for the city of Stockholm. They concluded that proactive measures such as spatial urban planning can provide a constructive tool for sustainable urban development on a regional as well as national and international scale. However, their findings depend on the geographical context as well as the ecosystem services' scale of operation (Daily and Matson, 2008). Their comprehensive and freely available database ensures that the outcomes of the project Susta-Info are easily available to stakeholders all over the world.

The project Susta-Info has developed a relevant support tool for local authorities and experts which could help decision-makers mould future policies towards attaining sustainable development. Further reflections included the fact that there were many databases that provided the same re-processed information, and that dissemination of information was more important and should be converted into messages that people could understand.

3.1.4 Rural Management Supporting Sustainable Agriculture and Rural Livelihoods

Rural management policies, strategies and models concerned with global change and addressing key agricultural issues (Schröter et al., 2005) contrast with those developed for urban areas. An economic and a biophysical model for Europe was presented by van Meijl et al. (2006). The potentially negative impact of the liberalization of agricultural policies on European agricultural land use is small, because the decline in Europe's competitiveness has led partly to extensification rather than land abandonment, and the European agricultural reforms (Ackrill et al., 2008) at the beginning of this millennium focused on income support rather than market protection, which has led to reduced agricultural production.

Global climate change is likely to alter the supply of ecosystem services that are vital for human well-being (Daily and Matson, 2008). Schröter et al. (2005) used a range of ecosystem models and scenarios of climate and land use change to conduct a Europe-wide assessment. Findings show that large changes in climate and land use typically result in considerable changes in ecosystem service supply. Many changes increase vulnerability because of a decreasing supply of ecosystem services such as declining soil fertility and water availability and increasing risk of forest fires (Daily and Matson, 2008).

Models have been developed to estimate the carbon fluxes from agricultural soils. Vleeshouwers and Verhagen (2002) produced a model that was parameterized for grassland and several arable crops. Data from agricultural, meteorological, soil and land use databases were input into the model, which was used to evaluate the effects of different carbon dioxide mitigation measures on soil organic carbon in agricultural areas in Europe. The study revealed considerable regional differences in the effectiveness of carbon dioxide abatement measures, resulting from the interaction between crop, soil and climate. Considerable differences between the spatial patterns of carbon fluxes resulting from different measures were also noted.

Large European projects have subsequently been funded to further develop rural management strategies. The projects System for Environmental and Agricultural Modelling Linking European Science and Society (SEAMLESS), Sustainability Impact Assessment: Tools for Environmental Social and Economic Effects of Multifunctional Land Use in European Regions (SENSOR), Knowledge Assessment and Sharing on Sustainable Agriculture (KASSA) and Rural Future Networks (RUFUS) cover agricultural modelling, sustainability impact assessment, sustainable agriculture and rural networks issues, respectively.

The SEAMLESS (2005–2006) consortium was funded by the EU Framework 6 to assess the impact of policy and behaviour changes in agro-forestry and agriculture;

it was established in response to a research and policy need from the European Commission. Its objective was to develop a framework which integrates approaches from economics, and environmental and social sciences to enable assessment of the impact of policy and behavioural changes as well as innovations in agriculture and agro-forestry. The contributions of agriculture to sustainable development and multi-functionality were assessed at different spatial scales, from individual farms to widespread scenarios, allowing consideration of both top-down and bottom-up approaches to land management change (SEAMLESS, 2010).

Qualitative tools were integrated into SEAMLESS to consider institutional and social contexts. SEAMLESS used practical test cases to evaluate and improve tools and to assess their utility (SEAMLESS, 2010). Through clarification of the benefits, costs and externalities associated with farming system management, SEAMLESS achieved ex-ante analyses of the impacts of policy and behavioural changes. Interactions between the European Union, associated candidate countries and other countries were assessed by incorporating appropriate models. SEAMLESS could be viewed as essential for the integrated assessment of agricultural systems in the context of agro-ecological innovations, rural development, sustainability, agricultural policy reform, European Union enlargement and world trade liberalization (SEAMLESS, 2010; Wacziarg and Horn Welch, 2008).

van Ittersum et al. (2008) presented the rationale, design and illustration of the component-based SEAMLESS integrated framework for agricultural systems to assess, ex-ante, agricultural and agri-environmental policies and technologies across a range of scales, from a single field belonging to one farm to an entire region within the European Union, as well as some global interactions. One illustrative example assesses the effects of a trade liberalization (Wacziarg and Horn Welch, 2008) proposal on Europe's agriculture and indicates how SEAMLESS addresses the challenges for integrated assessment tools (i.e. linking micro and macro analysis, and assessing economic, environmental, social and institutional indicators), using standalone model components for field, farm and market analysis, linking them to their conceptual and technical features. Project SEAMLESS adopted a communicative and holistic approach involving economic, environmental and social science issues throughout the project while also ensuring engagement of stakeholders from a wide variety of sectors.

The EU project SENSOR (2010) was concerned with the sustainability of land use in European regions. The consortium developed tools for the assessment of different scenario impacts on environmental and socio-economic sustainability. The technical objective of SENSOR was to build, validate and implement sustainability impact assessment tools (Helming et al., 2008; see below) such as databases and spatial reference frameworks for the analysis of land and human resources in the context of agricultural, regional and environmental policies (SEAMLESS, 2010).

The scientific challenge was to establish relationships between different environmental and socio-economic processes as characterized by indicators considered to be quantitative measures of sustainability. Scenario techniques were used within an integrated modelling framework, reflecting various aspects of multi-functionality and their interactions (SEAMLESS, 2010).

The focus was on sensitive European regions, particularly those in accession countries, since accession poses significant questions for policymakers regarding the socio-economic and environmental effect of existing and proposed land use policies. SENSOR delivered novel solutions for integrated modelling, spatial and temporal scaling and aggregation of data, selection of indicators, database management, analysis and prediction of trends, education and implementation (SEAMLESS, 2010).

Helming et al. (2008) developed an ex-ante Sustainability Impact Assessment Tool (SIAT) as part of the SENSOR project for land use in European regions. This meta-modelling toolkit translates global economic trend and policy scenarios into land use changes at 1-km^2 grid resolution for Europe. Based on qualitative and quantitative indicator analyses, impacts of simulated land use changes on social, environmental and economic sustainability issues were assessed at regional scale. Valuation of these impacts was based on the concept of multi-functionality of land use and conducted through expert valuations leading to the determination of sustainability choice spaces for regions.

Further to the SIAT tool, SENSOR developed a Framework for Participatory Impact Assessment (FoPIA). This tool complements SIAT and, as suggested by the SENSOR (2010) "should make a valuable contribution to the European Sustainable Development Strategy". The discussion on biofuels and the subsequent directive meant that the political relevance of biofuel changed, and SENSOR was found to contribute to this debate. In terms of a decision support tool, SENSOR has led to greater quality at the science-policy boundary. As SENSOR is used to predict the consequences of policy change that will affect land use zones, its use globally will be of great relevance to scientists and policymakers working in a number of disciplines. One concern may be that in the long term, the research SENSOR delivers will become outdated, but the large consortium size and all the specialized institutes from numerous disciplines that have developed because of it will help to ensure that it continues to deliver up-to-date research.

The KASSA consortium is a European Commission's Sixth Framework programme aiming at the capitalization of results from past and on-going research on sustainable agriculture. The KASSA proposal aimed to build up a comprehensive knowledge base on sustainable agricultural practices, approaches and systems in support of stakeholders such as farmers and professionals, researchers and policymakers at local, regional, national, European and global levels (KASSA, 2010).

The KASSA project involved a critical mass of skilled partners divided between four rather undifferentiated platforms: Europe (excluding Mediterranean regions), the Mediterranean, Asia and Latin America. Aims were achieved through successive work sequences, which started with a comprehensive inventory of existing results followed by progressive refinement of the findings, alternating critical analysis and the sharing of the results of each platform (KASSA, 2010). Knowledge generation and sharing were a key concern for this project.

Lahmar (2010) states that, according to KASSA findings, soil conserving cultivation has been adopted less in Europe than in the other platforms (see above). One example is that reduced tillage is more common in Europe than no-tillage and cover crops. The lack of understanding regarding soil-conserving cultivation systems makes it difficult and socio-economically risky for European farmers to give up ploughing practices. In a few member states, the adoption of soil-conserving cultivation practices has been encouraged and subsidised to mitigate soil erosion. However, in other countries within Europe, the adoption process is mainly driven by farmers seeking to achieve cost reductions in machinery, fuel and labour. Large-scale farms are the most likely to adopt sustainable agricultural practices.

Land use change is characterized by a high diversity of change trajectories depending on the local conditions, regional context and potentially harmful external influences. Policy intervention aims to counteract the negative consequences of these changes and provide incentives for positive developments. The project RUFUS aimed to provide policymakers and stakeholders with a better theoretical and practical understanding of how European agriculture policy measures interact with other forms of public intervention in rural development, and it focused specifically on applied topics, which have a high relevance for policy (RUFUS, 2010). The consortium assessed how policy regimes can be combined to ensure more sustainable development. An interdisciplinary methodology included qualitative analysis of the social dimension and endogenous potentials alongside economic and ecological variables.

Verburg et al. (2010) have undertaken research in the context of the RUFUS (2010) project. They provided a typology of land use change in Europe at a high spatial resolution based on a series of different land use change scenarios covering the period 2000–2030. A series of simulation models ranging from the global to the landscape level were used to translate scenario conditions in terms of demographic, economic and policy change into changes in European land use patterns. The findings indicate that the typologies based on current landscape and rural areas are poor indicators of the land use dynamics simulated for the regions. It is suggested that typologies based on (simulated) future dynamics of land change should be more appropriate for identifying regions with potentially similar policy needs. Similar to many other EU projects, RUFUS aimed to bridge the gap between science and policy by combining the findings from this project with other research and placing it in the context of policy problems and goals that it can then transpose into meaningful and practical action.

3.1.5 Sustainable Water Management

Modern water management has to address sustainability issues associated with problems such as flooding (Scholz, 2006, 2010) and desertification. Sustainable (urban) drainage systems (also called best management practices) are typical examples of adaptive measures that can be retrofitted to reduce the pressure on traditional combined sewer systems during periods of high rainfall. Figure 3.2a shows an example of a derelict site in Glasgow, which would be suitable for retrofitting a swale and pond system (Figure 3.2b), as discussed by Scholz (2006). Projects labelled Managing Water for the City of the Future (SWITCH), Strategic Alliance for Integrated Water Management Actions (SAWA) and Surveillance System for Assessing and Monitoring Desertification (DeSurvey) cover adaptive sustainable water management (Hedelin, 2007) techniques such as sustainable drainage systems, flood control and desertification assessments, respectively.

FIGURE 3.2 Belvidere Hospital grounds in Glasgow, Scotland: (a) site photograph taken on 15 May 2004, and (b) artist impression of proposed site development.

SWITCH is the name of a major research programme (budget exceeding 20 million Euros) which has been implemented and co-funded by the European Union. SWITCH aimed to bring about a paradigm shift in urban water management away from existing ad hoc solutions to urban water management and towards a more coherent and integrated approach (SWITCH, 2010).

The SWITCH consortium represented academics, urban planners, water utilities and consultants, thereby adopting a communicative and holistic approach involving economic, environmental and social sciences. This network of researchers and practitioners worked very successfully directly with civil societies through new 'learning alliances' in ten cities located in the partner countries. Learning alliances are platforms that brought city stakeholders (utilities, planners, non-governmental organizations and finance departments) together with applied researchers. Other methods used to accelerate the sharing and adoption of more sustainable urban water solutions across different geographical, climatic and socio-cultural settings were presentations, research, training and knowledge sharing (SWITCH, 2010).

Wagner and Zalewski (2009) reported as part of the SWITCH project that intensive, unsustainable development of urban areas has been generally related to landscape degradation, which is partly due to the high density of city infrastructure, excessive increase in impermeable surfaces, reduction of green areas and river canalization and simplification. Furthermore, Wagner and Zalewski (2009) confirmed for the partner cites that decreased storm water infiltration causes increased flooding during rain events, while accelerated runoff via traditional drainage systems results in excessive decreases in air humidity during periods of dry weather.

The project SWITCH helped to alter the way urban water is managed, 'switching' from an engineered approach to a more integrated system philosophy that encourages a sustainable approach by forming a global network of stakeholders and scientists who have the expertise to change the way water is managed in cities. For example, in Birmingham (UK), brown roofs are now being incorporated into more areas of the city, while in Beijing, rainwater collection tanks are being encouraged with support from guideline information and installations. The project is actively engaging the public and is therefore successful in practical terms compared to many projects where only a decision support tool has been developed.

The SAWA consortium receives funding under the European Union Interreg IVb North Sea Region Programme. The consortium aims to develop a strategy which adheres to the European Water Frame Directive and meets the requirements of the existing Flood Directive. At the same time, the consortium is addressing challenges arising from climate change issues (SAWA, 2010).

The Flood Directive focuses on the quantitative aspects of flood risks, whereas in the case of the European Water Frame Directive, water quality and good water conditions are pivotal. Each directive deals with water management. Therefore, it is necessary for the SAWA consortium to consider both aspects even though this might lead to a conflict of interests (SAWA, 2010).

The consortium developed and tested adaptive flood risk management plans, and identified and deployed cost-effective local-scale adaptive measures such as sustainable flood retention basins (SFRBs) (Scholz, 2010; Scholz and Yang, 2010). Flood risk management plans for rural and urban locations are being prepared. Furthermore, a series of fully functional locally run measures, which not only mediate flood risks but furthermore have no negative effect on water quality, are being set up. An SFRB is a good example of an adaptive local and regional structural measure to control flooding and diffuse pollution (SAWA, 2010; Scholz and Yang, 2010). For example, Figure 3.3 shows an adaptive SFRB that has a clearly multi-functional land use purpose: in this case, flood control, recreation and agricultural production.

Yang et al. (2010) applied three feature selection techniques (Information Gain, Mutual Information and Relief) to an SFRB dataset to identify the importance of variables in terms of classification and to determine the minimum number of variables required to efficiently classify these basins. Four benchmark classifiers (Support Vector Machine, K-Nearest Neighbours, C4.5 Decision Tree and Naïve Bayes) were used to verify the effectiveness of the classification using the identified variables. The results of the statistical analysis using these benchmark classifiers and comparing corresponding errors showed that only 9 out of 40 collected variables were optimal to accurately classify SFRB. The proposed approach produced a simple, rapid, effective and efficient classification tool for water resources managers and planners.

The trans-national SAWA project developed a common understanding with respect to the implementation of ideas and concepts together with the establishment of cultural and social aptitudes. Moreover, the formulation of messages directed at policy- and decision-makers at the EU level is an important outcome of the project.

DeSurvey is a project funded by the European Commission under the Sixth Framework Programme. The corresponding consortium contributed to the implementation of proposed European actions labelled 'Mechanisms of desertification'

FIGURE 3.3 Sustainable flood retention basin near Freiburg, Germany: site photograph taken on 15 July 2010.

and 'Assessment of the vulnerability to desertification and early warning options' within the 'Global Change and Ecosystems' priority. The consortium consists of 39 organizations with a wide range of skills from 10 EU Member States and 6 developing nations (DeSurvey, 2010).

Despite the relevance of appropriate actions to counter desertification in Europe (Hill et al., 2008), there is a lack of standardized procedures to perform them on an operational scale. The DeSurvey project offers a contribution towards filling this gap by complementing assessments of desertification status with early warning and vulnerability evaluation of the land use systems involved. In this context, the interactive effects of climatic and human drivers of desertification are considered in a dynamic way (DeSurvey, 2010).

The DeSurvey consortium delivered a set of integrated procedures of desertification assessment and forecasting (e.g., an operational prototype that includes flexible procedures) with generic and case-specific components to which users can adapt their biophysical and socio-economic environments and their data availability accordingly. DeSurvey continues to support international, European, national and regional authorities, organizations and institutions in fulfilling their monitoring, surveillance and reporting obligations, and to help them increase the efficiency of desertification treatment policies (DeSurvey, 2010).

As part of the DeSurvey project, Martinez et al. (2007) analysed the consistency of regional scale mapping products derived from sensor systems used for vegetation monitoring, especially over hot spot areas affected by land degradation processes. The results demonstrated the competitiveness of products derived from the Fraction of Vegetation Cover and Leaf Area Index compared to similar satellite products at local scales and the suitability of using the former to reproduce the temporal variability over dry land condition areas.

The project has developed a dynamic methodology for determining desertification early and allowing for the assessment of vulnerable land prior to any adverse impacts affecting it. It has also contributed to the efficiency of desertification treatment policies by supporting European, national and regional organizations and authorities in assessing and monitoring desertification, and in this respect has been successful in having sufficient geographic coverage while ensuring engagement of stakeholders from a wide variety of sectors.

3.1.6 Sustainable Coastal Development and Management

Important coastal management issues are predominantly the large-scale concerns traditionally associated with land use planning, estuary management, natural resource and landscape protection and water quality (O'Connor et al., 2010). A lot of European coastal regions are under threat from the impacts of, for example, flooding and environmental degradation. However, corresponding social interaction problems are often neglected. Therefore, projects such as Science and Policy Integration for Coastal System Assessment (SPICOSA) look at policy options to address a combination of physical and social issues.

SPICOSA is a European Union integrated project, which has created a self-evolving, operational research approach framework for the assessment of policy options for the sustainable management of coastal zone systems. SPICOSA has contributed to stakeholders' understanding of social interactions within coastal zone systems and of how this impact on the environment and future policies. This project aimed to support the implementation of existing directives, guidelines and good practices. It is therefore of high relevance to Europe's integrated coastal zone management policies (SPICOSA, 2010).

The 'systems thinking' approach inherent in SPICOSA's framework, and its practical application, is likely to have a positive influence on understanding sustainability transitions in the future. Key to this approach is the science-policy interface to improve and enhance closer and more effective integration and common deliberation over coastal issues. The project outcomes will guide and support European coastal policymakers by providing them with the means to understand the effects of their policies (SPICOSA, 2010).

Vernier et al. (2010) assessed the environmental scenarios of changes in agriculture using the spatial distribution model called Soil and Water Assessment Tool. The effectiveness of each scenario was considered in terms of relative reduction in water uptake for irrigation, nitrates, pesticides and suspended matter compared to the current situation. The authors used a representative case study in France (Charente watershed) as part of the SPICOSA project.

This project aimed to improve the use of science and technology for the benefit of society. Because of the lack of policy, SPICOSA was developed to address increasing negative impacts on coastal zones because of human influence. The idea was that it would act as an early warning system that could alert decision-makers and help to improve the way research information was communicated. Understanding social aspects and knowledge improvement was viewed as important to improving debate between scientist and policymakers. An important 'take-home' message was that it should be made easier for society to grasp the significance of science-policy research and its application for the management of coastal areas.

3.1.7 Case Study Projects of International Importance

Hundreds of research projects with relevance to European urban and land management issues were scrutinized, categorized and assessed as part of the groundwork for this review chapter. The following projects were identified as potentially significant regarding their likely long-term impact on land and urban management practices in Europe (Table 3.1): SUME (2010), BRIDGE (2010), SPICOSA (2010), SEAMLESS (2010), SENSOR (2010), PLUREL (2010), KASSA (2010), SUSTA-INFO (2010), SWITCH (2010), DeSurvey (2010) and RUFUS (2010).

Section 3.1 emphasizes projects with a high-impact potential such as those above funded by the European Union under Framework Programmes 6 and 7. Most of the projects are associated with large consortia that received several millions of Euros in funding. Key publications associated with the research outlined in Table 3.1 are presented in Table 3.3. Some of the project reports are very detailed, summarizing research findings while others do not disseminate enough information to make meaningful judgements to the success. A sub-set of research publications that are not directly related to the highlighted case study projects, but which are highly relevant

TABLE 3.3
Sub-set of European Union Research Publications Selected as Resource Data According to Subject Area

Reference	Subject Area	Publication Type	Relevance
URBS PANDENS (2010)	Urban	Project report	Framework 5
LUTR (2010)	Urban	Project report	Framework 5
SUME (2010)	Urban	Project report	Framework 7
BRIDGE (2010)	Urban	Project report	Framework 7
PLUREL (2010)	Urban and rural	Project report	Framework 6
Susta-Info (2010)	Urban and rural	Project report	Framework 6
Jansson and Colding (2007)	Urban and rural	Journal	Framework 6; Susta-Info (2010)
SEAMLESS (2010)	Rural	Project report	Framework 6
van Ittersum et al. (2008)	Rural	Journal	Framework 6; SEAMLESS (2010)
SENSOR (2010)	Rural	Project report	Framework 6
Helming et al. (2008)	Rural	Book chapter	Framework 6; SENSOR (2010)
KASSA (2010)	Rural	Project report	Framework 6
Lahmar (2010)	Rural	Journal	Framework 6; KASSA (2010)
RUFUS (2010)	Rural	Project report	Framework 7
Verburg et al. (2010)	Rural	Journal	Frameworks 6 and 7; RUFUS (2010)
SWITCH (2010)	Water	Project report	Framework 6
Wagner and Zalewski (2009)	Water	Journal	Framework 6; SWITCH (2010)
DeSurvey (2010)	Water	Project report	Framework 6
Martinez et al. (2007)	Water	Conference	Framework 6; DeSurvey (2010)
SAWA (2010)	Water	Project report	Interreg IV
Scholz (2010)	Water	Book	Interreg IV, SAWA (2010)
Scholz and Yang (2010)	Water	Journal	Interreg IV, SAWA (2010)
Yang et al. (2010)	Water	Journal	Interreg IV, SAWA (2010)
SPICOSA (2010)	Coastal	Project report	Framework 6
Vernier et al. (2010)	Coastal	Conference	Framework 6; SPICOSA (2010)

research publications providing important background information to the selected projects is shown in Table 3.4. This information is not complete, but provides a starting point for further studies.

3.1.8 Conclusions and Recommendations

This review chapter has highlighted high-impact European research in the management of urban, rural and coastal areas that should make a significant contribution to policy developments. The clear trend in work output favours the holistic and multidisciplinary assessment approach of the total environment, leading to decision-making planning tools that should be of practical value for many stakeholders. The projects foster knowledge transfer from universities to councils and from developed to less developed nations. However, practitioners participating even in successful

TABLE 3.4
Sub-set of Non-European Union-Funded Research Publications Selected as Resource Data According to Subject Area

Reference	Publication Type	Subject Area	Relevance
Camagni et al. (2002)	Journal	Urban	Analysis of costs associated with urban mobility and urban form
Heynen et al. (2006)	Book	Urban	Urban political ecology and the politics of urban metabolism
Jaeger et al. (2010)	Journal	Urban	Suitability criteria for measures of urban sprawl
Simeonova and van der Valk (2009)	Journal	Urban	Assessment of the need for a communicative approach to improve environmental policy integration in urban land use planning
Ackrill et al. (2008)	Journal	Urban and rural	Analysis of the European Common Agricultural Policy and its reform
Bella and Irwin (2005)	Journal	Urban and rural	Spatial modelling of land use change at the rural-urban interface
de Groot (2006)	Journal	Urban and rural	Assessment tool for land use conflicts in planning for sustainable, multi-functional landscapes
Daily and Matson (2008)	Journal	Urban, rural, coastal and water	Introduction to the ecosystem services concept
Elliot (2006)	Book	Urban, rural, coastal and water	General introduction to sustainable development
Lafferty and Hovden (2003)	Journal	Urban, rural, coastal and water	Analytical framework for environmental policy integration
UN-HABITAT (2010)	Report	Urban, rural, coastal and water	Promotion of socially and environmentally sustainable human settlement development
Wacziarg and Horn Welch (2008)	Journal	Urban, rural, coastal and water	Introduction to world trade liberalization
Schröter et al. (2010)	Journal	Rural	Evaluation of ecosystem service supply and vulnerability to global change
Vleeshouwers and Verhagen (2002)	Journal	Rural	Critical review and model of carbon emission and sequestration by agricultural land use
Hedelin (2007)	Book	Water	Introduction to sustainable water management
Hill et al. (2008)	Journal	Water	Mediterranean desertification and land degradation
Scholz (2006)	Book	Water	Introduction to runoff management control
O'Connor et al. (2010)	Journal	Coastal	Analysis of coastal issues and conflicts

projects such as SWITCH very rarely integrate models and decision support tools into their day-to-day work practices, usually because the corresponding methods are too academic and not site-specific.

Most of the research output is intended to be practical with a strong integrated community focus regarding the management of urban, semi-urban and rural areas, having strong stakeholder involvement in decision-making on various levels. However, the real value of the relatively soft project output is very difficult to assess academically in a quantitative way. The evidence-based project output in terms of practical and novel solutions to real problems should be much higher for most European projects considering the significant funds invested (often over 1 million Euros per project). In terms of 'money well-spent', the author estimates that the ratio of good value for money projects to less valuable projects is rather low. However, a quantitative assessment is difficult since the success of projects is centrally assessed with the help of tick lists and milestone completion dates, which very much rely on self-assessments of often vague objectives by lead partners rather than on independent and critical peer reviews.

The most common reasons for failed projects are well known: (a) important stakeholders are sometimes not involved from the planning stage onward and (b) and the target audience and suitable dissemination pathways for the project outcomes are not identified in advance. It follows that some projects did not generate many new findings but were rather 're-inventing the wheel'.

On a more practical level, many consortia are very large, and there is the tendency for individual partners to focus on their own projects rather than on trans-national project outcomes. This became apparent when analysing project outcomes based on paper publications. Only a limited number of papers provided clear evidence for the achievement of trans-national project objectives. For example, the study on SFRB as part of SAWA does have a trans-national application, but virtually all work was undertaken by only 1 out of 22 project partners.

This review has highlighted representative European case study projects of relevance to policymakers. Holistic decision-making tools supported by numerical models seem to be most promising in influencing future policy. For example, the SWITCH approach, in combination with other proposed related and compatible methodologies by SAWA, SUME and BRIDGE, is likely to lead to cities being more sustainable in terms of water management. However, there is no mechanism in the current European funding scheme that links projects of a similar nature with each other, thereby avoiding duplication and generating better quality research.

Further research should focus on how to improve the uptake of decision-making tools in practice. This could be achieved by linking funding provisions to the international rather than national (or even regional) project deliverables. Moreover, funding should be directed toward substantive consortia with reasonable transaction costs, a proven track record and a strong project management team. However, innovative research based on young academics and practitioners without a significant track record but excellent skills to tackle major scientific and societal challenges should not be overlooked. There is also a clear need to increase project output by avoiding vague project objectives that cannot be assessed with quantitative assessment criteria.

The project assessment should be undertaken by peer review and part of the funding should be withheld if trans-national objectives have not been met.

3.2 ECOLOGICAL EFFECTS OF WATER RETENTION

3.2.1 INTRODUCTION

Section 3.2 summarizes the findings published in recent high-impact journal papers to enable further research into the assessment of floodplain and retention basin functions. The main purpose is to provide the scientific foundation for future retention basin classification systems and models. The review concentrates on the River Rhine (Rhein) valley, which is representative for other rivers in temperate climates.

3.2.2 RIVER REGULATION AND FLOODING

In the northern part of Europe, the total annual amount of rainfall significantly increased in the last century, which, at a probability of 95%, may not be attributed to the natural variability of the climate (Hegerl et al., 1994). A statistically significant increase of precipitation was observed particularly during winter and spring (Bardossy and Caspary, 1990; Schönwiese et al., 1993). This finding is supported, for example, by the observation that the mean water flow in the largest German rivers has increased by approximately 26% since 1933 (Schumann, 1993). Further changes of precipitation patterns are predicted independently from different climatic models (ICPR, 1998; IPCC, 1997), which will lead to an enhanced increase of precipitation during winter and spring, causing increased risk of flooding in Central and Northern Europe.

Land use change such as urbanization in headwater areas can lead to flooding during storms (Pfister et al., 2004). Future changes in peak flow will depend on the variability of extreme rainfall in combination with land use management. Therefore, the sustainable development of a floodplain strategy, which promotes sustainable floodplains in the River Rhine valley by finding a compromise between flood prevention and nature development, has been proposed elsewhere (Nijland, 2005).

As a flood prevention measure (Molles et al., 1998; LfU, 1999) or to restore floodplain forests (Middleton, 2002; Rood et al., 2005), water retention basins, which are often located in forest ecosystems, were created. These forests were often exposed to flooding as never before, or, in the case of former alluvial floodplains, have adapted to the new hydrological conditions by changes in structure and floristic composition of the forest (Trémolières et al., 1998), and in the soil chemistry (Trémolières et al., 1993). Therefore, the construction of flood retention areas in such ecosystems usually leads to damage and changes of some vegetation communities. In areas where the history of human use of natural areas is old and intense, the creation of retention areas should not only consider ecological objectives but also social aspects (Winkel, 2000).

The topsoil of flooded forested retention basins is relatively high in phosphorus and relatively low in nitrogen and potassium, if compared to areas that were not regularly flooded. Previous findings show that flooding is also responsible for an increase of the bioavailability of nutrients (Trémolières et al., 1999).

Finally, the species richness of existing vegetation is likely to increase with the duration of interruption of the floods in the riparian forest. This phenomenon can be observed because of the introduction of species that are intolerant of the floods (Deiller et al., 2001).

The River Rhine was subject to drastic environmental changes (e.g., regulation in the 19th century, canalization between 1930 and 1977, and discharge of wastewater), which caused reductions of the total surface area of the alluvial floodplains and of the hydrological connectivity (i.e. disconnection of the former lateral arms and of the alluvial forests), as well as a deterioration of the water quality. At present, only small areas are still subject to annual floods (Deiller et al., 2001).

Due to the regulation works within the upper River Rhine valley, the ecosystem of the floodplain forest has been drastically modified. Most of the riparian forest has been disconnected from the river to prevent flooding (Deiller et al., 2001).

The disappearance of alluvial hardwood forest in the upper River Rhine valley through centuries of river regulation and flood protection measures is well documented. The importance of conserving the last remaining alluvial hardwood forest communities, which represent the highest level of organization in river plains at the end of succession, has been highlighted in the literature (Schnitzler, 1994). Despite efforts to protect the river valley habitat, the Rhine is the most damaged large fluvial system in Europe. Nowadays, most of the remaining hardwood forests along the Rhine are young, semi-natural and not flooded.

The wilderness (i.e. areas unaffected by men) in terms of forest communities has not existed in the River Rhine valley for approximately 800 years (Volk, 1998). The present riparian forest developed on land reclaimed from previous riverbeds. *Quercus* spp. (oak), *Salix* spp. (willow), *Carpinus betulus* (common hornbeam), *Populus* spp. (poplar), *Ulmus* spp. (elm), *Alnus* spp. (alder), *Fagus* spp. (beech) and various wild fruit trees were preferentially planted. Over thousands of years, timber production and settlement increases reduced the distribution of *Fraxinus* spp. (ash) to a few isolated pockets. Most of the current *Fraxinus* spp. (ash) forests are, however, man-made (Volk, 2002).

River regulation also led to increased shipping and subsequently to bank erosion in the upper valley (France and Germany) and subsequent sedimentation in the lower valley (The Netherlands). Therefore, the use of *Salix* spp. (willow) for riverbank protection as a sustainable ecological engineering technique has been proposed (Van Splunder et al., 1994).

Furthermore, the degradation (e.g., silting up due to lowering of the water table) of cut-off channels of the Rhine after river regulation has been observed (Klein et al., 1994). Therefore, the re-establishment of the connectivity of these channels with surface water flows to regain the most natural possible ecosystem functioning has often been proposed.

Because of large-scale decreases in the groundwater table due to river regulation, intensive peat mineralization has been observed (Kazda, 1995). The peat of the drained areas showed reductions in nitrogen, organic carbon, calcium and magnesium, but an increase of exchangeable aluminium together with a decrease of pH.

Moreover, a decline of *Carex acutiformis* (lesser pond-sedge; indicator for wet soil) was noted.

3.2.3 Restoration of Floods in Forests and Meadows

The technique of ecological flooding (i.e. artificial flooding to enhance the ecological habitat value of retention basins and wet meadows) assists in the regeneration of floodplain biotopes, and promotes a typical floodplain biotic community, and the readaption of the bioceonosis (characterises interacting organisms living together in a habitat) to a regular flooding regime. Floodplain biotopes of early succession in terms of their associated biotic community compositions are created predominantly through geomorphologic processes (Rood et al., 2005).

The restoration of floods in hardwood forests is likely to contribute to the preservation of the alluvial vegetation composition and structure, and even of the ecosystem. Moreover, biological growth and degradation processes can be increased by flooding. This should lead also to benefits to the associated water quality. Eutrophication is likely to be less of a problem, because of shading of open water surfaces due to tree canopies (Brettar et al., 2002; Trémolières et al., 1998). The recent creation of forested retention areas (i.e. polder) in, for example, the upper Rhine valley has been outlined in the literature (Brettar et al., 2002). These retention areas attenuate floods and improve water quality.

The rehabilitation of wet meadows has also received recent attention. In theory, great differences from a biological point of view are likely to be found between the flora of old and new meadows. Research has shown that typical plant species for flooded meadows were not observed even between 15 and 20 years after restoration attempts of wet meadows, which had been dominated by the originally sown plant species, and a high proportion of ruderal and arable species (Bissels et al., 2004). This can be explained only partly by the limited dispersal and recruitment of seeds, which usually restrain restoration successes. Findings have also shown that the restoration of natural flooding regimes alone is not sufficient to guarantee a high success rate (Sanchez-Perez et al., 1991).

3.2.4 Nutrient Control in the River Valley

Trees and other vegetation on floodplains are efficient in purifying flood water and shallow groundwater contaminated by diffuse pollution resulting from agricultural and urban activities (Sanchez-Perez et al., 1991). The efficiency of self-purification of nutrients on floodplains depends on the vegetation type and the textural and soil hydro-morphological characteristics. Regulating or at least limiting nutrients are nitrate, phosphate and potassium. The concentrations of nutrients are a function of season, vegetation and flood water quality.

Self-purification is difficult to measure, because stocks of nutrients (particularly N, P, K, Mg, Ca and S) in alluvial deciduous forest topsoil are much higher than corresponding annual inputs from the atmosphere (usually between 10 and 100 times) and from floods (usually between 5 and 30 times) according to Sanchez-Perez et al. (1993). Except for nitrogen, the nutrient concentrations in old and new flood banks are usually similar. Nutrient concentrations are likely to decrease with succession, and the groundwater below hardwood forests is generally low.

Relatively low concentrations of nutrients such as nitrate and phosphate in groundwater were observed under forests in comparison to meadows (Takatert et al., 1999). This observation can be explained by more efficient uptake by woody in comparison to herbaceous plants. Overall assessments show frequently that hydrological variations are much more important than those concerning substrate and type of vegetation.

As for flooded riparian forests, leaf litter production correlates positively with an indication of flood duration and high nutrient (e.g., nitrogen and phosphorus) concentrations within the leaves regardless of the nutrient content in the soil (Trémolières et al., 1998). It can be assumed that the 'extra nutrients' originate from eutrophic flood waters.

Furthermore, the creation of many small and shallow freshwater lakes in the lower River Rhine valley will result in the dominance of submerged macrophytes, which can be deliberately used to reduce nutrients. In contrast, large and frequently dry retention basins in the upper valley are less likely to be colonized by submerged vegetation (Van Geest et al., 2003).

The nitrogen variations in alluvial *Fraxinus* spp. (ash) and *Ulmus* spp. (elm) forests have been studied elsewhere (Sanchez-Perez et al., 1991). The researchers found out that these variations relate strongly to the seasonal cycle, forest typology and individual flood events. Infiltration of eutrophic flood waters in forests was identified as an important natural purification process for river water contaminated with nitrogen. Furthermore, nitrogen concentrations usually decrease with an increase of the age of the flooded habitats (Sanchez-Perez et al., 1993).

With the exception of forested riparian floodplains, high concentrations of nitrate and ammonia can usually be measured in shallow groundwater on current and former floodplains, which are used for agricultural purposes. Nitrate concentrations are a function of the soil type as well. Nitrate concentrations are usually low in silty and clayey soils. The frequencies and durations of flooding events impact also on the nitrate concentrations of loamy soils as shown elsewhere (Sanchez-Perez and Trémolières, 1997).

In comparison, dry forests have generally low leaf litter production rates with low nitrogen contents. Research has shown that flooding has no significant influence on the leaf content of potassium, magnesium and calcium regardless of the concentrations within the soil and flood waters (Trémolières et al., 1998).

The nitrate from alluvial groundwater influenced by agricultural and urban practices may be denitrified by certain bacteria communities in the presence of organic carbon (Sanchez-Perez et al., 2003) from river water flooding riparian wetlands of floodplains (Scholz, 2006; Scholz and Lee, 2005). Successful denitrification requires virtually oxygen-free conditions; for example, river water during the night (Venterik et al., 2003). Oxygen levels are usually too high during the day to promote denitrificaton. Moreover, denitrification is also regulated by nitrate diffusion from water into sediment, which usually only contains low concentrations of dissolved oxygen.

Furthermore, floodplains are sinks for nitrogen, and denitrification often drives nitrogen losses (Forshay and Stanley, 2005; Van der Lee et al., 2004). Floodplain restoration is therefore likely to enhance the overall retention of nitrogen and reduce nitrogen exports from upstream catchment runoff via regulated main channels and rivers such as the River Rhine. The retention of nitrogen within the floodplain sediment of the lower River Rhine is insignificant in comparison to the total dissolved load (i.e. predominantly nitrate) in the main river (Van der Lee et al., 2004).

Flooded new floodplains contain frequently more phosphorus in comparison to former (i.e. old) floodplains that have not been flooded for more than a century. Plants on floodplains release phosphorus during autumn via leaf fall (Weiss et al., 1991). Flooded sites offer the best nutrition in terms of phosphorus availability for genera such as *Quercus* spp. (oak) and *Cornus* spp. (dogwood). These woody plants have a specific adaptive behaviour in terms of phosphorus release (Weissschmitt and Trémolières, 1993).

Research has shown that phosphate variations in alluvial forests (planted with *Fraxinus* spp. (ash) and *Ulmus* spp. (elm)) relate strongly to the seasonal cycle, forest typology and specific flood events. Infiltration of eutrophic flood waters into such habitats has also been identified as an important natural purification process for waters contaminated with nutrients (Sanchez-Perez et al., 1993).

Except for forested floodplains, high concentrations of phosphate can be measured in shallow groundwater of current and former floodplains used by the agricultural industry (Sanchez-Perez and Trémolières, 1997). This can be explained with the significant phosphorus retention potential of sediment within such flooded floodplains (Van der Lee et al., 2004).

3.2.5 Vegetation Communities

Submerged macrophytes in temporary flooded lakes are usually either scarce or abundant (Van Geest et al., 2003). There is no correlation between macrophyte abundance and phosphorus in flooded lake water. The longer the flooded periods, the less macrophyte numbers can be observed. The macrophyte dominance decreases with surface area, depth and age of the lake. In comparison, Nymphaeid cover increases with lake depth and age. Helophytes increases with lake age, but decreases with lake size, and the presence of trees, cattle grazing, surface area, use of manure and lake depth.

In contrast to large lakes, small lakes with even relatively large depths are dominated by submerged macrophytes. But flooding is likely to have little influence on the vegetation within lakes located within the floodplain, if the lakes are not flooded within the growth season, and if currents are low (Van Geest et al., 2003).

Furthermore, the vegetation composition is strongly correlated with the age of the lake and the occurrence of drawdown periods, and only weakly correlated with the amount of time the river was flooding (Van Geest et al., 2005). Particularly concerning older lakes, drawdown during periods of low river flow is prevented due to the accumulation of silt, which seals the lake sediment. This process is a major driving force for aquatic vegetation succession. The main transition is a shift from desiccation-tolerant species to desiccation-sensitive species as the lakes mature. Low water levels are likely to enhance the ecological status of floodplain lakes.

After the Rhine was straightened and canalized, a subsequent change in vegetation was apparent. For example, the moss flora in flooded forested floodplains has been assessed (Van der Poorten et al., 1995). The researchers looked at bryological dynamics and assessed vegetation community changes. Bryophyte flora that is like that of good quality waters in the uplands has disappeared from the main channel but can still be found in some of the groundwater-fed cut-off channels (Van der Poorten and Klein, 1999). Therefore, the researchers promote the protection of these areas in future flood management programs.

Furthermore, the regulation of the Rhine allowed the appearance and subsequent spread of strict hydrophytes that are predominantly tolerant of pollutants. In comparison, the species communities indicative of variable water levels decreased (Van der Poorten and Klein, 1999).

The tree density of the areas, which were not flooded, increased due to an increase in sapling density than stem density. Sapling density increased between two and three times depending on sites. But density increases correlated positively with an increase of time duration with respect to the last flood (Trémolières et al., 1998).

The disturbances caused by floods usually allow for a more equal coexistence of several species. In contrast, the areas unaffected by flooding are often dominated by only a small number of woody species that are representative for the regeneration phase (Deiller et al., 2001).

Flooding affects the abundance and distribution of plants, because many plant species are damaged or even killed by flooding events due to the associated oxygen depletion within the plants. Shoot elongation to restore the contact of the leaves with the atmosphere above the water surface is therefore stimulated by flooding. This improves the inward diffusion of oxygen and the rate of photosynthesis. Therefore, the capacity to elongate is a selective trait of plants in flood-prone environments with a shallow water depth for prolonged periods of time. The absence of this phenomenon is often an indication of deep floods of short duration (Voeseneck et al., 2004).

Finally, the availability of underwater light, which is a function of water depth and suspended solids load, determines the distribution of riparian plant species during flooding. Extreme floods carrying high loads of suspended solids are likely to change the plant distribution pattern for the following years (Vervuren et al., 2003).

Serratula tinctoria (saw-wort) is an indicator plant indicative of endangered species-rich floodplain meadows. The presence of this species correlates positively with bare soil and bryophytes. In comparison, the Ellenberg-N value of the established vegetation correlates negatively with the density and proportion of seedlings. Nutrient-rich sites in the River Rhine valley, which are usually mown in June, stimulate the establishment of *S. tinctoria* (saw-wort) according to Bissels et al. (2004).

Researchers have established the link between macrophyte community structure and diversity with habitat variability, and in turn with the hydrological conditions and fluvial dynamics of cut-off river channels. They claim that it is necessary to preserve the fluvial dynamics to maintain the aoristic and phytosociological diversity of these endangered ecosystems (Robach et al., 1997).

The Charophyte richness and abundance usually decreases when nutrient concentrations increase. The ammonium concentration of the water is negatively correlated to the species richness. *Chara globularis* (fragile stonewort) amongst other species occurred in cut-off channels characterized by low levels of flood disturbances. In comparison, *Chara vulgaris* (common stonewort) and *Nitella confervacea* (least stonewort) are usually present in cut-off channels that are frequently disturbed by floods. In contrast to *Nitella* spp. (stonewort), which has a limited distribution and high variability in abundance over time, *Chara major* (sometimes called *waterdrieblad*) should not be considered as a pioneer fugitive species, because this species is able to colonize cut-off channels that were undisturbed for more than 15 years (Bornette and Arens, 2002).

Finally, research indicates that large gradients of connection and disturbance of networks of cut-off channels in the floodplains induce high diversities within plant communities. The functioning of habitat can subsequently be predicted with the help of biological traits (Trémolières, 2004).

The increase of flood duration correlates positively with the increase of seed density within soil samples of flooded meadows. Flooding has, however, no impact on the proportion of species not occurring in the vegetation composition and the overall species richness. An increase in disturbance indicators due to flooding is responsible for the corresponding increase in seed density within the aboveground vegetation.

It is likely that a long-term persistent seed bank will be established in a flooded area. The aboveground vegetation composition within meadows and their corresponding adjacent fields and the associated land use management strategies have a greater influence on the seed bank composition than floods (Hölzel and Otte, 2001).

Alluvial grassland communities in flooded floodplains of the upper Rhine are usually ordered along an elevated gradient of increasing flood frequency. But no general relationship between species-richness and productivity has been found in the literature (Donath et al., 2004).

Furthermore, the seed bank composition variations of various flooded meadows of the northern stretch of the upper Rhine have been studied (Hölzel and Otte, 2004). Soil seed bank depletion was attributed, for example, to post-flood germination flushes due to very favourable growth conditions during the drawdown period observed during early summer floods. It follows that the promotion of persistent seed banks as an example of effective management strategies is important to avoid seed bank depletion through unforeseen variability of hydraulic regimes.

Finally, ephemeral wetland species can be conserved by allowing irregular flooding of arable fields, avoiding of engineered drainage and landfill, and ploughing of fields as late as realistically possible. Moreover, the successful emergence of ephemeral wetland vegetation after flooding depends on the presence of a strong corresponding seed bank in the topsoil (Bissels et al., 2005).

Neither regular flooding nor haymaking, autumnal grazing or wind dispersal had an impact on the recolonization of newly created wet meadows with typical plant species for alluvial meadows (Donath et al., 2003). The vegetation of old meadows is likely to be much more differentiated along prevailing environmental gradients in comparison to the vegetation composition of new meadows. Restoration successes in terms of the reestablishment of alluvial meadows are usually rare; that is, indicator organisms for these flooded meadows are typical for regularly disturbed ruderal and arable habitats.

Concerning the artificial creation of flooded meadows, limited seed dispersal has been identified as the main reason for a slow restoration success rate with respect to plant communities typical for traditionally flooded meadows. A potentially successful method in speeding up this process is the removal of nutrient-rich and seed-rich topsoil, and its replacement with mown plant material containing the seeds of desired plant species. This method is likely to be most successful in regularly flooded areas, because of the suppression of competitors as well as the creation of ideal moisture conditions for most seedlings (Hölzel and Otte, 2003; Volk, 1998).

The success of this method is also a function of the quality of the transferred mown plant matter, and the degree of competition between new plants and plants already established in the vicinity of the targeted site (Hölzel and Otte, 2003). Moreover, a relatively low-cost-to-benefit ratio is another required variable that would benefit such a strategy.

3.2.6 Plankton and Fish

The study of plankton and fish communities is particularly relevant for the lower River Rhine valley because of the large abundance of shallow freshwater lakes. The plankton species richness depends on the vegetation type of the floodplains (Van den Brink et al., 1994a). The composition of both phytoplankton and zooplankton communities on flooded floodplains is a function of the local hydrology, habitat type and available resources of nutrients. The nutrient status depends on resources that were already available before flooding, and new ones that are a function of the nitrogen and phosphorus content of the eutrophic flood water from the main channel. This is particularly the case for lowland rivers (e.g., lower River Rhine), which are usually more polluted with nutrients than their corresponding upstream stretches (Van den Brink et al., 1994a).

Frequently flooded floodplains are associated with high densities of filter feeders. This is likely to be due to the high plankton biomass and insufficiently developed aquatic vegetation. In comparison, shredders are most abundant in watercourses unaffected from floods. These waters such as ponds and lakes have usually a mature aquatic vegetation and low biomass content with respect to phytoplankton (Van den Brink et al., 1994b).

Furthermore, hydrological parameters such as flood pulse duration and water level are important for the management of flooded floodplains. In comparison, watercourses rarely influenced by floods are dependent on vegetation coverage and water temperature (Grift et al., 2003).

Rehabilitated water bodies and retention basins within floodplains of the River Rhine provide, for example, more suitable habitat for 0-group fish than groyne fields. The total fish density increases with decreasing water flow rates, whereas the rheophilic species number decreases. In general, flow velocity and water depth are the most important variables when considering habitat for fish production (Grift et al., 2003).

Finally, flooded floodplains provide an ideal habitat for the larvae of fish. This is especially true where the floodplains have a complex structure (e.g., irregular embankment and shoreline with flat slopes), and if they have varying water flow rates and large proportions of flooded terrestrial vegetation during the growing seasons (Yoshida and Dittrich, 2002).

3.2.7 Flow and Biological Simulations

A one-dimensional unsteady-state flow simulation model for the upper River Rhine valley was developed elsewhere (Yoshida and Dittrich, 2002). The model takes the roughness of the channel and the corresponding floodplains into account. The floodplains have the greatest impact on the model output. It is surprising that the type of riparian vegetation was only of minor importance. But a comparison between measured and predicted hydrograph samples highlighted errors of up to just <10%.

Hydrological modelling indicated that only the first 300 m of a floodplain adjacent to the River Rhine functions as a hydrological buffer, which increases the denitrification potential of the floodplain. This strip of land increases in width if the river has meanders (i.e. an unregulated river) according to Weng et al. (2003).

An unsteady model, which estimates velocities, friction factors and the components of discharge in the corresponding main channel and partly vegetated floodplains, has been proposed (Helmio, 2005). This model showed that a significant amount of discharge was transported via the floodplains despite vegetation in some river cross-sections particularly through remarkably high flows.

Finally, the BIO-SAFE model, which quantifies the relevance of species and ecotypes, as well as channel and floodplain characteristics for sustainable river management, has been developed. This model can be used to assess the biodiversity for various river design and management options in the planning process. Tests have shown that qualitative and quantitative information can be obtained with respect to the biodiversity compliance of river designs with international guidelines (De Nooij et al., 2004).

3.2.8 Classification Concepts

Classifications of natural and engineered habitats are important to aid communication between stakeholders. For example, trophic levels and various levels of biological organization can be used as techniques to classify wetlands according to their degradation status. The following rapid assessment techniques have been proposed as early warning key indicators of wetland degradation (Van Dam et al., 1998): (a) phytoplankton indicators for toxicity assessments, (b) rapid toxicity bioassays using invertebrates and vertebrates, (c) biomarkers as early warning indicators for specific pollutants and (d) rapid methods of monitoring aquatic community assemblages. These indicators could also be used as variables for a wetland classification scheme.

Furthermore, floodplains could be classified according to plant species (i.e. the corresponding life-form, place in the successional stage, phenology and abundance), which usually segregate according to their habitat (Schnitzler et al., 1992): (a) geochemistry such as the lime content and texture, (b) water stress including water level and (c) fluvial dynamics such as flooding.

In general, vegetation, soil and geomorphologic data can be used to classify floodplains according to their degree of succession. The corresponding classes or groups could be applied as a basis for rehabilitation plans (Schoor, 1994).

The hierarchical classification of factors responsible for the distribution of vegetation has been proposed (Carbiener et al., 1995). The following list shows the levels in order of decreasing importance: mineralization, trophy status (particularly phosphate and ammonia), rheology, sedimentology and morphology. Moreover, aquatic macrophyte communities can be used as biological indicators to classify floodplain areas according to their spatial-temporal trophic quality status. The reasoning behind this approach is that biological indicators respond well to river-groundwater interchanges and have a 'memory' of hydrological events such as floods.

There is a strong relationship between the quotient of measured nutrient load versus the estimated sum of point and diffuse emissions of nutrients and the aerial runoff. This correlation could be used for the formulation of a parameter that might be helpful for the classification of floodplains (Behrendt, 1996).

The Rhine basin is divided into eight different action areas with corresponding action plans according to the following criteria (Bohm et al., 2004): (a) prevailing flood danger, (b) geophysical situation and (c) potential effects of retention measures. Based on this concept, priority zones to prevent any risk increases, and reserve zones, which demand precautionary measures concerning construction only, have been proposed (Bohm et al., 2004).

Finally, a classification of floodplains based on their compartmentalization is also possible. Such a classification would need to be dynamic, because dykes of secondary importance are frequently either removed or restored without trans-national agreements being arranged. Neither a simple removal nor a total restoration of secondary dykes would reduce the risks of overall flood damage, but that this could be achieved by a strategic compartment plan, which would certainly be a useful tool for a multi-purpose floodplain classification scheme at the same time (Alkema and Middelkoop, 2005).

Forests on floodplains of the River Rhine, for example, can be classified into the following four groups (Schnitzler, 1995): (a) *Salici-Populetum nigrae* (softwood) dominated by *Salix alba* (white willow) and *Populus nigra* (black poplar); (b) *Ligustro-Populetum nigrae* (softwood) dominated by *Ligustrum* spp. (privet) and *Populus nigra* (black poplar); (c) *Fraxina-Populetum albae* (softwood) dominated by *Fraxinus* spp. (ash) and *Populetum albae* (white poplar) and (d) *Querco-Ulmetum minoris* (hardwood) dominated by *Quercus* spp. (oak) and *Ulmus minor* (golden elm).

Salix spp. (willow) and *Populus* spp. (poplar) are good indictors for the presence of virtual bare sediment, and of forests <100 years of age depending on hydrological and edaphic gradients. Allogenic processes of flooding are fundamental in determining the temporal and spatial patterns of species. With an increase in succession (i.e. hardwood replaces softwood), the forest species and community diversity increases (Schnitzler, 1995).

Furthermore, the alluvial hardwood forest (*Querco-Ulmetum minoris*) in the upper Rhine valley is principally composed of three vegetation units (Trémolières et al., 1998): (a) unprotected wet riparian forest, which is frequently flooded by calm waters; (b) dry riparian forest, which was last flooded before river canalization (about 1967), and (c) dry riparian forest, which was last flooded before river straightening and embankment work (about 1867).

The findings of conventional land use mapping were compared with those of the maximum likelihood classification method using Landsat-TM data (Heinzmann and Zollinger, 1995). The corresponding groups of both classification systems were revised to achieve common classification groups; that is, cropland, forest, pasture, urban area and vineyard. The classes forest and pasture of both systems corresponded well with each other, regardless of the methodologies used and different scales. Correlations between land use type and geomorphologic variables compared also well with each other. But the data for the classes cropland, urban area and vineyard did not correspond well with each other.

Previous sub-chapters have identified gaps in scientific knowledge. These gaps need to be filled before someone can build a solid foundation for the classification of retention basins. For example, it is unclear to what extent flooding may cause damage to different plant species in potentially economically and ecologically important forest and meadow ecosystems. The processes responsible for the occurrence of damages to the vegetation due to frequent water level variations require further research. Therefore, the age and corresponding ecological developmental stage of any flood retention basin requires assessment.

The benefit and shortcomings for vegetation (e.g., increase of production) and groundwater quality by improvement of biogeochemical cycling (e.g., reduction of nutrient transfer) require further assessment. This should be integrated into environmental impact assessments of water retention areas.

Flood retention basin classification and decision support tools for both practitioners and experts should be developed. For example, a hierarchical classification of flood water retention basins according their hydraulic, vegetation and water quality characteristics should be undertaken. Retention basins should be connected to flooding regime variables (e.g., water height, duration and frequency of flooding, seasonal dependencies and flow velocity of water) for classification purposes.

There are many different retention basins in Germany including examples for forested ones, which are flooded during extreme rainfall events. The great diversity of basins is apparent, and it is therefore surprising that retention basins have not yet been classified to aid communication among different stakeholders.

3.2.9 Conclusions and Recommendations

Historical river regulation strategies for the River Rhine to control flooding risks have been associated with serious ecological problems, including the disappearance of ecosystems such as flooded meadows and alluvial forests. Nowadays, flood retention areas, which should be located as far upstream as possible, are used to control flooding, improve the water quality and to enhance habitats.

The response of vegetation to ecological flooding has been discussed. There are serious problems associated with the restoration of floods in forests and alluvial meadows because typical vegetation communities are difficult to establish.

Flooded floodplains play a key role in the control of nutrients such as nitrogen and phosphorus. These nutrients can be successfully reduced by the floodplain vegetation and the active biomass within flooded surfaces. Flood retention basins can therefore function also as important water quality control techniques.

Section 3.2 has also highlighted problems with vegetation community changes. A serious loss of typical indicator species for succession of flooded floodplains and cut-off channels has been identified. The depletion of seed banks of flooded meadows has also been discussed in this context. Novel restoration strategies to promote endangered plants have been proposed and tested with limited success.

Changes in the plankton and fish communities of floodplains have been highlighted. Some species can be used successfully as indicator organisms for change. But plankton species are difficult to detect and are therefore only of limited relevance for floodplain management tools.

Hydraulic simulation modelling can be helpful to assist engineers and planners in river management and flood forecasting. But floodplain roughness is difficult to estimate accurately. This dynamic variable would need to be approximated, if it should be applied in any decision support tools.

Various concepts for classifying and managing floodplains and retention basins have been discussed. There is a gap in knowledge and understanding of characterizing floodplains based on a holistic and multi-disciplinary approach.

3.3 CLASSIFICATION OF SUSTAINABLE FLOOD RETENTION BASINS

3.3.1 Introduction

3.3.1.1 Climate Change, Flooding Risk and Landscape Planning

In the northern part of Europe, the total annual amount of rainfall significantly increased in the last century, which cannot be explained by the natural variability of the climate (Scholz, 2006). In accordance with Bardossy and Caspary (1990), a statistically significant increase of precipitation has been observed, particularly during winter and spring. This finding is supported by the observation that the mean flow rate in the largest German rivers, for example, has increased by approximately 25% since 1930. Further changes of precipitation patterns are predicted independently from different climatic models, which will lead to more precipitation (predominantly rain) during winter and spring, causing an increased risk of flooding in Central and Northern Europe (Scholz, 2006).

Pfister et al. (2004) claim that land use change such as urbanization in headwater areas can lead to flooding during storms. Future changes in peak flow will depend on the variability of extreme rainfall in combination with land use management. High urban runoff and diffuse pollution from agricultural sources are increasingly becoming a serious problem for the management of retention basins. These problems need to be addressed by civil and environmental engineers, and urban and landscape planners.

Hooijer et al. (2004) have outlined the European Union-funded IRMA-SPONGE project that aimed to assess the impact of flood risk reduction measures, and changes to land use and climate change on the international River Rhine catchment planning process. The project focused specifically on climate change, land use planning and management (particularly retention basins) and ecological quality of rivers and floodplains.

Most recently, Nijland (2005) has proposed a sustainable floodplain development strategy, which is based on a project funded by the European Interregional IIIb Programme. The aim is to promote the sustainable development of floodplains in the River Rhine valley by finding a compromise between flood prevention, and the protection of nature and the natural landscape. However, it is too early to assess any long-term impact of revised landscape design on the river runoff.

3.3.1.2 General Classification of Floodplains and Retention Basins

Schnitzler et al. (1992) indicated that floodplains could be classified according to plant species (i.e. the corresponding life-form, place in the successional stage, phenology and abundance), which usually segregate according to their habitat: (a) geochemistry such as the lime content and texture, (b) water stress including the depth of water and (c) fluvial dynamics such as flooding.

Moreover, vegetation, soil and geomorphologic data can be used to classify floodplains and possibly also SFRBs according to their degree of succession. The corresponding classes or groups could be used as a basis for rehabilitation strategies (Schoor, 1994).

Carbiener et al. (1995) proposed a hierarchical classification of factors responsible for the distribution of vegetation. The following list shows the levels in order of decreasing importance: water mineralization, trophic status (particularly phosphate and ammonia), rheology, sedimentology and morphology. Moreover, aquatic macrophyte communities can be used as bio-indicators to classify floodplain areas and potentially also SFRB according to their spatial and temporal trophic quality status. The reasoning behind this approach is that bio-indicators respond well to river-groundwater interchanges and have a 'memory' of hydrological events such as floods.

Haase (2003) examined indicators such as the expansion of loam due to flooding, groundwater table, relief and land use to characterize floodplain functionality in urban areas. However, he used only one case study to test his classification variables. Unless this case study is typical for a vast majority of cases (unlikely but not proven), the research is only incremental.

Furthermore, the Rhine basin, for example, is divided into eight different action areas with corresponding action plans to combat floods according to the following criteria (Bohm et al., 2004): prevailing flood danger, geophysical situation and potential effects of retention measures. Priority zones to prevent any risk increases, and reserve zones, which demand precautionary measures concerning construction only, have been recommended by Bohm et al. (2004). The level of risk is reflected in the retention basin location and selection of the corresponding engineering measures. This approach is logical and important for flood risk and landscape managers, but offers little help for civil and environmental engineers, and landscape and urban planners in characterizing different types of SFRB with the aim of identifying their individual function and defining corresponding policies and detailed guidelines for their management.

A classification of floodplains based on their compartmentalization is also possible. However, such a classification would need to be dynamic because dykes of secondary importance (predominantly located in lowland areas) are frequently being either removed or restored without trans-federal and/or trans-national agreements being arranged. Alkema and Middelkoop (2005) claim that neither a simple removal nor total restoration of secondary dykes would reduce the risk of overall flood damage, but that this could be achieved by a strategic compartment plan, which would certainly be a useful tool for a multi-purpose floodplain classification scheme at the same time. Any classification of retention basins needs to be simple, transparent and dynamic, and should consider the entire catchment.

Furthermore, any classification system or model should be tested on a real data set containing enough case studies (i.e. SFRB). Different degrees of system complexity and their impact on the correctness of results need to be judged (e.g., Van Lienden and Lund, 2004).

3.3.1.3 Need for a New Classification System

German flood retention basin guidelines describe different components (e.g., dam height and basin volume) of basins but fall short of classifying flood retention basins according to types (Gesellschaft zur Förderung der Abwassertechnik (ATV-DVWK), 2001). Virtually all retention basins featuring in official German databases are classified as flood retention basins including drinking water reservoirs, basins used for hydropower generation, off-line polders for active river regulation, covered reservoirs, fishponds and even some traditional village ponds. Such a loose grouping of different retention basin types into one category frequently leads to legal disputes and misunderstandings between practitioners and the public. A classification scheme for SFRB is therefore timely and urgently required to assist communication.

Moreover, a rapid classification methodology is relevant for stakeholders such as local authorities and non-governmental organizations, and it will greatly assist them with urban and landscape planning. For example, an SFRB that was initially built for flood protection purposes only might have become a site of scientific interest where the flora and fauna should now be under legal protection. Finally, the mathematical relationships among the classification variables will help engineers and scientists to identify cause and effect phenomena, especially, for example, when an SFRB requires upgrading.

The aim of Section 3.3 is to define sub-classes (i.e. types) for SFRB. The key objectives are (a) to determine all relevant variables, particularly the key classification ones; (b) to assess the certainty of accurate determination associated with each classification variable; (c) to determine weightings for all classification variables according to their relative importance depending on various basin applications; (d) to provide civil and environmental engineers with a rapid expert classification system; (e) to apply the classification system on a large and detailed case study data set and (f) to illustrate and discuss examples of the most relevant SFRB types for civil and environmental engineers. Scholz (2007a, 2007b) provided the basis for this chapter.

3.3.2 Expert System for Classifying Basins

3.3.2.1 Classification Variables and Corresponding Certainty

The most important classification variables (i.e. 34 in total) for various types of SFRB to control runoff during floods in a temperate climate were identified and grouped based on a literature review, various recent site visits in Germany, UK, Ireland and Denmark and group discussions among British, German, French, Irish and U.S. engineers, scientists and landscape and urban planners. The identified classification variables and their corresponding identification numbers are as follows (Tables 3.5 and 3.6): Naturalness (1), Dam Height (2), Dam Length (3), Outlet Flexibility (4), Animal Passage (5), Floodplain Elevation (6), Connectivity (7), Wetness (8), Channel System (9), Flooding Depth (10), Flooding Duration (11), Flood Frequency (12), Slope (13), Velocity (14), Embankment Length (15), Flood Water Volume (16), Water Surface Area (17), Rainfall (18), Drainage (19), Clay (20), Season (21), Elevation (22), Vegetation (23), Algae (24), Pollution (25), Sediment Depth (26), Sediment Composition (27), Flotsam (28), Catchment Size (29), Urban (30), Arable (31), Pasture (32), Forest (33) and Groundwater (34). The qualitative and quantitative characterization (i.e. grouping in bins) of these variables for flood retention basins regardless of their purpose is shown in Table 3.5. A bin number (i.e. point between 1 and 5) and a numerical value were assigned to each quantitative variable. In contrast, qualitative variables (i.e. 5, 21, 23, 25 and 28) received points based on their corresponding bin number (Table 3.5).

Furthermore, the experienced and trained user should be able to estimate most variables during a desktop study, which should take <50 minutes and a site visit of <40 minutes; that is, total investigation time of <1.5 hours. A certainty value (e.g., very low = 0%; medium = 50% and very high = 100%) should be assigned to each variable to reflect the estimated risk of selecting a false value. The assessment should always be carried out by the same team comprising at least one engineer and one scientist to reduce the risk of bias and inconsistency. In this case study, the author (chartered engineer, chartered scientist and chartered environmentalist) was joined, for example, by a scientist.

3.3.2.2 Selection of Weightings

Table 3.6 summarizes the classification variables and associated weightings with a bias toward hydraulics, sustainable drainage, environmental protection, recreational activities and landscape aesthetics, respectively. The use of sums of weightings implied the assumption of additivity of variable importance, which is obviously a simplification of reality and does not hold true for all scenarios. This is a simplification of a complex reality and there are obviously exceptions, which the experienced user needs to identify. The scores associated with the weightings should be seen as recommendations, which could be changed by the expert user depending on regional and national differences reflected in federal, national and/or international guidelines.

The weightings were chosen based on a literature review, and comprehensive discussions between civil, environmental and water engineers, environmental scientists, sociologists and landscape managers. Representatives from different subject disciplines have provided the author with suggested relative weightings based on their own experience. The findings were then collected, and mathematical relationships were identified. For example, the arrangement of weightings for individual classification variables were based on linear and second-order polynomial relationships between opposites such as deep versus shallow, or presence or

TABLE 3.5
Variable Determination Template: Qualification and Quantification of 34 Classification Variables for Sustainable Flood Retention Basins to Control Runoff During Typical Floods

No.	Bin 1	Bin 2	Bin 3	Bin 4	Bin 5
1	Old, permanently filled and man-made structure with full engineering control; that is, similar to a traditional flood retention basin (>60%)	Mostly permanently filled and man-made traditional structure (>50%–60%) with some natural elements, but automatically controlled	Engineered structure (40%–50%) fitted well into the natural landscape with passive control	Aesthetically pleasing natural and occasionally dry formation with some engineered (30% to <40%) features, but relatively natural channel base	Almost entirely natural and occasionally dry formation including a wide natural channel base (<30% engineered; e.g., inlet and outlet)
2	Relatively high dam (>10 m) as the dominant feature of the engineered structure; local climate change	Part of engineered structure features a dam (height between >8 and 10 m)	Small dam (height between 5 and 8 m) and natural containment combined	Ecologically engineered and aesthetically pleasing structure with a small dam (height between 2 and <5 m)	Insignificant dam (<2 m embankment height) but natural containment of water (e.g., in a valley)
3	Very long dam (>2,000 m)	Long dam (1,000–2,000 m)	Normal dam length (>100 to <1,000 m)	Short dam (50–100 m)	Very short dam (<50 m)
4	Combined or separate very flexible outlets (>90% flexibility)	Combined or separate flexible outlets (>70%–90% flexibility)	Combined and partly flexible outlets (30%–70% flexibility)	Separate outlets; at least one outflow is fixed (10% to <30% flexibility)	Separate and fixed baseflow and/or flood flow outlets (<10% flexibility)
5	Very problematic animal passage	Problematic animal passage	Standard animal passage	Easy animal passage	Very easy animal passage
6	Retention basin predominantly elevated (>3 m; e.g., valley in the highlands with normal slopes)	Basin elevated (>2–3 m) and not well integrated into the landscape	Basin elevated (1.5–2 m), but well-integrated into the landscape	Retention basin partly elevated (1 to <1.5 m) and structure perfectly integrated into the landscape	Retention basin not elevated (<1 m) and virtually level with the river; for example, lowland stream

(Continued)

TABLE 3.5 (*Continued*)
Variable Determination Template: Qualification and Quantification of 34 Classification Variables for Sustainable Flood Retention Basins to Control Runoff During Typical Floods

No.	Bin 1	Bin 2	Bin 3	Bin 4	Bin 5
7	Basin directly connected (<5 m) with the watercourse (i.e. basin and stream are virtually on-line)	Short connectivity (5 to <10 m), but well integrated into the landscape	Short connectivity (10–20 m), but elements are clearly separated from each other	Long connectivity (>20–50 m)	Very long connectivity (i.e. separate structures are far away (>50 m) from each other)
8	Very wet basin (virtually entirely submerged; i.e. >70% wet area); similar to a pond	Wet basin with minor natural components (>60%–70% wet area)	Partly wet basin with minor natural components (40%–60% wet area)	Dry basin or pond, but with natural components (30% to <40% wet area)	Very dry area with natural components; that is, dry basin or pond (<30% wet area)
9	Permanently flooded main channel (i.e. on-line basin with no by-pass); that is, >95% of water volume	Permanently flooded main channel (>90%–95% of water volume) and very occasionally flooded by-pass	Partly flooded main channel (80%–90% water volume) and occasionally partly flooded by-pass	Main channel (70% to <80% water volume) and partly flooded by-pass	Main channel (<70% water volume) and by-pass (i.e. off-line basin) taking most of the flood water
10	Mostly very deep flooding depth as in reservoirs (>3 m); virtually, permanently wet	Deep flooding depth (>2–3 m)	Partly flooded; normal flooding depth (1–2 m)	Partly flooded; shallow flooding depth (0.5 to <1 m)	Only occasionally and partly flooded; shallow flooding depth (<0.5 m); virtually, a dry basin
11	Very prolonged flooding event due to intense and long storms (>70 d); virtually, a wet basin	Prolonged flooding event due to long storms (>50–70 d)	Occasional flooding event (30–50 d)	Mostly short flooding event (7 to <30 d)	Mostly very short flooding events (<7 d); virtually, a permanently dry basin or pond

(Continued)

TABLE 3.5 (*Continued*)

Variable Determination Template: Qualification and Quantification of 34 Classification Variables for Sustainable Flood Retention Basins to Control Runoff During Typical Floods

No.	Bin 1	Bin 2	Bin 3	Bin 4	Bin 5
12	Very high flood frequency due to high storm frequencies (>15 per a) leads to landscape destruction	High flood frequency due to high storm frequencies (>12–15 per a) leads to landscape damage	Normal flood frequency (9–12 per a) controlled by management strategies	Low flood frequency, which does not lead to any not aesthetically sights (5 to <9 per a)	Very low flood frequency (<5 per a); rarely monitored
13	Very low gradient (<1%)	Low gradient (1% to <3%)	Normal gradient (3%–6%)	High gradient (>6%–9%)	Very high gradient (>9%)
14	Mostly very high velocity due to intense storms (>80 cm/s)	High velocity (>60–80 cm/s)	Variable velocity (40–60 cm^{-1})	Low velocity (25 to <40 cm/s)	Mostly very low velocity, but never stagnant (<25 cm/s)
15	Very short embankment (<200 m)	Short embankment (200 to <350 m)	Normal embankment length (350–500 m)	Aesthetically pleasing long embankment (>500–700 m)	Very long embankment (>700 m); very variable topography
16	Very high volume (>10,00,000 m^3)	High volume (>100,000–10,00,000 m^3)	Standard volume (50,000–100,000 m^3)	Low volume (5,000 to <50,000 m^3)	Very low volume (<5,000 m^3)
17	Very large surface area reflecting a very large basin (>100,000 m^2)	Large surface area reflecting a large catchment (>20,000 to 100,000 m^2)	Normal surface area covered with water (10,000–20.000 m^2)	Small aesthetically pleasing surface area (2,000 to <10,000 m^2)	Very small surface area reflecting a very small catchment area (<2,000 m^2)
18	Very high precipitation due to intense and long storms (>1,500 mm/a) in the catchment; highland climate	High precipitation due to long storms (>1,200–1,500 mm/a) in the catchment; partly highland climate	Normal precipitation (800–1,200 mm/a) in the catchment; wet temperate climate	Relatively low precipitation (600 to <800 mm/a) in the catchment; temperate lowland climate	Very low precipitation (<600 mm/a) in the catchment; sometimes semi-arid climate

(*Continued*)

TABLE 3.5 (*Continued*)
Variable Determination Template: Qualification and Quantification of 34 Classification Variables for Sustainable Flood Retention Basins to Control Runoff During Typical Floods

No.	Bin 1	Bin 2	Bin 3	Bin 4	Bin 5
19	Very well-drained with a high infiltration rate (>50 cm/d)	Well-drained with a relatively high infiltration rate (>30–50 cm/d)	Drained with a moderate infiltration rate (10–30 cm/d)	Not well-drained with a low infiltration rate (2 to <10 cm/d)	Mostly not drained (<2 cm/d); usually a boggy area
20	Very high proportion of clay present (>40%)	High proportion of clay (>20%–40%)	Variable proportions of clay present (10%–20%)	Some clay present (5% to <10%)	Clay mostly absent (<5%)
21	Very weakly seasonally influenced (i.e. insignificant seasonal variation)	Weakly seasonally influenced	Moderately seasonally influenced	Mostly seasonally influenced (e.g., area around Freiburg in Baden, Germany)	Strongly seasonally influenced (e.g., typical temperate climate)
22	Very high elevation (>500 m); for example, highlands	High elevation (>300–500 m)	Typical elevation (150–300 m)	Low elevation (50 to <150 m)	Very low elevation (<50 m); for example, marshes
23	Predominantly not vegetated area (e.g., only short, mowed lawn)	Partly not vegetated area (e.g., some grassland or submerged aquatic vegetation)	Vegetated area with low roughness (e.g., bushes and some reeds); early succession of plants	Vegetated area with moderate roughness (e.g., predominantly reeds, bushes and some trees)	Vegetated area with high roughness (e.g., reeds and mature trees such as willows)
24	Virtually no algae present (<5% cover)	Low numbers of algae detected (5% to <10% cover)	Normal algal presence (10%–30% cover)	Dominant algal bloom (>30%–90% cover)	Very dominant algal bloom (>90% cover)
25	Polluted (high organic and inorganic solids content; occasionally sewage)	Occasionally polluted with solids and/or organics	Occasionally partly polluted	Minor occasional pollution; environment may still be rich in species	Rarely minor pollution; environment may be rich in species

(*Continued*)

TABLE 3.5 (*Continued*)

Variable Determination Template: Qualification and Quantification of 34 Classification Variables for Sustainable Flood Retention Basins to Control Runoff During Typical Floods

No.	Bin 1	Bin 2	Bin 3	Bin 4	Bin 5
26	Occasional presence of sediment (<3 cm) due to passive and/or occasional basin management	Occasional presence of sediment (3 to <6 cm) due to passive and/or occasional basin management	Aesthetically pleasing sediment layer (6–8 cm); occasional basin management	Deep sediment layer (>8–10 cm) present (i.e. mature system)	Very deep and potentially natural sediment layer (>10 cm) present (i.e. mature and stable system)
27	Deep inorganic sediment (>10 cm) requiring regular removal	Inorganic sediment layer (>5–10 cm), which is sometimes managed	Mixed sediment layer (≤5 cm of each type)	Organic and predominantly man-made sediment layer (>5–10 cm)	Deep organic and potentially natural sediment layer (>10 cm)
28	Virtually no flotsam	Occasional presence of flotsam	Typical presence of flotsam	Partly covered with flotsam; regular removal required	Dominant cover with flotsam causing problems; immediate removal required
29	Very large catchment (>50 km^2)	Large catchment (>20–50 km^2)	Standard catchment size (10–20 km^2)	Small catchment (0.5 to <10 km^2)	Very small catchment (<0.5 km^2)
30	Very high urban catchment proportion (>50%); for example, city or town	High urban catchment proportion (>30%–50%); for example, town or village	Significant urban catchment proportion (20%–30%); for example, village	Insignificant urban catchment proportion (10% to <20%); for example, small village	Not urbanized catchment (<10%); for example, only individual houses and/or farms
31	Not agriculturally used; some isolated arable fields (<7%)	Insignificant agriculture; arable fields (7% to <10%)	Significant arable land use (10%–15%); typical for Central Europe	Highly intensively used (mostly cash crops) agricultural catchment (>15%–20%)	Very highly intensively used (mostly cash crops) agricultural catchment (>20%)

(*Continued*)

TABLE 3.5 (*Continued*)
Variable Determination Template: Qualification and Quantification of 34 Classification Variables for Sustainable Flood Retention Basins to Control Runoff During Typical Floods

No.	Bin 1	Bin 2	Bin 3	Bin 4	Bin 5
32	Not agriculturally used; some isolated pastures used for grazing (<7%)	Insignificant agriculture; predominantly pastures used for grazing (7% to <10%)	Significant pasture (10%–15%); typical for central Europe	Highly intensively pasture catchment proportion (>15%–20%)	Very highly intensively used pasture (>20%)
33	Not intensively used for forestry purposes (<8%)	Virtually an insignificant forested catchment proportion (8% to <15%)	Significant forested catchment proportion (15%–20%)	High forested catchment proportion (>20%–30%)	Very high forested catchment proportion (>30%)
34	Not significantly groundwater-fed (<10%), but may exfiltrate into the local groundwater	Minor groundwater infiltration (10% to <20%)	High groundwater infiltration (20%–30%)	Very high groundwater infiltration (>30%–50%)	Most of the system depends on groundwater infiltration (>50%)

Bin number for each variable corresponds to a score present in Table 3.6.

absence of clay. However, the application of alternative weighting methodologies based on fuzzy logic, case-based reasoning and neural networks could also be considered (Lee et al., 2005).

3.3.2.3 Definition of Basin Types

The variable determination template (Table 3.5) and classification matrix (Table 3.6) are based on similar principles such as the classification key for ditches in wetlands (Scholz, 2006; Scholz and Lee, 2005; Scholz and Trepel, 2004a, 2004b) and the Sustainable Urban Drainage Decision Support Key, Matrix and Model (Scholz, 2006; Scholz et al., 2005).

The purpose of the variable determination template (Table 3.5) is to guide the user in working through the list of classification variables. The corresponding weighting matrix (Table 3.6) can be used individually, if only one purpose of the retention basin dominates (e.g., 90% purpose of enhancing the landscape aesthetically), or collectively, depending on their individual proportion of use (e.g., 40%, 25%, 25% and 10% use for sustainable drainage, environment protection, recreation and aesthetics, respectively). Table 3.6 also recommends weightings for SFRB where their purpose is unclear or not obvious.

Finally, Table 3.7 defines and characterizes six different types of SFRB as a function of their predominant purpose based on data collected during desktop studies and field visits. The determination of characteristics for all SFRB types is based on experience, and therefore on the expert judgement of the author and his multi-disciplinary research group. In theory, some SFRB types could be combined or alternatively more types could be identified. However, the proposed classification system is practical and was supported by observations on site.

3.3.2.4 Worked Example

Step 1: For a particular SFRB (e.g., Figure 3.4), use the variable determination template in Table 3.5. Go through each row and note the most appropriate bin number for each variable. For example, bin 3 for variable 1 (40%), bin 1 for variable 2 (12 m), (...) and bin 1 for variable 34 (5%).

Step 2: Decide on the most appropriate SFRB application. Each selected bin number (Table 3.1) corresponds to a score in Table 3.2. Write all scores down and sum them up. For example, the main application bias is toward hydraulic optimization; therefore, $15+25+(\ldots)+10=277$.

Step 3: Define the SFRB type with Table 3.3. For example, select the column hydraulics bias based on Step 2 and identify the type and name of the SFRB as 1 and Hydraulic Flood Retention Basin, respectively.

3.3.3 Results and Discussion

3.3.3.1 Classification Variables

Voeseneck et al. (2004) discussed the following habitats that are important for a decision support system: river, riverbanks, natural levees, floodplain, floodplain channel, back swamp areas (with or without flood channels), areas protected by

TABLE 3.6
Classification Matrix: Classification Variables and Associated Weightings for Different Bins (i.e. 1–5 per Application) with and without Bias toward Different Applications of Sustainable Flood Retention Basins

	Hydraulics Bias					Sustainable Drainage Bias					Environmental Protection Bias				
No.	1	2	3	4	5	1	2	3	4	5	1	2	3	4	5
1	25	20	15	10	5	15	12	9	6	3	5	10	15	20	25
2	25	20	15	10	5	15	12	9	6	3	5	10	15	20	25
3	5	4	3	2	1	1	2	3	4	5	4	8	12	16	20
4	25	20	15	10	5	2	4	6	10	8	1	2	3	4	5
5	10	8	6	4	2	2	4	6	8	10	5	10	15	20	25
6	5	4	3	2	1	2	4	6	8	10	4	8	12	16	20
7	10	8	6	4	2	1	2	3	4	5	3	6	9	12	15
8	2	4	6	8	10	4	8	12	16	20	3	6	9	12	15
9	10	8	6	4	2	1	2	3	4	5	2	4	6	8	10
10	2	4	6	8	10	3	6	9	12	15	3	6	9	12	15
11	2	4	6	8	10	2	4	6	8	10	2	4	6	8	10
12	1	2	3	4	5	2	4	6	8	10	3	6	9	12	15
13	10	8	6	4	2	4	5	3	2	1	5	4	3	2	1
14	1	2	3	4	5	2	4	6	8	10	2	4	6	8	10
15	20	16	12	8	4	1	2	3	4	5	3	6	9	12	15
16	25	20	15	10	5	1	2	3	4	5	2	4	6	8	10
17	1	2	3	4	5	3	6	9	15	12	3	6	9	12	15
18	1	2	3	4	5	3	6	9	12	15	5	4	3	2	1
19	1	2	3	4	5	25	20	15	10	5	1	2	3	4	5
20	2	4	6	8	10	3	6	9	12	15	1	2	3	4	5
21	5	4	3	2	1	10	8	6	4	2	2	4	6	8	10
22	10	8	6	4	2	2	4	5	3	1	1	2	3	4	5
23	1	2	3	4	5	2	4	6	8	10	3	6	9	12	15
24	10	8	6	4	2	15	12	9	6	3	2	4	5	3	1
25	1	2	3	4	5	1	2	3	4	5	5	10	15	20	25
26	25	20	15	10	5	25	20	15	10	5	10	8	6	4	2
27	5	4	3	2	1	5	4	3	2	1	1	2	3	4	5
28	20	16	12	8	4	10	8	6	4	2	3	6	9	12	15
29	15	12	9	6	3	5	4	3	2	1	1	2	3	4	5
30	1	2	3	4	5	5	4	3	2	1	3	6	9	12	15
31	10	8	6	4	2	5	4	3	2	1	10	8	6	4	2
32	5	4	3	2	1	5	4	3	2	1	10	8	6	4	2
33	2	4	6	8	10	2	4	6	8	10	1	2	3	4	5
34	10	8	6	4	2	15	12	9	6	3	1	2	3	4	5

(Continued)

TABLE 3.6 (*Continued*)

Classification Matrix: Classification Variables and Associated Weightings for Different Bins (i.e. 1–5 per Application) with and without Bias toward Different Applications of Sustainable Flood Retention Basins

	Recreational Bias					Landscape Aesthetics Bias					No Obvious Bias				
No.	**1**	**2**	**3**	**4**	**5**	**1**	**2**	**3**	**4**	**5**	**1**	**2**	**3**	**4**	**5**
1	2	4	6	8	10	4	8	16	20	12	16	14	13	12	9
2	5	4	3	2	1	4	8	12	20	16	16	14	12	11	9
3	4	5	3	2	1	5	10	15	20	25	4	5	6	6	7
4	5	4	3	2	1	5	4	3	2	1	14	12	9	7	4
5	2	4	6	8	10	1	2	3	4	5	6	7	7	8	8
6	3	6	15	12	9	3	6	12	15	9	4	5	7	8	7
7	2	4	6	10	8	1	2	3	5	4	6	6	6	6	5
8	3	6	9	15	12	4	8	10	6	2	3	5	8	10	11
9	2	4	5	3	1	4	5	3	2	1	6	6	5	4	3
10	2	4	6	10	8	2	4	6	8	10	2	5	7	9	11
11	2	4	6	8	10	2	4	6	8	10	2	4	6	8	10
12	3	6	9	12	15	3	6	9	12	15	2	4	6	8	9
13	5	4	3	2	1	5	4	3	2	1	7	6	5	3	2
14	3	6	9	12	15	1	2	3	4	5	2	3	5	6	8
15	3	6	9	12	15	3	6	9	12	15	11	11	10	9	8
16	8	10	6	4	2	12	15	9	6	3	15	14	10	8	5
17	2	4	8	10	6	8	10	6	4	2	2	4	6	7	7
18	2	6	10	8	4	1	2	3	5	4	2	3	5	5	5
19	2	4	5	3	1	5	4	3	2	1	4	4	5	4	4
20	2	4	5	3	1	1	2	3	5	4	2	4	5	7	8
21	4	5	3	2	1	4	8	12	20	16	5	5	5	5	4
22	1	2	3	4	5	8	10	6	4	2	6	6	5	4	3
23	3	6	12	15	9	3	6	12	15	9	2	4	6	8	8
24	15	12	9	6	3	20	16	12	8	4	11	9	7	5	2
25	5	10	15	20	25	2	4	6	8	10	2	5	7	9	11
26	15	12	9	6	3	20	6	12	8	4	21	16	12	8	4
27	1	2	3	5	4	5	4	3	2	1	4	3	3	3	2
28	5	4	3	2	1	16	20	12	8	4	14	12	10	7	5
29	10	8	6	4	2	4	5	3	2	1	10	8	6	5	3
30	3	6	12	15	9	1	2	3	5	4	2	3	5	7	7
31	1	2	4	5	3	4	8	10	6	2	8	7	6	4	2
32	1	2	4	5	3	4	6	10	8	2	5	5	4	3	2
33	2	4	8	10	6	1	3	5	4	2	2	4	6	7	8
34	1	2	3	5	4	4	5	3	2	1	7	6	5	4	3

The calculation of the sum must be undertaken with the help of Table 3.5, and the sum should be compared with the corresponding entry in Table 3.7.

TABLE 3.7
Definition of the Sustainable Flood Retention Basin (SFRB) Types According to Their Predominant Purposes with Total Scores Obtained from Tables 3.6 after Applying Table 3.5

Type	Name	Definition of SFRB Type	Hy-draulics Bias	Sustain-able Drainage Bias	Environmen-tal Protection Bias	Recre-ational Bias	Land-scape Aesthetics Bias	No Obvious Bias
1	Hydrau-lic Flood Retention Basin	Managed, traditional and large SFRB that is hydraulically optimized (or even automated) and captures sediment	>260	<130	<160	<110	<140	>240
2	Traditional Sustainable Flood Retention Basin	Usually, large retention basin used for active flood protection adhering to best management practices	>240–260	>170–190	160 to <190	110 to <130	>140 to <160	>220–240
3	Sustainable Flood Retention Wetland	Aesthetically pleasing retention and treatment wetland used for passive flood protection adhering to sustainable drainage and best management practices	>220–240	>210	190–220	130–150	160–180	>200–220
4	Aesthetic Flood Treatment Wetland	Treatment wetland for the retention and treatment of contaminated runoff, which is aesthetically pleasing and integrated into the landscape, and has some social and recreational benefits	200–220	>190–210	>220–250	>170–190	>220	180–200
5	Integra-ted Flood Retention Wetland	Integrated semi-natural flood retention wetland for passive treatment of runoff, passive flood retention and enhancement of recreational benefits	180 to <200	150–170	>250–280	>190	>200–220	160 to <180
6	Natural Flood Retention Wetland	Passive natural flood retention wetland that became a site of scientific interest requiring protection from adverse human impacts	<180	130 to <150	>280	>150–170	>180–200	<160

FIGURE 3.4 Sustainable Flood Retention Basin in Merzhausen (near Freiburg, Baden) on 4 April 2006. Applying Tables 3.5 and 3.6, the score was 277 (i.e. Hydraulic Flood Retention Basin).

summer flooding embankments and areas protected by flood dykes. In addition, structures such as SFRB are likely to develop their own habitats after years or decades depending on management, flooding characteristics and climate. These habitats are indirectly reflected in the selection and characterization of the classification variables (Table 3.4).

Van den Brink et al. (1994a, 1994b) highlighted the importance of topographical, geomorphologic, hydrological and habitat characteristics. They discussed variables including location (i.e. hydrological habitat; e.g., site in open connection, and isolated site in river foreland or behind the main dyke), water surface area, water depth, distance from river (i.e. Connectivity; variable number 7 in Table 3.5), flood duration, sediment composition (sandy, clayey and organic) and vegetation (submerged, nymphaeid and emergent). Table 3.5 summarizes most of these variables either directly or indirectly.

Van den Brink et al. (1994a) related phytoplankton and zooplankton to their feeding characteristics (i.e. predator, filter feeder, scraper and sucker) and their habitat (i.e. open water, sediment and macrophyte). These criteria are relevant if the retention basins should have ecological benefits. However, most of these ecological variables are too specific for civil engineers and time-consuming to determine for the purpose of a global SFRB classification system. Nevertheless, Animal Passage (variable number 5; Table 3.6) accounts for the functioning of the SFRB (i.e. particularly the dam and its outlet arrangement) as a barrier in the landscape restricting animal movement.

Concerning the final variable on groundwater infiltration (i.e. Groundwater; variable 7 in Table 3.6), Trémolières et al. (1993) and Scholz and Trepel (2004a, 2004b) indicated that this variable is important, if a watercourse is being fed by groundwater. However, reservoir water exfiltration can be a more frequent problem in practice.

3.3.3.2 Application of the Classification System

The author estimates that there are >3,000 areas within the River Rhine catchment of the federal state of Baden-Württemberg (South-west Germany), which could be classified as retention basins using the widest possible definition in agreement with German guidelines (ATV-DVWK, 2001). The author revised and applied an active database provided by the Federal Environmental Ministry of Baden-Württemberg, and updated by the Institute for Landscape Management, which characterizes approximately 660 of the known important and strongly man-made retention basins. It was obvious that there is a bias toward large engineered and permanent retention basins in this data base. However, the author estimates that the volume of these known retention basins is likely to be >95% of the total basin volume in the River Rhine catchment (i.e. for Baden-Württemberg only). The remaining retention basins are dominated by watercourses such as wet meadows, ditches and semi-natural ponds.

For the purpose of Section 3.3, an SFRB is defined as an "aesthetically pleasing retention basin predominantly used for flood protection adhering to sustainable drainage and best management practices". The adjective aesthetic means visually pleasing and is defined in the context of landscaping as "pertaining to the appreciation of beauty and good taste". The noun that corresponds to aesthetic is aesthetics, which means "the study of the appreciation of beauty and good taste". Landscape design and management is concerned both with aesthetic and functional elements of landscaping. For example, a long, large and grey concrete dam is usually seen as less aesthetically pleasing than a short, small and planted dam. An SFRB has therefore to be visually attractive. This obviously goes beyond a characterization based purely on traditional engineering variables such as dam height, retention volume and surface area.

An overwhelming majority of 92% of all recorded retention basins in Baden-Württemberg had a clear flood protection purpose. With respect to the approximately 660 flood retention basins, 77% were classified as SFRB. However, this was expected considering the dominant flood retention purpose of the basins.

There is a relatively great variability among the variables summarizing the key characteristics of SFRB (Tables 3.3 and 3.4, which are used predominantly for hydraulic purposes such as water retention and sedimentation, and which are in the same area (i.e. South Baden). Therefore, the variable determination template (Table 3.5) and the corresponding classification matrix (Table 3.6) can be used to sub-classify SFRB (Table 3.7).

For example, Figures 3.4 and 3.7 show four closely related engineered SFRB sub-classified as types 1–4 (Table 3.3), respectively. The selected SFRB examples had all a dominant hydraulic bias (Table 3.2). The first example (Figure 3.4) shows a hydraulically optimized basin, which captures large amounts of sediment during floods. Figure 3.5 shows a retention basin used predominantly for flood protection but adhering to best management practices including water treatment. Figure 3.6 shows an aesthetically pleasing retention and treatment wetland, which has an important treatment

FIGURE 3.5 Sustainable Flood Retention Basin in Müllheim (Baden) on 26 April 2006. Applying Tables 3.5 and 3.6, the score was 250 (i.e. Traditional Sustainable Flood Retention Basin).

FIGURE 3.6 Sustainable Flood Retention Basin near Denzlingen (Baden) on 19 April 2006. Applying Tables 3.5 and 3.6, the score was 232 (i.e. Sustainable Flood Retention Wetland).

FIGURE 3.7 Sustainable Flood Retention Basin near Königschaffhausen (Baden) on 18 April 2006. Applying Tables 3.5 and 3.6, the score was 212 (i.e. Aesthetic Flood Treatment Wetland).

and infiltration function, and is used for passive flood protection adhering to sustainable drainage and best management practices. The final example (Figure 3.7) is an Aesthetic Flood Treatment Wetland. This is a permanently wet treatment wetland for the retention and active purification of contaminated runoff, which is not hydraulically optimized, but perfectly integrated into the landscape, and has some social, recreational and ecological benefits.

A careful estimation based on a randomly (i.e. 1 in 3) selected sub-sample of 56 thoroughly researched SFRB in South Baden shows that the predominant use is for hydraulic purposes (48 basins), followed by environmental protection (3 basins) and recreational (3 basin), and finally by sustainable drainage (1 basin) and landscape aesthetics (1 basin) purposes. The mean hydraulic, drainage, environmental, recreational and landscape bias estimated values for all 56 SFRB are 50%, 11%, 16%, 13% and 10%, respectively. The corresponding standard deviations were 17.8%, 8.1%, 8.1%, 8.9% and 6.2%, respectively.

Moreover, 15, 21, 9, 8, 1 and 2 of the thoroughly researched SFRB belong to subclasses 1–6, respectively (Table 3.7). This bias toward engineered SFRB is likely to be representative for the remaining SFRB in Baden-Württemberg.

3.3.3.3 Prioritization of Classification Variables

A correlation analysis was performed between all classification variables. Some variables such as Outlet Flexibility, Flood Duration and Connectivity can be seen as independent due to their lack of correlation with other variables. In contrast, the variables

Forest, Flotsam and Flood Water Volume, for example, correlated very well with other variables, and should therefore be regarded as dependent variables (Table 3.8). However, this does not necessarily mean that they could be replaced by other variables or that they may replace other variables, because their relationships among each other and with other variables are also functions of the retention basin use and specific local boundary conditions such as storm characteristics and local management policies not captured by the classification system.

Table 3.8 presents an unbiased attempt to prioritize the classification variables based on their correlation with other variables (i.e. concerning the sum of absolute correlation coefficients, the more the better) and the certainty (i.e. confidence of the expert team in choosing the data point) associated with the correctness of the corresponding numerical value. The variables Rainfall (mm), Dam Height (m), Flood Water Volume (m^3), Elevation (m), Dam Length (m), Flotsam (–), Floodplain Elevation (m), Forest (%) and Animal Passage (–) were the most promising variables. A cut-off point at 500 priority points based on expert judgement (i.e. high correlation value for Forest and high certainty value for Animal Passage) was chosen.

Tables 3.2 and 3.3 were also applied for only the top nine variables listed in Table 3.3 and for the 48 SFRB with a clear hydraulic application bias. However, the relative importance of the nine selected variables based on their corresponding proportion of summed-up scores in relation to the maximum possible score was applied to adapt Tables 3.2 and 3.3. The relative adjusted mean absolute error between the original scores (based on 34 variables including the key variables) and the new scores (based on the 9 key variables only) was 8.6%. The corresponding standard deviation was 5.86%, which is relatively low considering the rather high diversity of SFRB.

It follows that this variable reduction exercise (i.e. 9 instead of 34 variables) would result in a mean shift of only 0.7 of a type of SFRB (Table 3.5). For example, type 2 or type 4 would have been selected instead of type 3. This loss in SFRB identification accuracy might be justified in some cases by the reduction in assessment time to approximately 20 minutes, considering that all these key variables (except for flotsam) can be easily determined during a desktop study (Figures 3.4 and 3.5).

3.3.4 Conclusions and Recommendations

Section 3.3 provides clear definitions for different types of SFRB associated with different applications including hydraulic purposes, sustainable drainage, environmental protection, recreational activities and landscape aesthetics. The following new types of SFRB were defined: Hydraulic Flood Retention Basin, Traditional Sustainable Flood Retention Basin, Sustainable Flood Retention Wetland, Aesthetic Flood Treatment Wetland, Integrated Flood Retention Wetland and Natural Flood Retention Wetland.

The definitions for the different types of SFRB are based on a set of tables with weightings for 34 classification variables. Weightings can be obtained with the support of the variable determination template containing qualitative and frequently also quantitative descriptions for the classification variables.

TABLE 3.8
Prioritization Example for Classification Variables (See Table 3.5 for Details) Used for a Detailed Data set of 56 Sustainable Flood Retention Basins in South Baden (Germany)

Variable	No.	PP[a]	Correlation[b]	Certainty[c]	Mean	SD[d]
		Nine Key Classification Variables				
Rainfall (mm)	18	648	7.72	84	731	99.3
Dam height (m)	2	624	7.77	80	4	3.1
Flood water volume (m^3)	16	622	7.92	78	54,100	82,900
Elevation (m)	22	603	6.56	92	237	42.1
Dam length (m)	3	601	7.16	84	286	426.8
Flotsam (–)	28	594	7.92	75	1.9	1.25
Floodplain elevation (m)	6	593	7.47	79	0.5	1.52
Forest (%)	33	535	8.82	61	23	23
Animal passage (–)	5	515	6.63	78	1.8	0.92
		Other Classification Variables				
Algae (%)	24	490	5.60	88	5	16.5
Channel system (%)	9	470	5.27	89	93	21.0
Water surface area (m^2)	17	464	7.53	62	83,100	52,100
Embankment length (m)	15	459	7.67	60	420	438.9
Pollution (–)	25	449	7.18	63	2.8	1.13
Wetness (%)	8	441	5.68	78	22	28.0
Pasture (%)	32	422	6.76	63	29	26.0
Naturalness (%)	1	417	5.99	70	52	20.8
Connectivity (m)	7	401	4.24	95	0.9	2.69
Season (–)	21	397	4.59	87	3.9	0.49
Arable (%)	31	387	6.20	63	27	26.6
Sediment composition (cm)	27	385	5.91	65	3.9	5.99
Vegetation (–)	23	377	4.80	79	3.1	1.26
Groundwater (%)	34	375	5.68	66	4	8.7
Sediment depth (cm)	26	370	5.77	64	5.2	7.93
Catchment size (km^2)	29	363	7.52	48	7.6	10.78
Flooding depth (m)	10	358	5.90	61	1.7	0.83
Slope (%)	13	341	4.84	71	2.7	2.02
Urban (%)	30	330	4.68	71	12	13.1
Outlet flexibility (%)	4	262	3.71	71	20	24.4
Flood frequency (a–1)	12	218	4.79	46	7	6.1
Flood duration (d)	11	209	3.90	54	27	75
Clay (%)	20	156	4.86	32	11	13.1
Drainage (cm/d)	19	144	4.62	31	14	13.5
Velocity (cm/s)	14	126	4.29	29	49	56.8

[a] Priority points = (column 4) × (column 5).
[b] Sum of all absolute correlation coefficients for one particular variable with all other variables.
[c] Certainty of a correct value expressed in percentage by the author and his research team.
[d] Standard deviation.

The River Rhine catchment with particular bias toward the upper river catchment in Baden has been used as a case study. A bias toward engineered SFRB in South Baden was obvious. The nine most important variables were Rainfall, Dam Height, Flood Water Volume, Elevation, Dam Length, Flotsam, Floodplain Elevation, Forest and Animal Passage, considering a correlation analysis and estimating the certainty of numerical correctness based on the confidence of the research team in the field. A variable reduction exercise has shown that 9 instead of 34 variables could also be used but that the SFRB identification accuracy would decrease too much to justify the gain in assessment time reduction.

The characterization methodology outlined could be implemented world-wide. However, the specific variable determination template, the classification matrix and the associated SFRB type definitions are likely to be only applicable for Europe, Northern America and other regions with a temperate and oceanic climate.

REFERENCES

Ackrill, R., Kay, A. and Morgan, W., 2008. The common agricultural policy and its reform: The problem of reconciling budget and trade concerns, *Canadian Journal of Agricultural Economics – Revue Canadienne D Agroeconomie*, **56**(4), 393–411.

Alkema, D. and Middelkoop, H., 2005. The influence of floodplain compartmentalization on flood risk within the Rhine-Meuse delta, *Natural Hazards*, **36**, 125–145.

ATV-DVWK, 2001. *Hochwasserrückhaltebecken – Probleme und Anforderungen aus wasserwirtschaftlicher und ökologischer Sicht* (Hennef: Deutsche Vereinigung für Wasserwirtschaft, Abwasser und Abfall e.V. (ATV-DVWK), Gesellschaft zur Förderung der Abwassertechnik (in German).

Bardossy, A. and Caspary, H.J., 1990. Detection of climate change in Europe by analyzing European atmospheric circulation patterns from 1881 to 1989, *Theoretical Applied Climatology*, **42**(3), 155–167.

Behrendt, H., 1996. Inventories of point and diffuse sources and estimated nutrient loads – a comparison for different river basins in Central Europe, *Water Science and Technology*, **33**(4–5), 99–107.

Bella, K.P. and Irwin, E.G., 2005. Spatially explicit micro-level modelling of land use change at the rural–urban interface, *Agricultural Economics*, **27**(3), 217–232.

Bissels, S., Hölzel, N., Donath, T.W. and Otte, A., 2004. Evaluation of restoration success in alluvial grasslands under contrasting flooding regimes, *Biological Conservation*, **118**(5), 641–650.

Bissels, S., Hölzel, N., Donath, T.W. and Otte, A., 2005. Ephemeral wetland vegetation in irregularly flooded arable fields along the northern upper Rhine: The importance of persistent seed banks, *Phytocoenologia*, **35**(2–3), 469–488.

Bohm, H.R., Haupter, B., Heiland, P. and Dapp, K., 2004. Implementation of flood risk management measures into spatial plans and policies, *River Research and Applications*, **20**(3), 255–267.

Bornette, G. and Arens, M.F., 2002. Charophyte communities in cut-off river channels – the role of connectivity, *Aquatic Botany*, **73**(2), 149–162.

Brettar, I., Sanchez-Perez, J.M. and Trémolières, M., 2002. Nitrate elimination by denitrification in hardwood forest soils of the upper Rhine floodplain – correlation with redox potential and organic matter, *Hydrobiologica*, **469**, 11–21.

BRIDGE, 2010. Official project web site of Sustainable Urban Planning Decision Support Accounting for Urban Metabolism (BRIDGE). European Union project.

Camagni, R., Gibelli, M.C. and Rigamonti, P., 2002. Urban mobility and urban form: the social and environmental costs of different patterns of urban expansion, *Ecological Economics*, **40**(2), 199–216.

Carbiener, R., Trémolières, M. and Muller, S., 1995. Vegetation of running waters and water quality: Thesis, debates and prospects, *Acta Botanica Gallica*, **142**(6), 489–531 (in French).

Couch, C., Karecha, J., Nuissl, H. and Rink, D., 2005. Decline and sprawl: An evolving type of urban development observed in Liverpool and Leipzig, *European Planning Studies*, **13**(1), 117–136.

Daily, G.C. and Matson, P.A., 2008. Ecosystem services: From theory to implementation, *Proceeding of the National Academy of Sciences of the United States of America*, **105**(28), 9455–9456.

de Groot, R., 2006. Function-analysis and valuation as a tool to assess land use conflicts in planning for sustainable, multi-functional landscapes, *Landscape and Urban Planning*, **75**(3–4), 175–186.

De Nooij, R.J.W., Lenders, H.J.R., Leuven, R.S.E.W., De Blust, G., Geilen, N., Goldschmidt, B., Muller, S., Poudevigne, I. and Nienhuis, P.H., 2004. BIO-SAFE: Assessing the impact of physical reconstruction on protected and endangered species, *River Research and Applications*, **20**(3), 229–313.

DeSurvey, 2010. Official project website of Surveillance System for Assessing and Monitoring Desertification (DeSurvey). European Union project.

Deiller, A.F., Walter, J.M.N. and Trémolières, M., 2001. Effects of flood interruption on species composition of woody regeneration in the upper Rhine alluvial hardwood forest, *Regulated Rivers – Research and Management*, **17**(4–5), 393–405.

Donath, T.W., Hölzel, N., Bissels, S. and Otte, A., 2004. Perspectives for incorporating biomass from non-intensively managed temperate flood-meadows into farming systems, *Agriculture Ecosystems and Environment*, **104**, 439–451.

Donath, T.W., Hölzel, N. and Otte, A., 2003. The impact of site conditions and seed dispersal on restoration success in alluvial meadows, *Applied Vegetation Science*, **6**, 13–22.

Elliot, J.A., 2006. *An Introduction to Sustainable Development* (Abingdon: Taylor and Francis).

Forshay, K.J. and Stanley, E.H., 2005. Rapid nitrate loss and denitrification in a temperate river floodplain, *Biogeochemistry*, **75**(1), 43–64.

Grift, R.E., Buijse, A.D., Van Densen, W.L.T., Machiels, M.A.M., Kranenbarg, J., Breteler, J.P.K. and Backx, J.J.G.M., 2003. Suitable habitats for 0-group fish in rehabilitated floodplains along the lower River Rhine, *River Research and Applications*, **19**(4), 353–374.

Haase, D., 2003. Holocene floodplains and their distribution in urban areas – functionality indicators for their retention potentials, *Landscape and Urban Planning*, **66**(1), 5–18.

Hedelin, B., 2007. *Criteria for the Assessment of Sustainable Water Management* (Berlin: Springer Verlag).

Hegerl, G.C., von Storch, H., Hasselmann, K., Santer, B.D., Cubasch, U. and Jones, P.D., 1994. Detecting Anthropogenic Climate Change with an Optical Fingerprint Method. Report 142 (Hamburg: Max-Planck-Institut für Meteorologie).

Heinzmann, U. and Zollinger, G., 1995. Validation of representativeness with relief parameters based on comparison of two land use classifications, *Catena*, **24**(1), 69–87.

Helming, K., Tscherning, K., König, B., Sieber, S., Wiggering, H., Kuhlman, T., Wascher, D.M., Pérez-Soba, M., Smeets, P.J.A.M., Tabbush, P., Dilly, O., Hüttl, R.F. and Bach, H., 2008. Ex ante impact assessment of land use changes in European regions - The SENSOR approach. In: K. Helming, M. Perez-Soba and P. Tabbush (Eds.) *Sustainability Impact Assessment of Land Use Changes* (Berlin: Springer), Part I, pp. 77–105.

Helmio, T., 2005. Unsteady 1D flow model of a river with partly vegetated floodplains – application to the Rhine river, *Environmental Modelling and Software*, **20**(3), 361–375.

Heynen, N.C., Kaika, M. and Swyngedouw, E., 2006. *In the Nature of Cities: Urban Political Ecology and the Politics of Urban Metabolism* (Abingdon: Routledge, Taylor and Francis Group).

Hill, J., Stellmes, M., Udelhoven, T., Röder, A. and Sommer, S., 2008. Mediterranean desertification and land degradation: Mapping related land use change syndromes based on satellite observations, *Global and Planetary Change*, **64**(3–4), 146–157.

Hölzel, N. and Otte, A., 2001. The impact of flooding on the soil seed bank of flood-meadows, *Journal of Vegetation Science*, **12**(2), 209–218.

Hölzel, N. and Otte, A., 2003. Restoration of a species-rich flood meadow by topsoil removal and dispore transfer with plant material, *Applied Vegetation Science*, **6**(2), 131–140.

Hölzel, N. and Otte, A., 2004. Inter-annual variation in the soil seed bank of flood-meadows over two years with different flooding patterns, *Plant Ecology*, **174**(2), 279–291.

Hooijer, A., Klijn, F., Pedroli, G.B.M. and Van Os, A.G., 2004. Towards sustainable flood risk management in the Rhein and Meuse basins: Synopsis of the findings of IRMA-SPONGE, *River Research and Applications*, **20**(3), 343–357.

ICPR, 1998. Plan on Flood defence. Action International Commission for the Protection of the River Rhine (ICPR). https://www.iksr.org.

IPCC, 1997. R. T. Watson, M. C. Zinyowera and R. H. Moss (Eds) *The Regional Impacts of Climate Change: An Assessment of Vulnerability*. A Special Report of the Intergovernmental Panel of Climate Change (IPCC) Working Group II. (Cambridge: Cambridge University Press).

Jaeger, J.A.G., Bertiller, R., Schwick, C. and Kienast, F., 2010. Suitability criteria for measures of urban sprawl, *Ecological Indicators*, **10**(2), 397–406.

Jansson, Å. and Colding, J., 2007. Tradeoffs between environmental goals and urban development: The case of nitrogen load from the Stockholm County to the Baltic Sea. *AMBIO: A Journal of the Human Environment*, **36**(8), 650–656.

KASSA, 2010. Official project web site of Knowledge Assessment and Sharing on Sustainable Agriculture (KASSA). European Union project.

Kazda, M., 1995. Changes in alder fens following a decrease in the ground-water table – Results of a geographical information-system application, *Journal of Applied Ecology*, **32**(1), 100–110.

Klein, J.P., Maire, G., Exinger, F., Lutz, G., Sanchex-Perex, J.M., Trémolières, M. and Junod, P., 1994. The restoration of former channels in the Rhine alluvial forest – The example of the Offendorf-Nature-Reserve (Alsace, France), *Water Science and Technology*, **29**(2), 301–305.

Lafferty, W. and Hovden, E., 2003. Environmental policy integration: Towards an analytical framework, *Environmental Politics*, **12**(3), 1–22.

Lahmar, R., 2010. Adoption of conservation agriculture in Europe: Lessons of the KASSA project, *Land Use Policy*, **27**(1), 4–10.

Lee, B.-H., Scholz, M., Horn, A. and Furber, A.M., 2005. Prediction of constructed treatment wetland performance with case-based reasoning (Part B), *Environmental Engineering Science*, **23**(2), 203–211.

LfU, 1999. *Auswirkungen der Ökologischen Flutungen des Polder Altenheim, Ergebnisse des Untersuchungsprogramms 1993–1996*. Landesamt für Umwelt (LfU) (in German).

LUTR, 2010. Official project web site of Land Use and Transportation Research: Policies for the City of Tomorrow (LUTR). European Union project.

Martinez, B., Verger, A., Garcia-Haro, F.J., Gilabert, M.A. and Melia, J., 2007. Procedure for the regional scale mapping of FVC and LAI over land degraded areas in the DeSurvey project. In: IEEE International, ed. *Proceedings of the Geoscience and Remote Sensing Symposium*. Barcelona, pp. 3452–3455.

Middleton, B.A. (Ed), 2002. *Flood Pulsing in Wetlands: Restoring the Natural Hydrological Balance* (Hoboken, NJ: John Wiley & Sons).

Molles, M.C., Crawford, C.S., Ellis, L.M., Valett, H.M. and Daham, C.N., 1998. Managed flooding for riparian ecosystem restoration – Managed flooding reorganizes riparian forest ecosystems along the middle Rio Grande in New Mexico, *Bioscience*, **48**(9), 749–756.

Nijland, H.J., 2005. Sustainable development of floodplains (SDF) project, *Environmental Science and Policy*, **8**(3), 245–252.

O'Connor, M.C., McKenna, J. and Cooper, J.A.G., 2010. Coastal issues and conflicts in North West Europe: A comparative analysis, *Ocean and Coastal Management*, **53**(12), 727–737.

Pfister, L., Kwadijk, J., Musy, A., Bronstert, A. and Hoffmann, L., 2004. Climate change, land use change and runoff prediction in the Rhine-Meuse basins, *River Research and Applications*, **20**(3), 229–241.

PLUREL, 2010. Official project website of Peri-urban Land Use Relationships - Strategies and Sustainability Assessment Tools for Urban-Rural Linkages (PLUREL). European Union project.

Robach, F., Eglin, I. and Trémolières, M., 1997. Species richness of aquatic macrophytes in former channels connected to a river: A comparison between two fluvial hydrosystems differing in their regime and regulation, *Global Ecology and Biogeography Letters*, **6**, 267–274.

Rood, S.B., Samuelson, G.M., Braatne, J.H., Gourley, C.R., Hughes, F.M.R. and Mahoney, J.M., 2005. Managing river flows to restore floodplain forests, *Frontiers in Ecology and the Environment*, **3**(4), 193–201.

RUFUS, 2010. Official project website of Rural Future Networks (RUFUS). European Union project.

Sanchez-Perez, J.M. and Trémolières, M., 1997, Variation in nutrient levels of the groundwater in the Upper Rhine alluvial forests as a consequence of hydrological regime and soil texture, *Global Ecology and Biogeography Letters*, **6**(3–4), 211–217.

Sanchez-Perez, J.M., Trémolières, M., Schnitzler, A. and Carbiener, R., 1991. Evolution of physicochemical quality of shallow groundwater as a function of seasonal cycles and succession of Rhine alluvial forests (Querco-Ulmetum Minoris ISSL 24), *Acta Oecologica – International Journal of Ecology*, **12**, 581–601 (in French).

Sanchez-Perez, J.M., Trémolières, M., Schnitzler, A. and Carbiener, R., 1993. Nutrient content in alluvial soils submitted to flooding in the Rhine alluvial deciduous forest, *Acta Oecologica – International Journal of Ecology*, **14**, 371–387.

Sanchez-Perez, J.M., Vervier, P., Grabetian, F., Sauvage, S., Loubet, M., Rols, J.L., Briac, T. and Weng, P., 2003. Nitrogen dynamics in the shallow groundwater of a riparian wetland zone of the Garonne, South-west France: Nitrate inputs, bacterial densities, organic matter supply and denitrification measurements, *Hydrology and Earth System Sciences*, **7**(1), 97–107.

SAWA, 2010. Official project web site of Strategic Alliance for Integrated Water Management Actions (SAWA). European Union project.

Schnitzler, A., 1994. Conservation of biodiversity in alluvial hardwood forests of the temperate zone – the example of the Rhine valley, *Forest Ecology and Management*, **68**(2–3), 385–398.

Schnitzler, A., 1995. Successional status of trees in gallery forest along the river Rhine, *Journal of Vegetation Science*, **6**(4), 479–486.

Schnitzler, A., Carbiener, R. and Trémolières, M., 1992. Ecological segregation between closely related species in the flooded forests of the upper Rhine plain, *New Phytologist*, **121**(2), 293–301.

Scholz, M., 2006. *Wetland Systems to Control Urban Runoff* (Amsterdam: Elsevier).

Scholz, M., 2007a. Ecological effects of water retention in the River Rhine Valley: A review assisting future retention basin classification, *International Journal of Environmental Studies*, **64**(2), 171–187.

Scholz, M., 2007b. Expert system outline for the classification of sustainable flood retention basins (SFRBs), *Civil Engineering and Environmental Systems*, **24**(3), 193–209.

Scholz, M., 2010. *Wetland Systems – Storm Water Management Control*. Series: Green Energy and Technology (Berlin: Springer Verlag).

Scholz, M. and Lee, B.-H., 2005. Constructed wetlands: A review, *International Journal of Environmental Studies*, **62**(4), 421–447.

Scholz, M. and Trepel, M., 2004a. Water quality characteristics of vegetated groundwater-fed ditches in a riparian peatland, *The Science of the Total Environment*, **332**(1–3), 109–122.

Scholz, M. and Trepel, M., 2004b. Hydraulic characteristics of groundwater-fed open ditches in a peatland, *Ecological Engineering*, **23**(1), 29–45.

Scholz, M. and Yang, Q., 2010. Guidance on variables characterising water bodies including sustainable flood retention basins, *Landscape and Urban Planning*, **98**(3–4), 190–199.

Scholz, M., Hedmark, Å. and Hartley, W., 2012. Recent advances in sustainable multifunctional land and urban management in Europe: A review, *Journal of Environmental Planning and Management*, **55** (7), 833–854.

Scholz, M., Morgan, R. and Picher, A., 2005. Stormwater resources development and management in Glasgow: Two case studies, *International Journal of Environmental Studies*, **62**(3), 263–282.

Schoor, M.M., 1994. The relation between vegetation, hydrology and geomorphology in the Gemenc Floodplain Forest, Hungary, *Water Science and Technology*, **29**(3), 289–291.

Schönwiese, C.-D., Rapp, J., Fuchs, T. and Denhard, M., 1993. Klimatrend Atlas Europa 1891–1990. Berichte des Zentrums für Umweltforschung (Nummer 20), in German.

Schröter, D., Cramer, W., Leemans, R., Prentice, I.C., Araújo, M.B., Arnell, N.W., Bondeau, A., Bugmann, H., Carter, T.R., Gracia, C.A., de la Vega-Leinert, A.-C., Erhard, M., Ewert, F., Glendining, M., House, J.I., Kankaanpää, S., Klein, R.J.T., Lavorel, S., Lindner, M., Metzger, M.J., Meyer, J., Mitchell, T.D., Reginster, I., Rounsevell, M., Sabaté, S., Sitch, S., Smith, B., Smith, J., Smith, P., Sykes, M.T., Thonicke, K., Thuiller, W., Tuck, G., Zaehle, S. and Zierl, B., 2005. Ecosystem service supply and vulnerability to global change in Europe, *Science*, **310**(5752), 1333–1337.

Schumann, A.H., 1993. Changes in hydrological time series - a challenge for water management in Germany. In: *Extreme Hydrological Events: Precipitation, Floods and Droughts, Proceedings of the Joint IAMAP-IAHS Meeting Yokohama*, July 1993, Japan, IAHS Publication No. 213, pp. 95–102.

SEAMLESS, 2010. Official project web site of System for Environmental and Agricultural Modelling; Linking European Science and Society (SEAMLESS).

SENSOR, 2010. Official project web site of Sustainability Impact Assessment: Tools for Environmental Social and Economic Effects of Multifunctional Land Use in European Regions (SENSOR). European Union project.

Simeonova, V. and van der Valk, A., 2009. The need for a communicative approach to improve environmental policy integration in urban land use planning, *Journal of Planning Literature*, **23**(3), 241–261.

SPICOSA, 2010. Official project website of Science and Policy Integration for Coastal System Assessment (SPICOSA). European Union project.

SUME, 2010. Official project website of Sustainable Urban Metabolism for Europe (SUME). European Union project.

Susta-Info, 2010. Official project website of Global Knowledge for Local Sustainability (Susta-Info). European Union project.

SWITCH, 2010. Official project website of Managing Water for the City of the Future (SWITCH). European Union project.

Takatert, N., Snchez-Perez, J.M. and Trémolières, M., 1999. Spatial and temporal variations of nutrient concentration in the groundwater of a floodplain: Effect of hydrology, vegetation and substrate, *Hydrological Processes*, **13**(10), 1511–1526.

Trémolières, M., 2004. Plant response strategies to stress and disturbance: The case of aquatic plants, *Journal of Biosciences*, **29**(4), 461–470.

Trémolières, M., Eglin, I., Roeck, U. and Carbiener, R., 1993. The exchange progress between river and groundwater on the central Alsace floodplain (Eastern France). The case of the canalised river Rhine, *Hydrobiologia*, **254**, 133–148.

Trémolières, M., Sanchez-Perez, J.M., Schnitzler, A. and Schmitt, D., 1998. Impact of river management history on the community structure, species composition and nutrient status in the Rhine alluvial hardwood forest, *Plant Ecology*, **135**, 59–78.

Trémolières, M., Schnitzler, A., Sanchez-Perez, J.M. and Schmitt, D., 1999. Changes in foliar nutrient content and resorption in Fraxinus excelsior L., Ulmus minor Mill. and Clematis vitalba L. after prevention of floods, *Annals of Forest Science*, **56**(8), 641–650.

UN-HABITAT, 2010. Official website of United Nations-HABITAT (UN-HABITAT). European Union project.

URBS PANDENS, 2010. Official project web site of Urban Sprawl: European Patterns, Environmental Degradation and Sustainable Development (URBS PANDENS). European Union project.

Van Dam, R.A., Camilleri, C. and Finlayson, C.M., 1998. The potential of rapid assessment techniques as early warning indicators of wetland degradation: A review, *Environmental Toxicology and Water Quality*, **13**(4), 297–312.

Van den Brink, F.W.B., Beljaards, M.J., Botts, N.C.A. and van der Velde, G., 1994b. Macrozoobenthos abundances and community composition in three lower Rhine floodplain lakes with varying inundation regimes, *Regulated Rivers – Research and Management*, **9**(4), 279–293.

Van den Brink, F.W.B., van Katwijk, M.M. and van der Velde, G., 1994a. Impact of hydrology on phytoplankton and zooplankton community composition in floodplain lakes along the lower Rhine and Meuse, *Journal of Plankton Research*, **16**(4), 351–373.

Van der Poorten, A. and Klein, J.P., 1999. Aquatic bryophyte assemblages along a gradient of regeneration in the river Rhine, *Hydrobiologia*, **410**, 11–16.

Van der Poorten, A., Klein, J.P. and DeZuttere, P., 1995. Bryological evaluation of a flood project by the Rhine floods: The Erstein nature reserve (Alsace, France), *Belgian Journal of Botany*, **128**, 139–150 (in French).

Van Geest, G.J., Coops, H., Roijackers, R.M.M., Buijse, A.D. and Scheffer, M., 2005. Succession of aquatic vegetation driven by reduced water-level fluctuations in floodplain lakes, *Journal of Applied Ecology*, **42**(2), 251–260.

Van Geest, G.J., Roozen, F.C.J.M., Coops, H., Roijackers, R.M.M., Buijse, A.D., Peeters, E.T.H.M. and Scheffer, M., 2003. Vegetation abundance in lowland floodplain lakes determined by surface area, age and connectivity, *Freshwater Biology*, **48**(3), 440–454.

van Ittersum, M.K., Ewert, F., Heckelei, T., Wery, J., Olsson, J.A., Andersen, E., Bezlepkina, I., Brouwer, F., Donatelli, M., Flichman, G., Olsson, L., Rizzoli, A.E., van der Wal, T., Wien, J.E. and Wolf, J., 2008. Integrated assessment of agricultural systems – A component-based framework for the European Union (SEAMLESS), *Agricultural Systems*, **96**(1–3), 150–165.

Van der Lee, G.E.M., Venterik, O. and Asselman, N.E.M., 2004. Nutrient retention in floodplains of the Rhine distributaries in The Netherlands, *River Research and Applications*, **20**(3), 315–325.

Van Lienden, B.J. and Lund, J.R., 2004. Spatial complexity and reservoir optimization results, *Civil Engineering and Environmental Systems*, **21**(1), 1–17.

van Meijl, H., van Rheenen, T., Tabeau, A. and Eickhou, B., 2006. The impact of different policy environments on agricultural land use in Europe, *Agriculture, Ecosystems and Environment*, **114**(1), 21–38.

Van Splunder, I., Coops, H. and Schoor, M.M., 1994. Tackling the bank erosion – problem – (re-)introduction of willows on riverbanks, *Water Science and Technology*, **29**(3), 379–381.

Venterik, O., Hummelink, E. and Van den Hoorn, M.W., 2003. Denitrification potential of a river floodplain during flooding with nitrate-rich water: Grassland versus reedbeds, *Biogeochemistry*, **65**(2), 233–244.

Verburg, P.H., van Berkel, D.B., van Doorn, A.M., van Eupen, M. and van den Heiligenberg, H.A.R.M., 2010. Trajectories of land use change in Europe: A model-based exploration of rural futures, *Landscape Ecology*, **25**(2), 217–232.

Vernier, F., Bordenave, P., Chavent, M., Leccia, O. and Petit, K., 2010. Modelling scenarios of agriculture changes on freshwater uses and water quality at a large watershed scale – the case of the Charente watershed (France). In: D. A. Swayne, W. Yang, A. A. Voinov, A. Rizzoli, and T. Filatova, eds. *Proceedings of the International Environmental Modelling and Software Society (iEMSs) 2010 International Congress on Environmental Modelling and Software Modelling for Environment's* Sake, Fifth Biennial Meeting, Ottawa. Available from: https://www.iemss.org/iemss2010/index.php?n=Main.Proceedings (Accessed 2 April 2024).

Vervuren, P.J.A., Blom, C.W.P.M. and de Kroon, H., 2003. Extreme flooding events on the Rhine and the survival and distribution of riparian plant species, *Journal of Ecology*, **91**(1), 135–146.

Vleeshouwers, L.M. and Verhagen, A., 2002. Carbon emission and sequestration by agricultural land use: A model study for Europe, *Global Change Biology*, **8**(6), 519–530.

Voeseneck, L.A.C.J., Rijnders, J.H.G.M., Peeters, A.J.M., Van de Steeg, H.M.V. and De Kroon, H., 2004, Plant hormones regulate fast shoot elongation under water: From genes to communities, *Ecology*, **85**(1), 16–27.

Volk, V.H., 1998. Contributions towards a new evaluation of nature protection of the riparian forests of the upper Rhine, *Forstwisenschaftliches Zentralblatt*, **117**(5), 289–304.

Volk, V.H., 2002. Is ash (Fraxinus excelsior L.) native to central European floodplains? *Forstwissenschaftliches Zentralblatt*, **121**(3), 128–137.

Wacziarg, R. and Horn Welch, K., 2008. Trade liberalization and growth: new evidence. *World Bank Economic Review*, **22**(2), 187–231.

Wagner, I. and Zalewski, M., 2009. Ecohydrology as a basis for the sustainable city strategy, *Reviews in Environmental Science and Bio/Technology*, **8**(3), 209–217.

Weiss, D., Carbiener, R. and Trémolières, M., 1991. Effect of alluvial deposit geochemistry and flood frequency on phosphorus bioavailability in three alluvial Rhine forests on the Alsace Plain, *Comptes Rendus de L'Academie des Sciences Serie III – Sciences de la Vie - Life Sciences*, **313**(6), 245–251 (in French).

Weissschmitt, D. and Trémolières, M., 1993. Effect of flood frequency on the phosphorus bioavailability in two alluvial forests on the Alsace Plain (France), *Comptes Rendus de L'Academie des Sciences Serie III – Sciences de la Vie – Life Sciences*, **316**, 211–218 (in French).

Weng, P., Sanchez-Perez, J.M., Sauvage, S., Vervier, P. and Giraud, F., 2003. Assessment of the quantitative and qualitative buffer function of an alluvial wetland: Hydrological modelling of a large floodplain (Garonne River, France), *Hydrological Processes*, **17**(12), 2375–2392.

Winkel, R., 2000. Wald, Wasser oder Wildnis? Die Implementationsproblematik des "Integrierten Rheinprogramms" auf der kommunalen Ebene. Eine vergleichende Untersuchung am baden-württembergischen Oberrhein, Diplomarbeit am Institut für Forstpolitik (Freiburg: Universität Freiburg), in German.

Yang, Q., Shao, J., Scholz, M. and Plant, C., 2010. Feature selection methods for characterizing and classifying adaptive sustainable flood retention basins, *Water Research*, **45**(3), 993–1004.

Yoshida, H. and Dittrich, A., 2002. 1 D unsteady-state flow simulation of a section of the upper Rhine, *Journal of Hydrology*, **269**(1–2), 79–88.

4 Water and Wastewater Treatment Technology

4.1 CONSTRUCTED WETLANDS

4.1.1 Introduction and Definitions

Wetlands have been recognized as a natural resource throughout human history. Their importance is appreciated in their natural state by people such as the Marsh Arabs around the confluence of the rivers Tigris and Euphrates in southern Iraq, as well as in managed forms, for example, rice paddies, particularly in Southeast Asia (Mitsch and Gosselink, 2000).

The water purification capability of wetlands is now being recognized as an attractive option in wastewater treatment. For example, the Environment Agency has spent more than £1M on a reed bed scheme in South Wales, UK. This system is designed to clean up mine water from the colliery on which the CW and associated community park is being built.

Reed beds provide a useful complement to traditional sewage treatment systems. They are often a cheap alternative to expensive wastewater treatment technology such as trickling filters and activated sludge processes (Bannister, 1997; Cooper et al., 1996; Hammer, 1989; Kadlec and Knight, 1996; Robinson et al., 1999; Scholz and Xu, 2001). Vertical-flow and horizontal-flow wetlands based on soil, sand and/or gravel are used to treat domestic and industrial wastewater (Cooper et al., 1996; Decamp and Warren, 1998; Decamp et al. 1999; Green et al., 1999; Rivera et al., 1995). They are also applied for passive treatment of diffuse pollution including mine wastewater drainage (Hammer, 1989; Kadlec and Knight, 1996; Mungur et al., 1995, 1997) and urban and motorway water runoff after storm events (Green et al., 1999; Kadlec and Knight, 1996; McNeill and Olley, 1998; Scholes et al., 1999). Furthermore, wetlands serve as a wildlife conservation resource and can be seen as natural recreational areas for the local community (Hawke and José, 1999). The functions of macrophytes within CWs have been reviewed previously. *Phragmites* spp., *Typha* spp. and other swamp plants are widely used in Europe and Northern America (Brix, 1999; Kadlec and Knight, 1996).

A considerable amount of work on CWs has already been done in the UK by universities, the water authorities and the Natural History Museum. In particular, WRc Swindon, Severn Trent Water and Middlesex University have made an important textbook contribution to CW research (Cooper et al., 1996).

Defining wetlands has long been a problematic task, due to not only the diversity of environments that are permanently or seasonally influenced by water but also the specific requirements of the diverse groups of people involved with the study and management of these habitats.

DOI: 10.1201/9781003509370-4

The Ramsar Convention, which brought wetlands to the attention of the international community, proposed the following (Convention on Wetlands of International Importance Especially as Waterfowl Habitat, 1971): "Wetlands are areas of marsh, fen, peatland or water, whether natural or artificial, permanent or temporary, with water that is static or flowing, fresh, brackish or salt, including areas of marine water the depth of which at low tide does not exceed six metres".

Another, more succinct, definition is as follows (Smith, 1980): "Wetlands are a half-way world between terrestrial and aquatic ecosystems and exhibit some of the characteristics of each".

This complements the Ramsar description, since wetlands are at the interface between water and land. This concept is particularly important in areas where wetlands may only be 'wet' for relatively short periods of time in a year, such as in areas of the tropics with marked wet and dry seasons.

These definitions put an emphasis on the ecological importance of wetlands. However, the natural water purification processes occurring within these systems have become increasingly relevant to those people involved with the practical use of wetlands for water treatment. There is no single accepted ecological definition of wetlands. Wetlands are characterized by (U.S. Army Corps of Engineers, 2000) the presence of water, unique soils that differ from upland soils and vegetation adapted to saturated conditions.

Whichever definition is adopted, it can be seen that wetlands encompass a wide range of hydrological and ecological types, from high-altitude river sources to shallow coastal regions, in each case being affected by prevailing climatic conditions. For this review chapter, however, the main emphasis will be upon constructed treatment wetlands in a temperate climate. Section 4.1 is based on a review article by Scholz and Lee (2005). The original article has been revisited and updated where appropriate.

4.1.2 Hydrology of Wetlands

The biotic status of a wetland is intrinsically linked to the hydrological factors by which it is affected. These affect the nutrient availability and physicochemical parameters such as soil and water pH and anaerobiosis within soils. In turn, biotic processes will have an impact upon the hydrological conditions of a wetland.

Water being the hallmark of wetlands, it is not surprising that the input and output of water, the water budget, of these systems determine the biochemical processes occurring within them. The net result of the water budget, the hydroperiod, may show great seasonal variations but ultimately delineates wetlands from terrestrial and fully aquatic ecosystems.

From an ecological standpoint, as well as an engineering one, the importance of hydrology cannot be overstated, as it defines the species diversity, productivity and nutrient cycling of specific wetlands. Hydrological conditions must be considered if one is interested in the species richness of flora and fauna or if the interest lies in utilizing wetlands for pollution control.

The stability of wetlands is directly related to their hydroperiod – that is the seasonal shift in surface and sub-surface water levels. The terms flood duration and flood frequency refer to wetlands that are not permanently flooded and give some indication of the time involved in which the effects of inundation and soil saturation will be most pronounced.

Of relevance to riparian wetlands is the concept of flooding pulses as described elsewhere (Junk et al., 1989). These pulses cause the greatest difference in high and low water levels and benefit wetlands by the input of nutrients and washing out of waste matter that these sudden high volumes of water provide on a periodic or seasonal basis. It is particularly important to appreciate this natural fluctuation and its effects since wetland management often attempts to control the level by which waters rise and fall. Such manipulation might be due to the overemphasis placed on water and its role in the lifecycles of wetland flora and fauna, without considering that such species have evolved in such an unstable environment (Fredrickson and Reid, 1990).

The balance between the input and output of water within a wetland is called its water budget, which is summarized by Equation 4.1.

$$\Delta V/\Delta t = P_n + S_i + G_i - \mathrm{ET} - S_o - G_o \pm T \tag{4.1}$$

where V is the volume of water storage in wetlands; $\Delta V/\Delta t$ is the change in volume of water storage in wetland per unit time (t); P_n is the net precipitation; S_i is the surface inflow including flooded streams; G_i represents groundwater inflows; ET is evapotranspiration; S_o represents surface outflows; G_o indicate groundwater outflows and T either means tidal inflow (+) or outflow (–).

In general terms, wetlands are most widespread in those parts of the world where precipitation exceeds water loss through evapotranspiration and surface runoff. The contribution of precipitation to the hydrology of a wetland is influenced by several factors. Precipitation such as rain and snow often passes through a canopy of vegetation before it becomes part of the wetland. The volume of water retained by this canopy is termed the interception. Factors such as precipitation intensity and vegetation type will affect interception, for which median values of several studies have been calculated as 13% for deciduous forests and 28% for coniferous woodland (Dunne and Leopold, 1978).

The precipitation that remains to reach the wetland is termed the through fall. This is added to the stem flow, which is the water running down vegetation stems and trunks, generally considered a minor component of a wetland water budget, such as 3% of through fall in cypress dome wetlands in Florida (Heimburg, 1984). Thus, through fall and stem flow form Equation 4.2, the most commonly used precipitation equation for wetlands.

$$P_n = \mathrm{TF} + \mathrm{SF} \tag{4.2}$$

where P_n is the net precipitation; TF represents the throughfall and SF is the stem flow.

4.1.3 Wetland Chemistry

Because wetlands are associated with waterlogged soils, the concentration of oxygen within sediments and the overlying water is of critical importance. The rate of oxygen diffusion into water and sediment is slow and this coupled with microbial and animal respiration leads to near anaerobic sediments within many wetlands (Moss, 1998). These conditions favour rapid peat build-up, since decomposition rates and inorganic content of soils are low. Furthermore, the lack of oxygen in such conditions affects the aerobic respiration of plant roots and influences plant nutrient availability. Wetland plants have consequently evolved to be able to exist in anaerobic soils.

Whilst deeper sediments are generally anoxic, a thin layer of oxidized soil usually exists at the soil-water interface. The oxidized layer is important since it permits oxidized forms of prevailing ions to exist. This contrasts with the reduced forms occurring at deeper levels of soil. The state of reduction or oxidation of iron, manganese, nitrogen and phosphorus ions determines their role in nutrient availability and also toxicity. The presence of oxidized ferric iron (Fe^{3+}) gives the overlying wetland soil a brown coloration, while reduced sediments have undergone gleying, a process by which ferrous iron (Fe^{2+}) gives the sediment a blue-grey tint.

The level of reduction of wetland soils is, therefore, important in understanding the chemical processes that are most likely to occur in the sediment and influence the above water column. The most practical way to determine the reduction state is by measuring the redox potential, also called the oxidation-reduction potential, of the saturated soil or water. The redox potential quantitatively determines whether a soil or water sample is associated with a reducing or oxidizing environment. Reduction is the release of oxygen and gain of an electron (or hydrogen), while oxidation is the reverse; that is, the gain of oxygen and loss of an electron as shown in Equation 4.3 (Mitsch and Gosselink, 2000):

$$E_H = E^0 + 2.3[RT/nF]\log[\{ox\}/\{red\}] \tag{4.3}$$

where E_H is the redox potential on the hydrogen scale; E^0 is the potential of reference (mV); R represents the gas constant, which is 81.987 cal/deg/mol; T stands for temperature; n is the number of moles of electrons transferred and F is the Faraday constant that is 23,061 cal/mole-volt.

Oxidation (and therefore decomposition) of organic matter (a very reduced material) occurs in the presence of any electron acceptor, particularly O_2, although NO^{3-}, Mn^{2+}, Fe^{3+} and SO_4^{2-} are also commonly involved in oxidation, but the rate will be slower in comparison with O_2. A redox potential range between +400 and +700 mV is typical for environmental conditions associated with free dissolved oxygen (DO). Below +400 mV, the oxygen concentration will begin to diminish, and wetland conditions might become increasingly more reduced (>−400 mV). It should be noted that redox potentials are affected by pH and temperature, which influence the range at which particular reactions occur.

The following thresholds are therefore not definitive: (a) Once wetland soils become anaerobic, the primary reaction at approximately +250 mV is the reduction of nitrate (NO_3^-) to nitrite (NO_2^-) and finally to nitrous oxide (N_2O) or free nitrogen gas (N_2); (b) at about +225 mV, manganese is reduced to manganous compounds. Under further reduced conditions, ferric iron becomes ferrous iron between approximately +100 and −100 mV and sulphates become sulphides between approximately −100 and −200 mV and (c) under the most reduced conditions (<−200 mV), the organic matter itself or carbon dioxide will become the terminal electron acceptor, resulting in the formation of low-molecular-weight organic compounds and methane gas ($CH_4\uparrow$).

Organic matter within wetlands is usually degraded by aerobic respiration or anaerobic processes (e.g., fermentation and methanogenesis). Anaerobic degradation of organic matter is less efficient than decomposition occurring under aerobic conditions.

Fermentation is the result of organic matter acting as the terminal electron acceptor (instead of oxygen as in aerobic respiration). This process forms low-molecular-weight acids (e.g., lactic acid), alcohols (e.g., ethanol) and carbon dioxide. Therefore, fermentation is often central in providing further biodegradable substrates for other anaerobic organisms in waterlogged sediments.

The sulphur cycle is linked with the oxidation of organic carbon in some wetlands, particularly in sulphur-rich coastal systems. Low-molecular-weight organic compounds that result from fermentation (e.g., ethanol) are utilized as organic substrates by sulphur-reducing bacteria in the conversion of sulphate to sulphide (Mitsch and Gosselink, 2000).

Previous work suggests that methanogenesis is the principal carbon pathway in fresh water. Between 30% and 50% of the total benthic carbon flux has been attributed to methanogenesis (Boon and Mitchell, 1995).

The prevalence of anoxic conditions in most wetlands has led to them playing a particularly important role in the release of gaseous nitrogen from the lithosphere and hydrosphere to the atmosphere through denitrification (Mitsch and Gosselink, 2000). However, the various oxidation states of nitrogen within wetlands are also important to the biogeochemistry of these environments.

Nitrates are important terminal electron acceptors after oxygen, making them relevant in the process of oxidation of organic matter. The transformation of nitrogen within wetlands is strongly associated with bacterial action. The activity of bacterial groups is dependent on whether the corresponding zone within a wetland is aerobic or anaerobic.

Within flooded wetland soils, mineralized nitrogen occurs primarily as ammonium (NH_4^+). Ammonium is formed through ammonification, the process by which organically bound nitrogen is converted to ammonium nitrogen under aerobic or anaerobic conditions. Soil-bound ammonium can be absorbed through plant root systems and be reconverted to organic matter, a process that can also be performed by anaerobic microorganisms.

The oxidized top layer present in many wetland sediments is crucial in preventing the excessive build-up of ammonium. A concentration gradient will be established between the high concentration of ammonium in the lower reduced sediments and the low concentration in the oxidized top layer. This may cause a passive flow of ammonium from the anaerobic to the aerobic layer, where microbiological processes convert the ion into further forms of nitrogen.

Within the aerobic sediment layer, nitrification of ammonium, firstly to nitrite (NO_2^-) and subsequently to nitrate (NO_3^-), is shown in Equations 4.4 and 4.5, preceded by the genus of bacteria involved in each process step. Nitrification may also take place in the oxidized rhizosphere of wetland plants.

$$\text{Nitrosomonas: } 2NH_4^+ + 3O_2 \rightarrow 2NO_2^- + 2H_2O + 4H^+ + \text{energy} \quad (4.4)$$

$$\text{Nitrobacter: } 2NO_2^- + O_2 \rightarrow 2NO_3^- + \text{energy} \quad (4.5)$$

A study in southern California indicates that denitrification was the most likely pathway for nitrate loss from experimental macrocosms and larger CWs (Bachand and Horne, 1999). Very high rates of nitrate nitrogen removal were reported (2,800 mg $\times$ N $\times$ m^2/d). Furthermore, nitrate removal from inflow (waste)water is generally lower in CWs compared to natural systems (Spieles and Mitsch, 2000). There is considerable interest in enhancing bacterial denitrification in CWs to reduce the level of eutrophication in receiving waters such as rivers and lakes (Bachand and Horne, 1999).

An investigation into the seasonal variation of nitrate removal showed maximum efficiency to occur during summer (Spieles and Mitsch, 2000). This study also indicated a seasonal relationship in the pattern of nitrate retention, in which nitrate assimilation and denitrification are temperature-dependent.

Further evidence supporting the importance of denitrification is presented elsewhere (Lund et al., 2000). The proportion of nitrogen removed by denitrification from a wetland in southern California was estimated by analysing the increase in the proportion of the nitrogen isotope ^{15}N found in the outflow water. This method is based on the tendency of the lighter isotope ^{14}N to be favoured by the biochemical thermodynamics of denitrification, thus reducing its proportion in water flowing out of wetlands in which denitrification is prevalent. Denitrification seems to be the favoured pathway of nitrate loss from a treatment wetland, as this permanently removes nitrogen from the system, compared to sequestration within algal and macrophyte biomass.

In some wetlands, nitrogen may be derived through nitrogen fixation. In the presence of the enzyme nitrogenase, nitrogen gas is converted to organic nitrogen by organisms such as aerobic or anaerobic bacteria and cyanobacteria (blue-green algae). Wetland nitrogen fixation can occur in the anaerobic or aerobic soil layer, overlying water, rhizosphere of plant roots and on leaf or stem surfaces. Cyanobacteria may contribute significantly to nitrogen fixation. In northern bogs, which are often too acidic for large bacterial populations, nitrogen fixation by cyanobacteria is particularly important (Etherington, 1983). However, it should be noted that while cyanobacteria are adaptable organisms, they are affected by environmental stresses. For example, cyanobacteria are particularly susceptible to ultraviolet radiation, whereby their nitrogen metabolism (along with other functions) is impaired (Donkor and Häder, 1996).

In wetland soils, phosphorus occurs as soluble or insoluble, organic or inorganic complexes. Its cycle is sedimentary rather than gaseous (as with nitrogen) and predominantly forms complexes within organic matter in peatlands or inorganic sediments in mineral soil wetlands. More than 90% of the phosphorus load in streams and rivers may be present in particulate inorganic form (Overbeck, 1988).

Soluble reactive phosphorus is the analytical term given to biologically available orthophosphate, which is the primary inorganic form. The availability of phosphorus to plants and microconsumers is limited due to the following main effects: (a) Under aerobic conditions, insoluble phosphates are precipitated with ferric iron, calcium and aluminium; (b) phosphates are adsorbed onto clay particles, organic peat and ferric/aluminium hydroxides and oxides and (c) phosphorus is bound up in organic matter through incorporation in bacteria, algae and vascular macrophytes.

There are three general conclusions about the tendency of phosphorus to precipitate with selected ions (Reddy et al., 1999): In acid soils, phosphorus is fixed as aluminium and iron phosphates; in alkaline soils, phosphorus is bound by calcium and magnesium and the bioavailability of phosphorus is greatest at neutral to slightly acid pH.

Under anaerobic wetland soil conditions, phosphorus availability is altered. The reducing conditions that are typical of flooded soils do not directly affect phosphorus. However, the association of phosphorus with other elements that undergo reduction has an indirect effect upon phosphorus in the environment. For example, as ferric iron is reduced to the more soluble ferrous form, phosphorus as ferric phosphate (reductant-soluble phosphorus) is released into solution (Faulkner and Richardson, 1989; Gambrell and Patrick, 1978). Phosphorus may also be released into solution by a pH change brought about by organic, nitric or sulphuric acids produced by chemosynthetic bacteria. Phosphorus sorption to clay particles is greatest under strongly acidic to slightly acidic conditions (Stumm and Morgan, 1996).

Great temporal variability in phosphorus concentrations of wetland influent in Ohio has been reported (Nairn and Mitsch, 2000). However, no seasonal pattern in phosphorus concentration was observed. This was explained by precipitation events and river flow conditions. Dissolved reactive phosphorus levels peaked during floods and on isolated occasions in late autumn. Furthermore, sedimentation of suspended solids appears to be important in phosphorus retention within wetlands (Fennessy et al., 1994).

The physical, chemical and biological characteristics of a wetland system affect the solubility and reactivity of different forms of phosphorus. Phosphate solubility has been shown to be regulated by temperature (Holdren and Armstrong, 1980), pH (Mayer and Kramer, 1986), redox potential (Moore and Reddy, 1994), interstitial soluble phosphorus level (Kamp-Nielson, 1974) and microbial activity (Gächter and Meyer, 1993; Gächter et al., 1988).

Where agricultural land has been converted to wetlands, there can be a tendency in solubilization of residual fertilizer phosphorus, which results in a rise of the soluble phosphorus concentration in floodwater. This effect can be reduced by physicochemical amendment, applying chemicals such as alum and calcium carbonate to stabilize the phosphorus in the sediment of these new wetlands (Ann et al., 1999a, 1999b).

The redox potential has significant effects on dissolved reactive phosphorus of chemically amended soils (Ann et al., 1999a, 1999b). The redox potential can alter with fluctuating water-table levels and hydraulic loading rates (HLRs). Dissolved phosphorus concentrations are relatively high under reduced conditions and decrease with increasing redox potential. Iron compounds (e.g., $FeCl_3$) are particularly sensitive to the redox potential, resulting in the chemical amendment of wetland soils. Furthermore, alum and calcium carbonate are suitable to bind phosphorus even during fluctuating redox potentials.

Macrophytes assimilate phosphorus predominantly from deep sediments, thereby acting as nutrient pumps (Carignan and Kaill, 1980; Mitsch and Gosselink, 2000; Smith and Adams, 1986). The most important phosphorus retention pathway in wetlands is via physical sedimentation (Wang and Mitsch, 2000).

Model simulations on CWs in north-eastern Illinois (U.S.) showed an increase in total phosphorus in the water column in the presence of macrophytes mainly during the non-growing period, with little effect during the growing season. Most phosphorus taken from sediments by macrophytes is reincorporated into the sediment as dead plant material and therefore remains in the wetland indefinitely. Macrophytes can be harvested to enhance phosphorus removal in wetlands. By harvesting macrophytes at the end of the growing season, phosphorus can be removed from the internal nutrient cycle within wetlands. Moreover, the model showed a phosphorus removal potential of three-quarters of that of the phosphorus inflow. Therefore, harvesting would reduce phosphorus levels in upper sediment layers and drive phosphorus movement into deeper layers, particularly the root zone. In deep layers of sediment, the phosphorus sorption capacity increases along with a lower desorption rate (Wang and Mitsch, 2000).

In wetlands, sulphur is transformed by microbiological processes and occurs in several oxidation stages. Reduction may occur between −100 and −200 mV on the redox potential scale. Sulphides provide the characteristic 'bad egg' odour of wetland soils.

Assimilatory sulphate reduction is accomplished by obligate anaerobes such as *Desulfovibrio* spp. Bacteria may use sulphates as terminal electron acceptors (Equation 4.6) in anaerobic respiration at a wide pH range, but highest around neutral (Mitsch and Gosselink, 2000).

$$4H_2 + SO_4^{2-} \uparrow H_2S^- + 2H_2O + 2OH^- \tag{4.6}$$

The greatest loss of sulphur from freshwater wetlands to the atmosphere is via hydrogen sulphide ($H_2S\uparrow$). In oceans, however, this is through the production of dimethyl sulphide from decomposing phytoplankton (Schlesinger, 1991).

Oxidation of sulphides to elemental sulphur and sulphates can occur in the aerobic layer of some soils and is carried out by chemoautotrophic (e.g., *Thiobacillus* spp.) and photosynthetic microorganisms. *Thiobacillus* spp. may gain energy from the oxidation of hydrogen sulphide to sulphur and further, by certain other species of the genus, from sulphur to sulphate (Equations 4.7 and 4.8)

$$2H_2S + O_2 \rightarrow 2S + 2H_2O + \text{energy} \tag{4.7}$$

$$2S + 3O_2 + 2H_2O \rightarrow 2H_2SO_4 + \text{energy} \tag{4.8}$$

In the presence of light, photosynthetic bacteria, such as purple sulphur bacteria of salt marshes and mud flats, produce organic matter as indicated in Equation 4.9. This is similar to the familiar photosynthesis process, except that hydrogen sulphide is used as the electron donor instead of water.

$$CO_2 + H_2S + \text{light} \rightarrow CH_2O + S \tag{4.9}$$

Direct toxicity of free sulphide in contact with plant roots has been noted. There is a reduced toxicity and availability of sulphur for plant growth if it precipitates with trace metals. For example, the immobilization of zinc and copper by sulphide precipitation is well known.

The input of sulphates to freshwater wetlands, in the form of Aeolian dust or as anthropogenic acid rain, can be significant. Sulphate deposited on wetland soils may undergo dissimilatory sulphate reduction by reaction with organic substrates (Equation 4.10).

$$2CH_2O + SO_4{}^{2-} + H^+ \rightarrow 2CO_2 + HS^- + 2H_2O \tag{4.10}$$

Protons consumed during this reaction generate alkalinity. This is illustrated by the increase in pH with depth in wetland sediments (Morgan and Mandernack, 1996). It has been suggested that this 'alkalinity effect' can act as a buffer in acid rain affected lakes and streams (Rudd et al., 1986; Spratt and Morgan, 1990).

The sulphur cycle can vary greatly within different zones of a particular wetland. The stable isotope $\delta^{34}S$ within peat, the$^{35}SO_4{}^{2-}$:Cl^- ratio and the stable isotopes $\delta^{18}O$ and $\delta^{34}S$ of sulphate within different waters were analysed, previously. The variability in the sulphur cycle within the watershed can affect the distribution of reduced sulphur stored in soil. This change in local sulphur availability can have marked effects upon stream water over short distances (Mandernack et al., 2000).

The estimation of generated alkalinity may be complicated due to the potential problems associated with the use of the$^{35}SO_4{}^{2-}$: Cl^- ratio and/or $\delta^{34}S$ value to estimate the net sulphur retention. These problems may exist because ester sulphate pools can be a source of sulphate availability for sulphate reduction and as a $\delta^{34}S$ sulphate buffer within stream water.

4.1.4 Wetland Mass Balance

The general mass balance for a wetland, in terms of chemical pathways, uses the following main pathways: inflows, intrasystem cycling and outflows. The inflows are mainly through hydrologic pathways such as precipitation, surface water runoff and groundwater. The photosynthetic fixation of atmospheric carbon nitrogen is important biological pathway. Intrasystem cycling is the movement of chemicals in standing stocks within wetlands, such as litter production and remineralization. Translocation of minerals within plants is an example of physical movement of chemicals. Outflows involve hydrologic pathways but also include the loss of chemicals to deeper sediment layers, beyond the influence of internal cycling (although the depth at which this threshold occurs is not certain). Furthermore, the nitrogen cycle plays an important role in outflows, such as nitrogen gas lost because of denitrification. However, respiratory loss of carbon is also an important biotic outflow.

There is great variation in the chemical balance from one wetland to another. However, the following generalizations may be made:

- Wetlands act as sources, sinks or transformers of chemicals depending on wetland type, hydrological conditions and length of time the wetland has received chemical inputs. As sinks, the long-term sustainability of this function is associated with hydrologic and geomorphic conditions as well as the spatial and temporal distribution of chemicals within wetlands.
- Particularly in temperate climates, seasonal variation in nutrient uptake and release is expected. Chemical retention will be greatest in the growing seasons (spring and summer) due to higher rates of microbial activity and macrophyte productivity.
- The ecosystems connected to wetlands affect and are affected by the adjacent wetland. Upstream ecosystems are the sources of chemicals, while those downstream may benefit from the export of certain nutrients or the retention of chemicals.
- Nutrient cycling in wetlands differs from that in terrestrial and aquatic systems. More nutrients are associated with wetland sediments than with most terrestrial soils, while benthic aquatic systems have autotrophic activity which relies more on nutrients in the water column than in the sediments.
- The ability of wetlands to remove anthropogenic waste is not limitless.

Equation 4.11 indicates a general mass balance for a pollutant within a treatment wetland. Within this equation, transformations and accretion are long-term sustainable removal processes, while storage does not serve in long-term average removal but can lessen or accentuate the cyclic activity.

$$\text{In-Out} = \text{Transformation} + \text{Accretion} + \text{Biomass Storage} + \text{Water/Soil Storage} \tag{4.11}$$

4.1.5 Macrophytes

Wetland plants are often central to wastewater treatment wetlands. The following requirements of plants should be considered for use in such systems (Tanner, 1996): (a) ecological adaptability (no disease or weed risk to the surrounding natural ecosystems); (b) tolerance of local conditions in terms of climate, pests and disease; (c) tolerance of pollutants and hypertrophic waterlogged conditions; (d) ready propagation, rapid establishment, spread and growth and (e) high pollutant removal capacity, through direct assimilation or indirect enhancement of nitrification, denitrification and other microbial processes.

Interest in macrophyte systems for sewage treatment by the UK water industry dated back to 1985 (Parr, 1990). The ability of macrophyte species and their assemblages within systems to treat wastewater most efficiently has been examined, previously (Kuehn and Moore, 1995). The dominant species of macrophyte varies from locality to locality. The number of genera (e.g., *Phragmites* spp., *Typha* spp. and *Scirpus* spp.) common to all temperate locations is high. The improvement of water quality with respect to key parameters including biochemical oxygen demand (BOD), chemical oxygen demand (COD), total suspended solids (TSS), nitrates and phosphates has been studied (Turner, 1995). Relatively little work has been conducted on the enteric bacteria removal capability of macrophyte systems (Perkins and Hunter, 2000).

There have been many studies to determine the primary productivity of wetland macrophytes, although estimates have generally tended to be high (Mitsch and Gosselink, 2000). The estimated dry mass production for *Phragmites australis* (Cav.) Trin. ex Steud. (common reed) is 1,000–6,000 g/m^2 × a in the Czech Republic (Kvet and Husak, 1978), 2,040–2,210 g/m^2 × a for *Typha latifolia* L. (cattail) in Oregon, U.S. (McNaughton, 1966), and 943 g/m^2 × a for *Scirpus fluviatilis* (Torr.) A. Gray [JPM][H&C] (river bulrush) in Iowa, U.S. (van der Valk and Davis, 1978). Little of this plant biomass is consumed as live tissue; it rather enters the pool of particulate organic matter following tissue death. The breakdown of this material is consequently an important process in wetlands and other shallow aquatic habitats (Gessner, 2000). Litter breakdown has been studied along with extensive work on one of the most widespread aquatic macrophyte; *P. australis* (Wrubleski et al., 1997).

There has been an emphasis on studying the breakdown of aquatic macrophytes in such a way as most closely resembles that of natural plant death and decomposition, principally by not removing plant tissue from macrophyte stands. Many species of freshwater plants exhibit so-called standing-dead decay, which describes the observation of leaves remaining attached to their stems after senescence and death (Kuehn et al., 1999). Different fractions (leaf blades, leaf sheaths and culms) of *P. australis* differ greatly in structure and chemical composition and may exhibit different breakdown rates, patterns and nutrient dynamics (Gessner, 2000).

Phragmites australis (Cav.) Trin. ex Steud. (common reed), formerly known as *Phragmites communes* (Norfolk reed), is a member of the large family Poaceae (roughly 8,000 species within 785 genera). Common reed occurs throughout Europe to 70° north and is distributed worldwide. It may be found in permanently flooded soils of still or slowly flowing water. This emergent plant is usually firmly rooted in wet sediment but may form lightly anchored rafts of ‘hover reed’. It tends to be replaced by other species at drier sites. The density of this macrophyte is reduced by grazing (e.g., by waterfowl) and may then be replaced by other emergent species such as *Phalaris arundinacea* L. (reed canary grass).

P. australis is a perennial, with shoots emerging in spring. Hard frost kills these shoots, illustrating the tendency for reduced vigour towards the northern end of its range. The hollow stems of the dead shoots in winter are important in transporting oxygen to the relatively deep rhizome (Brix, 1989).

Reproduction in closed stands of this species is mainly by vegetative spread, although seed germination enables the colonization of open habitats. Detached shoots often survive and regenerate away from the main stand (Preston and Croft, 1997).

Common reed or Norfolk reed is most common in nutrient-rich sites and absent from the most oligotrophic zones. However, the stems of this species may be weakened by nitrogen-rich water and are subsequently more prone to wind and wave damage, leading to an apparent reduction in density of this species in Norfolk (England) and elsewhere in Europe (Boar et al., 1989; Ostendorp, 1989).

T. latifolia L. (cattail, reedmace and bulrush) is a species belonging to the small family Typhaceae. This species is widespread in temperate parts of the Northern Hemisphere but extends to South Africa, Madagascar, Central America and the West Indies and is naturalized in Australia (Preston and Croft, 1997). It is typically found in shallow water or on exposed mud at the edge of lakes, ponds, canals and ditches and less frequently near fast flowing water. This species rarely grows at water depths below 0.3 m, where it is frequently replaced by *P. australis*. Reedmace is a shallow-rooted perennial producing shoots throughout the growing season, which subsequently die in autumn. Colonies of this species expand by rhizomatous growth at rates of 4 m/a, while detached portions of rhizome may float and establish new colonies (Hammer, 1989).

In contrast, colony growth by seeds is less likely. Seeds require moisture, light and relatively high temperatures to germinate, although this may occur in anaerobic conditions. Where light intensity is low, germination is stimulated by temperature fluctuation.

4.1.6 Physical and Biochemical Processes and Parameters

The key physicochemical parameters relevant for wetlands include the BOD, turbidity and the redox potential. The BOD is an empirical test to determine the molecular oxygen used during a specified incubation period (usually five days), for the biochemical degradation of organic matter (carbonaceous demand) and the oxygen used to oxidize inorganic matter (e.g., sulphides and ferrous iron). An extended test (up to 25 days) may also measure the amount of oxygen used to oxidize reduced forms of nitrogen (nitrogenous demand), unless this is prevented by an inhibitor chemical (Scholz, 2004a, 2004b). Inhibiting the nitrogenous oxygen demand is recommended for secondary effluent and pollution samples (American Public Health Association, 1995).

The European Union freshwater fisheries directive sets an upper BOD limit of 3 mg/L for salmonid rivers and 6 mg/L for coarse fisheries. A river is deemed polluted if the BOD exceeds 5 mg/L. Municipal wastewater values are usually between approximately 150 and 1,000 mg/L (Kiely, 1997).

Turbidity is a measure of the cloudiness of water, caused predominantly by suspended material, such as clay, silt, organic and inorganic matter, plankton and other microscopic organisms, scattering and absorbing light. Turbidity in wetlands and lakes is often due to colloidal or fine suspensions, while in fast flowing waters the particles are larger and turbid conditions are prevalent during flood times (Kiely, 1997).

The redox potential is another key parameter for monitoring wetlands. The reactivities and mobilities of elements such as Fe, S, N, C and a number of metallic elements depend strongly on the redox potential conditions. Reactions involving electrons and protons are pH and redox potential dependent. Chemical reactions in aqueous media can often be characterized by pH and redox potential together with the activity of dissolved chemical species. The redox potential is a measure of intensity and does not represent the capacity of the system for oxidation or reduction (American Public Health Association, 1995). The interpretation of redox potential values measured in the field is limited by several factors, including irreversible reactions, 'electrode poisoning' and multiple redox couples.

4.1.7 Natural and Constructed Wetlands

4.1.7.1 Riparian Wetlands

Riparian wetlands are ecosystems under the influence of adjacent streams or rivers (Scholz and Trepel, 2004). A succinct definition is as follows (Gregory et al., 1991):

> Riparian zones are the interface between terrestrial and aquatic ecosystems. As ecotones, they encompass sharp gradients of environmental factors, ecological processes and plant communities. Riparian zones are not easily delineated but are composed of mosaics of landforms, communities, and environments within the larger landscape.

There are four main reasons as to why the periodic flooding, which is typical of riparian wetlands, contributes to the observed higher productivity compared to adjacent upland ecosystems: (a) there is an adequate water supply for vegetation; (b) nutrients are supplied and coupled with a favourable change in soil chemistry (e.g., nitrification, sulphate reduction and nutrient mineralization); (c) in comparison to stagnant water conditions, a more oxygenated root zone follows flooding and (d) waste products (e.g., carbon dioxide and methane) are removed by the periodic 'flushing'.

Nutrient cycles in riparian wetlands can be described as follows: (a) nutrient cycles are 'open' because of river flooding, runoff from upslope environments or both (depending on season and inflow stream or river type) and (b) riparian forests have a great effect on the biotic interactions within intrasystem nutrient cycles. The seasonal pattern of growth and decay often matches available nutrients; (c) water in contact with the forest floor leads to important nutrient transformations. Therefore, riparian wetlands can act as sinks for nutrients that enter as runoff and groundwater flow; and (d) riparian wetlands have often appeared to be nutrient transformers, changing a net input of inorganic nutrients to a net output of their corresponding organic forms.

The nitrogen cycle within a temperate stream-floodplain environment is of particular interest to ecological engineers. During winter, flooding contributes to the accumulation of dissolved and particulate organic nitrogen that is not assimilated by the trees due to their dormancy. This fraction of nitrogen is retained by filamentous algae and through immobilization by detritivores on the forest floor. As the waters warm in spring, nitrogen is released by decomposition and by shading of the filamentous algae by the developing tree canopy. Nitrate is then assumed to be immobilized in the decaying litter and gradually made available to plants. As vegetation increases, the plants take up more nitrogen and water levels fall due to evapotranspiration.

Ammonification and nitrification rates increase with exposure of the sediments to the atmosphere. Nitrates produced in nitrification are lost when denitrification becomes prevalent as flooding later in the year creates anaerobic conditions.

In terms of reducing the effects of eutrophication of open water by runoff, the use of riparian buffer zones, particularly of *Alnus incana* (grey alder) and *Salix* spp. (willow) in conjunction with perennial grasses, has been recommended (Mander et al., 1995). Riparian zones are also termed riparian forest buffer systems (Lowrance et al., 1979). Such zones were found to reduce the nutrient flux into streams.

The role of riparian ecosystems in nutrient transformations is specifically important in relation to the production of the greenhouse gas nitrous oxide (N_2O). Due to the inflow of excess agricultural nitrogen into wetland systems, the riparian zones are likely 'hot spots' for nitrous oxide production (Groffman et al., 2000).

The control of non-point source pollution can be successfully achieved by riparian forest buffers in some agricultural watersheds and most effectively if excess precipitation moves across, in or near the root zone of the riparian forest buffers. For example, between 50% and 90% retention of total nitrate loading in both shallow groundwater and sediment subject to surface runoff within the Chesapeake Bay watershed (U.S.) was observed. In comparison, phosphorus retention was found to be generally much less (Lowrance et al., 1979).

4.1.7.2 Constructed Wetlands

Natural wetlands usually improve the quality of water passing through them, acting in effect as ecosystem filters. Constructed wetlands are artificially created wetlands used to treat water pollution in its variety of forms. Therefore, they fall into the category of treatment wetlands. Treatment wetlands are solar-powered ecosystems. Solar radiation varies diurnally, as well as on an annual basis (Kadlec, 1999).

Constructed wetlands have the purpose to remove bacteria, enteric viruses, suspended solids, BOD, nitrogen (as ammonia and nitrate), metals and phosphorus (Pinney et al., 2000). Two general forms of CWs are used in practice: surface-flow (horizontal-flow) and sub-surface-flow (vertical-flow). Surface-flow CWs most closely mimic natural environments and are usually more suitable for wetland species because of permanent standing water. In sub-surface-flow wetlands, water passes laterally through a porous medium (usually sand and gravel) with a limited number of macrophyte species. These systems have often no standing water.

Constructed treatment wetlands can be built at, above or below the existing land surface if an external water source is supplied (e.g., wastewater). The grading of a particular wetland in relation to the appropriate elevation is important for the optimal use of the wetland area in terms of water distribution. Soil type and groundwater level must also be considered if long-term water shortage is to be avoided. Liners can prevent excessive desiccation, particularly where soils have a high permeability (e.g., sand and gravel) or where there is limited or periodic flow.

Rooting substrate is also an important consideration for the most vigorous growth of macrophytes. A loamy or sandy topsoil layer of 20–30 cm in depth is ideal for most wetland macrophyte species in a surface-flow wetland. A sub-surface-flow wetland will require coarser material such as gravel and/or coarse sand (Kadlec and Knight, 1996).

The use of flue-gas-desulphurization by-products from coal-fired electric power plants in wetland liner material was investigated previously (Ahn et al., 2001). These by-products are usually sent to landfill sites. This is now recognized as an increasingly unsuitable and impractical waste disposal method. Although this study was short (two years), no detrimental impact on macrophyte biomass production was reported. Moreover, flue-gas-desulphurization material may be a good substrate and liner for phosphorus retention in CWs.

The understanding of chemical transformations in constructed treatment wetlands has become a main research focus. Dissolved organic carbon is an important parameter in potable water treatment due to its reaction with disinfectants (e.g., chlorine) to form carcinogenic by-products, such as trihalomethanes. The transformations of dissolved organic carbon through a CW were observed, previously. The following conclusions with implications for treatment wetland design were made (Pinney et al., 2000):

- High levels of dissolved organic carbon may enter water supplies where soil aquifer treatment is used for groundwater recharge, as the influent for this method is likely to come from long hydraulic retention time (HRT) wetlands. There is consequently a greater potential for the formation of disinfection by-products.
- Shorter HRT will result in less dissolved organic carbon leaching from plant material compared to longer HRT in a wetland.
- Dissolved organic carbon leaching is likely to be most significant in wetlands designed for ammonia removal, which requires long HRT.

4.1.7.3 Constructed Storm Water Treatment Wetlands

Most CWs in the U.S. and Europe are soil- or gravel-based horizontal-flow systems planted with *T. latifolia* and/or *P. australis*. They are used to treat storm runoff as well as agricultural, domestic and industrial wastewater (Cooper et al., 1996; Kadlec and Knight, 1996; Scholz, 2003; Scholz et al., 2002, 2007), and have also been applied for passive treatment of mine wastewater drainage (Mays and Edwards, 2001; Mungur et al., 1997).

Storm runoff from urban areas has been recognized as a major contributor to pollution of the receiving urban watercourses. The principal pollutants in urban runoff are BOD suspended solids, heavy metals, de-icing salts, hydrocarbons and faecal coliforms (Scholz, 2004a, 2004b; Scholz and Martin, 1998).

Although various conventional methods have been applied to treat storm water, most technologies are not cost-effective or too complex. CWs integrated into a best management practice concept are a sustainable means of treating storm water and prove to be more economical (e.g., construction and maintenance) and energy efficient than traditional centralized treatment systems (Kadlec et al., 2000; Scholz et al., 2004). Furthermore, wetlands enhance biodiversity and are less susceptibility to variations of loading rates (Cooper et al., 1996; Scholz and Trepel, 2004).

Contrary to standard domestic wastewater treatment technologies, storm water (gully pot liquor and effluent) treatment systems must be robust to highly variable flow rates and water quality variations. The storm water quality depends on the load of pollutants present on the road, and the corresponding dilution by each storm event (Scholz, 2003; Scholz and Trepel, 2004).

In contrast to standard horizontal-flow constructed treatment wetlands, vertical-flow wetlands are flat, intermittently flooded and drained, allowing air to refill the soil pores within the bed (Cooper et al., 1996; Gervin and Brix, 2001; Green et al., 1998). When the wetland is dry, oxygen (as part of the air) can enter the top layer of debris and sand. The following flow of runoff will absorb the gas and transport it to the anaerobic bottom of the wetland. Furthermore, aquatic plants such as macrophytes transport oxygen to the rhizosphere. However, this natural process of oxygen enrichment is not as effective as the previously explained engineering method (Kadlec and Knight, 1996; Karathanasis et al., 2003).

While it has been recognized that vertical-flow CWs have usually higher removal efficiencies with respect to organic pollutants and nutrients in comparison to horizontal-flow wetlands, denitrification is less efficient in vertical-flow systems (Luederitz et al., 2001).

Heavy metals within storm water are associated with fuel additives, car body corrosion and tire and brake wear. Common metal pollutants from cars include copper, nickel, lead, zinc, chromium and cadmium. The quality standards of freshwater are commonly exceeded by copper (Cooper et al., 1996; Tchobanoglous et al., 2003; Scholz et al., 2002).

Metals occur in soluble, colloidal or particulate forms. Heavy metals are most bioavailable when they are soluble, either in ionic or weakly complexed form (Cheng et al., 2002; Cooper et al., 1996; Wood and Shelley, 1999).

There have been many studies on the specific filter media within CWs to treat heavy metals economically, such as limestone, lignite, activated carbon (Scholz and Martin, 1998), peat and leaves. Metal bioavailability and reduction are controlled by chemical processes including acid volatile sulphide formation and organic carbon binding and sorption in reduced sediments of CWs (Kadlec, 2002; Obarska-Pempkowiak and Klimkowska, 1999; Wood and Shelley, 1999). It follows that metals usually accumulate in the top layer (fine aggregates, sediment and litter) of vertical-flow and near the inlet of horizontal-flow constructed treatment wetlands (Cheng et al., 2002; Scholz and Xu, 2002; Vymazal and Krasa, 2003).

Physical and chemical properties of the wetland soil and aggregates affecting metal mobilization include particle size distribution (texture), redox potential, pH, organic matter, salinity and the presence of inorganic matter such as sulphides and carbonates (Backstrom, 2004).

The cation-exchange capacity of maturing wetland soils and sediments tend to increase as texture becomes finer because more negatively charged binding sites are available. Organic matter has a relative high proportion of negatively charged binding sites. Salinity and pH can influence the effectiveness of the cation-exchange capacity of soils and sediments because the negatively charged binding sites will be occupied by a high number of sodium or hydrogen cations (Knight et al., 1999).

Sulphides and carbonates may combine with metals to form relatively insoluble compounds. Especially the formation of metal sulphide compounds may provide long-term heavy metal removal because these sulphides will remain permanently in the wetland sediments as long as they are not re-oxidized (Cooper et al., 1996; Kadlec and Knight, 1996).

4.1.8 Case Study Evaluation

4.1.8.1 Purpose

The aim is to assess the treatment efficiencies for gully pot liquor of experimental vertical-flow CW filters containing *P. australis* and filter media of different adsorption capacities. Six out of twelve filters received inflow water spiked with metals. For two years, hydrated nickel and copper nitrate were added to sieved gully pot liquor to simulate contaminated primary treated storm runoff.

For those six CW filters receiving heavy metals, an obvious breakthrough of dissolved nickel was recorded after road salting during the first winter. However, a breakthrough of nickel was not observed since the inflow pH was raised to eight after the first year of operation. High pH facilitated the formation of particulate metal compounds such as nickel hydroxide.

During the second year, reduction efficiencies of heavy metal, BOD and suspended solids improved considerably. Concentrations of BOD were frequently <20 mg/L, an international threshold for secondary wastewater treatment. This is likely due to biomass maturation, and the increase of pH.

The case study is based on data and observations from a research project (since September 2002). Only the methodology and preliminary data (less than six months) have been published previously (Scholz, 2004a). The following sections focus on the principle observations relevant for this review chapter and not on specific data that may be only of importance for this particular experimental set-up.

The major purpose of this case study is to improve the design, operation and management guidelines of constructed treatment wetlands to secure a high wastewater treatment performance during all seasons particularly in cold climates. The main processes that have been discussed as part of the literature review will be looked at from a practical point of view. The objectives are to assess (a) the performance of vertical-flow constructed treatment wetland filters; (b) the compliance with water quality standards in terms of the reduction efficiencies of experimental CWs treating gully pot liquor receiving high loads of BOD, suspended solids, nickel, copper, nitrate and turbidity; (c) the impact of environmental conditions, such as variations of salt concentrations, pH an d temperature on the treatment performance of CWs during all seasons; (d) heavy metal leaching at low and high pH levels and (e) the overall role of adsorption media and *P. australis*.

4.1.8.2 Location, Materials and Methods

Twelve wetland filters (Figure 4.1) were located outdoors at The King's Buildings campus (The University of Edinburgh, Scotland) to assess the system performance (09/09/02 to 21/09/04). The 12 first days of operation were not analysed because the water quality was not representative. Inflow water, polluted by road runoff, was collected from randomly selected gully pots on the campus, the nearby predominantly housing estates and two major roads. After mixing both the sediment and the water phase within the gully pot, water was collected by manual abstraction with a 2-L beaker.

Round drainage pipes were used to construct the filters. All 12 vertical-flow wetland filters (Figure 4.1) were designed with the following dimensions: height of 83 cm

FIGURE 4.1 Constructed treatment wetland rig (The King's Buildings campus; The University of Edinburgh) in May 2004.

and diameter of 10 cm. In September 2002, the calculated empty filter bed volumes were approximately 6.2, 6.4, 4.0, 4.1, 3.8, 4.1, 3.8, 4.0, 3.8, 4.0, 4.0 and 4.0 l for Filters 1–12, respectively. The filter volume capacities were measured by draining the filters entirely.

Different packing order arrangements of filter media and plant roots were used in the wetland filters (Tables 4.1and 4.2). The outlet of each CW comprised a valve at the bottom of each filter.

The inflow waters of Filter 2 and Filters 7–12 were dosed with hydrated copper nitrate ($Cu(NO_3)_2.3H_2O$) and hydrated nickel nitrate ($Ni(NO_3)_2.6H_2O$). Filters 1 and 2 (controls) are like wastewater stabilization ponds or gully pots (extended storage) without a significant amount of filter media (Table 4.1). In comparison, Filters 3, 5, 7 and 9 are similar to gravel and slow sand filters, and Filters 4, 6, 8 and 10 are typical reed bed filters. The reed bed filters contain gravel and sand substrate and native *P. australis*, all of similar total biomass weight during planting and from the same local source. However, Filters 5, 6, 9 and 10 also contain adsorption media. Additional natural adsorption media (Filtralite and Frogmat) were used. Filtralite (containing 3% of calcium oxide (CaO)) with diameters between 1.5 and 2.5 mm is associated with enhanced metal and nutrient reduction (Brix et al., 2001; Scholz and Xu, 2002). Furthermore, Frogmat (natural product based on raw barley straw) has a high adsorption area and is therefore likely to be associated with a high heavy metal reduction potential. The use of other filter media with high adsorption capacities such as activated carbon (Scholz and Martin, 1998; Scholz et al., 2002) and oxide-coated sand (Sansalone, 1999) has been discussed elsewhere.

Filters 11 and 12 are more complex in their design and operation (Table 4.1). The top water layer of both filters is aerated (with air supplied by air pumps) to enhance oxidation (minimizing zones of reducing conditions) and nitrification (Cheng et al., 2002; Green et al., 1998; Obarska-Pempkowiak and Klimkowska, 1999). Filter 12 receives about 153% of Filter 11's mean annual inflow volume and load (Table 4.1). The hydraulic regime of Filter 12 differs from that of Filters 1–11 to identify the best filtration performance. A higher hydraulic load should result in greater stress on *P. australis* and biomass.

TABLE 4.1
Systematic and Stratified Experimental Set-up of Filter Content and Operation

Filter	Planted	Media Type[a]	Plus Metals[b]	Aerated	High Loading[c]
1	No	1	No	No	No
2	No	1	Yes	No	No
3	No	2	No	No	No
4	Yes	2	No	No	No
5	No	3	No	No	No
6	Yes	3	No	No	No
7	No	2	Yes	No	No
8	Yes	2	Yes	No	No
9	No	3	Yes	No	No
10	Yes	3	Yes	No	No
11	Yes	3	Yes	Yes	No
12	Yes	3	Yes	Yes	Yes

[a] 1, No media; 2, standard; 3, addition of Filtralite (light expanded clay product) and Frogmat (barley straw), see also Table 4.2;

[b] addition of hydrated copper and nickel nitrate;

[c] Filter 12 received approximately 153% additional inflow in comparison to, for example, Filter 11.

The filtration system was designed to operate in batch flow mode to reduce pumping and computer control costs. All filters were periodically inundated with pre-treated inflow gully pot liquor and partially drained (50%) or fully drained (100%) to encourage air penetration through the aggregates (Cooper et al., 1996; Gervin and Brix, 2001; Scholz, 2004a; Scholz and Xu, 2002). The theoretical hydraulic residence time (about two to seven days) was variable. It follows that the hydraulic conductivity is also highly variable due to frequent wetting and drying cycles.

Since 22 September 2003, the pH value of the inflow has been artificially raised by addition of sodium hydroxide (NaOH) to the sieved gully pot liquor. It follows that the inflow pH was therefore increased from a mean pH 6.7 to pH 8.1. The analytical methods have been described previously (Scholz et al., 2002; Scholz, 2004a).

4.1.8.3 Findings and Discussion

The pH of the inflow was artificially raised to assess its influence on the treatment performance and particularly on the potential breakthrough of heavy metals during the second winter. Raw gully pot liquor was sieved (pore size of 2.5 mm) to simulate preliminary treatment (Tchobanoglous et al., 2003). Sieving resulted in a mean annual reduction of the BOD and suspended solids by approximately 12% and 22%, respectively.

The inflow data set was divided into two sub-sets (winter and summer) to assess the effect of seasonal variations (e.g., temperature) and road management (e.g., road gritting and salting) on the water quality. Most variables including BOD (except for the first year of operation), suspended solids, total solids, turbidity and conductivity were relatively high in winter compared to summer (Scholz, 2004a).

Concerning BOD removal, the performances of all filters (except for Filters 1 and 2; extended storage) improved greatly over time. The reductions in BOD were also satisfactory for most filters, if compared to minimum American and European standards (<20 mg/L) for the secondary treatment of effluent.

Furthermore, the artificial increase of pH after the first year of operation had no apparent influence on the treatment performance of the BOD. There is no obvious difference in performance between Filters 8 and 11, indicating that aeration did not contribute significantly to the removal of the BOD. This has been confirmed by an analysis of variance.

In contrast to previous researchers who reported the worst seasonal performance for BOD removal during winter (Karathanasis et al., 2003), all filters with exception of Filters 1 and 2 showed high BOD removal figures (>94%) in the second winter. This suggests that soil microbes still have the capacity to decompose organic matter in winter.

Concerning other variables, reduction rates for suspended solids increased also in the second year although outflow concentrations exceeded frequently the threshold of 30 mg/L throughout the year except for summer. Turbidity values of the outflow decreased greatly over time. In contrast, conductivity removal deteriorated. Despite the artificial increase of pH in the inflow, the pH of the outflow was approximately neutral and comparable to the first year of operation. Moreover, the pH of the outflow was relatively stable in the second year.

TABLE 4.2
Packing Order of Vertical-Flow Wetland Filters

Height (cm)	Filter 1	Filter 2	Filter 3	Filter 4	Filter 5	Filter 6
61–83	W	W	W	W	W	W
56–60	W	W	6	6+P	7	7+P
51–55	W	W	6	6+P	6	6+P
36–50	W	W	4	4+P	5	5+P
31–35	W	W	3	3	5	5
26–30	W	W	3	3	4	4
21–25	W	W	2	2	3	3
16–20	W	W	2	2	2	2
11–15	2	2	2	2	2	2
0–10	1	1	1	1	1	1
Height (cm)	**Filter 7**	**Filter 8**	**Filter 9**	**Filter 10**	**Filter 11**	**Filter 12**
61–83	W	W	W	W	AW	AW
56–60	6	6+P	7	7+P	7+P	7+P
51–55	6	6+P	6	6+P	6+P	6+P
36–50	4	4+P	5	5+P	5+P	5+P
31–35	3	3	5	5	5	5
26–30	3	3	4	4	4	4
21–25	2	2	3	3	3	3
16–20	2	2	2	2	2	2
11–15	2	2	2	2	2	2
0–10	1	1	1	1	1	1

W, Water; P, *Phragmites australis* (Cav.) Trin. ex Steud. (common reed); AW, aerated water; 1, stones; 2, large gravel; 3, medium gravel, 4, small gravel; 5, Filtralite (light expanded clay product); 6, sand (0.6–1.2 mm); 7, Frogmat (barley straw).

Heavy metal removal efficiencies improved during the second year of operation. However, the reduction in metals was not sufficient to comply with American standards for secondary wastewater treatment. Dissolved nickel and copper concentrations should not exceed 0.0071 and 0.0049 mg/L, respectively (Tchobanoglous et al., 2003).

It has been known that the decomposition of aquatic plants after fall, reducing soil conditions, road gritting and salting during periods of low temperatures and acid rain contribute to increases of metal concentrations in the outflow (Norrstrom and Jacks, 1998; Sasaki et al., 2003). For example, high levels of conductivity were recorded in the filter inflow and outflows, and the breakthrough of dissolved nickel was observed during the first winter.

Concerning the effect of retention time on the treatment efficiency of metals, the heavy metal outflow concentrations of Filter 12 (higher loading rate) were slightly higher than the corresponding concentrations for the other filters (e.g., Filters 7 and 8). According to previous studies (Kadlec and Knight, 1996; Wood and Shelley, 1999), metal removal efficiencies for wetlands are highly correlated with influent concentrations and mass loading rates. Moreover, it was suggested that the formation of metal sulphides was favoured in wetlands with long retention times. This may lead to a more sustainable management of constructed treatment wetlands.

After the increase of the inflow pH, mean reduction efficiencies for nickel increased during the second winter compared to the first winter; for example, 90% and 65%, respectively, for Filter 7. Moreover, an obvious breakthrough of nickel was not observed during the second winter despite the presence of high salt concentrations in the inflow. This is likely due to the artificial increase of pH. A high pH facilitates nickel precipitation. For example, nickel hydroxide ($Ni(OH)_2$) may precipitate at pH 9.1 if the corresponding metal concentration is 1 mg/L (Tchobanoglous et al., 2003), which is similar to the inflow concentrations of spiked filters. Moreover, dissolved copper did not break through any filters throughout the study.

All filters acted as pH buffers after pH increase, and pH levels were subsequently reduced. It can be assumed that this buffering capacity is greatly enhanced by the presence of active biomass rather than macrophytes (Kadlec and Knight, 1996; Sasaki et al., 2003). However, the outflow pH values for the planted filters recorded were slightly lower than those for the unplanted filters. For example, the overall mean pH value for Filter 7 (unplanted filters) is 7.31, and the corresponding value for Filter 8 (planted filters) is 6.98 during the second year of operation.

Nitrogen is used by *P. australis* and microorganisms for developing new biomass. This explains the higher reduction of ammonia and nitrate for planted in comparison to unplanted filters. Moreover, storage of nutrients in plant-derived debris is another sustainable mechanism removing nitrogen as well as phosphate (Kadlec and Knight, 1996; Scholz et al., 2002).

Ammonification is slower in anaerobic than in aerobic soils because of the reduced efficiency of heterotrophic decomposition in anaerobic environments. Nitrification requires oxygen that was provided during the drawdown periods and/or during artificial aeration (Filters 11 and 12). Denitrification is supported by facultative anaerobes. These organisms can break down oxygen-containing compounds such as nitrate-nitrogen to obtain oxygen in an anoxic environment that was dominant during the long periods of filter flooding.

Some research indicates that most of the observed variation in ammonia-nitrogen removal could be attributed to fluctuations of the residence time in most wetlands (Kadlec and Knight, 1996). In comparison, this system also showed a decreased capability to treat ammonia-nitrogen in filters with low retention times. Therefore, it might be a beneficial approach to use planted and intermittently loaded systems with long retention times to obtain high ammonia-nitrogen removal.

As like previous findings (Kadlec and Knight, 1996), the reduction rates of ammonia-nitrogen decreased during the winter. Such temperature-dependent processes may result in the ammonia-nitrogen removal target to become the determining design component for constructed treatment wetlands in cold climate. In contrast, ortho-phosphate-phosphorus concentrations were low in winter but high in spring and summer. However, concentrations were always relatively low (<0.35 mg/L).

Ammonia-nitrogen reduction efficiencies for planted filters were higher than for comparable unplanted filters. Similar findings have been reported elsewhere (Kadlec and Knight, 1996). It can be inferred that the higher removal rate in the planted filters could be attributed to plant uptake of ammonia-nitrogen and increase in nitrification near the rhizomes and roots of *P. australis*.

While the ammonia-nitrogen outflow concentrations of the unplanted filters (e.g., Filter 7) decreased in summer, nitrate-nitrogen concentrations increased. It can be assumed that ammonia-nitrogen removal is mostly due to the increase of the nitrification rate in summer.

The nitrate-nitrogen concentration in planted filters (e.g., Filter 8) was much lower than that of unplanted filters (e.g., Filter 7) in summer. The filter with higher loading rates and shorter HRT (Filter 12) experienced a breakthrough of nitrate-nitrogen during low temperature periods. It can be concluded that the transfer of ammonia to nitrate and plant uptake attributed greatly to the overall nitrogen removal during periods of high temperature.

The ortho-phosphate-phosphorus reduction efficiencies for unplanted filters were slightly higher than for comparable planted filters. However, phosphorus concentrations were relatively low indicating that most wetlands are phosphorus and not nitrogen limited.

The contribution of macrophytes to the overall treatment performance is assumed to largely vary depending on the wetland design. Previous work indicates that macrophytes are likely to affect considerably the removal of pollutants in horizontal sub-surface constructed treatment wetlands, while their role is minor in pollutant removal for periodically loaded vertical-flow wetlands (Brix, 1997). Nevertheless, the secondary role of macrophytes concerning oxygen transport, clogging prevention and provision of an energy source for microorganism can influence positively the treatment performance of wetlands (Cooper et al., 1996).

4.1.9 Conclusions and Recommendations

The critical review sub-chapters highlighted the importance of hydrological and biochemical processes in natural and CWs. The latter are either dominated by *T. latifolia* in the U.S. or *P. australis* in Europe. Most research work has been performed in the U.S. and northern countries of Western Europe. There is a gap of knowledge and understanding relating to natural and CWs in the tropics and arid areas of South America, Africa and Asia.

The importance of BOD, suspended solids, nitrogen and phosphorus as performance indicator variables for engineering applications was stressed. The main quantitative and qualitative relationships between these variables have been summarized. This includes mass balance equations and nutrient transformation processes. Moreover, a high variability of particularly the heavy metal removal efficiencies in CWs has been reported.

The review was subsequently evaluated with the help of a case study in urban water. Despite the highly variable water quality of road runoff, the wetland filters showed great treatment performances particularly with respect to the BOD reduction in a cold climate. Removal efficiencies for suspended solids and nitrogen improved considerably over time and dissolved copper was removed satisfactorily in comparison to values obtained from the literature.

A breakthrough of dissolved nickel during the first winter of the first year of operation was observed. After creating an artificially high inflow pH of approximately 8 after 1 year of operation, nickel was successfully treated despite vulnerability to leaching when exposed to a high salt concentration during the second winter.

A high pH was apparently also linked to high suspended solids and ammonia-nitrogen removal efficiencies. The elevated pH had no apparent negative effect on the biomass including macrophytes. Moreover, filters showed a great pH buffering capacity. Findings indicate that conventional pH adjustment can be successfully applied to CW systems for storm water treatment.

The presence of Filtralite (adsorption filter media) and *P. australis* did not result in an obvious reduction of metal concentrations in outflow waters. Operational conditions such as inflow pH and retention time were more important for the heavy metal treatment.

This analytical chapter has shown the great popularity of natural and CW research. The general governing processes within natural wetlands have been covered adequately by the scientific literature. However, information on detailed processes within CWs contributing to the treatment performance of specific wastewater streams is lacking.

For example, the literature review has shown that a diverse microorganism community dominated by bacteria, fungi, algae and protozoa is present in the aerobic and anaerobic zones of wetlands. The diverse microbial ecology and plant community structure within complex wetland ecosystems has not yet been fully reported. The author recommends further research into wetland microbial ecology to perform sound temporal and spatial modelling of microbial pollution and treatment performance indicators.

4.2 METAL REMOVAL IN WETLANDS TREATING WATERS FOR POTABLE WATER PRODUCTION

4.2.1 Introduction

The exposure of metals in drinking water is a worldwide problem because some of them like arsenic (As) are hazardous for human consumption. In some developing countries (e.g., Bangladesh, India and China), As concentrations in drinking water exceed guidelines for human health protection, causing serious poisoning and even death (Srivastava et al., 2011). In China, more than two million people are exposed to As through contaminated drinking water (Guo et al., 2007). The National Drinking Water Standard (GB 5749-2006) decreased the maximal As concentration to 0.01 mg/L (Ministry of Health, 2006), which is equal to the permissible limit set by the World Health Organization (WHO) for safe drinking water (World Health Organization, 2004). To avoid carcinogenic and many other adverse health effects, it is important to develop highly efficient, easy-to-operate, cost-effective and environmentally benign As removal techniques.

A variety of conventional and non-conventional techniques (e.g., coagulation and filtration, reverse osmosis, ion exchange, oxidation and precipitation, adsorption and photocatalysis), their modifications and/or combinations have been tested to remove As from potential drinking water resources (Liao et al., 2011; Litteret al., 2010; Mohan and Pittman, 2007). These As treatment choices, however, are greatly limited to field practice with their inherent disadvantages such as relatively high costs, huge energy consumption rates and high chemical reagent utilizations (Litteret al., 2010). In contrast, less energy-consuming and solar-driving 'green' treatment systems like phytoremediation and CWs may be alternatives to remove As from potential drinking water resources due to their cost-effective and sustainable nature.

CWs are engineered filter systems that have been designed and constructed to utilize natural processes involving wetland vegetation, substrate and their associated microbial assemblages in treating wastewater (Kadlec and Wallace, 2009; Tsihrintzis, and Gikas, 2010; Scholz, 2010; Vymazal, 2007; Zhang et al., 2009). These biological filters take advantage of physical, chemical and biological processes occurring in natural wetlands, but do so within a semi-controlled environment (Scholz, 2010). Arsenic removal in CWs takes place because of plant uptake (Marchand et al., 2010), accretions of wetland soils (Rahman et al., 2011), microbial immobilization (Arroyo et al., 2013), adsorption and retention by substrates (Rahman et al., 2011) and precipitation in the water column (Henken, 2009). Although macrophytes, soil, detritus and biomass are important sinks for As in the short term, substrate is the main sink for As in the long term (Rahman et al., 2011). In sub-surface-flow wetlands, pollutants including As are in direct contact with the substrate, and adsorption and retention are, therefore, the main pathways of As removal.

Because of the great importance of substrate in As removal, different media including limestone, zeolite, cocopeat and gravel have been studied to assess As removal in CW filters. Moreover, several industrial by-products or wastes such as blast furnace slag, red mud, fly ash and sanding wastes have been examined as potential As adsorbents in the view of supporting waste recycling or reutilization strategies (Altundogan et al., 2002; Gupta et al., 2005; Kanel et al., 2006; Lim et al., 2009). The common belief is that aggregates rich in iron, aluminium, manganese and copper have a strong affinity for As, because they are more prone to form different compounds besides physical adsorption (Chakravarty et al., 2002; Maji et al., 2011). Though adsorption processes of As to various adsorbent materials have been studied in detail, the efficiency of sub-surface-flow wetlands packed with different substrates, or their combinations, has not been sufficiently examined in the scientific literature. Since wetlands with conventional soil or gravel media have commonly been used to treat acid mining wastewater (Lizama et al., 2012), little is known about the performance of using alternative substrates purifying As-polluted drinking water resources.

Wetland plants play a critical role with regard to As removal in CW filtration systems. Macrophytes, such as *P. australis* (Cav.) Trin. ex Steud. (common reed) and *Juncus effuses*, can accumulate As and other heavy metals in roots and shoots (Rahman et al., 20111). In addition to common aquatic plants, some terrestrial plants called As 'hyperaccumulators' such as *Pteris vittata* and *Pityrogramma calomelanos* L. (Silver Fern) can also remove a formidable quantity of As from soil and store it in their fronds, which are large, divided leaves (Gulz et al., 2005; Ma et al., 2011). However, the growth and As removal performance of hyperaccumulators is questionable in sub-surface CWs because they have different humidity, water content, oxygen availability and nutrient supply conditions when compared to soil. More attention should be paid to determine whether the traditional macrophytes or hyperaccumulators are suitable for the removal of As.

Previous research indicated that As(III) is thermodynamically unstable and easily converted to As(V) in aerobic environments (Van Halem et al., 2009). Therefore, it is reasonable to consider only As(V) compounds when evaluating As removal in water treatment (Li et al., 2012). Given this context, As(V) has been chosen as the target As species to evaluate the As adsorption characteristics to different wetland substrates including gravel, zeolite, ceramsite (rather small, typically multi-coloured, smooth and round balls of clay) and manganese sand. After assessing As adsorption results, tap water spiked with As(V) has been used to test the As removal efficiency in wetland filters packed with ceramsite and manganese sand, and planted with *J. effuses* and *P. vittata*, respectively.

Due to the benefits of small land occupation and relatively good oxygen availability, this study used vertical wetland filters to test their performance concerning As removal, and to assess the roles of different substrates and plants involved in As reduction. The major objectives of Section 4.2 are to (a) assess the As adsorption capacities and process mechanisms involving different substrates; (b) compare As removal performances for different substrates in column experiments with each other; (c) evaluate the effects of traditional macrophytes and hyperaccumulators on As removal using column experiments; (d) identify the main approaches to As removal in wetland filters based on an annual mass balance calculation and (e) assess the roles of other water quality variables including pH, DO and nutrients in As removal. Section 4.2 is based on the original article by Wu et al. (2014), which has been revisited and updated.

4.2.2 Materials and Methodology

4.2.1.1 Materials

Commercially obtained gravel, zeolite, ceramsite and manganese sand were tested as potential wetland substrates with measured porosity values of 45.0%, 38.6%, 42.8% and 54.0%, respectively. The corresponding media compositions are shown in Table 4.3. It was found that silicium dioxide (SiO_2) dominates the compositions for gravel, zeolite and ceramsite. Manganese sand contained mainly

TABLE 4.3
Major Mineral Composition of the Different Tested Substrates

	Contents (%)							
Substrates	SiO_2	Al_2O_3	MnO_2	Fe_2O_3	CaO	MgO	K_2O	Others
Gravel	79.52	7.36	–	1.86	3.84	0.95	3.41	3.06
Zeolite	72.01	10.25	–	2.02	3.58	0.98	1.96	–
Ceramsite	62.16	16.32	–	7.84	3.26	2.04	3.22	6.42
Manganese sand	19.41	13.28	43.93	20.74	1.84	0.42	–	–

magnesium dioxide (MnO_2), iron (III) oxide (Fe_2O_3) and SiO_2 (Table 4.3). The corresponding ranking order of the percentage contents of total metal oxides (mainly MnO_2, Al_2O_3 and Fe_2O_3) for different substrates was as follows: manganese sand > ceramsite > zeolite > gravel.

All chemical reagents were of analytical grade and used without further purification. All aqueous solutions were prepared using ultra-pure water. An As(V) stock solution with an As concentration of 1,000 mg/L was prepared using $Na_2HAsO_4.7H_2O$. Parts of the stock solution were subsequently diluted to the required concentrations for conducting As adsorption and column experiments. Nutrient stock solutions containing 1,000 mg/L ammonia-nitrogen (NH_4-N) and 300 mg/L ortho-phosphate-phosphorus (PO_4-P) were prepared using NH_4Cl and KH_2PO_4, respectively. These solutions were used to supply the required nutrients to allow for wetland plant growth after adequate dilution. *J. effuses* and *P. vittata* were both bought from a local botanical garden.

4.2.1.2 Batch Adsorption

Batch adsorption experiments were performed to obtain kinetic and isotherm data for all substrates with grain sizes of less than 0.25 mm. The Brunauer–Emmett–Teller surface area was determined by a Nova 4200e surface area analyser obtained from Quantachrome (http://www.quantachrome.co.uk). Regardless of kinetic or isotherm adsorption, each experiment was conducted in triplicates. For all adsorption kinetic experiments, 0.5 g of substrate was placed in a 50-mL polyethene bottle containing 20 mL of 1,000 μg/L As solution. Each solution was continuously shaken at 200 rpm ($r=2$) within a water batch at 25°C for 0.25, 0.50, 1.00, 2.00, 4.00, 8.00, 12.00, 24.00 and 48.00 h. Suspensions were centrifuged for 10 minutes at a speed of 3,500 rpm ($r=1,895$). The As remaining in the supernatant was determined using the atomic fluorescence spectrometer AF-610A (Beijing Rayleigh, Beijing, China) (Ministry of Water Resource, 2005).

For all adsorption isotherm experiments, 0.5 g of substrate was placed in a 50-mL polyethene bottle containing 20 mL of a solution with different initial As concentrations (0, 50, 100, 200, 400, 600, 800 and 1,000 μg/L). Each solution was shaken at 200 rpm ($r=2$) within a water bath at 25°C for 24 hours. Arsenic in suspension was centrifuged and subsequently determined using the same procedure and method mentioned above. The adsorption capacity (μg/g) at equilibrium was calculated using Equation 4.12.

$$q_e = \frac{(C_0 - C_e)V}{W} \tag{4.12}$$

where C_0 and C_e are the initial and equilibrium concentrations (μg/L) of As(V) in the solution, respectively; V (L) is the volume of solution; and W (g) is the mass of substrate used.

Data collected from the batch adsorption experiments were expressed as means and standard deviations (SD). The data were further fitted with various adsorption kinetic and isotherm models to help understand the As adsorption process and mechanisms to different tested substrates.

Pseudo-first order and pseudo-second-order adsorption kinetic models were applied to evaluate the kinetic order of the adsorption process. The specific rate equations are shown in Equations 4.13 and 4.14 (Ho and McKay, 1999).

$$\log\left(q_e - q_t\right) = \log q_e - k_1 t \tag{4.13}$$

$$\frac{t}{q_t} = \frac{1}{k_2 q_e^2 + q_e} \frac{t}{} \tag{4.14}$$

where q_t (μg As(V)/g) is adsorbed As at time t (h); and k_1 and k_2 are the rate constants of the pseudo-first-order adsorption (L/h) and pseudo-second-order rate constant (g/(μg h)).

The rate-limiting step of the substrate adsorption process can be calculated using first-order kinetic data (Maji et al., 2008). Assuming spherical geometry of the adsorbent, the calculated pseudo-first-order rate constant was utilized to correlate with the pore diffusion (Equation 4.15) and film diffusion coefficients (Equation 4.16). $t_{\frac{1}{2}}$ is the time (s) required to bring down the As concentration to half the initial concentration.

$$t_{1/2} = 0.03 \frac{r^2}{D_p} \tag{4.15}$$

$$t_{1/2} = 0.23 \frac{2\delta}{D_f} \times \frac{C_s}{C_p} \tag{4.16}$$

where r is the mean geometric radius of the substrate particle (cm); D_p and D_f are the pore diffusion and film diffusion coefficients (cm^2/s), respectively; C_s is the concentration of As on the adsorbent (μg/g); and δ is the film thickness of 0.001 cm.

The relationship between $t_{\frac{1}{2}}$ and k_1 (overall reaction rate constant) can be described in Equation 4.17 (Asher and Pankow, 1991). Values of $t_{\frac{1}{2}}$ can be calculated with k_1 obtained from Equation 4.13.

$$t_{\frac{1}{2}} = -\frac{1\text{n}(0.5)}{k_1} \tag{4.17}$$

Langmuir and Freundlich isotherms were used to simulate the adsorption isotherms of each wetland media used in the experiment. The linear forms of Langmuir and Freundlich equations can be expressed in Equations 4.18 and 4.19:

$$\frac{C_e}{q_e} = \frac{C_e}{q_m} + \frac{1}{k_L q_m} \tag{4.18}$$

$$1\text{n}\, q_e = 1\text{n}\, k_F + \frac{1}{n} 1\text{n}\, C_e \tag{4.19}$$

where q_m is the theoretical maximal capacity (μg/g), k_L is the Langmuir sorption equilibrium constant (L/μg) related to the adsorption energy, k_F is the equilibrium constant indicative of adsorption capacity, and n is the adsorption equilibrium constant whose reciprocal is indicative of adsorption intensity (Li et al., 2012).

To characterize the type of As adsorption to different substrates, the data were applied to the Dubinin-Radushkevich (D-R) isotherm (Kundu and Gupta, 2006), which can be expressed as Equations 4.20 and 4.21.

$$\ln q_e = \ln q_m - k_{DR}\varepsilon^2 \tag{4.20}$$

$$\varepsilon = \mathrm{RT}\ln\left(1+\frac{1}{C_e}\right) \tag{4.21}$$

where k_{DR} is the constant related to the adsorption energy (mol^2/kJ^2), R is the universal gas constant of 8.3145 kJ/mol K and T is the study temperature of 298 K.

The constant k_{DR} indicates the mean free energy of adsorption per molecule when it is transferred to the surface of the solid from the solution, and can be calculated using Equation 4.22, where E is the mean sorption energy (kJ/mol) providing important information about the physical and chemical nature of the adsorption process.

$$E = \frac{1}{\sqrt{2k_{DR}}} \tag{4.22}$$

4.2.1.3 Column Experiments

After selecting the wetland substrates, ten analogous wetland filters were designed, constructed and operated predominantly to assess As removal as a function of nutrient supply between May 2012 and May 2013. All experimental wetlands were located outside on the open balcony (3 m above the ground level) of the Changjiang River Scientific Research Institute (Wuhan, China) to allow exposure to natural climatic conditions. The rig was constructed using polyethylene columns of 120-mm diameter and 750-mm height. The outlet valves were located at the centre of the bottom plate of each wetland column and connected to 12-mm internal diameter vinyl tubing. This arrangement was used to allow for manual flow adjustment, and collection of outflow sample water at the same time.

The ten experimental CW columns were labelled either A, B, C, D, E or F. Wetlands A, B, C and D were planted and operated in duplicates, while wetlands E and F were operated as unplanted controls without duplicates. The packing orders of all wetlands were the same. Wetland columns A, B and E were filled with 0.2-m deep gravel (diameter between 10 and 50 mm) representing the bottom layer, and 0.4-m deep ceramsite (diameter between 8 and 10 mm) at the top. The compositions for columns C, D and F were the same except for the replacement of ceramsite by manganese sand (diameter between 3 and 4 mm).

The roles of *J. effuses* and *P. vittata* in removing As were also assessed. CWs A and C were planted with *J. effuses*, while wetlands B and D were vegetated with *P. vittata*. Each tested wetland column was arranged with similar plant biomass (approximately 150 g in wet weight) of equal viability and strength.

After nearly one month of As exposure (acclimatization period) to the wetland plants, all wetlands received 3.5 L of simulated As contaminated drinking water solution (i.e. tap water spiked with 500 μg/L As(V) and nutrients) every three days. This methodology may reduce media clogging and enhance oxygenation. The selected influent As(V) concentration of 500 μg/L is equal to the Chinese wastewater As(V) discharge concentration limit and falls within the moderate As(V) concentration range set for drinking water sources contaminated with As(V) in China (Ministry of Environmental Protection, 1996). The previously prepared nutrient stock solution was used as the predominant source of nutrient supply to enhance plant and microbial growth, and to improve the As removal efficiency of the wetland systems. The controlled supply of nutrients gave the research team the opportunity to assess nutrient requirements for As removal in wetland filters for the first time. The previously prepared nutrient stock solution was added to tap water spiked with As. The solution contained 2 mg/L NH_4-N and 0.6 mg/L PO_4-P, which was seen as reasonable for good plant growth and As removal.

Since June 2012, all wetland filters were fully saturated and flooded to a depth of approximately 5 and 0.5 cm above the top level of the packing media (ceramsite and manganese sand) to reduce the bed media clogging probability and improve the overall oxygen availability. An As(V) loading rate of 583.3 μg/d was used for all systems. The wetlands were fully and quickly drained within a brief period of time (less than 5 minutes) and subsequently refilled in a batch flow mode with a residence time of three days. All samples were collected from the drained water and further analysed on the same day when they were taken for the following parameters: As, NH_4-N, PO_4-P, pH and DO.

Arsenic was determined using the atomic fluorescence spectrometer AF-610A (Beijing Rayleigh, Beijing, China) (Ministry of Water Resource, 2005). The nutrients NH_4-N and PO_4-P were analysed according to American standard methods (APHA, 1998); ammonia F phenate method and automated ascorbic acid reduction method, respectively. A YSI 52 DO meter and a HANNA portable pH meter were used for DO and pH measurements.

Data collected from all CWs were used for statistical analysis to assess the effects of different substrates and plants on As removal and other water quality variables. All statistical tests including a one-way analysis of variance (ANOVA) were performed using the software SPSS (2003). Multiple comparisons were undertaken using the least significant difference (LSD) test, homogeneity of variance test and Duncan's multiple range test for differences between means. The selected level of significance was $p < 0.05$. The statistical tests were applied to assess the differences between wetland effluent As, NH_4-N, PO_4-P and other water quality variables such as pH and DO.

4.2.1.4 Annual Arsenic Mass Balance

All aboveground and belowground macrophyte biomass were harvested after one year of operation to examine the role of plant uptake on As removal in wetland filters. After dividing the biomass into roots, stems, leaves (none for *J. effuses*) and seeds (none for *P. vittata*), they were subsequently air-dried to determine the dry weight and prepare sub-samples for further As concentration analysis. Sub-samples of dried fractional biomass were powdered, wet digested with a mixed solution of nitric acid (HNO_3) and hydrochloric acid (HCl) at a ratio of 4 to 1, and analysed for As content according to the standardized atomic fluorescence spectrometric method with a detection limit of 0.4 μg/L.

No obvious As(V) saturation phenomenon was identified during the first year of operation. To briefly assess the annual As retention capacity within different filter media, substrate segments were collected at 15-cm intervals and then fully mixed to obtain a homogeneous sample at the end of the experiment. All of the substrate samples were air-dried, powdered, wet-digested and measured according to Chinese standard methods (Standardization Administration of the People's Republic of China, 2008).

An As mass balance was calculated for each wetland by considering the total As-mass input, the total As-mass output including effluent discharge, substrate retention, plant uptake and other unaccountable parts such as microbial assimilation and detritus adsorption. The findings of the mass balance calculation were then used to identify the role of substrates and plants in As reduction.

4.2.3 Results

4.2.3.1 Arsenic Adsorption Kinetics

The adsorption rate of As(V) was found to be time-dependant as seen in Figure 4.2. The uptake of As(V) increased with reaction time. The adsorption of As(V) was rapid in the first four hours and then slowed down as an equilibrium was reached (Figure 4.2.), which corresponds to physical and chemical adsorption processes, respectively. The substrate type impacted significant the adsorption process; gravel and zeolite were much faster in approaching the adsorption equilibrium in contrast to ceramsite and manganese sand. After 48 hours, the amounts of As(V) removal were 7.58, 15.06, 26.78 and 38.67 μg/g for gravel, zeolite, ceramsite and manganese sand, respectively.

Further kinetic simulation results indicated that the pseudo-first-order equations were suitable for describing As(V) adsorption to all four tested substrates (Table 4.4). Moreover, the ranking order of the calculated As(V) adsorption rate constants based on pseudo-first-order fitting were as follows: gravel ($0.50\,h^{-1}$) < zeolite ($1.56\,h^{-1}$) < ceramsite ($3.42\,h^{-1}$) < manganese sand ($4.55\,h^{-1}$). The coefficients for determining the As(V) adsorption rate limiting steps are shown in Table 4.5. The pore diffusion constants for different substrates ranged from 0.36×10^{-8} to $3.27 \times 10^{-8}\,cm^2/s$, while film diffusion constants ranged between 0.23×10^{-7} and $1.02 \times 10^{-7}\,cm^2/s$.

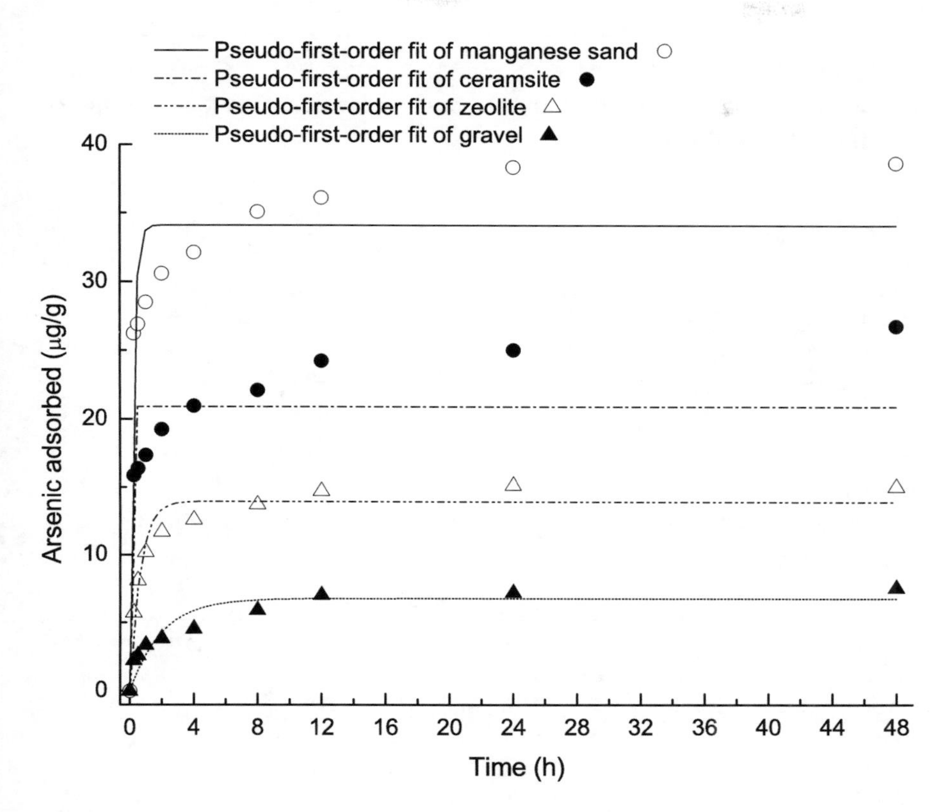

FIGURE 4.2 Pseudo-first-order kinetic model fitted for arsenic(V) adsorption to gravel, zeolite, ceramsite and manganese sand as a function of time.

4.2.3.2 Arsenic Adsorption Isotherm

The adsorption of As(V) was found to be concentration-dependant as seen in Figure 4.3. With the initial As(V) concentration of less than 1,000 μg/L, the uptake of As(V) increased with As in solution. The findings of the As(V) adsorption isotherm experiments showed that the adsorption capacities of different substrates varied considerably (Table 4.6); for example, the maximum As(V) adsorption capacities (obtained by the Langmuir equation) were 42.37 μg/g for manganese sand in comparison to 12.7 μg/g for gravel.

The ranking order for the As(V) adsorption capacity (Table 4.6) was as follow: manganese sand > ceramsite > zeolite> gravel. After comparing the coefficients of determination (R^2) values of the three adsorption isotherms, it was found that As(V) adsorption capacities of all tested substrates were best explained by the Freundlich adsorption isotherm. The corresponding correlation coefficients were 0.95, 0.92, 0.93 and 0.98 for gravel, zeolite, ceramsite and manganese sand, respectively (Table 4.6).

TABLE 4.4
Comparison of the Pseudo-First- and Second-Order Reaction Rate Constants for Different Tested Substrates

	Pseudo-First Order			Pseudo-Second Order		
Substrate	k_1	q_e	R^2	k_2	q_e	R^2
Gravel	0.50	6.78	0.85	0.04	4.92	0.31
Zeolite	1.56	13.96	0.95	0.21	11.88	0.54
Ceramsite	3.42	22.40	0.84	0.37	20.89	0.73
Manganese sand	4.55	34.10	0.89	0.48	32.52	0.82

TABLE 4.5
Calculated Pore Diffusion and Film Diffusion Constants for Different Tested Substrates

	Substrates			
Parameters	**Gravel**	**Zeolite**	**Ceramsite**	**Manganese Sand**
C_e (mg/L)	0.83	0.65	0.44	0.15
k_1 ($\times10^{-3}$/s)	0.14	0.43	0.95	1.26
$t_{1/2}$ ($\times10^{-3}$/s)	4.95	1.61	0.73	0.55
r (cm)	0.0245	0.0245	0.0245	0.0245
D_p ($\times10^{-8}$cm^2/s)	0.36	1.12	2.47	3.27
D_f ($\times10^{-7}$cm^2/s)	0.23	0.28	0.30	1.02

The magnitude of E of As(V) adsorption was calculated (Table 4.6) using the Dubinin-Radushkevich (D-R) isotherm. The values of E varied greatly between different substrates; for example, 184.43 kJ/mol for manganese sand in contrast to 4.66 kJ/mol for gravel.

4.2.3.3 Arsenic Removal

During the whole experimental period, the recorded maximum precipitation intensity was no more than 200 mm. Natural water loss was due to evaporation. A surplus height of 10–14.5 cm for each column avoided the occurrence of rainwater overflowing accidentally. The monthly mean As(V) removal ratios for wetlands with different substrates and plants are shown in Figure 4.4. Throughout the entire experiment, mean removal rates were higher for wetlands with manganese sand (approximately 90%) compared to those containing ceramsite (between about 30% and 80%) as shown in Figure 4.4.

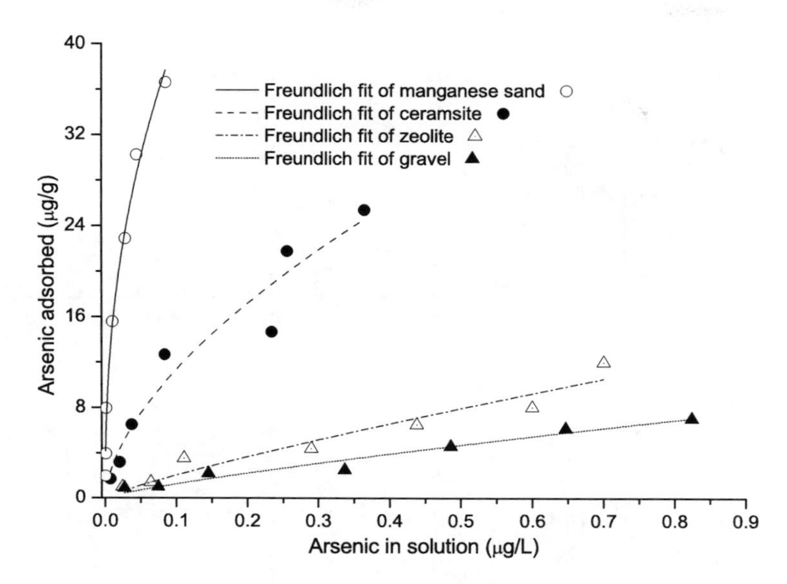

FIGURE 4.3 Freundlich isotherm equations fitted for arsenic(V) adsorption to gravel, zeolite, ceramsite and manganese sand as a function of the equilibrium aqueous arsenic concentrations.

TABLE 4.6
Comparison of the Correlation Coefficients of the Langmuir, Freundlich and Dubinin-Radushkevich (D-R) Isotherm for Different Tested Substrates

	Langmuir Isotherm			Freundlich Isotherm			Dubinin-Radushkevich (D-R) Isotherm			
Substrates	q_m (μg/g)	k_L (L/μg)	R^2 (–)	N (–)	k_F (μg/g)	R^2 (–)	q_m (μg/g)	k_{DR} ($\times 10^{-5}$ mol²/kJ²)	R^2 (–)	E (kJ/mol)
Gravel	12.70	1.44	0.95	1.23	8.25	0.95	8.58	230	0.78	4.66
Zeolite	17.32	1.69	0.91	1.19	14.16	0.92	8.46	200	0.65	15.81
Ceramsite	34.75	5.56	0.92	1.67	44.87	0.93	20.60	32.8	0.81	39.04
Manganese sand	42.37	56.37	0.95	2.33	108.22	0.98	32.04	1.47	0.84	184.43

Negligible effects of operation time, seasonal temperature variation and vegetation type on As (V) removal were observed for wetland columns packed with manganese sand (Table 4.7). As (V) was removed with mean efficiency ranges of 88%–92%, 88%–97% and 80%–91% for columns C (planted with *J. effuses*), D (planted with *P. vittata*) and F (unplanted), respectively. Wetlands packed with manganese sand and planted with As hyperaccumulators had the most stable and efficient As(V) removal capacity in comparison to other wetland designs.

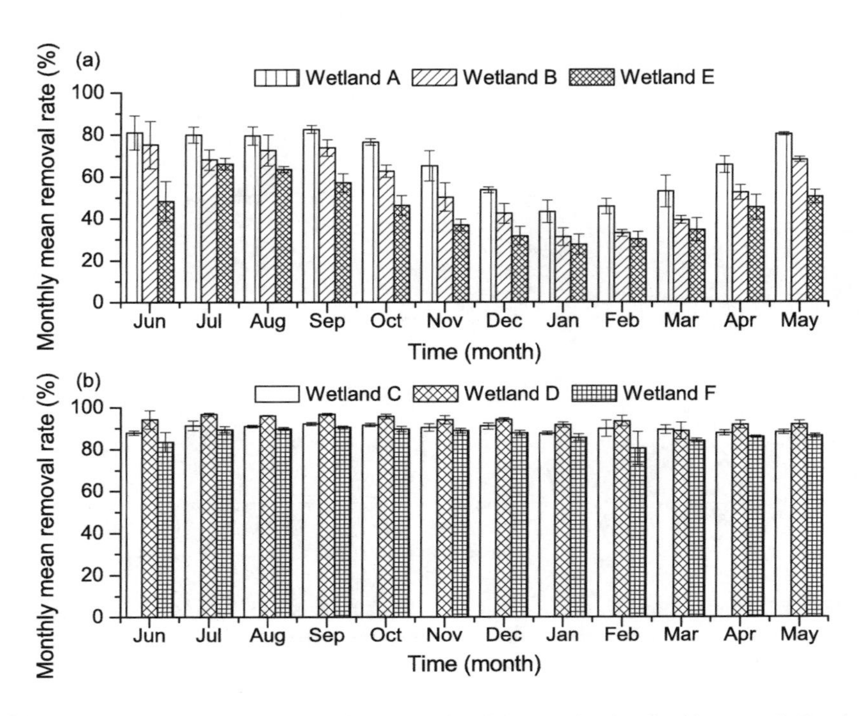

FIGURE 4.4 Variation in monthly mean arsenic(V) removal ratios for (a) ceramsite-packed wetland filters A (*Juncus effuses* planted), B (*Pteris vittata* L. planted) and E (unplanted), and (b) manganese sand packed wetland filters C (*Juncus* effuses planted), D (*Pteris vittata* L. planted) and F (unplanted) between June 2012 and May 2013.

In contrast to wetlands filled with manganese sand, the presence of *J. effuses* and *P. vittata* leads to an increased mean removal of As(V) by approximately 21% and 10% for wetlands A and B, respectively, if compared to the unplanted wetland E. Moreover, As(V) removal rates increased or decreased with the fluctuations of atmospheric temperature; the highest and lowest As(V) removal rates of 83% and 43% for wetland A occurred in warm September and cold January, respectively.

4.2.3.4 Nutrient Removal

Effluent NH_4-N and PO_4-P concentrations were evaluated to obtain an indication of the nutrient removal efficiency for wetlands treating As(V). Effluent concentrations of NH_4-N and PO_4-P are summarized in Table 4.7. The concentrations in the effluent were significantly lower than the corresponding ones in the influent, which can be expected.

Effluent NH_4-N and PO_4-P concentrations were significantly lower in *J. effuses* and *P. vittata* planted wetlands A, B, C and D than in the unplanted wetlands E and F (Table 4.7). Moreover, effluent NH_4-N and PO_4-P concentrations were both significantly higher in wetland F packed with manganese sand than in wetland E containing ceramsite. Regardless of the presence of ceramsite or manganese, NH_4-N removal

TABLE 4.7
Annual Mean Concentrations ±SD and Pollutant Removal Efficiencies for Arsenic (As (V)), Ammonia-Nitrogen (NH_4-N) and Ortho-Phosphorus- Phosphate (PO_4-P) and Other Variables Including Water Temperature Dissolved Oxygen (DO) and pH in the Influent and Effluent Waters of Experimental Ceramsite Packed Wetlands A (*Juncus effuses* planted) and B (*Pteris vittata* L planted) and E (unplanted) , and Manganese Sand Packed Wetlands C (*Juncus effuses* planted), D (*Pteris vittata* L planted) and F(unplanted)

		Effluent					
Variables	**Influent**	**Wetlands A**	**Wetlands B**	**Wetlands C**	**Wetlands D**	**Wetlands E**	**Wetlands F**
As (V)							
Concentrations (μg/L)	498.16 ± 8.03^a	147.28 ± 69.76^d	205.54 ± 81.04^c	48.73 ± 10.49^e	27.61 ± 15.04^f	253.54 ± 81.04^b	61.83 ± 17.56^e
Removal (%)		70.48 ± 13.89^c	58.80 ± 16.11^d	90.22 ± 2.10^b	94.46 ± 3.02^a	49.16 ± 14.31^e	87.58 ± 3.55^b
NH_4-N							
Concentrations (mg/L)	2.12 ± 0.12^a	0.35 ± 0.08^d	0.37 ± 0.09^{cd}	0.73 ± 0.13^{bc}	0.60 ± 0.17^{bcd}	0.67 ± 0.12^{cd}	0.76 ± 0.20^b
Removal (%)		83.08 ± 16.93^a	71.51 ± 18.71^b	65.51 ± 11.43^c	66.45 ± 18.05^c	64.02 ± 12.38^c	52.30 ± 18.62^d
PO_4-P							
Concentrations (mg/L)	0.60 ± 0.03^a	0.18 ± 0.06^e	0.28 ± 0.06^d	0.39 ± 0.08^c	0.35 ± 0.03^f	0.39 ± 0.07^c	0.44 ± 0.07^b
Removal (%)		69.62 ± 9.40^a	53.28 ± 9.67^b	35.39 ± 9.19^c	39.40 ± 12.99^c	35.41 ± 9.19^c	26.29 ± 9.79^d
DO (mg/L)	5.69 ± 0.57^a	3.29 ± 0.27^b	3.14 ± 0.28^c	3.85 ± 0.30^b	3.23 ± 0.19^c	2.72 ± 0.14^d	2.83 ± 0.23^d
pH (-)	7.25 ± 0.39^b	7.05 ± 0.22^c	7.28 ± 0.22^b	7.20 ± 0.27^b	7.44 ± 0.24^a	7.42 ± 0.18^a	7.41 ± 0.20^a

Values with a different superscript letter (i.e. a, b and c) indicate significant difference at $p < 0.05$ based on Tukey's HSD. Sampling number: 54 for each variable.

performed well; annual mean removal efficiencies were between 50% and 80%, if approximately 2 mg/L NH_4-N were added. The removal of PO_4-P in wetlands treating As varied considerably. Unplanted wetlands were associated with less than 40% removal, if compared to the influent phosphorus load. Meanwhile, the presence of *J. effuses* resulted in the maximum PO_4-P mean removal of 70% for wetland A packed with manganese sand (Table 4.7).

4.2.3.5 Other Variables

Changes in the online measured parameters pH and DO were also reported in Table 4.7. In general, the effluent pH was significantly higher in unplanted wetlands E and F than in those wetlands (A, B and C) planted with *J. effuses* and *P. vittata*. Wetlands packed with ceramsite had significantly lower pH values than corresponding wetlands packed with manganese ($p < 0.05$; Table 4.7). The DO concentrations were significantly higher in planted wetlands than in unplanted ones. However, minor differences in effluent DO concentrations were noted for wetlands planted with the same plant species whether ceramsite or manganese sand was present.

4.2.3.6 Annual Arsenic Mass Balance

Total As biomass production and As removal by plant harvesting is summarized in Table 4.8. Substrate types had little importance in wetland plant uptake of As. *J. effuses* seeds accumulated the highest As concentrations compared to the corresponding roots and stems, while As levels in *P. vittata* leaves were between 57 and 61 times and between 4.6 and 5.6 times higher compared the levels in roots and stems (Table 4.8), respectively. Total As removal showed great variability with different substrate and plant combinations (Table 4.8). Values of As uptake by wetland plants reached the minimum and maximal level of 41.56 and 804.01 mg/m^2 for CWs C (*J. effuses* planted in manganese sand) and D (*P. vittata* planted in manganese sand), respectively.

Annual As mass balance calculations were conducted for each wetland unit. As shown in Figure 4.5, substrate adsorption and retention contributed most to As removal in manganese sand-packed wetlands with removals of 82.92%, 75.93% and 82.47% for wetlands C, D and F, respectively. Wetland plants played an insignificant role in As reduction. The highest and lowest removal rates of 5.87% and 0.33% were recorded for *P. vittata*-planted wetlands C and D, respectively. Except for effluent discharge, substrate retention and plant accumulation, unaccountable As removal caused by processes including microbial assimilation and detritus adsorption accounted for between 15.01% and 23.12% and between 5.12% and 5.89% of the total reduction in ceramsite and manganese packed wetlands, respectively.

A brief As mass balance analysis for CWs with different substrates and plant designs was made in this study. Environmental field managers may benefit from results obtained for different filter media types, packing orders and depths, macrophyte types and planting intensities, harvesting seasons and operation parameters such as hydraulic residence time as discussed in the next section.

TABLE 4.8

Total Arsenic (As) Biomass Production in Ceramsite Packed Wetlands A (*Juncus effuses* planted) and B (*Pteris vittata* L planted), Manganese Sand Packed Wetlands C (*Juncus effuses* planted) and D (*Pteris vittata* L planted) and Total As Accumulation by Plant Uptake after 12 months Running

	Dry Weight (kg/m²)				As Content (mg/kg)				
Wet-land	Root	Stem	Leaf	Seed	Root	Stem	Leaf	Seed	Total As Uptake (mg/m²)
A	1.20 ± 0.20	2.65 ± 0.48	-	0.47 ± 0.23	1.63 ± 0.17	7.91 ± 0.31	-	181.36 ± 13.45	108.16 ± 13.28
B	3.29 ± 0.90	1.95 ± 0.09	0.23 ± 0.03	-	7.18 ± 0.39	76.98 ± 4.89	431.36 ± 21.27	-	272.95 ± 19.38
C	0.75 ± 0.08	1.01 ± 0.16	-	0.27 ± 0.09	1.18 ± 0.08	4.37 ± 0.23	-	134.29 ± 9.32	41.56 ± 3.88
D	8.59 ± 0.63	4.04 ± 0.23	0.53 ± 0.14		8.97 ± 0.45	112.68 ± 7.35	512.69 ± 35.17	-	804.01 ± 26.90

Note: In the present study, As content in wetland plant biomass before planting is negligible and was not taken into consideration because all of the wetland candidate plants are cultivated without As exposure before the test run.

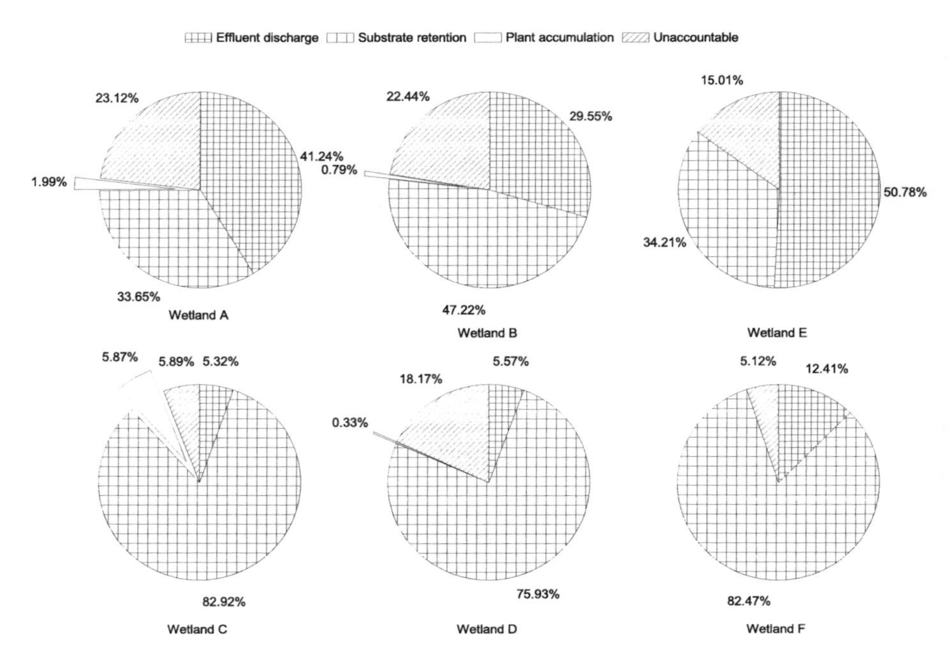

FIGURE 4.5 Mass balance of arsenic (As) evaluated as a percentage of the inflowing total As mass in the experimental wetland filters A, B, C, D, E and F.

4.2.4 Discussion

4.2.4.1 Arsenic Adsorption Processes

From the available literature, it appears that the use of suitable specific sorbent media to enhance the removal of As within CWs has been poorly studied, although various media have been investigated for As adsorption (Altundogan et al., 2002; Chakravarty et al., 2002; Gupta et al., 2005; Kanel et al., 2006; Lim et al., 2009; Maji et al., 2011; Rahman et al., 2011). The first criterion in comparing the As removal performances of different substrates was usually the maximum As adsorption capacity, which can be obtained from batch adsorption experiments. In this study, calculated maximum As adsorption capacities (obtained from Langmuir isotherms; Table 4.6) for different materials ranged between 12.70 and 42.37 μg/g. These values were comparable to a number of adsorbents such as zeolite of 8 μg/g (Jimenez-Cedillo et al., 2009), $FeCl_3$-treated clinoptilolite of 9.2 μg/g (Baskan and Pala, 2011), ferruginous manganese ore of 15.38 μg/g (Chakravarty et al., 2002), granular ferric hydroxide of 160 μg/g (Thirunavukkarasu et al., 2003) and laterite soil of 180 μg/g (Roberts et al., 2004).

The Brunauer–Emmett–Teller surface areas measured in this study ranged between 0.89 and 1.02 m^2/g. No significant difference in the outer surface area for each media assessed for As adsorption was recorded. Note that a 0.25-mm mesh sieve was applied for all media. With a similar outer surface, concentrations of iron, aluminium and manganese may greatly determine As adsorption (Chakravarty et al., 2002; Maji et al., 2011). The order of metal oxides in Table 4.3 mirrored the sequence for maximum As adsorption capacity. It follows that iron, aluminium and manganese contents can be used as effective indicators to help with a preliminarily evaluation of the As adsorption capacity.

During the As adsorption process, iron, aluminium and manganese are prone to form different compounds besides physical adsorption. Amorphous hydrous ferric oxide (FeOOH), hydrous aluminum oxide (AlOOH) and hydrous manganese oxide (MnOOH) are promising effective adsorptive materials for As removal from water (Mohan and Pittman, 2007; Roberts et al., 2004). The iron(III) oxide surface had a high affinity for As(V) and was capable of forming inner-sphere bidentate binuclear As(V)-Fe (III) (Nguyen et al., 2010). Arsenic adsorption by iron complexes occurred by ligand exchange of the As species for $(OH)_2$ and OH^- in the coordination spheres of surface structural Fe atoms (Macedo-Miranda and Olguin, 2007). Manganese dioxide such as MnO_2 can oxidize As(III) to As(V), and then adsorb the As(V) reaction product onto its solid phase. The most likely As(V)-MnO_2 complex is a bidentate binuclear corner-sharing (bridged) complex occurring at MnO_2 crystallite edges and interlayer domains. There is a potential advantage of using manganese dioxide to treat waters contaminated by As(III) and As(V) (Manning et al., 2002).

Adsorption kinetics and isotherms show large dependencies on the physical and/or chemical characteristics of the sorbent material, which also influenced the adsorption mechanism (Ranjan et al., 2009). As shown in Tables 4.4 and 4.5 the observed good correlation coefficients indicate that the As uptake process can be approximated by the pseudo-first-order kinetic model and Freundlich isotherm equation. For film diffusion to be the adsorption rate-limiting step, the value of the film diffusion co-efficient (D_f) should be in the range of 10^{-6} to 10^{-8} cm²/s, and for pore diffusion to be rate-limiting, the pore diffusion coefficient (D_p) should be in the range of 10^{-11} to 10^{-13} cm²/s (Michelson et al., 1975). In the present study, film diffusion appeared to be the rate-limiting step for the As adsorption kinetic process (Table 4.5).

The mean adsorption energy findings provide important information about the physical and chemical nature of the adsorption process. If $E < 8$ kJ/mol, physical adsorption dominates the adsorption process. If $E > 16$ kJ/mol, chemical adsorption is the dominant factor. If E is between 8 and 16 kJ/mol, the adsorption process is dominated by particle diffusion (Argun et al., 2007). Based on the judgement rule mentioned above, As adsorption to ceramsite and manganese sand was dominated by chemical forces. Physical processes determined the As adsorption onto the gravel surface, while particle diffusion was the main mechanism for As adsorption onto zeolite (Table 4.6). When relating As maximum adsorption capacity to the mean adsorption energy, it can be shown that substrates with bigger mean adsorption energy have higher maximum adsorption capacities.

4.2.4.2 Role of Substrates and Plants

Substrates played a key role in As removal within CWs. Adsorption, precipitation and co-precipitation of As on hydrous oxides of metals was a major sink for As fixation. Filter media with a relatively high porosity usually have larger outer surface areas and therefore higher As adsorption capacities (Zhang et al., 2013). The relatively high porosity of manganese sand (54%) may partly improve As removal due to adsorption when compared to ceramsite. In oxidizing environments with high levels of As(V), precipitation of As(V) with Ca, Mg, Al and Fe(III) may occur (Henken, 2009). Trapping within porous filter media and trapping with Fe and Mn on the substrate surface are the major As removal mechanisms in CWs (Singhakant et al., 2009a). Gravel is the most commonly used wetland aggregate supporting As removal. However, the As adsorption capacity is low; that is, in the range of up to 4.3 μg/g (Singhakant et al., 2009a). This value is smaller than 12.70 μg/g, which was obtained in this study. However, a relatively smaller grain size was applied (Table 4.6).

Iron (Fe) can act as a co-precipitation agent for As, particularly in the oxic zones (Buddhawong et al., 2005). In CWs, As usually adsorbs onto the surface of substrates, mineral particles, (oxy)(hydr)oxides and organic matter (Lizama et al., 2012). Iron oxide tezontle has been found to remove As. The total As mass removal efficiency during the first three months was 57.7% in unplanted sub-surface-flow CWs packed with tezontle (Zurita et al., 2012). In comparison to this study, ceramsite and manganese sand contributed between 33.65% and 47.22% and between 75.93% and 82.92%, respectively, to total As removal. High As removal performances for manganese sand during the whole running period of this study confirmed that As precipitation and adsorption onto most metal (hydr)oxides (especially onto Fe and Mn oxyhydroxides) were likely the main As removal mechanisms (Kneebone et al., 2002). Materials rich of Fe, Al and Mn (oxy)(hydr)oxides had great potential in removing As.

The main role attributed to macrophytes in terms of As removal has been the presence of a highly diverse microbial community within the root zone, which mediates a variety of removal mechanisms. The removal efficiency of As is significantly higher ($p < 0.001$) in CWs planted with *P. australis*. This is particularly the case for gravel-packed wetlands (Zhang et al., 2009). Wetlands planted with *J. effuses* have a substantially higher As retention capacity (59%–61% of the total As inflow) than wetlands without plants (only 44%) (Tsihrintzis and Gikas, 2010). This study shows that experiments also indicate that the presence of *J. effuses* can improve As removal efficiencies by nearly 2.64% and 21.32% (Table 4.7) for wetlands packed with ceramsite and manganese sand, respectively. However, the introduction of the As hyperaccumulator *P. vittata* resulted in no more than 9.64% and 6.88% As(V) reductions when compared to corresponding unplanted wetlands. Moreover, annual As mass balance analysis results further revealed that *P. vittata* contributed no more than 1% to the total As removal regardless if the wetland was packed with ceramsite or manganese. Findings imply that terrestrial As hyperaccumulators cannot exhibit their best capacity in As accumulation within fully water-saturated CWs. Although, *P. vittata* survived and grew normally at sufficient As exposure.

4.2.4.3 Relationship between Effluent and Arsenic Removal

Planted wetlands removed NH_4-N and PO_4-P significantly from As-polluted influent, because these nutrient species are easily removed by wetland plants and microbes (Tang et al., 2009, 2011). Furthermore, linear regression analysis results for representative wetlands D indicated that As removal was significantly and positively

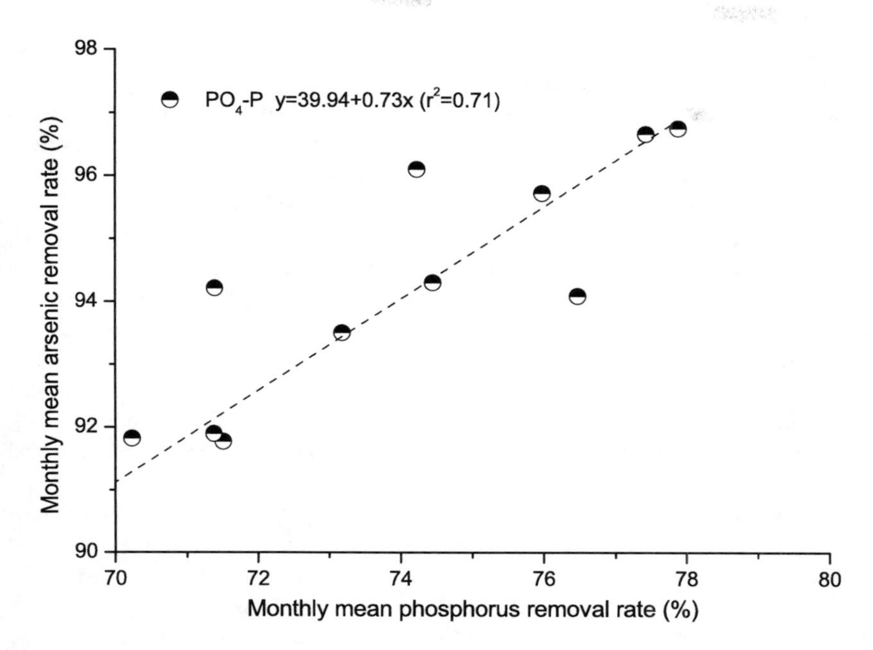

FIGURE 4.6 Higher monthly mean arsenic(V) removal rates observed at higher orthophosphate-phosphorus (PO_4-P) removal rates for the selected representative wetland D.

correlated to PO_4-P reduction rates (Figure 4.6). Adequate phosphorus supply is of particular importance to maintain the normal growth of wetland plants and associated microorganisms (Buddhawong et al., 2005). In this study, phosphorus addition may encourage plant and microbial growth, and thus indirectly enhanced As removal. Competitive As(V) adsorption to substrate is widely reported (Lizama et al., 2012).

The major role of wetland plants in heavy metal removal is not indirect uptake but substrate stabilization and bed media oxidation (Stefanakis and Tsihrintzis, 2012). Root oxygen release led to elevated oxygen availability within the planted wetland (Table 4.7), which favoured the immobilization of As compared to the unplanted wetlands (Tsihrintzis and Gikas, 2010). Significantly higher DO levels clearly indicated that planted wetlands exhibited more oxidized conditions than unplanted wetlands since the plants were directly involved in the diffusion of oxygen via the aerechyma (spongy tissue that creates spaces or air channels in some plants) to the rhizomes (Brix, 1997, although the detailed processes involving macrophytes are still debated (Wießer et al., 2002). Precipitation of As with metal oxides resulted in a pH decline of wetland effluent when compared to the influent (Table 4.7). Similar findings were also reported by Tsihrintzis and Gikas (2010).

4.2.4.4 Arsenic Mass Balance and Saturated Media

Compared to unplanted wetlands, annual As mass balance calculations indicated that the presence of wetland plants improved As-bonding capacity (Tsihrintzis and

Gikas, 2010), and thus reduced the amount of As flushed out through effluent discharge (Figure 4.6). Moreover, substrate retention contributed to the majority of As removal in wetland filters whether planted or not (Figure 4.6). Similar findings were also reported elsewhere (Singhakant et al., 2009b) indicating that As complexation with Fe and Mn on the media surface was 31% and 38%, respectively. Moreover, As trapping within wetland substrate was 42% and 52%, respectively, of total As.

In the present assessment, the As content in aboveground parts including seeds, leaves and stems was higher than in the corresponding belowground parts such as roots (Table 4.8). This finding disagrees with previous studies (Singhakant et al., 2009b; Stefanakis and Tsihrintzis, 2012), which indicate that the belowground biomass accumulated significantly more heavy metals compared to the corresponding aboveground biomass. A possible reason could be that *J. effuses* and *P. vittata* were harvested in the growing season (May) in the present study, which is associated with a high translocation efficiency from belowground biomass to aboveground parts (Tang et al., 2009).

Compared to gravel, excellent As adsorption performance of ceramsite and manganese sand may reduce the importance of wetland plants in As removal. Wetland plants only contributed to less than 6% of the total As removal, which was significantly lower than reported for gravel bed wetlands, where figures were around 50% (Rahman et al., 2011). Similarly, low plant uptake rates can also be found elsewhere (Ye at al., 2003), indicating that only 2% of the As accumulation in the plants was obtained in an experimental CW treating electric utility wastewater. Finally, processes including microbial assimilation, detritus adsorption, atmospheric volatilization and pore water storage may cause the unaccountable parts of As removal (Rahman et al., 2011).

No obvious As saturation phenomenon was detected during the first year of operation. However, it is likely that wetland media will eventually get exhausted and require change either by replacement with fresh media or through regeneration by reactivation of their adsorption capacity. Replacement is usually the common choice for exhausted wetland filter media while regeneration is very much in an experimental stage. Replacement is more attractive and cost-effective compared to regeneration, because wetlands are usually packed with a large amount of cheap and portable filter media. The relatively small available outer surface area of media such as gravel is indicative of a limited adsorption capacity, particularly if compared to detritus originating from long-term wetland operation. However, a dilute NaOH solution can successfully be used to regenerate media saturated with nanoparticles at reasonable cost (Ghosh and Gupta, 2012). For wetland media saturated with As(V), however, this technique may be economically unjustified.

4.2.5 Conclusions and Recommendations

Due to the small-scale system and relatively short running period, the results obtained in this experimental wetland study cannot directly represent the real performance of a large-scale field operation. Nevertheless, Section 4.2 discussed the As removal performance in wetlands with different substrates and plants, and provides an initial indication of the overall performance for wetland designers. Moreover, the findings provide a strong indication that the use of wetland filters as a low-cost option for decontaminating As-polluted drinking water resources in developing countries is a promising alternative to conventional high-cost treatment options. Results from batch adsorption kinetics and isotherm experiments indicated that aggregates containing iron, aluminium and manganese show excellent As adsorption performances. For manganese sand, its easy availability, low cost (common industrial waste), fast adsorption rate and excellent adsorption capacity cuts down operational expenditures and make it a good wetland substrate.

During the whole running period, wetlands packed with manganese sand showed more than 90% removal of influent As(V) regardless if planted or unplanted. Differences in water content, microbial community and nutrient supply processes within the root zone micro-environment of wetland filters may be the cause for the unsatisfactory removal of As from the traditional hyperaccumulator *P. vittata*. Therefore, *J. effuses* should be considered as an aquatic plant for As(V) removal in the future.

Annual As mass balance analysis indicated that substrate adsorption was the major contributor to As removal in the wetland systems. Direct wetland plant uptake played a negligible role in total As removal, but its indirect function of substrate stabilization and bed media oxidation can significantly improve the As removal performance when compared to unplanted controls.

Wetland plants showed positive contributions to total As removal. However, further detailed research work should be conducted to identify the optimal plant harvesting season when the above biomass As concentration has reached the maximum value. Adsorption was the main As removal mechanism in the CWs. Further studies on As saturation are recommended to test the substrate lifetime with a combination of As batch adsorption experiments. The presence of As(V) and different substrates and vegetation change the microbial composition of the biomass, As species as well as their fate and distribution. Studies of the microbial population dynamics, and As translocation, accumulation and reduction mechanism should therefore be conducted. In addition, future works require the evaluation of the effect of loading rate and retention time on As removal in wetland filters.

4.3 REDUCTION OF OCHER IN GROUNDWATER WELLS

4.3.1 Question and Goals

The aim of Section 4.3 is to determine whether both chemical fouling and biological fouling by iron bacteria such as *Gallionella ferruginea* can be reduced in groundwater wells, raw water pumps and raw water pipes (Figures 4.7 and 4.8). For this purpose, the literature was reviewed and problems with fouling were explained using a representative Oldenburgisch-Ostfriesischer Wasserverband (OWVV) case study. Solutions were identified to help water suppliers.

FIGURE 4.7 Typical example of a pump affected by ochre formation. (Picture: Oldenburgisch-Ostfriesischer Wasserverband.)

4.3.2 Method

A review of the German and international literature was carried out with no time limit. The following search terms (also in German) and related forms of these were used in the correct context: biological fouling, well management, chemical fouling, iron bacteria and flow velocity. For the case study, almost exclusively unpublished Oldenburgisch-Ostfriesischer Wasserverband documents (not shown) that were written in German were viewed.

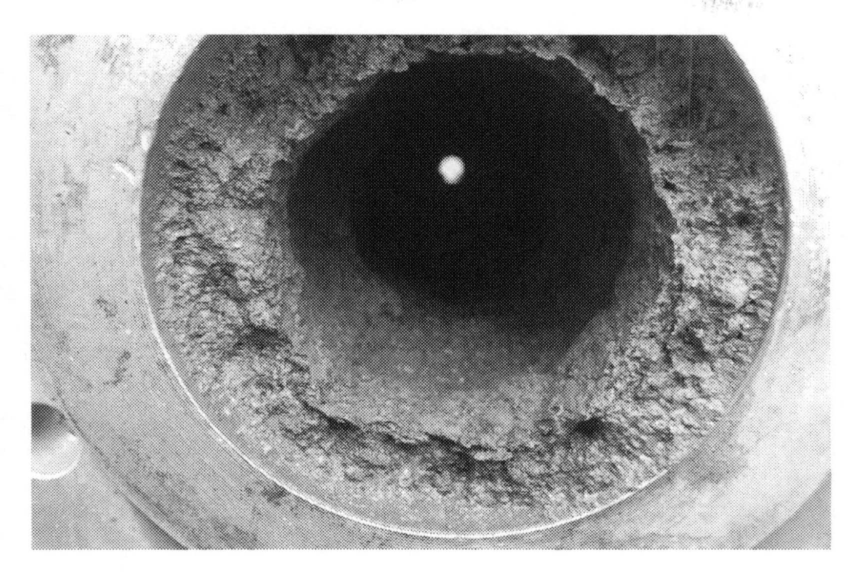

FIGURE 4.8 Representative example of clogging due to ochre within a raw water pipe. (Picture: Oldenburgisch-Ostfriesischer Wasserverband.)

4.3.3 Background and Challenge

Ocher is a natural clay pigment. It consists of a mixture of iron oxide and varying amounts of sand and clay. The colour of ocher ranges from yellow to deep orange or even brown. Ocher, which contains a large amount of hematite (Fe_2O_3) or dehydrated iron oxide, produces a reddish hue. Clays coloured with iron oxide that are produced during the extraction of copper and tin are also referred to as ocher (Houben and Treskatis, 2019).

When iron ions dissolved in water find favourable electrochemical conditions for oxidation and precipitation, ocher formation occurs. The shape of the iron element depends on purely chemical aspects, the electrochemical potential, the oxygen partial pressure, the pH value, temperature and the general pressure (Van Beek et al., 2009; Houben and Treskatis, 2019).

Deep groundwater does not contain oxygen. However, enrichment and downward spread of oxygen due to water extraction via groundwater pumps can usually be observed at all well locations. Although the absolute oxygen concentrations vary between drilling locations depending on various technical and natural conditions, there is often enough oxygen to form ocher (Henkel et al., 2012; Menz, 2016; Houben and Treskatis, 2019).

Contamination in both surface water and groundwater can lead to problems in hydraulic engineering facilities. Tophof et al. (2022) systematically examined clogging on filter systems. In addition, the formation of incrustations by ocher leads to additional energy costs when pumping from the wells and in the raw water pipe system due to the resulting friction losses (Figures 4.8 and 4.9). In further raw water treatment, a reduction in process and energy efficiency as well as an increased use of chemicals and, if necessary, the use of a filtration stage can be observed (Houben and Treskatis, 2019).

Formulations were also found in water-bearing layers up to 4 m around boreholes. Such distant incrustations are difficult to remove using mechanical and chemical methods. Removing pipe and annulus material to install a replacement well in the same location could therefore impose a hydraulic disadvantage on the new well as the iron scales remain present in the aquifer (Henkel et al., 2012; Houben and Weihe, 2010).

Biological fouling must be distinguished from chemical fouling. *G. ferruginea* is an iron-oxidizing autotrophic (chemolithotrophic) bacterium (Figure 4.9). Chemolithotrophy is the oxidation of inorganic chemicals to produce energy (Hallneck and Pedersen, 1991). It has been known for over 150 years that this bacterium plays a significant role in the oxidation and fixation of iron. This bacterium can be easily recognized by its twisted stalks. At the end of a stem is the bean-shaped iron bacterium, which, however, can detach from the stem if living conditions become unfavourable (Figure 4.9).

In their work, Hallbeck and Pedersen (1995) proved for the first time that the stem has a protective function against the toxic oxygen radicals produced during chemical iron oxidation. The researchers also showed that the oxidation rate of iron, not its concentration, is harmful to cells. The stalk (Figure 4.9) gives *G. ferruginea* a unique opportunity to colonize and survive in high iron habitats that are inaccessible to bacteria without a defence system against the oxidation of iron (Cullimore, 2007).

FIGURE 4.9 *Gallionella ferruginea* in a drinking water pipeline. (Attribution: Koelle, CC BY-SA 3.0 <http://creativecommons.org/licenses/by-sa/3.0/>, via Wikimedia Commons.)

G. ferruginea was identified early on in many different habitats such as freshwater iron-bearing mineral vents, shallow brackish waters, shallow marine hydrothermal environments and active deep-sea hydrothermal vents, as well as iron-associated bottom environments (Halbach et al., 2001). In these places, they form biofilms by interacting in a network of different bacteria.

This bacterium can be responsible for the production capacity of a groundwater well decreasing over time. This also means that raw water flows can no longer be measured precisely, which is insufficient for control and planning work. The ocher deposited by *G. ferruginea* can be deposited in the filter pipe, in the gravel fill, in the aquifer and on the pump. These deposits can lead to a drop in the operating water level into the suction area of the pump (Thronicker and Szewzyk, 2013).

Optimal living conditions for *G. ferruginea* are often found in wells with neutral, slightly oxic and normal conditions. The pH values should be between 5.4 and 7.6 (optimally between 6.0 and 7.2). The redox potential should be between −10 and 320 mV. Note that 20–300 mV would be optimal. A cheap sour substance content is between 0.005 (better 0.1 mg/L) and 1 mg/L. The iron content should be between 0.2 and 25 mg/L. Carbon dioxide should also be present (Cullimore, 2007; Houben and Treskatis, 2019).

4.3.4 Critical Review of the Literature

4.3.4.1 Preventive Measures Against Ocher Formation

The use of preventive measures against ocher formation is more effective for a smooth technical process than subsequent measures such as pigging (use of a cleaning device in pipelines). Someone may distinguish between non-invasive and invasive concepts and measures (Menz, 2016).

Computer-aided classification systems to support decision-making are a clearly non-invasive measure to reduce the challenges associated with ocher formation. The classification of drilling sites considering their vulnerability to aging and the development of adapted operational plans can be of practical importance. A mathematical approach can be used to quantify well aging and identify factors that contribute to well performance loss (Thronicker and Szewzyk, 2013).

Most appropriate clogging indicators should be identified to analyse the worst and best site conditions for their impact on ochre formation. Accordingly, a well located at a large distance from the nearest surface body of water with a thick layer of groundwater above the well screen, located in a closed aquifer with a high redox potential, will have the lowest aging potential. Compared to unfavourable site conditions and based on the average lifespan of a typical drinking water well, this can mean a difference in well capacity of up to 90%. In addition, optimized renovation intervals could be determined for the identified well classes based on their aging potential (Menz, 2016).

The use of preventative invasive measures against ochre formation can also be effective. Therefore, preventing ochre formation using hydrogen peroxide (H_2O_2) is at least theoretically a suitable measure. However, preventative treatment could also be a potential source of oxygen in wells and filter packages (Menz, 2016; Houben and Treskatis, 2019).

Current treatment methods with hydrogen peroxide were evaluated by reviewing the literature and operator data as well as through laboratory and field studies (Menz, 2016). Studies have shown significant improvement potential for treatment with hydrogen peroxide. However, the effects of the treatment were small, especially when ochre encrustations were already present. Results of laboratory and field tests could not fully demonstrate the effectiveness of preventive treatment. However, with increased concentrated hydrogen peroxide solutions and an improved treatment process, ochre formation can be delayed, and the rehabilitation potential of well and raw water pipes can be improved.

4.3.4.2 Chemical Ochre Formation

At high oxygen concentrations, iron hydroxide can form chemically (Henkel et al., 2012). However, nitrate can also serve as an electron receiver (Hedrich et al., 2011; Scholz, 2015). Centimetre-thick deposits form over the years. These plaques are soft at first, but become harder as they age and become difficult to remove.

However, chemical curing processes are quite well known (Henkel et al., 2012). This review therefore focuses on the biological investigation of curing. In practice, no distinction is often made between iron and manganese curing, nor between biological and chemical curing. This makes it difficult to examine the literature, as some articles often only talk about curing. However, the author assumes that at least a large proportion of the iron is chemically deposited (Scholz, 2015).

Since iron encrustations are attached to iron-oxidizing bacteria and their biomass, one cannot assume that a low biomass content indicates lower biological encrustation. The ratio between biomass and iron oxide in young cherts (rock composed of silica with a fine-grained texture) is often only 1–20 to 1–10. In older cherts, the ratio is up to 1–60. This may be due to bacterial migration because nutrients can no longer easily penetrate through the hardened crusts (Houben and Treskatis, 2019). It follows that in practice it is difficult to distinguish between biological and chemical curing, as the age of curing and the nutrient conditions in the groundwater also play a key role.

The chemical curing risk reduction processes are discussed in the following paragraphs: oxygen reduction, pH reduction and pigging. A high oxygen content within the upper well shaft must be prevented to reduce the risk of chemical curing. Therefore, the filter section above the underwater pump should be filled with water. This can be achieved by reducing the pump speed.

The curing rate depends on the pH value of the raw water. At lower pH values, fewer signs of curing can usually be observed. In addition, the speed of ochre decomposition increases if an ochre layer is already present on the well materials (Henkel et al., 2012; Houben and Treskatis, 2019).

The aging of incrustations must be stopped by countermeasures such as pigging so that encrustations can be removed before they harden. The hardening of curing can be delayed for years and decades by inhibitors such as phosphate, silicate and organic components in the groundwater. A low pH value in groundwater is also conducive to delay (Henkel et al., 2012; Houben and Treskatis, 2019).

The removal of deposits such as ochre through maintenance measures such as pigging is necessary (Thronicker and Szewzyk, 2013; Scholz, 2015). However, optimizing pigging through a better understanding of chemical and biological curing would be desirable.

4.3.4.3 Biological Ochre Formation

The role of iron bacteria in well aging is well documented (Cullimore, 2007). In groundwater with neutral pH values, the group of neutrophilic iron and manganese bacteria often predominates, although this group is very heterogeneous. The individual bacteria sometimes have very different metabolic properties (Houben and Treskatis, 2019).

For example, *G. ferruginea* oxidizes dissolved iron, therefore removing it from water and producing an insoluble precipitate of iron hydroxide. Over the years, the iron bacteria form centimetre-thick deposits. These deposits are initially soft, but as they age, they become harder and become more difficult to remove. The removal of the deposits through maintenance measures is often necessary (Thronicker and Szewzyk, 2013).

The iron dissolved in the water serves *G. ferruginea* as an electron source for its metabolism during iron oxidation. This bacterium donates electrons to oxygen and dissolved divalent iron ions are transformed into insoluble iron hydroxide (Cullimore, 2007).

While it is important to learn more about the ecology of diverse groups of iron bacteria, the biofilm in which they live should also be better understood as an ecosystem. Some bacteria attach to surfaces using fimbriae (cell appendages) (Mikkelsen et al., 2011). Extracellular polymeric substances (long-chain compounds) are then secreted by bacteria and contribute to biofilm formation (Sutherland, 2001). These long-chain compounds are positively charged and therefore attract negatively charged particles such as chemically formed ochre. The microorganisms within the biofilm have different metabolic processes and products and can therefore support each other in growth (Stams and Plugge, 2009). This creates a complex biofilm that slowly hardens and reduces the water flow cross-sections within wells and submersible pumps.

There are basically three groups of iron bacteria (Cullimore, 2007): autotrophic iron oxidizers, heterotrophic iron reducers and heterotrophic iron oxidizers. Autotrophic organisms produce food from their environment using light (photosynthesis; not relevant to this subchapter) or chemical energy (chemosynthesis). In comparison, heterotrophic organisms cannot synthesize their own food and rely on other organisms for nutrition.

Autotrophic iron oxidizers oxidize iron (Fe) from Fe^{2+} to Fe^{3+} to produce energy. They need carbon dioxide as a carbon source to build their cell substance. Traces of oxygen and nitrate (as an alternative source of oxygen) are sufficient for them to maintain their metabolic processes. These organisms are active primarily on the surface of the biofilm to obtain oxygen. Due to their position in the biofilm, autotrophic iron oxidizers serve as an indicator of the spatial extent of icing (Hallbeck and Pedersen, 1991; Cullimore, 2007).

Heterotrophic iron reducers reduce Fe^{3+} to Fe^{2+}. They do not need DO for their metabolism and can directly dissolve Fe^{3+} deposits. However, to maintain their metabolic processes, they also need organic components dissolved in the water. They are usually located within the biofilm and live on iron compounds that have already been deposited by the iron-oxidizing bacteria. The presence of iron reducers indicates that icing has already developed and can serve as an indicator of the thickness of the icing on the gravel within the production pipe. Increasing concentrations may indicate a possible change in the tendency to dry out (Cullimore, 2007).

Heterotrophic iron oxidizers convert Fe^{2+} dissolved in water into insoluble Fe^{3+}. These oxidizers also require organic substances dissolved in water for their metabolism and cannot meet their carbon requirements with carbon dioxide alone. They need additional oxygen or nitrate to maintain their metabolic processes. These organisms, like the autotrophic iron oxidizers (see above), reside on the surface of the ochre deposits and serve as an indicator of the extent of the ochre formation. In addition, they provide information about potentially relevant organic compounds in water (Cullimore, 2007).

Different iron bacteria often occur in groundwater and wells. However, under certain conditions such as stress, only a few or even just one species of iron bacteria such as *Gallionella*, *Leptothrix*, *Sphaerotilus*, *Kineosporia* or *Siderocapsa* can dominate. However, *Gallionella* is generally considered to be one of the most important primary iron sequesters (Figure 4.9; Ehrlich, 2002).

From a water hygiene perspective, iron bacteria present in drinking water pipes are not pathogenic. However, large iron deposits can cause biofilms to become so dense and thick that they provide a protected habitat for potentially pathogenic organisms. The presence of pathogens in drinking water can negatively influence the composition of the microbial community, the biotic interactions between individual non-pathogenic and pathogenic microorganisms, and the structure of the biofilm grown on the inner surface of the pipe (Felföldi et al., 2010).

The processes that lead to an increase in the risk of biological decomposition are complex and should therefore be examined in situ in the biofilm. The more iron the groundwater contains at the most appropriate pH ranges and DO and nutrient concentrations, the more likely ochre formation is in the presence of iron bacteria such as *G. ferruginea* (Houben and Treskatis, 2019).

The following biodegradation risk reduction processes are discussed below: ultrasonic application, flow rate reduction and uniform well water delivery, manipulation of oxygen levels, iron bacterial predation and iron-reducing organisms.

It should be possible to reduce clumping by using ultrasound. Ultrasound can tear the cell walls of microorganisms (Berlitz and Kögler, 1997). This method is theoretically more suitable for deep versus shallow wells. However, the curing problem is greater in deeper than in shallow wells, which is why ultrasound is used (see also above). In addition, the effect of ultrasound also depends on many other factors such as well design, pressure, material properties and water turbidity. In general, the following boundary conditions could be considered favourable: wells up to a depth of approximately 50 m, small diameter of both the borehole and the filter pipe inside it, filter pipe of low density and low filter area density (Houben and Treskatis, 2019).

Ochre is preferentially deposited in system parts with the highest flow, such as pumps. The attachment takes place with the excretion of so-called sticky extracellular polymeric substances (polysaccharides and proteins) (Thronicker and Szewzyk, 2013).

The faster the water flow, the more nutrients are available to the iron bacteria living on pipe walls. A reduction in flow velocity should therefore also lead to a reduction in biological clogging (Cullimore, 2007).

The literature also points to advantages such as the reduction of silting when uniform well water extraction is conducted to keep the water level stable (Van Beek et al., 2009). The mixing of oxic and reduced well water should therefore be prevented in practice to promote clogging.

G. ferruginea lives in oxygen-poor conditions. The bacterium oxidizes and fixes iron (Figure 4.9). To obtain energy from this process, *G. ferruginea* must live in a relatively specific environment that contains reduced iron, the right amount of oxygen, and enough carbon, phosphorus and nitrogen (Halbach et al., 2001). It follows that an environment with high oxygen, very little reduced iron and low concentrations of carbon, phosphorus and nitrogen should lead to a reduction of *G. ferruginea*. The study of biofilms in situ is necessary to better understand and control these complex processes.

However, since groundwater in some localities is rich in iron, the introduction of oxygen into the well water should be prevented to reduce early oxidation if the formation of ochre in aquifers should be avoided (Weihe, 2000). However, the relatively new technology of underground de-icing specifically leads to massive ochre formation in the aquifer and should therefore be assessed critically. The first industrial applications such as those in Bremen were therefore not successful (personal communication with Mr. Tompke, SWB Wasser, water utility in Bremen, 24 April 2023).

In underground de-icing, there is an opportunity to avoid aboveground oxidation of iron and manganese by carrying out underground iron and manganese oxidation with the help of pumping oxygen-enriched water into the aquifer (Houben and Threskatis, 2019). The ochre formed then remains underground and should make it more energy-intensive to pump water out of the ground in the future. There may also be other disadvantages that have not yet been identified when studying the literature.

Hallbeck and Pedersen (1995) speculated that both the braided twisted stems and the general encrustation due to ochre creation serve the iron bacterium as protection against predators, erosion and reactive oxygen species. Except for very large and some poisonous animals, most living things have some predators that control their numbers. It follows that there should be some small organisms that can live at least in the horizontal raw water pipes, and which should successfully reduce *G. ferruginea*. However, literature findings have not provided a clear answer to this.

On the other hand, it is easier to find organisms in the literature that can reduce iron (Greene et al., 2003). The most common genera found are *Geobacter* and *Geothrix*, which occur naturally in groundwater. *Geobacter metallireducens* is often used in biodegradation and bioremediation. In addition, *Geothrix fermentans* is known for its ability to use electron acceptors Fe(III) as well as other high potential metals. It also uses many substrates as electron donors.

4.3.5 Possible Solutions for a Representative Case Study

4.3.5.1 Specific Problem Statement

Many water associations in Central Europe have problems with silting, especially in their groundwater wells. Figure 4.10 shows an overview of wells in the

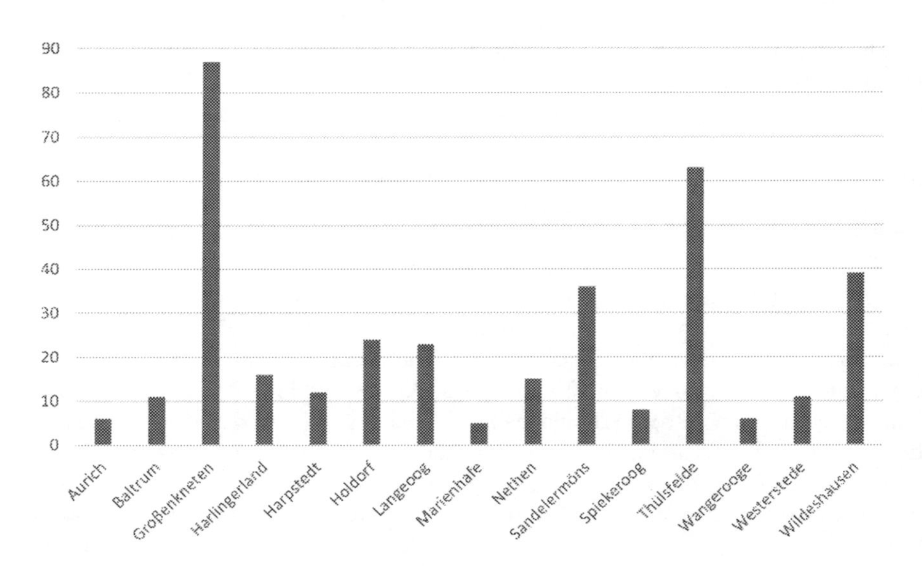

FIGURE 4.10 Number of well groups in the water board area of the Oldenburgisch-Ostfriesischen Wasserverbandes – wells in Grossenkneten are affected by clogging. (Attribution: Oldenburgisch-Ostfriesischer Wasserverband.)

Oldenburgisch-Ostfriesische Wasserverband (OOWV; water utility) association area. Many wells in Groβenkneten are affected by icing deposits. A pipe wall covering can lead to a cross-sectional narrowing of 40% within one year (Figure 4.10).

In practice, it has been observed by the OOWV that pumps fail much earlier due to clogging, before the end of their official service life (Figure 4.10). Even the use of pump washing systems was not successful. OOWV engineers replace pumps in some areas annually. A pump may currently cost around €6,500. In addition, there are €8000 for replacing the pump and around €4500 for general cleaning of the systems (personal communication with Julian Damröse, OOWV, March 2023).

The OOWV would like to know how much water it can extract from groundwater wells (personal communication with Ulrich Weihe, OOWV, January 2023). Unfortunately, the water flow rate and quantity cannot be accurately measured in some wells and the associated raw water pipes that lead to drinking water treatment stations (Houben, 2006). The reason for this is a hard deposit up to 11 mm thickness, which has settled in some pumps as well as in the neighbouring pipes (Figure 4.10) after several decades (Houben and Weihe, 2010).

It is reasonable to assume that this is due to chemical and biological clogging of the production well pumps. The OOWV water management information department assumes that these are iron(III) structures (ochre) in the well. These could be metabolites of *G. ferruginea*. The OOWV has received various external reports that contradict each other. One report confirms biological fouling by *G. ferruginea*, while another report highlights that there are few iron bacteria in the pipe water system.

However, only three groups of iron bacteria with different frequencies were distinguished by the OOWV: low occurrence of autotrophic iron oxidizers, low to medium number of heterotrophic iron reducers and very low occurrence of heterotrophic iron oxidizers in isolated samples. In general, the iron bacteria numbers determined indicate low curing. The variability is very high between different wells (personal communication with Jürgen Sander, OOWV, February 2023).

4.3.5.2 Possible Solutions

The following key approaches are discussed in this subchapter: flow rate manipulation, depth control, use of larger pumps and pipes, use of alternative piping materials, and the development of energy-efficient well management strategies. Other possible solutions are also briefly mentioned.

The OOWV has observed that clogging can be reduced if the pumps run continuously and evenly at reduced speed. A suction flow regulator can help to homogenize the inflow (Houben, 2006). However, this can also lead to increased entry losses if it is affected by incrustations themselves. Switching pumps on and off and fluctuating flow velocities, on the other hand, lead to more ochre formation (Thronicker and Szewzyk, 2013).

The more *G. ferruginea* is stressed, the harder the ochre deposits become. This could be observed particularly within the pumps. Some of these failed due to overheating as ochre deposits grew and severely restricted flows. It follows that more rather than fewer pumps at lower speed in the same area should reduce ochre formation and protect groundwater. However, this would require a much larger capital investment for new wells from the OOWV. A new well may currently cost around €400,000. Furthermore, such an approach has not yet been tested (personal communication with Jürgen Sander, OOWV, February 2023).

A lower flow rate should not only lead to a slowdown in curing but also to curing with a higher water content (Weihe, 2000). This would then also make it easier to clean pumps and pipes.

The OOWV has observed in practice that deeper (compared to shallower) extraction of groundwater leads to less ochre formation. Wells with filter edges that are closer to the surface are more prone to icing. This may be due to the presence of oxygen and nitrate (Houben and Treskatis, 2019). However, a deeper well also comes with higher capital and operating costs. Therefore, a cost-utility analysis should be conducted.

Large drilling and construction diameters with long filter sections and low extraction quantities have proven themselves in continuous operation. This can be explained by the fact that aging phenomena are highly dependent on changes in the natural flow conditions in the groundwater body (Weihe, 2000).

The failure of pumps and the increased frequency of raw water pipe cleaning in the OOWV due to silting could be reduced by choosing larger cross-sectional diameters for pumps, risers and smaller raw water pipes. This would also be beneficial in terms of reducing flow velocities (see also above), as high flow velocities lead to more clogging (Thronicker and Szewzyk, 2013).

Previous tests with various pipe materials and coatings in the OOWV association area have shown that copper pipes and pipes coated with copper and silver lead to a remarkable reduction in the biological ocher formation on these pipes. However, the challenge with this approach is that metal coatings such as copper dissolve slowly (within six months) in groundwater when the pH is below about 8. Therefore, pipes would have to be made from solid copper, increasing capital costs. The same is especially true for other more expensive precious metals with similar properties to silver. It would therefore be necessary to conduct a cost-use analysis (personal communication with Jürgen Sander, OOWV, March 2023).

Optimizing well management as part of an economic efficiency analysis includes the development of maintenance and flushing strategies, which should be carried out in a much more targeted manner and as needed based on targeted hydro-chemical system monitoring. The economic analysis should also be based on an energy efficiency assessment. Dynamic economic efficiency calculations taking energy efficiency into account should be carried out for well management optimization scenarios. The long-term savings potential through a change in well management will be compared with the potential investment costs. Short- to medium-term energy optimization should be achieved through maintenance-optimized control. Energy consumption and energy-price-oriented optimization must support well management (Houben and Treskatis, 2019).

The OOWV also looked at other alternatives to solve the problem:

a. Pigging (every 2 years according to plan) to clean pipes and improve flow (consequential problems due to biofilm destruction have been observed);
b. Chemical cleaning (controversial regarding environmental impact within wells and their environments);
c. Optimization of pumps in terms of pressure (in the network), operating speed, pump selection and cleaning (further research is planned);
d. Uniform water withdrawals (see above);
e. Assessment of the individual wells and, if necessary, initiation of well construction measures (cleaning, renovation and new construction) and geographical redefinition of the work area; and
f. Optimization of raw water pipes (risers, branch pipes and transitions between branch pipes and collecting pipes).

4.3.6 Summary and Outlook

There are various ways of reducing or eliminating clogging within pumps and pipelines. A constant and low flow rate and larger cross-sectional diameters of pumps and pipelines are advantageous. Deep versus shallow wells result in less ocher formation. Compared to modern pipes, conventional copper does not lead to any biological ocher creation. The development of energy-efficient well management strategies should be promoted.

It is recommended to study in situ (in the pipes) particularly the biological clogging processes. The quantification of chemical and biological fouling is important to propose the right approaches to reduce them.

4.4 GREYWATER

4.4.1 Introduction

4.4.1.1 Background

Researchers estimate that one-third of the world population could have insufficient water resources by 2025 (FAO, 2007). Therefore, recycling of wastewater for non-portable purposes has been considered as a new strategy to conserve conventional water resources (Jefferson et al., 2000). The most common practises of recycling treated wastewater and greywater can be found in the agricultural, industrial, urban and environmental sectors (Surendran and Wheatley, 1998).

Greywater is a major proportion of domestic wastewater (around 50%–80% according to Eriksson et al. (2002), which is generated from all household wastewater streams, except toilet discharge (Eriksson et al., 2002; Jefferson et al., 2000; Jeppesen, 1996). However, some literature has excluded the flow contributions of kitchen sinks, garbage disposal units and/or dishwashers from greywater (Al-Jayyousi, 2003; Christova-Boal et al., 1996; Emmerson, 1998; World Health Organization, 2006). High fluctuations in quality and a considerable overlap in characteristics between black and grey wastewater have been reported (Eriksson et al., 2-002). The compounds present in greywater vary from source to source and depend on different lifestyles, customs and installations as well as on the use of chemical household products (Jeppesen, 1996). Furthermore, there could be chemical and biological degradation of the chemical compounds within the transportation network and during storage affecting physical and chemical parameters (Eriksson et al., 2002; Nolde, 2000; Li et al., 2009).

Reported physiochemical parameters of relevance for greywater are summarized in Table 4.9. Food particles and raw animal fluids from kitchen sinks, soil particles as well as hair and fibres from laundry wastewater are examples of sources of solid material in greywater (Eriksson et al., 2002). High temperatures may be unfavourable since they enhance microbial growth and could induce precipitation in supersaturated solutions (Christova-Boal et al., 1996).

Section 4.4 is based on the methodology paper by Abed and Scholz (2016). This article has been revisited and upgraded.

Measurements of turbidity and suspended solids provide some information concerning the overall content of particles and colloids that could induce clogging of installations such as the piping used for greywater transportation as well as sand filters and CWs used for subsequent treatment (Eriksson et al., 2002). Measurements of the traditional wastewater parameters five-day biochemical oxygen demand (BOD_5), COD and nutrients such as nitrogen (N) and phosphorus (P) in form of ammonia-nitrogen (NH_4-N), nitrate-nitrogen (NO_3-N) and ortho-phosphate-phosphorus (PO_4-P) also give valuable information about the chemistry of greywater (Houshia et al., 2012). Ramona et al. (2004) argued that wastewater would be better classified as a function of pollution load rather than origin, and hence suggesting the notion of low (bath, shower and washbasin) and high (kitchen, washing machine and dishwasher) strength greywater.

TABLE 4.9
Characteristics of Real Greywater (GW)

Reference	Greywater Source	Temp. (°C)	pH(–)	Turbidity (NTU)	TSS (mg/L)	EC (μS/cm)	DO (mg/L)	BOD_5 (mg/L)	COD (mg/L)	NH_4–N (mg/L)	NO_3–N (mg/L)	PO_4–P (mg/L)
Eriksson et al. (2002)	Bathroom GW	29	6.4–8.1	60–240	54–200	82–250		76–2,001	100–633	≤0.1–15.0	0.28–6.30	0.94–48.80
	Laundry GW	28–32	8.1–10.0	14–296	120–280	190–1,400		48–380	12.8–725.0	0.04–11.30	0.4–2.0	4–171
	Kitchen GW	27–38	6.3–7.4		235–720		2.2–5.8	1,040–1,460	3.8–1,380	0.002–23.0	0.3–5.8	12.7–32.0
	Mixed GW	18–38	5.0–8.7	15.3 to ≥200.0		320–20,000		90–360	13–549	0.03–25.40	0.0–4.9	4–68
Al-Jayyousi (2003)	Hand basin							109	2,631	9.6		
	Combined			69				121	371	1		
	Single person			14				110	256			
	Single family			76.5						0.74		
	Block of flats			20				33	40	10		
	College			59				80	146	10		
	Large college			57				96	168	0.8		
Christova-Boal et al. (1996)	Bathroom GW	25	6.4–8.1	60–240	48–120	82–250		76–200		≤0.1–15.0		
	Laundry GW	25	9.3–10.0	50–210	88–250	190–1,400		48–290		≤0.1–1.9		
Li et al. (2009)	Bathroom GW		6.4–8.1	44–375	7–505			50–300	100–633			
	Laundry GW		7.1–10.0	50–444	68–465			48–472	231–2,950			
	Kitchen GW		5.9–7.4	298	134–1,300			536–1,460	26–2,050			
	Mixed GW		6.3–8.1	29–375	25–183			47–466	100–700			

(Continued)

TABLE 4.9 (*Continued*)
Characteristics of Real Greywater (GW)

Reference	Greywater Source	Temp. (°C)	pH(–)	Turbidity (NTU)	TSS (mg/L)	EC (μS/cm)	DO (mg/L)	BOD_5 (mg/L)	COD (mg/L)	NH_4–N (mg/L)	NO_3–N (mg/L)	PO_4–P (mg/L)
Al-Hamaiedeh and Bino (2010)	Real GW range		6.9–7.8		23–358	157–200		110–1,240	92–2,263		0.44–0.93	
	Real GW average		7.2		275	183		942	1,712		0.68	
Pidou et al. (2008)	Mixed GW LC		6.6–7.6	35				39	144	0.7	3.9	0.5
	Shower GW HC		7.3–7.8	42				166	575	1	7.5	1.3
	Real Raw GW			46.6				205	791	1.2	6.7	1.66
Ramona et al. (2004)	Shower GW1		7.5	23	29.8	1,317		78	170	1.5–3.0	0.05–1.70	0.02–0.19
March et al. (2004)	Raw GW		7.3–8.0	5–62					39–441			
Eriksson et al. (2010)	Raw GW 1		7.7–8.1		51–135		2.5–4.5	18–68		0.36–4.40		0.02–2.20
	Raw GW 2		8.2–8.3		67–390		9.3–9.5	≤3		0.07–0.13		0.25–0.28
Nghiem et al. (2006)	Real GW		5.0–10.9					33–1,460	3.8–1,380.0			
Houshia et al. (2012)	Raw GW		6.1			1,500		126.6			38	
Leal et al. (2012)	Raw GW		7.24			74.4			1,476		≤0.10	2.97

Temp., Temperature, NTU, nephelometric turbidity unit, TSS, total suspended solids, EC, electric conductivity, DO, dissolved oxygen, BOD_5, five-day biochemical oxygen demand, COD, chemical oxygen demand, NH_4–N, ammonia–nitrogen, NO_3–N,nitrate–nitrogen, and PO_4–P, ortho–phosphate–phosphorus.

A major difficulty when treating greywater is the considerable variation in its composition. Reported mean values of, for example, COD and BOD_5, vary from 40 to 371 mg/L and from 33 to 466 mg/L, respectively, between sites and with similar variations arising at an individual site (Al-Jayyousi, 2003; Eriksson et al., 2002; Gross et al., 2007a, 2007b; Ramona et al., 2004). This has been attributed to changes arising in the quantity and type of detergent products employed during washing. Moreover, significant chemical changes may take place over time periods of only a few hours (Jefferson et al., 2000). Among other pollutants, trace elements and heavy metals have been reported as important components to take into consideration for treatment, storage and recycling purposes as indicated in Tables 4.10 and 4.11 (Eriksson et al., 2010; Leal et al., 2012).

The BOD_5 and DO concentrations decrease during the sedimentation period when greywater is stored. Evidence has shown that 50% removal of BOD_5 could be achieved when greywater is stored over a 4-hour-period (Jefferson et al., 2000). However, extended storage may lead to the risk of odour increases and possibly health issues due to enhanced microorganism growth (Jefferson et al., 2001). Furthermore, the BOD_5 concentration in, for example, greywater washing hand basins has been reported as being slightly lower than the one generated from mixed resources as well it varies with different discharge patterns (Al-Jayyousi, 2003).

There has been considerable research into the quality processes of raw greywater occurring during the storage stage (Liu et al., 2010). For example, Dixon et al. (1999) indicated improvements in greywater quality during complex storage processes.

4.4.1.2 Synthetic Greywater

In general, recycling of greywater is widely accepted compared to blackwater due to the lack of urine and faeces in the former (Domenech and Sauri, 2010). So, the pathogens and nutrients occurring in greywater are present in much lower concentrations than in blackwater (Eriksson et al., 2002).

Greywater does not contain the right nutrient and trace element ratio required for standard biological treatment or advanced treatment by membrane bioreactors (Al-Jayyousi, 2003; Jefferson et al., 2001; Wichmann and Otterpohl, 2009). Furthermore, low concentrations of trace elements have been linked to greywater (Eriksson et al., 2002). Some synthetic greywaters have been created by mixing different recipes of chemical products that household use and/or analytical grade chemicals known to be present in real greywater. Consequently, these chemicals are expected to control the characteristics of the generated greywater in terms of water quality (Schäfer et al., 2006).

Nghiem et al. (2006) investigated the feasibility of submerged ultrafiltration technology applied for greywater recycling. The synthetic greywater solution contained kaolin, cellulose, humic acid, sodium hypochlorite, calcium chloride electrolyte and a sodium bicarbonate buffer. These materials were also used in combination with sodium dodecyl sulphate to represent synthetic greywater proposed by Schäfer et al. (2006).

Nazim and Meera (2013) studied the treatment ability of a synthetic greywater by adding different concentrations of an enzyme protein solution to examine the reduction of chemical variables including nutrients. The mixture of synthetic greywater

TABLE 4.10
Trace Element Concentrations (mg/L) of Real Greywater (GW)

Reference	Greywater Source	Aluminium	Boron	Calcium	Potassium	Magnesium	Sodium	Sulphur	Silicon	Phosphorus
Eriksson et al. (2002)	Bathroom GW	≤0.1	≤0.1	3.5–7.9	1.5–5.2	1.4–2.3	7.4–18.0	1.2–3.3	3.2–4.1	
	Laundry GW	≤1.0–21	0.1–0.5	3.9–14.0	1.1–17.0	1.1–3.1	44–480	9.5–40.0	3.8–49.0	
	Kitchen GW	0.67–1.8		13–30	19–59	3.3–7.3	29–180			
	Mixed GW	0.10–3.55		11–35	6.6	1.5–19.0	21–230			
Christova-Boal et al. (1996)	Bathroom GW	≤1.0		3.5–7.9	1.5–5.2	1.4–2.3	7.4–18.0	1.2–3.3	3.2–4.1	0.11–1.80
	Laundry GW	≤1.0–21.0		3.9–12.0	1.1–17.0	1.1–2.9	49–480	9.5–40.0	3.8–49.0	0.062–42.000
Li et al. (2009)	Bathroom GW	2.44		33.8	8.1	5.74		23.7		
	Laundry GW	0.49		60.79	11.20–23.28	6.15		19		
	Kitchen GW	0.003		47.9	5.79	5.29		16.3		
Ramona et al. (2004)	Shower GW	0.03	0.14	71.0–93.6	9.8–12.4	43.2–50.0	93.0–142.7			
Nghiem et al. (2006)	Real GW			3.6–200.0						
Houshia et al. (2012)	Raw GW			89.5	37.3	132.2				
Leal et al. (2012)	Raw GW			42.8	14.5	11.6	128			
Kariuki et al. (2012)	Kitchen GW1			4.9	23.4	4.8	15.38			
	Laundry GW1			1.3	26.9	2.54	39.23			
	Bath GW2			0.96	10	0.27	6.15			
	Kitchen GW2			0.93	16.9	0.28	9.89			
	Laundry GW2			0.32	31.8	1.14	35.38			
Jefferson et al. (2001)	Real GW	0.003		47.9	5.79	5.29		16.3		

TABLE 4.11
Heavy Metal Concentrations (mg/L) of Real Greywater (GW)

Reference	Greywater Source	Cadmium	Chromium	Copper	Iron	Manganese	Nickel	Lead	Zinc	Molybdenum
Eriksson et al. (2002)	Bathroom GW	0.00054–0.01000		0.06–0.12	0.34–1.10			0.003	0.059–6.300	
	Laundry GW	0.00036–0.03800	≤0.025	≤0.050–0.322	0.29–1.00	0.029	≤0.028	0.033 to ≤0.063	0.09–0.44	
	Kitchen GW	0.00052–0.00700	≤0.025–0.130	0.05–0.26	0.6–1.2	0.031–0.075	≤0.025	0.005–0.140	0.096–1.800	
	Mixed GW	≤0.006–0.030	≤0.01026–0.05000	0.018–0.230	<0.05–4.37	0.014–0.075	≤0.015–0.050	≤0.01–0.15	≤0.01–1.60	
Christova-Boal et al. (1996)	Bathroom GW	≤0.01		0.06–0.12	0.34–1.10				0.2–6.3	
	Laundry GW	≤0.01		≤0.05–0.27	0.29–1.00				0.09–0.32	
Li et al. (2009)	Bathroom GW			0.0618	0.36	0.0121			0.0644	
	Laundry GW			0.08	0.11	≤0.05			0.00	≤0.05
	Kitchen GW			0.006	0.017	0.04			0.03	0.00

(*Continued*)

TABLE 4.11 (*Continued*)
Heavy Metal Concentrations (mg/L) of Real Greywater (GW)

Reference	Greywater Source	Cadmium	Chromium	Copper	Iron	Manganese	Nickel	Lead	Zinc	Molybdenum
Al-Hamaiedeh and Bino (2010)	GW Range							1.00–1.31		
	GW Average	0.008						1.19		
Ramona et al. (2010)	Shower GW 1	≤0.02	≤0.02	≤0.02	0.19	≤0.02	≤0.02	≤0.02	0.18	≤0.02
	Shower GW 2	≤0.02	≤0.02	≤0.02	0.06	≤0.02	≤0.02	≤0.02	0.03	≤0.02
Eriksson et al. (2010)	Raw GW 1	0.0001		0.0087–0.0110			0.007–0.039	0.0025–0.0031		
	Raw GW 2	≤ 0.0001–0.0090		0.0085–0.0250			0.0055–0.0079	0.0018–0.0032		
Leal et al. (2012)	Raw GW			0.0906	0.29			≤0.010		
Kariuki et al. (2012)	Kitchen GW1	5.5	16.1	0.9	1.9	1.4		0.9	6.6	
	Laundry GW1	7	0.9	1	3.6	0.4		0.8	0.4	
	Bath GW2	10.7	11.1	2.6	3.8	0.3		0.2	0.2	
	Kitchen GW2	10	11.3	2.3	9.7	0.2		0.3	0.1	
	Laundry GW2	11.2	16.1	2.9	17.5	0.3		0.0	0.7	

contained glucose, sodium acetate trihydrate, ammonium chloride, disodium hydrogen phosphate, potassium dihydrogen phosphate, magnesium sulphate and cow dung.

Diaper et al. (2008) introduced a synthetic greywater recipe to simulate combined laundry and bathroom greywater from an Australian residential dwelling. The constituents of the greywater included a variety of personal hygiene and household products, some laboratory-grade chemicals (sodium dodecyl sulphate, sodium hydro carbonate, sodium phosphate, boric acid, and lactic acid) and secondary sewage effluent sourced from a local wastewater treatment plant.

Fenner and Komvuschara (2005) described a new approach to model the effect of factors influencing ultraviolet disinfection efficiency of real and synthetic greywaters. A range of synthetic greywater recipes has been developed for both soft and hard waters to ensure they were representative of the properties of real greywater samples. A typical synthetic greywater recipe comprised dextrin, ammonia chloride (NH_3Cl), yeast extract, soluble starch, sodium carbonate (Na_2CO_3), monosodium phosphate (NaH_2PO_4), potassium phosphate (K_2PO_4) and an *Escherichia coli* culture mixed with distilled water.

Surendran and Wheatley (1998) proposed a biological treatment process for greywater obtained from large buildings. The synthetic greywater used comprised a known amount of soap, detergent, starch yeast extract and cooking oil. Settled sewage was also added to provide appropriate bacteria counts.

Jefferson et al. (2001) dosed synthetic and real greywater with nutrient supplements. The synthetic greywater recipe comprised synthetic soap, hair shampoo, sunflower oil and tertiary effluent.

Gross et al. (2006) have developed a new small-scale vertical-flow CW for decentralized treatment of greywater. The removal of indicator and pathogenic microorganisms was investigated to assess the reuse of treated greywater for irrigation purposes. The focus was on the removal dynamics of *Escherichia coli*, *Staphylococcus aureus* and *Pseudomonas aeruginosa* in three different synthetic greywaters.

Each greywater was made by combining three waste stocks representing laundry, bath and kitchen wastes (Gross et al., 2006). The composition of synthetic greywater for each stock contained laundry soap, shampoo, cooking oil and kitchen effluent (comprising one egg and one tomato). All greywater types were supplemented with raw sink effluent from a large dining room. This effluent, which contained an inoculum of *E. coli* and other bacteria, was added in a small enough volume not to affect the composition of the synthetic greywater (Gross et al., 2007b).

In a controlled study, a recirculating vertical-flow CW has been investigated to assess the effect of irrigation with treated greywater on soil properties (Travis et al., 2010). The greywater was prepared according to a similar recipe used by Gross et al. (2007b). However, pulverized bar soap was applied instead of shampoo in the synthetic greywater.

Gross et al. (2007b) developed an economically sound, low-tech and easily maintainable combined vertical-flow CW and trickling filter system for greywater treatment and subsequent recycling. The greywater was prepared artificially by mixing laundry detergent, boric acid and raw kitchen effluents into tap water.

Comino et al. (2013) proposed a functional hybrid phytoremediation pilot platform for the treatment of greywater. The pilot plant was tested with and without vegetation for different design specifications as well as for various organic and hydraulic loads of synthetic greywater. This study by Comino et al. (2013) followed one by Gross et al. (2007b) in terms of the preparation of artificial greywater.

Pinto et al. (2010) conducted glasshouse experiments to understand the effects of greywater reuse for irrigation of plants. Changes in soil pH, electric conductivity and nutrient content (total nitrogen and total phosphorus) due to greywater irrigation were assessed. Synthetic greywater was prepared by mixing a commonly available local detergent with potable water.

Winward et al. (2008) evaluated the three treatment technologies CWs, membrane bioreactors and membrane chemical reactors for indicator microbial removal and greywater reuse potential under conditions of low- and high-strength greywater influents. A high-strength supplementary solution together with real greywater was pumped to the treatment systems. Real greywater was referred to as low- or high-strength solution based on a mixture of locally sourced shampoo diluted by tap water.

4.4.1.3 Chemicals Used in Greywater Simulation

The increased focus on the treatment and reuse of highly variable real greywater has driven some researchers to create greywater with stable properties artificially as indicated in Table 4.12a (Hourlier at al., 2010). The concentrations of the corresponding greywater pollutants (e.g., organic strength, nitrogen, phosphorus, surfactants and metals) as a result of mixing the ingredients listed have been published in the references shown in Table 4.12a and Table 4.12b shows the corresponding water quality. However, most recipes cannot be reproduced accurately because the environmental boundary conditions are variable or unreported. Moreover, some ingredients such as cow dung, shampoo and kitchen effluent are unspecified. A reproduction of the published water quality data is therefore of little use to the readers of Section 4.4. Nevertheless, a review of the most common chemicals used for artificial greywater recipes is summarized below.

Kaolin is a common clay mineral composed of alternating sheets of aluminium hydroxide and silicate (Essington, 2004). It is frequently selected as an artificial greywater component to represent suspended organic and inorganic solids in greywater, which may originate from natural clay containing various mineral components. These solids are often generated from kitchen and laundry effluents (Eriksson et al., 2002). Kaolin is also used in synthetic wastewater recipes (Fitria et al., 2014; Nghiem et al., 2006; Marfil-Vega et al., 2010; Schäfer et al., 2006).

Cellulose is the principal structural component of plant cells and leaves. Furthermore, most of the carbohydrates found in soils are derived from cellulose, which is one of many polymers found in nature (Essington, 2004). Cellulose is frequently chosen to mimic organic fibres in greywater, since kitchen sinks and dishwashers are common sources of organic fibres (Nghiem et al., 2006; Schäfer et al., 2006).

All natural waters contain humic (Essington, 2004) constituents as the result of biodegradation of animal and plant matter or might form in situ due to the presence of soils, nutrients and cellulosic substrates for microbial action in the waste (Wall and Choppin, 2003). Humic acid is often used to represent dissolved organic matter in greywater (Nghiem et al., 2006; Schäfer et al., 2006).

TABLE 4.12A
Recipes reported for different synthetic greywaters

Reference	Surendran and Wheatley (1998)		Diaper et al. (2008)		Nazim and Meera (2013)		Fenner and Komvuschara (2005)	
Country	UK		Australia		India		UK	
Treatment approach	Multi-stage bio-filter		• Biological with suspended media • Chemical flocculants, ultraviolet disinfection and filtration • Settling, biological with fixed media		Using garbage enzyme after filtration		Ultraviolet disinfection system	
Dextrin	85 mg/l		Sunscreen or moisturiser	15 or 10 mg/l	Glucose	300 mg/l	Dextrin	85 mg/l
Ammonium chloride	75 mg/l		Toothpaste	32.5 mg/l	Sodium acetate trihydrate	400 mg/l	Ammonium chloride	75 mg/l
Yeast extract	70 mg/l		Deodorant	10 mg/l	Ammonium chloride	225 mg/l	Yeast extract	70 mg/l
Soluble starch	55 mg/l		Sodium sulphate	35 mg/l	Sodium dihydrogen phosphate	150 mg/l	Soluble starch	55 mg/l
Sodium carbonate	55 mg/l		Sodium hydrogen carbonate	25 mg/l	Potassium dihydrogen phosphate	75 mg/l	Sodium carbonate	55 mg/l
Washing powder	30 mg/l		Sodium phosphate	39 mg/l	Magnesium sulphate	50 mg/l	Sodium dihydrogen phosphate	11.5 mg/l
Sodium dihydrogen phosphate	11.5 mg/l		Clay (unimin)	50 mg/l	Cow dung	225 ml/l	Potassium phosphate	4.5 mg/l
Potassium sulphate	4.5 mg/l		Vegetable oil	0.7 mg/l			Escherichia coli culture	15 ml/l

(Continued)

TABLE 4.12A (*Continued*)
Recipes reported for different synthetic greywaters

Settled sewage	10 ml/l	Shampoo/hand wash	720 mg/l				
Shampoo	0.1 ml/l	Laundry	150 mg/l				
Cooking oil	0.1 ml/l	Boric acid	1.4 mg/l				
Biochemical oxygen demand	approx. 200 ml/l	Lactic acid	28 mg/l				
		Secondary effluent	20 ml/l				
Reference	Gross et al. (2007b)/ Comino et al. (2013)	Nghiem et al. (2006)		Jefferson et al. (2001)]		Hourlier et al. (2010)	
Country	Israel/Italy	Australia		UK		France	
Treatment Approach	Vertical-flow constructed wetland/ Hybrid constructed wetland	Submerged ultrafiltration membranes		Membrane bioreactors and activated sludge systems		Direct membrane nano-filtration	
Laundry detergent	20 g	Humic Acid	20 mg/l	Synthetic soap	0.64g	Lactic acid	100 mg/l
Boric acid	0.86 g	Kaolin	50 mg/l	Hair shampoo	8.0 ml	Cellulose	100 mg/l
Kitchen effluent	400 ml	Cellulose	50 mg/l	Sunflower oil	0.1 ml	Sodium dodecyl sulphate	50 mg/l
Tap water	150 l	Calcium chloride	0.5 mM	Tertiary effluent	24 ml	Glycerol	200 mg/l
		Sodium chloride	10 mM	Tap water	10 l	Sodium hydrogen carbonate	70 mg/l
		Sodium hydrogen carbonate	1 mM			Sodium sulphate	50 mg/l
						Septic effluent	10 mg/l

TABLE 4.12B
Characteristics of different synthetic greywaters proposed in Table 4.12a

Parameter	Unit	Surendran and Wheatley (1998)	Diaper et al. (2008)	Nazim and Meera (2013)	Gross et al. (2007b)	Comino et al. (2013)	Nghiem et al. (2006)	Hourlier et al. (2010)
Biochemical oxygen demand	mg/l	215	146.7	192	28.0–688			58–75
Chemical oxygen demand	mg/l		276.7	290	702–984	77.4		391–505
Ammonia-nitrogen	mg/l	11		9.6	0.1–0.5			
Nitrate-nitrogen	mg/l		<0.2		0.0–5.8			
Nitrite–nitrogen	mg/l		<0.003		0.0–1.0			
Total nitrogen	mg/l				25.0–45.2			
Ortho-phosphate-phosphorus	mg/l	4.9		110				
Total phosphorus	mg/l		17.8		17.2–27.0			
pH	–		7.4	6.16	6.3–7.0	7.3	7.5–8.0	6.29–7.29
Redox potential	mV							
Turbidity	NTU	72	52.1				140	4–42
Total dissolved solids	mg/l	12.3		563		247.4		
Total suspension solid	mg/l	196	59		85–285			41–87
Total organic carbon	mg/l	81.8	62.2					
Dissolved organic carbon	mg/l							106-149
Electronic conductivity	μs/cm		322.2		1000–1300	495.1		159–212
Dissolved oxygen	mg/l							
Aluminium	mg/l		1.6					
Boron	mg/l				1.4–1.7			
Calcium	mg/l		7.6					
Magnesium	mg/l		1.3					
Sodium	mg/l		65.3					
Surfactants	mg/l				4.7–15.6			33.5–69.8
Salinity	–					0.1		

Notes: NTU = nephelometric turbidity unit.

Boric acid is frequently applied to represent boron ions in greywater. One source of boron is natural and the other is a result of human activities (e.g., extraction plant, industry and detergent containing sodium perborate). It follows that many water sources and wastewaters may contain boron in variable concentrations (Diaper et al., 2008; Gross et al., 2007b).

The following salts have been previously suggested as possible ingredients in synthetic greywater: sodium chloride (dissolved monovalent salt) is found as a common ingredient of soap solutions and dyes (McBain et al., 1912; Myers, 1988). Sodium hydrogen carbonate (natural buffer) and sodium dodecyl sulphate are used for the manufacture of detergents. Their greatest cleaning application is as filler in powdered home laundry detergents (Myers, 1988; Zhu et al., 2015). Sodium hydrogen carbonate, sodium dodecyl sulphate and sodium phosphate are important in the manufacture of textiles by reducing negative charges on fibres, so that dyes can penetrate evenly (Syafalni et al., 2012). Some of these salts have previously been used in synthetic grey and municipal wastewater recipes (Diaper et al., 2008; Fenner and Komvuschara, 2005; Fitria et al., 2014; Nghiem et al., 2006; Schäfer et al., 2006; Surendran and Wheatley, 1998).

Calcium nitrate and calcium chloride have been suggested as components in synthetic greywater. Calcium salts are chosen to provide calcium ions to artificial greywater. Previous research used calcium salts in synthetic greywater (Nghiem et al., 2006; Schäfer et al., 2006). Laboratory-grade chemicals such as potassium nitrate, mono-potassium phosphate and magnesium sulphate have been chosen in previous studies (Fenner and Komvuschara, 2005; Nazim and Meera, 2013) to resemble real greywater in terms of nutrients and macronutrients generated from laundry and kitchen effluents. Low suspended solids and turbidity linked to greywater indicate that a substantial proportion of pollutants are dissolved. Although organics present in greywater are relatively similar to domestic greywater, their chemical natures are quite different. So, the deficiency of nutrients and low values of biodegradable organic matter are limiting the effectiveness of biological treatment of greywater (Al-Jayyousi, 2003).

Iron(III) chloride, manganese(II) chloride, chromium(III) nitrate, zinc sulphate, copper sulphate, cadmium oxide, nickel oxide, and lead(II) oxide are commonly selected to provide heavy metals to artificial greywater, as discussed in publications reported in Table 4.11 Sources of heavy metals in real greywater may be from cosmetics (Eriksson et al., 2010), other products such as skin emulsions (creams, lotion and jelly), soap, shampoo, hair cream, henna dye (Bocca et al., 2014; Chauhan et al., 2010) and from body parts such as hair, nails and died skin cells (Chjnacka et al., 2012; Eriksson et al., 2002). Henna is a reddish dye obtained from the dried and powdered leaves of the tree called henna (*Lawsonia inermis*).

Ammonium molybdate tetrahydrate is used to provide molybdenum in artificial greywater. Molybdate is also known to enhance the biological treatment of wastewater (Jefferson et al., 2001). Sodium hydroxide and hydrochloride acid are widely used as buffers to adjust the pH value of a chemical solution.

Small quantities of secondary or tertiary effluent obtained from predominantly domestic wastewater treatment plants is frequently recommended as an additive to synthetic greywater to provide a source of pathogens and microorganisms in general

(Diaper et al., 2008; Fenner and Komvuschara, 2005; Fitria et al., 2014; Gross et al., 2007a, 2007b; Hourlier et al., 2010). However, the addition of microbes might not be necessary for experiments in non-sterile environments such as outdoor trials where a microbial population adjusted to the system assessed will establish naturally eventually. One target of this study is to evaluate the stability of chemical compositions of artificial greywater through specific storage time experiments, without the contribution of biological treatment, which is offered by microorganism. There are numerous papers in the peer-reviewed literature indicating greywater recipes that have no artificially introduced microorganism in the list of ingredients (Nghiem et al., 2006; Schäfer et al., 2006).

4.4.1.4 Aim and Objectives

There is a need to develop standard synthetic greywater recipes to allow for the easy comparison of similar experiments in the future. Original experiments and a detailed literature review have been performed to support the development of stable generic synthetic greywater recipes for both low and high concentrations.

The aim of Section 4.4 is to propose practical recipes to be used for the simulation of greywater, which can be used with confidence to assess different treatment technologies. The objectives are (a) to review previous greywater recipes and corresponding components; (b) to evaluate the quality of the new synthetic greywater and compare it with recipes found in the literature; (c) to examine the stability of synthetic greywater as a function of time and (d) to show that water quality changes are not caused by internal reactions of used chemicals.

The scope of this chapter is limited to weak and strong standard synthetic greywater recipe proposals being prepared under non-sterile conditions. It follows that specific greywater types, which are often a function of geographical region, cultural and religious practices as well as guidelines and legislation, are beyond the scope of Section 4.4.

4.4.2 Materials and Methodologies

4.4.2.1 Synthetic Greywater

Household greywater was created artificially by using analytical grade chemicals (Table 4.12) purchased from Fisher Scientific Co. Ltd. (Bishop Meadow Road, Loughborough, UK). The synthetic greywater was prepared under non-sterile conditions as a stock solution by mixing the selected chemicals with de-chlorinated public mains tap water at a temperature of around 25°C. The following water quality parameters of greywater were simulated: BOD, COD, ammonia-nitrogen, nitrate-nitrogen, ortho-phosphate-phosphorus, pH, redox potential, turbidity, total suspension solids and electronic conductivity. The resultant key pollutants of the proposed recipes are summarized in Table 4.13.

Two stock solutions were mixed separately to represent low (LC) and high (HC) greywater strengths and stirred by a magnetic stirrer (3.0 cm long and 0.5 cm wide) with rounded edges for 1 hour at 1,200 rpm (Schäfer at al., 2006). The two solutions were stored overnight at 4°C and stirred for a further 30 minutes before the start of

TABLE 4.13
Proposed Ingredients for Low- and High-Strength Synthetic Greywaters

Item	Chemical Name	Chemical Formula	Molar Mass (g/mol)	Low Concentration (mg/L)	High Concentration (mg/L)	Composition Percentages
1	Kaolin	$Al_2Si_2O_5(OH)_4$	258.16	15	100	Al (20.90%), H (1.56%), O (55.78%) and Si (21.76%)
2	Cellulose	$(C_6H_{10}O_5)_n$	162.14	15	100	C (44.45%), H (6.22%) and O (49.34%)
3	Humic acid	$C_{187}H_{186}O_{89}N_9S_1$	4015.55	5	20	C (55.90%), H (4.67%), O (35.46%), N (4.67%) and S (0.80%)
4	Sodium chloride	$NaCl$	58.44	10	120	Cl (60.66%) and Na (39.34%)
5	Sodium hydrogen carbonate	$NaHCO_3$	84.01	10	85	C (14.30%), H (1.20%), Na (27.37%) and O (57.14%)
6	Calcium chloride	$CaCl_2$	147.02	10	55	Ca (36.11%) and Cl (63.89%)
7	Potassium nitrate	KNO_3	101.10	0	90	K (38.67%), N (13.85%) and O (47.48%)
8	Calcium nitrate	$Ca(NO_3)_2$	164.09	0	150	Ca (24.43%), N (17.07%) and O (58.50%)
9	Magnesium sulphate	$MgSO_4$	120.37	2	240	Mg (20.19%), S (26.64%) and O (53.17%)
10	Monopotassium phosphate	KH_2PO_4	136.09	13	85	H (1.48%), K (28.73%), O (47.03%) and P (22.76%)
11	Iron(III) chloride	$FeCl_3$	162.20	0.3	50.0	Fe (34.43%) and Cl (65.57%)
12	Boric acid	H_3BO_3	61.83	0.6	3.0	H (4.89%), B (17.48%) and O (77.63%)
13	Manganese(II) chloride	$MnCl_2$	125.84	0.03	3.20	Cl (56.34%) and Mn (43.66%)

(*Continued*)

TABLE 4.13 (*Continued*)
Proposed Ingredients for Low- and High-Strength Synthetic Greywaters

Item	Chemical Name	Chemical Formula	Molar Mass (g/mol)	Low Concentration (mg/L)	High Concentration (mg/L)	Composition Percentages
14	Zinc sulphate	$ZnSO_4$	161.44	0.25	15.00	O (39.64%), S (19.86%) and Zn (40.50%)
15	Copper sulphate	$CuSO_4$	159.61	0.025	7.000	Cu (39.81%), O (40.10%) and S (20.09%)
16	Ammonium molybdate tetrahydrate	$(NH_4)_6Mo_7O_{24}$	1163.94	0.35	0.35	H (2.08%), Mo (57.71%), N (7.22%) and O (32.99%)
17	Cadmium oxide	CdO	128.41	0.02	12.50	Cd (87.54%) and O (12.46%)
18	Nickel oxide	NiO	74.69	0.02	0.06	Ni (78.58%) and O (21.42%)
19	Chromium(III) nitrate	CrN_3O_9	99.99	0.045	70.000	Cr (21.85%), N (17.65%) and O (60.50%)
20	Sodium sulphate	Na_2SO_4	142.04	2.60	25.00	Na (32.37%), O (45.06%) and S (22.57%)
21	Sodium phosphate monobasic	H_2NaPO_4	119.98	0.00	250.00	H (1.68%), Na (19.16%), O (53.34%) and P (25.82%)
22	Lead(II) oxide	Pb_3O_4	685.60	0.16	1.40	Pb (90.67%) and O (9.33%)
23	Secondary treatment effluent with microbial content (ml/l)	–	–	20.00	100.00	–

Al, Aluminium; H, hydrogen; O, oxygen; Si, silicon, C, , N, nitrogen; S, sulphur; Cl, chlorine; Na, sodium; Ca,calcium; K, potassium; Mg;magnesium; P = phosphorus; Fe, iron; B, boron; Mn, manganese; Zn, zinc; Cu, copper; Mo, molybdenum; Cd, Cadmium; Ni, nickel; Cr, chromium; Pb, lead, and item 23 was not considered in this study.

subsequent experiments. The concentration levels of the proposed synthetic greywater are shown in Table 4.13. These concentrations were subject to environmental conditions typical for Greater Manchester (temperate and oceanic climate) between November and May.

Sodium hydroxide (NaOH) and hydrochloride acid (HCl) were used to adjust the pH value of the solution (Nghiem et al., 2006). A wide range for pH values for real greywater has been reported in literature (Table 4.9). However, in this experiment, the pH values for both low and high-strength greywaters were adjusted at pH ranges of around 5–7 and 7–10, respectively.

4.4.2.2 Experimental Set-up, Data and Analysis

The set-up design includes two groups of black plastic buckets (volumes of 14 L each) selected to store 10 L of the prepared greywater for two days and seven days of residence storage times. The storage times selected represent typical ones reported in literature (Tables 4.13 and 4.14). Moreover, there are practical considerations of regular feeding of experimental set-ups avoiding weekends. Each group has two bucket replicates; the first group was used for storing low-concentration greywater and the second for keeping high-strength greywater.

The buckets were subjected to real weather conditions at a quiescent place on university grounds from 1 November 2014 to 30 April 2015. Samples were collected manually after the specific storage time (two and seven days) to conduct several analytical tests as outlined in the next section.

Water quality sampling was carried out according to the American Public Health Association (2005), unless stated otherwise, to monitor the properties of synthetic greywater. The spectrophotometer DR 2800 (Hach Lange, Germany) was used for standard water quality analysis concerning variables including COD (mg/L), ammonia-nitrogen (NH_4-N, mg/L), nitrate-nitrogen (NO_3-N, mg/L), orthophosphate-phosphorus (PO_4-P, mg/L), total suspension solids (TSS, mg/L) and colour (Pa/Co).

The five-day biochemical oxygen demand (BOD_5, mg/L) was determined in all water samples with the OxiTop IS 12-6 system, a mono-metric measurement device, supplied by the Wissenschaftlich–Technische Werkstätten (Weilheim, Germany). Turbidity was measured with a Turbicheck Turbidity Meter (Lovibond Water Testing, Tintometer Group, Dortmund, Germany). The redox potential (redox) was obtained with a sensION+benchtop multi-parameter meter (Hach Lange, Düsseldorf, Germany). The electric conductivity (EC, μs/cm) was determined by a conductivity Meter entitled METTLER TOLEDO FIVE GOTM (Keison Products, Chelmsford, Essex, England, United Kindom). DO (mg/L) for all samples was measured by an HQ30d Flexi meter (Hach Lange, Düsseldorf, Germany).

Microsoft Excel has been used for the general data analysis (e.g., mean, standard deviation, minimum and maximum values). The non-parametric Mann-Whitney test was computed using IBM SPSS Statistics Version 20 and applied to compare the variance in test results of two (unmatched) independent samples, since all sample data were not normally distributed.

TABLE 4.14
Water Quality Parameters After Two and Seven Days of Storage Time

Parameter	Unit	Number	Mean	Standard Deviation	Minimum	Maximum	Reduction (%)
				Inflow (LC)			
Biochemical oxygen demand	mg/L	33	15.2	7.45	5.0	30.0	na
Chemical oxygen demand	mg/L	31	25.2	9.99	8.2	48.3	na
Ammonia-nitrogen	mg/L	30	0.2	0.11	0.0	0.5	na
Nitrate-nitrogen	mg/L	32	1.4	1.61	0.1	7.6	na
Ortho-phosphate-phosphorus	mg/L	31	6.3	2.35	3.8	12.0	na
pH	–	33	6.9	0.37	6.0	7.9	na
Redox potential	mV	33	15.7	53.07	−190.2	65.7	na
Turbidity	NTU	33	22.6	7.95	9.8	41.6	na
Total suspension solids	mg/L	33	40.2	18.70	10.0	87.0	na
Electronic conductivity	µs/cm	33	150.8	61.89	98.7	452.0	na
Dissolved oxygen	mg/L	33	10.1	1.53	7.7	12.2	na
Colour	Pa/Co	24	199.9	71.30	26.0	332.0	na
Temperature	°C	33	17.3	6.37	6.7	27.0	na
				Two-Day Outflow (LC)			
Biochemical oxygen demand	mg/L	21	5.7	3.96	0.0	10.0	62.3
Chemical oxygen demand	mg/L	21	27.9	10.26	2.7	41.9	−10.8
Ammonia-nitrogen	mg/L	19	0.1	0.09	0.0	0.3	45.2
Nitrate-nitrogen	mg/L	19	1.3	0.80	0.1	3.1	10.4
Ortho-phosphate-phosphorus	mg/L	19	5.6	2.04	3.5	10.9	11.4
pH	–	48	7.2	0.70	6.3	10.1	na
Redox potential	mV	48	17.5	30.68	−116.1	51.0	na

(*Continued*)

TABLE 4.14 (*Continued*)
Water Quality Parameters After Two and Seven Days of Storage Time

Parameter	Unit	Number	Mean	Standard Deviation	Minimum	Maximum	Reduction (%)
Turbidity	NTU	48	21.3	7.81	2.9	35.4	5.5
Total suspension solids	mg/L	48	30.8	12.92	13.0	76.0	23.4
Electronic conductivity	µs/cm	48	128.4	23.57	79.0	215.0	na
Dissolved oxygen	mg/L	48	10.7	0.94	8.8	12.6	−6.3
Colour	Pa/Co	36	156.0	51.13	34.0	265.0	22.0
Temperature	°C	48	16.0	4.85	5.3	21.8	na
			Seven-Day Outflow (LC)				
Biochemical oxygen demand	mg/L	15	7.0	6.21	0.0	20.0	54.0
Chemical oxygen demand	mg/L	22	19.6	9.83	6.0	36.7	22.2
Ammonia-nitrogen	mg/L	18	0.1	0.07	0.0	0.3	45.2
Nitrate-nitrogen	mg/L	17	1.1	1.27	0.0	4.0	21.4
Ortho-phosphate-phosphorus	mg/L	17	8.2	6.03	2.6	25.7	−29.4
pH	–	44	7.2	0.60	6.4	8.9	na
Redox potential	mV	44	18.3	26.66	−56.4	53.2	na
Turbidity	NTU	44	20.1	5.71	12.6	34.1	11.1
Total suspension solids	mg/L	44	31.0	9.52	18.0	56.0	22.9
Electronic conductivity	µs/cm	48	143.0	38.83	97.7	263.0	na
Dissolved oxygen	mg/L	48	11.5	0.84	10.4	14.3	−13.9
Colour	Pa/Co	36	171.5	33.14	128.0	258.0	14.2
Temperature	°C	48	14.1	3.87	6.7	20.0	na

(*Continued*)

TABLE 4.14 (*Continued*)
Water Quality Parameters After Two and Seven Days of Storage Time

Parameter	Unit	Number	Mean	Standard Deviation	Minimum	Maximum	Reduction (%)
				Inflow (HC)			
Biochemical oxygen demand	mg/L	33	32.3	12.81	10.0	60.0	na
Chemical oxygen demand	mgLl	30	115.4	39.57	63.9	189.0	na
Ammonia-nitrogen	mg/L	30	0.4	0.18	0.1	0.8	na
Nitrate-nitrogen	mg/L	32	9.2	7.81	0.2	29.8	na
Ortho-phosphate-phosphorus	mg/L	30	50.6	13.06	30.7	92.6	na
pH	–	33	8.1	1.93	5.4	11.5	na
Redox potential	mV	33	–29.3	89.61	–182.1	97.9	na
Turbidity	NTU	33	184.6	50.34	18.3	285.0	na
Total suspension solids	mg/L	33	317.5	54.73	190.0	473.0	na
Electronic conductivity	µs/cm	33	936.8	156.16	617.0	1180.0	na
Dissolved oxygen	mg/L	33	10.0	1.69	6.9	12.6	na
Colour	Pa/Co	27	1427.3	444.54	787.0	2499.0	na
Temperature	°C	33	17.6	6.58	6.5	27.8	na
				Two-Day Outflow (HC)			
Biochemical oxygen demand	mg/L	19	14.5	8.48	0.0	30.0	55.2
Chemical oxygen demand	mg/L	21	110.7	28.63	43.3	164.0	4.1
Ammonia-nitrogen	mg/L	19	0.4	0.26	0.0	0.9	6.8
Nitrate-nitrogen	mg/L	20	6.2	4.18	0.5	15.0	32.8
Ortho-phosphate-phosphorus	mg/L	20	46.5	14.37	23.7	70.1	8.2
pH	–	48	8.3	1.35	5.6	9.8	na

(*Continued*)

TABLE 4.14 (*Continued*)

Water Quality Parameters After Two and Seven Days of Storage Time

Parameter	Unit	Number	Mean	Standard Deviation	Minimum	Maximum	Reduction (%)
Redox potential	mV	48	−28.4	60.63	−107.6	88.6	na
Turbidity	NTU	48	215.7	49.45	111.0	341.0	−16.9
Total suspension solid	mg/L	48	345.0	48.49	229.0	447.0	−8.7
Electronic conductivity	µs/cm	48	948.3	105.86	627.0	1196.0	na
Dissolved oxygen	mg/L	48	10.3	0.78	9.0	12.1	−3.0
Colour	Pa/Co	36	1697.0	292.83	1121.0	2311.0	−18.9
Temperature	°C	48	17.0	4.94	6.0	21.5	na
			Seven-Day Outflow (HC)				
Biochemical oxygen demand	mg/L	15	14.7	6.40	5.0	30.0	54.5
Chemical oxygen demand	mg/L	24	108.3	24.47	67.2	159.5	6.2
Ammonia-nitrogen	mg/L	16	0.4	0.19	0.0	0.8	0.01
Nitrate-nitrogen	mg/L	18	2.8	2.24	0.4	9.3	69.6
Ortho-phosphate-phosphorus	mg/L	17	45.8	18.23	20.3	79.4	9.5
pH	–	48	8.1	1.20	5.9	9.8	na
Redox potential	mV	48	−27.4	57.02	−108.3	78.1	na
Turbidity	NTU	48	209.3	38.14	122.0	281.0	−13.4
Total suspension solid	mg/L	48	322.5	73.45	3.1	434.0	−1.6
Electronic conductivity	µs/cm	48	1105.6	351.09	668.0	2460.0	na
Dissolved oxygen	mg/L	48	10.9	0.72	9.4	12.0	−9.0
Colour	Pa/Co	36	1882.8	409.34	1119.0	2889.0	−31.9
Temperature	°C	48	15.7	3.49	8.4	20.8	na

LC, Low-concentration synthetic greywater; NTU, nephelometric turbidity unit; na, not applicable; HC; high-concentration synthetic greywater.

4.4.3 Results and Discussion

4.4.3.1 Synthetic Greywater Characteristics

The inflow water parameters in Table 4.13 refer to characteristics of prepared synthetic greywater just before utilization in the experiment. These parameters were compared and discussed with published results of real greywater constituents obtained from previous research studies (Table 4.9).

The figures shown in Table 4.13 are based on outside (greywater systems exposed to the elements) experiments. The data variability is therefore high, resulting in some unexpected findings, which are, however, not statistically ($p > 0.05$) significant. For example, the mean COD of the inflow (LC greywater) was 25.2 mg/L. After two days of storage, the average outflow COD was 27.9 mg/L. Furthermore, the corresponding standard deviations are relatively high, and the sample numbers of both data sets are different.

There are very few reported data regarding colour of real greywater. The test results of synthetic greywater have shown ranges of colour from 26 to 332 Pa/Co and from 787 to 2,499 Pa/Co for LC and HC greywater concentrations, respectively. The temperature was around 6.5°C–37.0°C for both types of proposed greywater, which was similar to figures reported by Eriksson et al. (2002) and Christova-Boal et al. (1996). Depending on the sources of greywater, there is a wide range of pH for real greywater. Most of these waters were simulated by using LC synthetic greywater with a pH between 6.0 and 7.9, while the pH values for HC greywater were between 5.4 and 11.5, representing those real discharges, which were commonly generated from laundries (Christova-Boal et al., 1996; Eriksson et al., 2002; Nghiem et al., 2006; Wichmann and Otterpohl, 2009).

The reported ranges for turbidity and TSS as shown in Table 4.9 and were successfully simulated particularly by the ingredient kaolin (Table 4.12) for both greywater strengths (Table 4.13). Those values for simulated HC greywater (mean of 318 mg/L and range between 190 and 473 mg/L; Table 4.13) are particularly represented by the solids in the discharges from laundry, kitchen and mixed greywater sources as shown in Table 4.9 (Al-Hamaiedeh and Bino, 2010; Christova-Boal et al., 1996; Eriksson et al., 2002; Wichmann and Otterpohl, 2009), while the simulated LC greywater (mean of 40 mg/L and range between 10 and 87 mg/L; Table 4.13) is linked to waters from hand basins, showers and similar mixed greywater sources as indicated in Table 4.9 (Al-Jayyousi, 2003; Eriksson et al., 2010; March et al., 2004; Pidou et al. 2008; Ramona et al., 2004). Electric conductivity data for real greywater in literature have demonstrated high levels for laundry and mixed greywater sources (Christova-Boal et al., 1996; Eriksson et al., 2002; Houshia et al., 2012; Ramona et al., 2004). In contrast, low values are linked to bathroom fluxes (Al-Hamaiedeh and Bino, 2010; Christova-Boal et al., 1996; Eriksson et al., 2002; Leal et al., 2012). The DO was around the reported upper limits, especially in the absence of significant numbers of microorganisms in the synthetic greywater.

Numerous water quality parameters of the proposed greywaters (Table 4.12) have similar values in terms of averages or are at least within the published ranges (Tables 4.13 and 4.14). Although the concentrations of BOD_5 in low strength greywater, in particular, are less than some of the reported values for real greywater, but they agree with those indicated by Eriksson et al. (2010) and Winward et al. (2008).

The review on COD concentrations in literature reveals that there is a wide variation of greywater types and compositions (Table 4.9). This can be explained by a great variety of household chemicals used causing a high degree of fluctuation from sample to sample (Al-Jayyousi, 2003; Eriksson et al., 2002, 2010). Compared with those obtained from the analysis of synthetic greywater (Table 4.13), the LC greywater COD concentrations were similar to the lower limits of reported studies. Furthermore, the test results for synthetic greywater (Table 4.122) have shown appropriate simulations for reported values of ammonia-nitrogen (NH_4-N), nitrate-nitrogen (NO_3-N) and ortho-phosphate-phosphorus PO_4-P, in terms of mixed greywater regardless of the sources of origin (Eriksson et al., 2002, 2010; Pidou et al, 2008; Ramona et al., 2004).

In the literature, various recipes for synthetic greywater, which was utilized for different treatment technologies, have been proposed (Table 4.13a). This study illustrates how to choose analytical grade chemicals to create two strength solutions of synthetic greywater (Table 4.12). Organic and inorganic matter, dissolved and suspended solids, nutrients, macronutrients, trace elements and microorganisms were resembled carefully to simulate real greywater components and associated properties. Depending on data shown in Tables 4.12 and 4.13, synthetic greywater solutions represent reality well. The recipe was based on the molar weight of the chemical composition multiplied by the percentage of the specific element in that chemical. For example, 100 mg of Iron (III) chloride provides 34 mg/L of iron (Table 4.12).

4.4.3.2 Stability of Synthetic Greywater

Table 4.13 shows all water quality results of LC and HC synthetic greywaters after two and seven days of storage. For LC greywater, the pH has increased from 6.9 to 7.2 for a two-day storage period. There was no significant ($p > 0.05$) change after seven days of storage. However, data show a reduction in colour, turbidity and TSS for the outflow of two-day storage experiments by 22.0%, 5.5% and 23.4%, respectively. The percentages concerning the outflow for the seven-day storage experiments were 14.2%, 11.1%, and 22.9%, respectively. The number of colloids and particles is likely to reduce over time as physical (e.g., coagulation and flocculation) processes reduce turbidity and suspended solids. However, biochemical processes such as biodegradation will lead to an increase in microorganisms and debris contributing to an increase in turbidity and fine material (Christova-Boal et al., 1996; Dixon et al., 1999; Eriksson et al., 2002; Gross et al., 2006, 2007a; Wichmann and Otterpohl, 2009).

A statistical analysis has shown no significant ($p > 0.05$) changes in colour, pH, turbidity and TSS, when both synthetic greywaters are stored for two or seven days. This confirms previous findings (Diaper et al., 2008; Hourlier et al., 2010) showing that suspended solids and insoluble particle concentrations of chemical greywaters are highly stable because they originate from inert materials.

Figure 4.11a and b illustrates the variations in BOD_5 concentrations for both LC and HC synthetic greywater, respectively. The values for LC greywater have shown significant ($p < 0.05$) reductions in the averages from 15.2 to 5.7 mg/L and to 7.0 mg/L at two and seven days of storage time, respectively (Table 4.13 and Figure 4.12a). While for HC greywater, the BOD has dropped significantly ($p < 0.05$) from 32.3 to 14.5 mg/L after two days of storage with a reduction of 55.2%, and it was stable

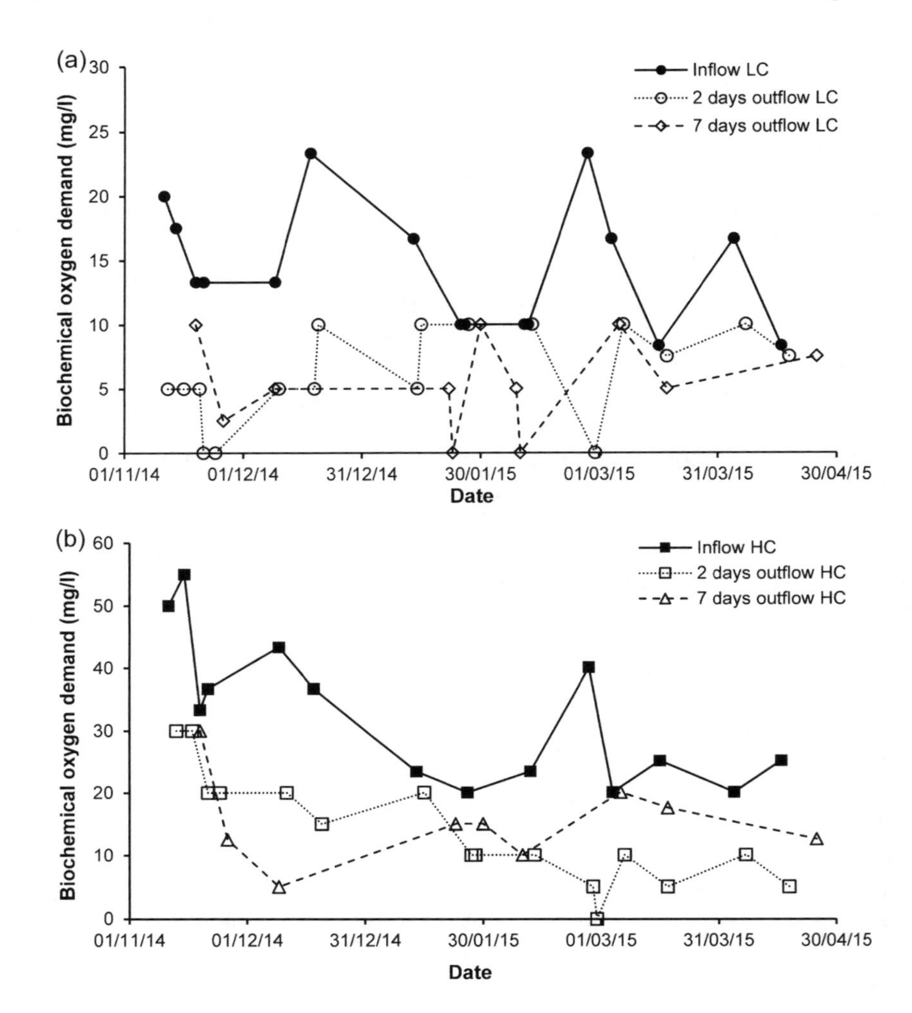

FIGURE 4.11 Effect of storage time on the variation of (a) five-day biochemical oxygen demand (BOD_5) of low-concentration synthetic greywater (LC); (b) BOD_5 of high-concentration synthetic greywater (HC); (c) chemical oxygen demand (COD) of LC, (d) COD of HC; (e) ammonia-nitrogen (NH_4–N) of LC, (f) NH_4–N of HC; (g) nitrate-nitrogen (NO_3–N) of LC, (h) NO_3–N of HC; (i) ortho-phosphate-phosphorus (PO_4–P) of LC and (j) PO_4–P of HC greywater.

(Continued)

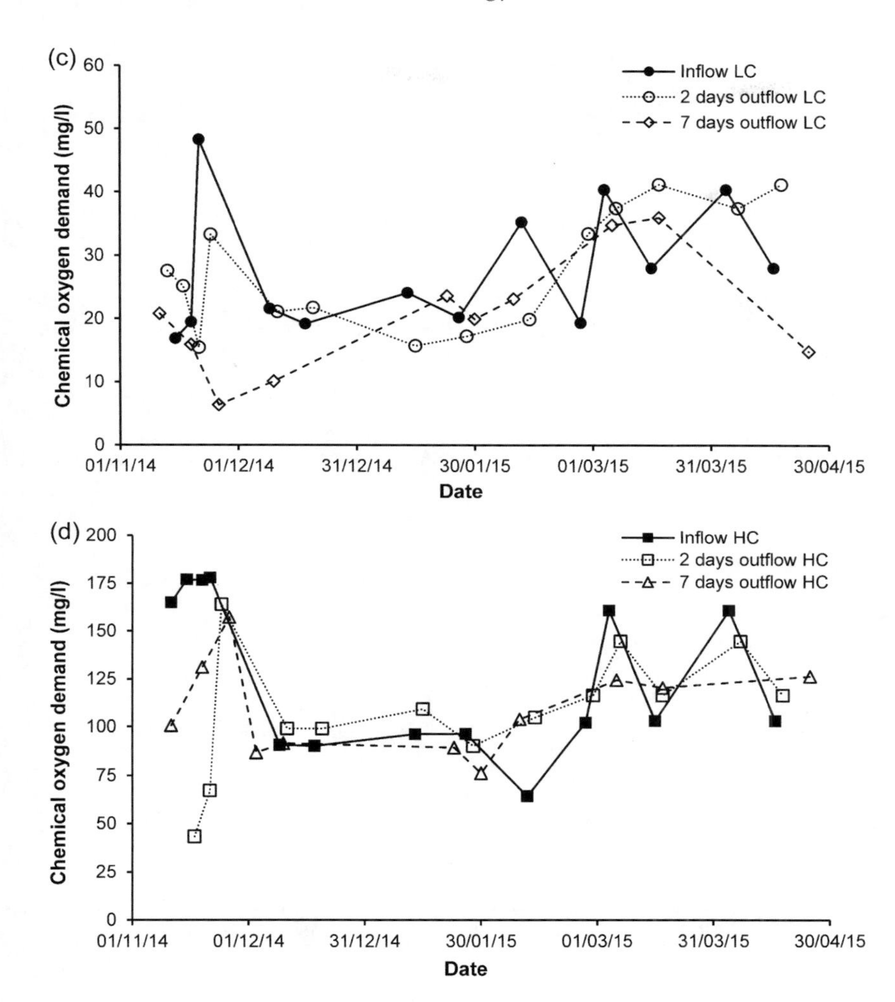

FIGURE 4.11 (*Continued*) Effect of storage time on the variation of (a) five-day biochemical oxygen demand (BOD_5) of low-concentration synthetic greywater (LC); (b) BOD_5 of high-concentration synthetic greywater (HC); (c) chemical oxygen demand (COD) of LC, (d) COD of HC; (e) ammonia-nitrogen (NH_4–N) of LC, (f) NH_4–N of HC; (g) nitrate-nitrogen (NO_3–N) of LC, (h) NO_3–N of HC; (i) ortho-phosphate-phosphorus (PO_4–P) of LC and (j) PO_4–P of HC greywater.

(Continued)

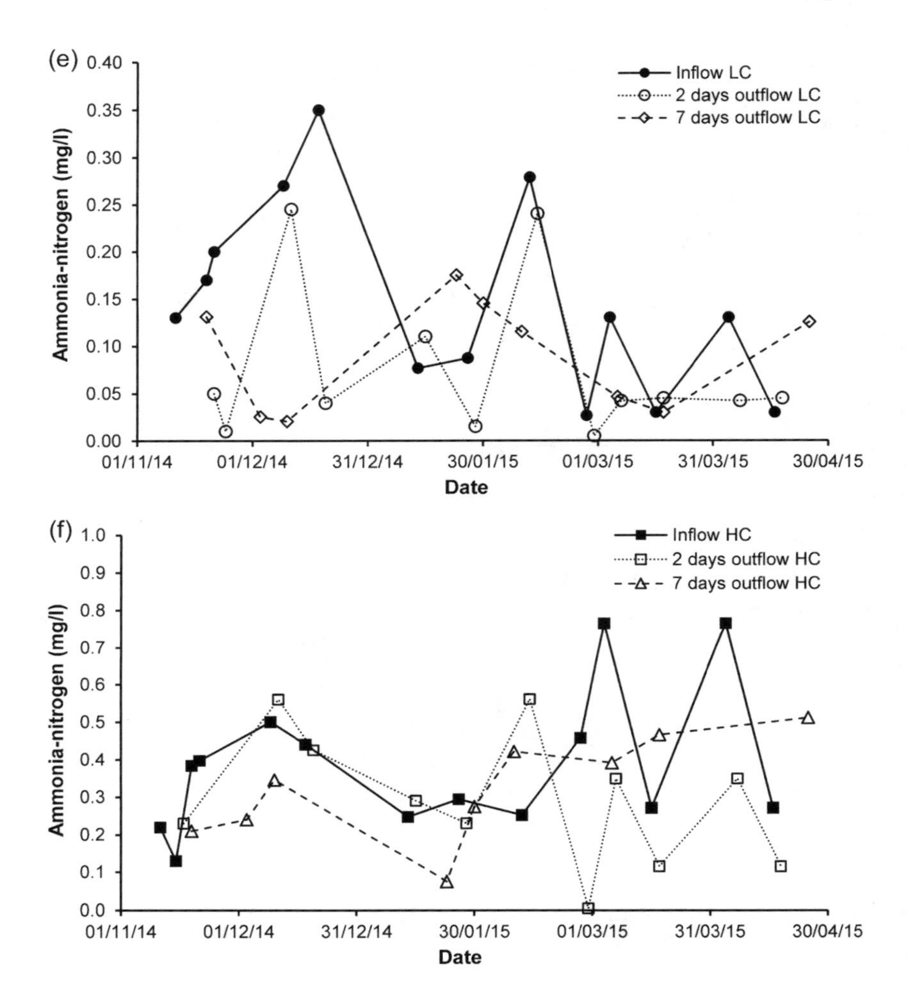

FIGURE 4.11 (*Continued*) Effect of storage time on the variation of (a) five-day biochemical oxygen demand (BOD_5) of low-concentration synthetic greywater (LC); (b) BOD_5 of high-concentration synthetic greywater (HC); (c) chemical oxygen demand (COD) of LC, (d) COD of HC; (e) ammonia-nitrogen (NH_4–N) of LC, (f) NH_4–N of HC; (g) nitrate-nitrogen (NO_3–N) of LC, (h) NO_3–N of HC; (i) ortho-phosphate-phosphorus (PO_4–P) of LC and (j) PO_4–P of HC greywater.

(*Continued*)

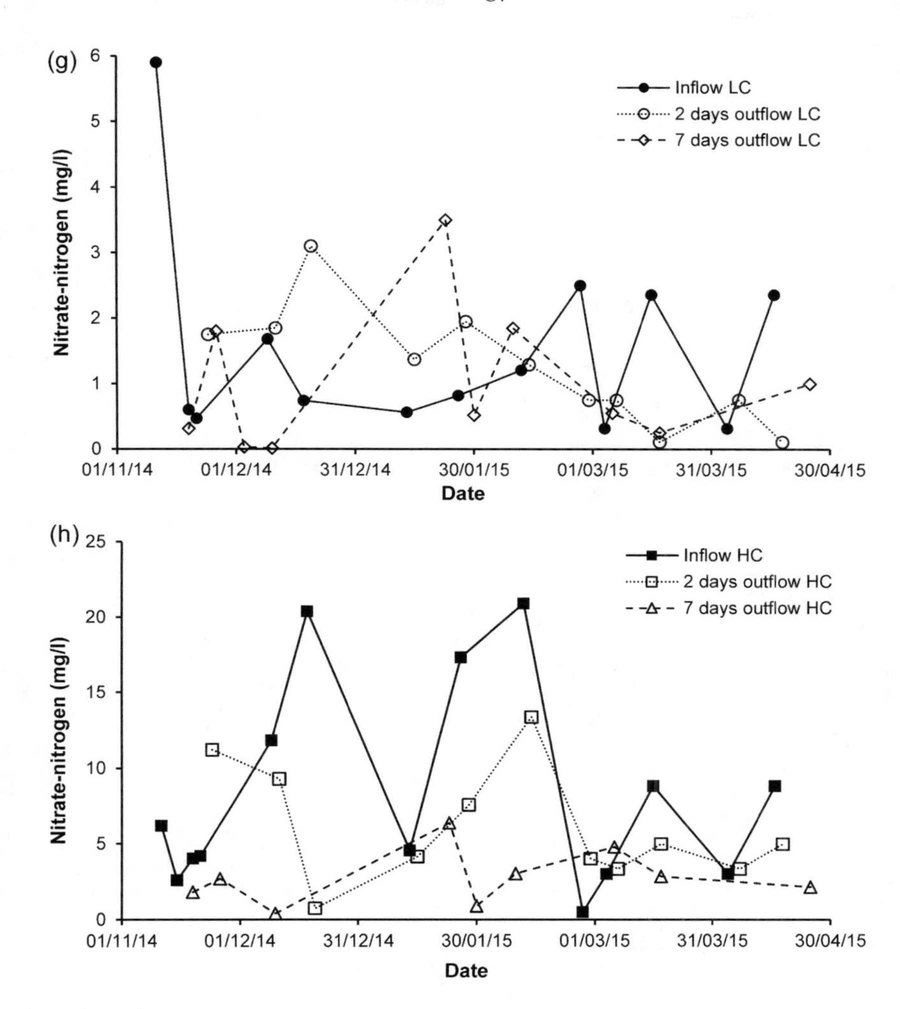

FIGURE 4.11 (*Continued*) Effect of storage time on the variation of (a) five-day biochemical oxygen demand (BOD_5) of low-concentration synthetic greywater (LC); (b) BOD_5 of high-concentration synthetic greywater (HC); (c) chemical oxygen demand (COD) of LC, (d) COD of HC; (e) ammonia-nitrogen (NH_4–N) of LC, (f) NH_4–N of HC; (g) nitrate-nitrogen (NO_3–N) of LC, (h) NO_3–N of HC; (i) ortho-phosphate-phosphorus (PO_4–P) of LC and (j) PO_4–P of HC greywater.

(*Continued*)

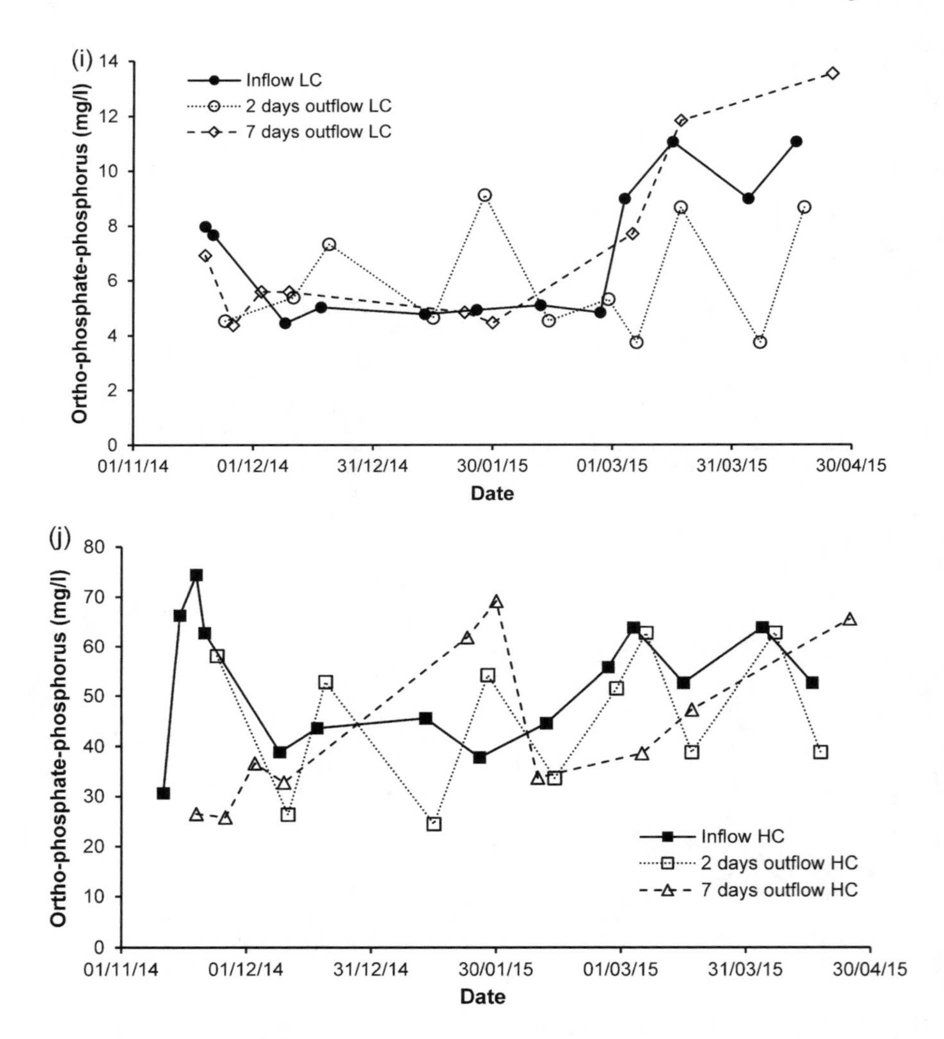

FIGURE 4.11 (*Continued*) Effect of storage time on the variation of (a) five-day biochemical oxygen demand (BOD_5) of low-concentration synthetic greywater (LC); (b) BOD_5 of high-concentration synthetic greywater (HC); (c) chemical oxygen demand (COD) of LC, (d) COD of HC; (e) ammonia-nitrogen (NH_4–N) of LC, (f) NH_4–N of HC; (g) nitrate-nitrogen (NO_3–N) of LC, (h) NO_3–N of HC; (i) ortho-phosphate-phosphorus (PO_4–P) of LC and (j) PO_4–P of HC greywater.

at around 14.7 mg/L for outflow water after seven days (Table 4.13, Figure 4.12a). This change has been confirmed by comparing available data evidence, which was reported by Jefferson et al. (2000). Microbial contamination is the reason for the drop in organic strength (Friedler et al., 2006; Maiga et al., 2014).

The COD in the LC greywater increased from 25.2 to 27.9 mg/L (not statistically significant ($p > 0.05$); see also above) for the two-day storage time experiment.

However, it decreased to 19.6 mg/L for the seven-day storage time test (Figure 4.12b). In contrast, the COD for HC greywater dropped from 115.4 to 110.7 mg/L (reduction by 4.1%) and to 108.3 mg/L (reduction by 6.2%) for two-day and seven-day storage times, respectively. The variations in test results are shown in Figure 4.11c and d in that order. Some of the COD data variations can be attributed to both experimental variability (see discussion in the previous section) and biodegradation of the fraction of the COD, which is biodegradable (Essington, 2004; Friedler et al., 2005).

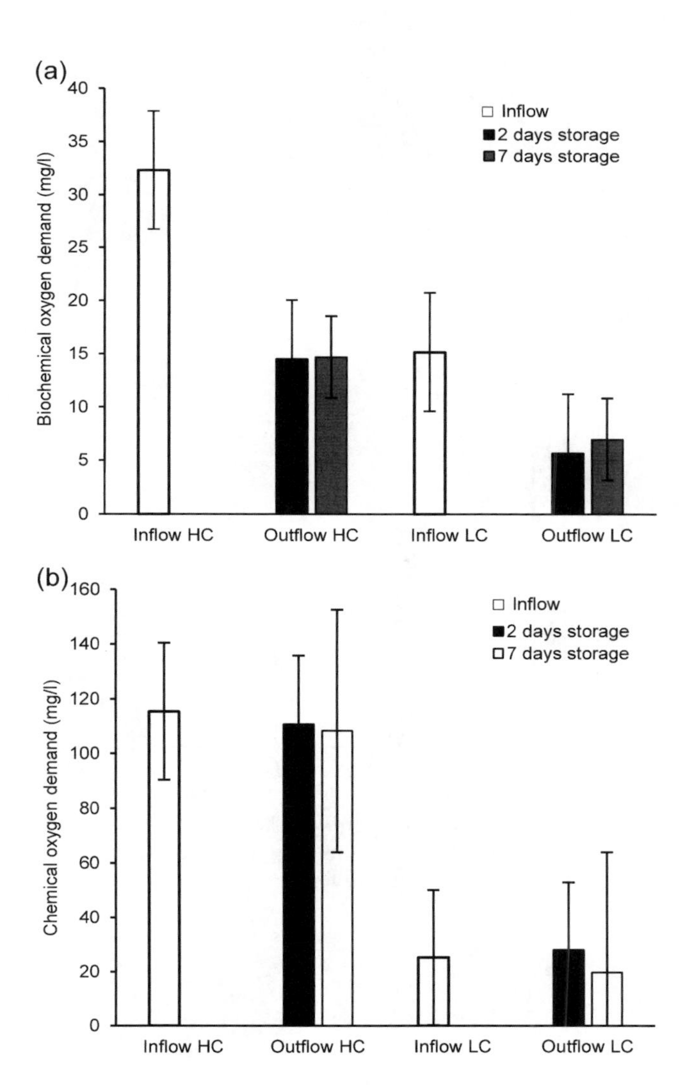

FIGURE 4.12 Effect of storage time on the synthetic greywater characteristics (a) five-day biochemical oxygen demand; (b) chemical oxygen demand; (c) ammonia-nitrogen; (d) nitrate-nitrogen and (e) ortho-phosphate-phosphorus.

(Continued)

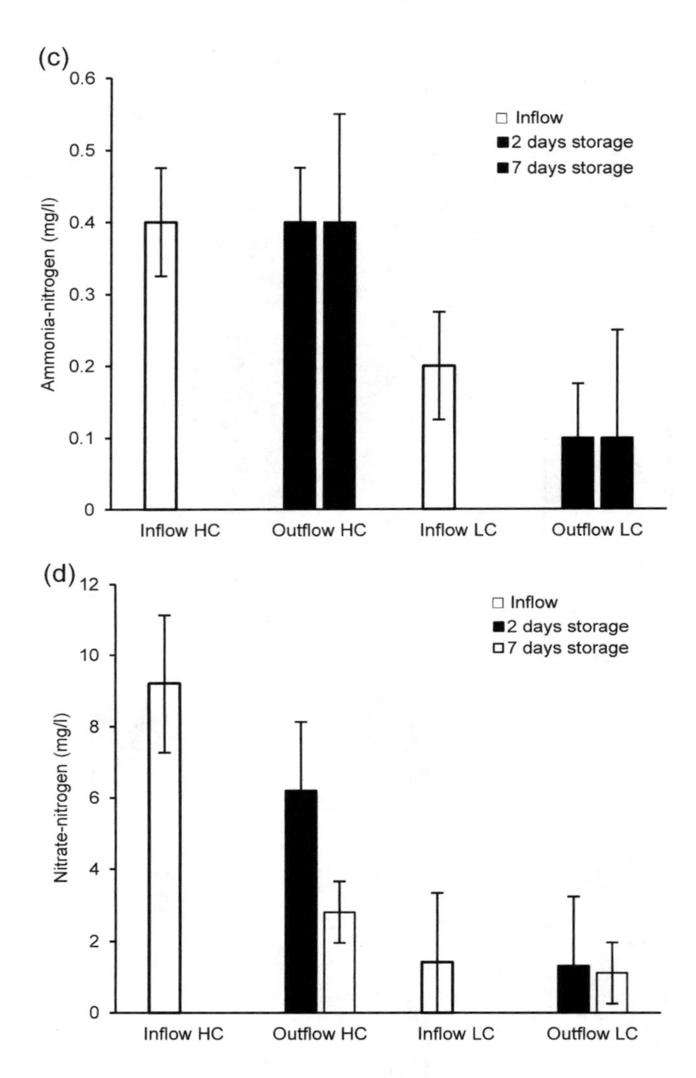

FIGURE 4.12 (*Continued*) Effect of storage time on the synthetic greywater characteristics (a) five-day biochemical oxygen demand; (b) chemical oxygen demand; (c) ammonia-nitrogen; (d) nitrate-nitrogen and (e) ortho-phosphate-phosphorus.

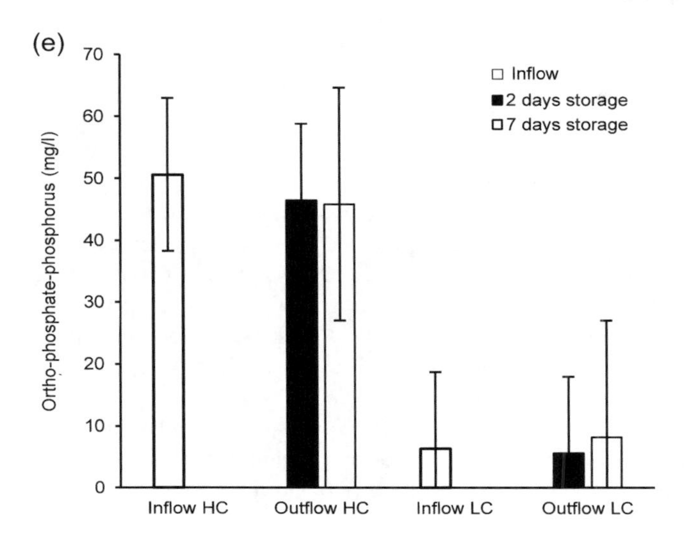

FIGURE 4.12 (*Continued*) Effect of storage time on the synthetic greywater characteristics (a) five-day biochemical oxygen demand; (b) chemical oxygen demand; (c) ammonia-nitrogen; (d) nitrate-nitrogen and (e) ortho-phosphate-phosphorus.

For HC greywater, the averages of ammonia-nitrogen show a stable behaviour with values of around 0.4 mg/L without change through storage (Figure 4.11e and f). The corresponding values for LC greywater have decreased from 0.2 to 0.1 mg/L after two days of storage. The results show no change for seven days outflow (Figure 4.12c). The measured values for ammonia-nitrogen are close to the detection limit. Therefore, the transformation of ammonia to nitrite and subsequently to nitrate cannot be evidenced in this experiment (Essington, 2004).

A considerable change was observed for the average values of nitrate-nitrogen after both storage times. The values dropped from 9.2 to 6.2 mg/L and 2.8 mg/L after storage times of tow and seven days, respectively (Table 4.13, Figure 4.11h). However, a significant ($p < 0.05$) reduction was noted for two days of storage regarding HC synthetic greywater. In contrast, the nitrate-nitrogen values of LC greywater decreased slightly from 1.4 to 1.3 mg/L and to 1.1 mg/L after two and seven days of storage time in this order (Figures 4.11g and 4.12d). The reduction of nitrate-nitrogen can be explained by denitrification (Essington, 2004).

Also, there are no significant ($p > 0.05$) changes in the reduction of ortho-phosphate-phosphorus for both storage times (Figure 4.11i and j). They decreased from 50.6 to 46.5 mg/L (reduction of 8.2%) for two-day storage and decreased to 45.8 mg/L (reduction of 26.4%) for seven-day storage of HC greywater. The ortho-phosphate-phosphorus concentrations also decreased from 6.3 to 5.6 mg/L for two-day storage experiments, and to 8.2 mg/L for seven-day storage of LC greywater (Figure 4.12e). Phosphorus is likely to be taken up by microbes developing in the outside systems (Friedler et al., 2006). However, considering that microbes were not deliberately added to the greywater recipe, microbial biomass development was rather slow. Therefore, changes in phosphorus concentrations were small.

4.4.4 Conclusions and Further Research

The proposed new synthetic greywater recipes mimic real greywater well in both composition and properties. Furthermore, they provide a good matrix for microorganisms to survive and contain compounds in detectable concentrations identified as having a potentially detrimental environmental impact.

The suggested recipes for LC and HC greywater loadings are easy to prepare and replicate by others in the future. All selected materials were of chemical analytical grade. High quantity stock solutions can be prepared and stored at 4°C without major concern.

Throughout monitoring of the synthetic greywater properties during storage, the water quality parameters concerning their average values are chemically relatively stable. It has been noticed that only significant ($p < 0.05$) fluctuations in the BOD_5 for both greywater concentrations may occur. In addition, it is not recommended to store the synthetic greywater for more than two days to avoid depletion of DO due to development of microorganisms. Furthermore, significant changes in nitrate-nitrogen content might be noticed after two days of storage.

4.5 NITROGEN REMOVAL IN WETLANDS TREATING DOMESTIC WASTEWATER

4.5.1 Introduction

Nitrogen (N) is an essential macronutrient in all ecosystems. Excess N, however, can be an important pollutant of receiving waters, and is a growing concern worldwide. Domestic wastewater contains high concentrations of N (Black and Veatch Corporation, 2010) and represents a predominant point source of N pollution to surface waters. Dissolved inorganic nitrogen species such as ammonia-nitrogen (NH_3-N) and nitrate-nitrogen (NO_3-N) in domestic wastewater can exacerbate eutrophication in open waters (Ault et al., 2000). Nitrogen pollution can also cause low DO conditions in surface waters (Beutel, 2006), either directly through the biological oxidation of NH_3-N, or indirectly through the decay of phytoplankton blooms initially stimulated by N pollution. In addition, a high level of NH_3-N is toxic to aquatic biota (Ault et al., 2000), while at elevated levels NO_3-N is toxic to infants (U.S. EPA, 2002). Significant treatment of domestic wastewater is, therefore, required to reduce N loading to open waters and protect water resources and consequently, public health.

CWs have emerged as a viable method for the treatment of various wastewaters worldwide because they are easy to operate, require low maintenance and have low investment costs (Machate et al., 1997). Indeed, the last decades have seen considerable development in the exploration of CW systems for the treatment of wastewater from several sources including industrial effluents, urban and agricultural stormwater runoff, domestic and animal wastewaters, landfill leachate, acid mine drainage and gully pot liquor (Kadlec et al., 2000; Moshiri, 1993; Scholz, 2004a, 2004b; Scholz and Lee, 2005; Scholz and Xu, 2002; Zhao et al., 2004). The treatment performance of these systems varies, depending on variables such as system type and design, retention time, hydraulic and pollutant mass loading rates, climate, vegetation and microbial communities (U.S. EPA, 1995). Generally, high efficiencies (>70%) are recorded in CW for parameters such as biochemical oxygen demand (BOD_5), COD,

TSS and faecal coliforms. The efficiency of nitrogen removal has been found to be lower and more variable (Kadlec and Wallace, 2009; Moshiri, 1993). Depending on several factors, the NH_3-N removal rate in free water surface (FWS) flow CW, for example, is known to typically range from −23% to 58% (Watson et al., 1989). In European systems, typical removal efficiency of ammonia-nitrogen (NH_4-N) in long-term operation is only 35%, or up to 50% after specific modifications are made to improve nitrogen removal (Luederitz et al., 2001; Verhoeven and Meuleman, 1999). Removal efficiencies of NH_4-N in Irish CW are also highly variable and classically range between 67% and 99.9% (Babatunde et al., 2008).

The concept of integrated constructed wetland (ICW) (Scholz et al., 2007; Harrington et al., 2007), promoted by the ICW Initiative of the Irish Department of Environment, Heritage and Local Government, is a specific design approach to constructed treatment wetlands. These FWS CWs, which employ the concept of restoration ecology, specifically mimic the structure of natural wetlands (Harrington and Ryder, 2002; Scholz et al., 2007). They are multi-celled with sequential through-flow and are based on the holistic and interdisciplinary use of land to control water quality. Usually, ICWs have shallow water depths and comprise many plant species that facilitates microbial and animal diversity (Jurdo et al., 2010; Nygaard and Ejrnæs, 2009), and are generally appealing, which enhances recreation and amenity values (Scholz et al., 2007). Previous applications of a specific type of ICW, namely, Constructed Farm Wetlands, defined by Carty et al. (2008) to treat farmyard runoff in the Annestown stream catchment (about 25 ha) in south County Waterford, Ireland, demonstrated very good treatment performance. Evaluation of the long-term performance of these systems by Mustafa et al. (2009) showed contaminant concentration removal efficiencies of BOD (97.6%), COD (94.9%), TSS (93.7%), NH_4-N (99%), NO_3-N (74%) and MRP (91.8%). Other studies such as Harrington et al. (2004), Dunne et al. (2005), Harrington and McInnes (2009) showed similar results. Such successful applications inspired the construction of a new industrial-scale ICW system which was commissioned in October 2007 to treat combined sewage from Glaslough village in County Monaghan, Ireland.

Pollutants removal in ICW systems can be achieved through a combination of physical, chemical and biological processes that naturally occur in wetlands and are associated with the vegetation, sediments and their microbial communities (Kadlec and Knight, 1996; Scholz, 2006; Vymazal, 2001). The N biogeochemical cycle within wetland ecosystems is complex and involves several transformation and translocation processes. These include ammonia volatilization, ammonification, N fixation, burial of organic N, ammonia sorption to sediments, nitrification, denitrification, anammox, and assimilation (Kadlec and Wallace, 2009; Vymazal, 2007). Commonly, N removal through bacterial transformations involves a sequential process of ammonification, nitrification and denitrification (Kadlec et al., 2000). Denitrification is believed to be the major N removal pathway, and typically accounts for more than 60% of the total N removal in constructed wastewater wetlands (Kadlec and Knight, 1996; Spieles and Mitch, 2000). This microbial process consists of the reduction of oxidized forms of N, mainly nitrate and nitrite, to the gaseous compounds nitrous oxide and dinitrogen. Anaerobic conditions are a prerequisite for the occurrence of denitrification (Kadlec and Knight, 1996). While nitrate availability often regulates denitrification, organic carbon content, pH and temperature also play important roles. Temperature affects denitrification by controlling rates of diffusion at the sediment-water interface in wetlands.

Denitrification rates in CW have been shown to increase dramatically with temperature, within a lower and upper bound of around 5°C and 70°C, respectively (Vymazal, 2007). The microbial activities related to nitrification and denitrification can decrease considerably at water temperatures below 15°C or above 30°C, and most microbial communities for nitrogen removal function at temperatures greater than 15°C (Kuschk et al., 2003). Nitrification involves the sequential biochemical oxidation of reduced N species such as ammonia (NH_3) to nitrite (NO_2^-) and nitrate (NO_3^-) under strict aerobic conditions, which may be present in the sediment-water interface of FWS CW. The nitrification process requires high oxygen concentrations and is highly sensitive to DO levels (Lee et al., 2009). Being an anaerobic process, denitrification is also sensitive to DO levels.

Hydraulic characteristics such as water depth, HLR and HRT are important factors for determining the treatment performance of CW (Kadlec and Knight, 1996; Kadlec and Wallace, 2009). At lower HLR and longer HRT, higher nutrient removal efficiencies are usually obtained (Sakadevan and Bavor, 1999). Most recent studies, however, have only focused on the system performance by comparing inlet and outlet concentrations of contaminants. There is limited information to quantify N removal in full-scale industry-sized CW based on wetland hydrology and corresponding pollutant concentration profiles. This section evaluates the N removal performance of a full-scale ICW applied as the main unit treating domestic wastewater in Ireland. Removal of two N species, namely, NH_3-N and NO_3-N were analysed, with the objective to (a) compare the annual and seasonal N removal efficiencies, (b) estimate the areal N removal rates and determine areal first-order kinetic coefficients for N removal and (c) assess the influence of water temperature on the N removal performance. Section 4.5 is based on an updated version of an original article by Dzakpasu et al. (2011).

4.5.2 Materials and Methods

4.5.2.1 Study Area

The studied ICW system is located within the walls of Castle Leslie Estate at Glaslough in County Monaghan, Ireland (06°53′37.94″ W, 54°19′6.01″N). Ireland has a relatively mild temperate maritime climate. Mean seasonal temperatures for Monaghan in 2009 were 10.7°C (spring), 14.9°C (summer), 7.9°C (autumn) and 2.9°C (winter). The mean annual rainfall is approximately 970 mm (Met Éireann, 2010). The site is surrounded by woodland and required sensitive development in terms of landscape fit, and biodiversity, amenity and habitat enhancement.

The ICW (Figure 4.13) comprises a small pumping station, two sludge cells, and five shallow vegetated cells. It was commissioned in October 2007 to treat combined sewage from Glaslough village and to improve the water quality of the Mountain Water River, which flows through the site. The design capacity of the ICW system is 1,750 per person equivalent and covers a total area of 6.74 ha. The total surface area of the CW cells is 3.25 ha. There is no artificial lining of the wetland cells. Excavated local soil material was used to construct the base of the wetland cells and

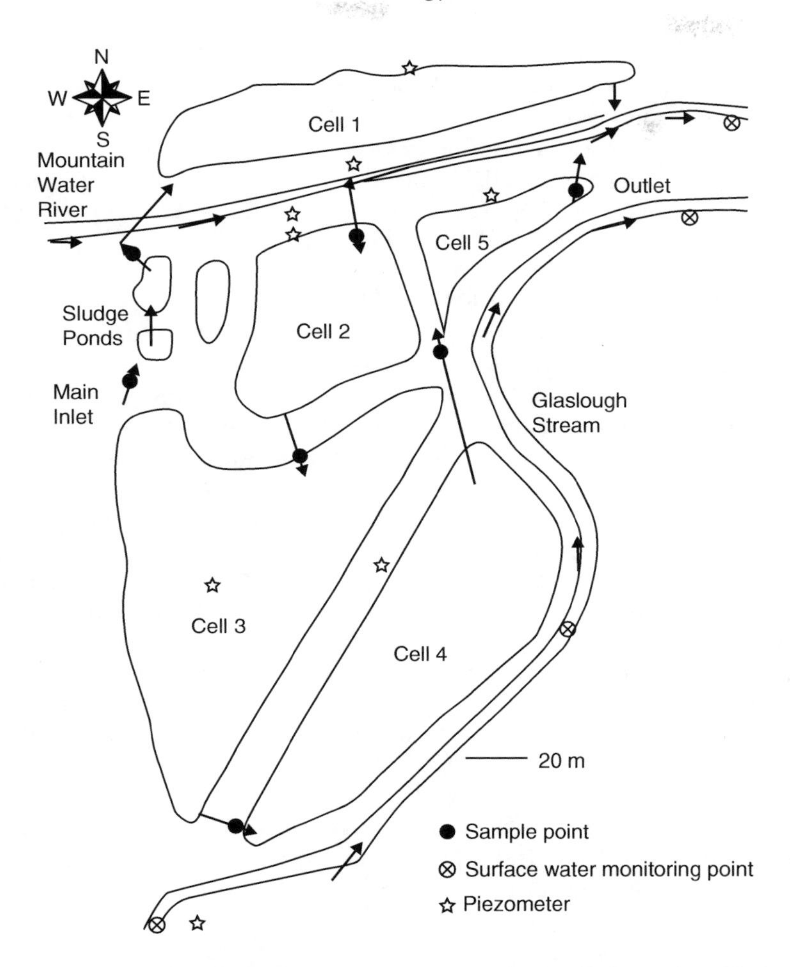

FIGURE 4.13 Schemata of integrated constructed wetland cells located at Glaslough in County Monaghan, Ireland showing all sampling locations.

compacted to a thickness of 500 mm to form a low permeability liner. A site investigation by the Geological Survey of Ireland (IGSL Ltd., Business Park, Naas, County Kildare, Ireland) in September 2005 indicated a soil coefficient of permeability of 9×10^{-11} m/s. The main ICW system is flanked by the Mountain Water River and the Glaslough Stream.

Untreated influent wastewater from the village is pumped directly into a receiving sludge cell. The system contains two sludge cells that can be used alternately so that one can be desludged without interrupting the process operation. The purpose of the sludge cell is to retain the suspended solids contained in the influent wastewater. In this way, the build-up of sludge in the wetland cells, which could degrease the capacity of the cells, is prevented. From the sludge cell, the wastewater subsequently flows by gravity sequentially through the five vegetated cells, and the effluent of the last cell discharges directly to the adjacent Mountain Water River.

The wetland cells, which were originally planted with *Carex riparia* Curtis, *P. australis* (Cav.) Trin. ex Steud., *T. latifolia* L., *Iris pseudacorus* L., and *Glyceria maxima* (Hartm.) Holmb., currently include a complex mixture of *Glyceria fluitans* (L.) R.Br., *Juncus effusus* L., *Sparganium erectum* L. emend Rchb, *Elisma natans* (L.) Raf., and *Scirpus pendulus* Muhl.

4.5.2.2 Water Quality Monitoring and Analysis

A suite of automated sampling and monitoring instrumentation such as the ISCO 4700 Refrigerated Automatic Wastewater Sampler (Teledyne Isco, Inc., NE., United States) has been used for weekly wetland water sampling from April 2008 to May 2010. These samplers take flow weighted composite water samples for the inlet and outlet of each wetland cell. Additionally, all water flows into and out of each ICW cell were measured and recorded with the Siemens Electromagnetic Flow Meters FM MAGFLO and MAG5000 (Siemens Flow Instruments A/S, Nordborgrej, Nordborg, Demark) and their allied computer-linked data loggers. Mean flows were recorded at 1-minute interval frequency. A weather station is located on site, beside the inlet pump sump to measure local temperature, precipitation and evapotranspiration.

The water samples were analysed weekly for NH_3-N and NO_3-N at the Monaghan County Council wastewater laboratory in Ireland, using the HACH Spectrophotometer DR/2010 49300-22. NH_3-N was determined by HACH Method 8038 (HACH Company, 2000), based on the Nessler method (adapted from Standard Methods for the Examination of Water and Wastewater). NO_3-N was determined by HACH Method 8171, based on the cadmium reduction method (using powder pillows) after HACH Company (2000). For the purpose of quality assurance, the water samples were also analysed monthly with the Lachat QuikChem 8500 Flow Injection Analysis System (Lachat Instruments, Loveland, CO, U.S.).

Removal rates for NH_3-N and NO_3-N, based on a two-year data set (April 2008–May 2010) were quantified using three common approaches for CW (Kadlec and Knight, 1996). The first approach estimated the mass removal efficiency (%) as shown in Equation 4.23:

$$\text{Removal efficiency} = \frac{Q_o C_o - Q_e C_e}{Q_o C_o} \times 100 \quad (4.23)$$

The second approach estimated the areal removal rate (mg-N/m^2/d) as indicated by Equation 4.24:

$$\text{Removal rate} = q \times (C_o - C_e) \quad (4.24)$$

The third approach estimated the area-based first-order removal rate constants for ammonia (K_A) and nitrate (K_N) using the K–C^* model, assuming plug flow conditions (Equation 4.25):

$$\ln\left(\frac{C_e - C^*}{C_o - C^*}\right) = -\frac{K}{q} \tag{4.25}$$

where Q_o and Q_e are the daily volumetric water inflow and outflow rates (m^3/d), C_o and C_e are influent and effluent concentrations, respectively, of NH_3-N or NO_3-N (mg N/L), C^* is the background concentration (mg N/L) and K is the areal first-order removal rate constant (m/yr). The K values were normalized to 20°C (K_{20}) based on Equation 4.26 using values estimated from Equation 4.27 (Kadlec and Knight, 1996). A C^* of 0 mg/L, recommended by Kadlec (2009), was used to calibrate the model.

The effect of temperature on the areal first-order removal rate constants for the N species was modelled using the modified Arrhenius relationship (Equation 4.26):

$$K_{(t)} = K_{(20)}\theta^{(t-20)} \tag{4.26}$$

where $K_{(t)}$ and $K_{(20)}$ are the first-order removal rate constants (m/yr), t is temperature (°C) and θ is an empirical temperature coefficient (Kadlec and Knight, 1996). A linear form of Equation 4.26 was used to estimate parameters of the model from the data set (Equation 4.27):

$$\log\left(K_{(t)}\right) = \log\theta(t-20) + \log\left(K_{(20)}\right) \tag{4.27}$$

Values of $\log(K_{(t)})$ versus $(t-20)$ were plotted and fit with a linear regression. The resulting slope and intercept were equal to $\log\theta$ and $\log(K_{(20)})$, respectively.

The hydraulic loading rate, q (m/yr), was calculated as shown in Equation 4.28:

$$q = \frac{Q}{A} \tag{4.28}$$

where Q is the total water inflow rate (m^3/d), and A is the total surface area for five wetland cells (m^2).

The overall dynamic wetland water budget was calculated with Equation 4.29.

$$Q_o - Q_e + Q_c + (P - \mathrm{ET} - I)A = \frac{dV}{dt} \tag{4.29}$$

where Q_c is catchment runoff rate (m^3/d), P is the daily precipitation rate (m/d), ET is the daily evapotranspiration rate (m/d), I is the daily infiltration rate (m/d) and $\frac{dV}{dt}$ is the net change in volume (m^3/d).

Data distributions were tested for normality. Data presentation uses means of actual measured values. Statistically significant differences were determined at $\alpha = 0.01$, unless otherwise stated. Comparisons of means were by paired student t-tests and analysis of variance (ANOVA). Regression analysis used the standard least squares fit. All statistical analyses were performed using Minitab 16 statistical software (Minitab Inc., Umited Kingdom).

4.5.3 Results and Discussion

4.5.3.1 Hydrology

Overall, surface flows from the sludge cell and precipitation were considered as the inflow sources to the ICW system, whereas evapotranspiration and water infiltration were assumed to be lost water. Precipitation and evapotranspiration were calculated as the amount of water falling on, or evaporating from the wetland cell surface, respectively. The HLR, HRT, and mean dimensions of each ICW cell are presented in Table 4.15. Furthermore, Table 4.16 shows the total water budget for the ICW.

During the study period, highest monthly rainfall (296 mm/month) was recorded in November 2009 and the lowest (5.6 mm/month) was recorded in June 2009 (Figure 4.14). There was no significant seasonal variation in daily rainfall; however, maximum daily rainfall between months was largely variable.

Domestic wastewater inflow to the ICW varied monthly (Figure 4.14), with individual system values ranging between 1.4 and 613 m^3/d. The average inflow rate (± SD) was 104 ± 106.1 m^3/d, yielding average hydraulic loading of 7 ± 10.5 mm/d,

TABLE 4.15
Dimensions and Hydraulic Characteristics of ICW System

ICW Section	Area (m^2)	Depth (m)	Volume (m^3)	HRT (d)	HLR (mm/day)
Pond 1	4,664	0.42	1958.9	18	24.4
Pond 2	4,500	0.38	1710.0	16	26.8
Pond 3	12,660	0.32	4051.2	32	10.7
Pond 4	9,170	0.36	3301.2	23	16.1
Pond 5	1,460	0.29	423.4	3	100.3
Total wetland	3,2454	–	11444.7	92	7.3

TABLE 4.16
Daily Water Fluxes and Distribution of Total Water Budget for ICW Between April 2008 and May 2010

Water Fluxes (m^3/d)	Cell 1		Cell 2		Cell 3		Cell 4		Cell 5		Total ICW	
	Mean	SD	Mean	SD	Mean	SD	Mean	SD	Mean	SD	Mean	SD
Inputs												
Wastewater inflow	104	106.1	112	132.4	109	134.3	128	158.5	143	191.6	104	106.1
Precipitation	21	46.1	20	44.5	57	125.1	41	90.6	7	14.4	139	65.7
Outputs												
Wetland effluent	112	132.4	109	134.3	128	158.5	143	191.6	131	179.4	131	179.4
Evapotranspiration	4	3.5	4	3.4	16	12.9	8	6.9	1	1.1	39	27.9
Infiltration	1.6		1.5		4.4		3.2		0.5		11.2	

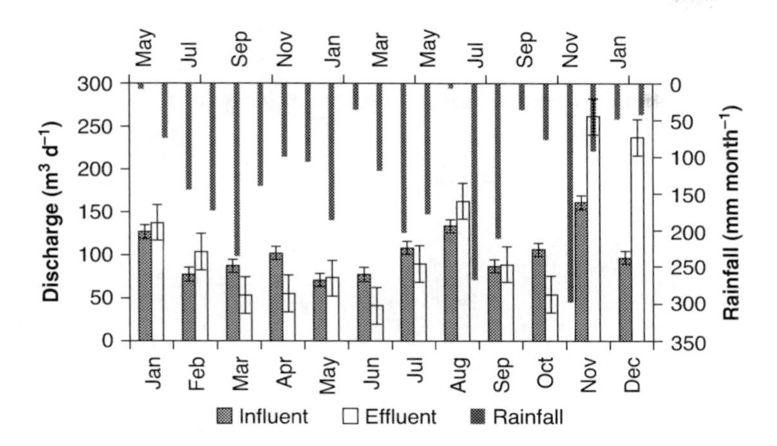

FIGURE 4.14 Average rainfall and water fluxes at influent and effluent points of integrated constructed wetlands between 2008 and 2010.

whereas the associated discharge at the effluent point ranged from 0 to 492 m^3/d with an average (± SD) of 131 ± 179.4 m^3/d. The average daily outflow volumes recorded for the ICW were higher than the average daily inflow volumes, probably due to precipitation inputs.

The net change in volume recorded during the study period (average ± SD) was 62 ± 371.3 m^3/d. Overall, precipitation represented approximately 56% of the total input to the ICW, and suggested that water inflow originated mainly from precipitation. Moreover, a strong linear correlation ($R^2 = 0.97$, $p < 0.01$, $n = 708$) was observed between precipitation and wetland volumetric flow rate and suggested that precipitation possibly had a significant influence on the hydraulic loading rate. Evapotranspiration and infiltration constituted about 25% and 5%, respectively, of water outflows from the ICW system, whereas the effluent accounted for nearly 50%. Catchment runoff and groundwater inflow were assumed to be negligible. The highest evapotranspiration rate (134 ± 18.4 m^3/d) was recorded in summer and the lowest (13.7 ± 4.4 m^3/d) in winter.

4.5.3.2 Nitrogen Concentrations

Overall, NH_3-N was recorded as the dominant species of N contained in the influent wastewater received by the ICW. Annual influent concentrations (average±SD) of 40 ± 13.6 and 5 ± 3.8 mg/L were recorded respectively for NH_3-N and NO_3-N, indicating a high variability of the influent domestic wastewater (Table 4.17). The NH_3-N concentrations received by the ICW over the study period were slightly higher than other FWS CW receiving primary domestic effluent, reported by Kadlec and Wallace (1996) and Boutilier et al. (2010) where concentrations varied depending on climate. Other studies such as Ran et al. (2004) have reported slightly higher influent concentrations as well. Average concentrations of N in the ICW effluent were consistently less than 1.0 mg/L and recorded an average of 0.8 ± 1.6 mg/L for NH_3-N and 0.3 ± 0.2 mg/L for NO_3-N. The effluent concentrations of both N species were significantly lower ($p < 0.01$, $n = 120$) than the influent.

TABLE 4.17
Influent and Effluent Nitrogen Concentrations at ICW Between April 2008 and May 2010

		Influent			Effluent		
Parameter	**Unit**	**Mean**	**SD**	***n***	**Mean**	**SD**	***N***
Ammonia	mg N/L	40	13.6	120	0.8	1.6	120
Nitrate	mg N/L	5	3.8	101	0.3	0.2	101

n, Sample number; SD, standard deviation.

Furthermore, influent concentrations of the two N species showed some seasonal variations (Table 4.18). Nevertheless, whereas the variations in concentrations of the influent NO_3-N was significant ($p<0.01$, $n=18$), variations of the influent NH_3-N were not. The highest (average $\pm$ SD) seasonal influent concentration of NH_3-N (42 ± 10.1 mg/L) and NO_3-N (8 ± 6.3 mg/L) was recorded in summer and spring, respectively (Table 4.18), and indeed the highest removal rate occurred in the same season. The effluent NH_3-N concentrations were slightly higher in winter (3 ± 3.1 mg/L) compared to the other seasons. No seasonal variations in the effluent NO_3-N were observed and were typically in the region of 0.3 mg/L. The effluent NH_3-N concentrations were highest during winter probably because of increased surface outflow rates (Dunne et al., 2005; Kadlec and Knight, 1996) caused by increases in precipitation-driven hydrological inputs, which subsequently decreased HRT. Other explanations for this increase may include vegetation senescence and subsequent nutrient release from vegetation to the overlying wetland water column during this period (Kadlec, 2003). Additionally, ice cover during the severe winter in late December 2009 through early January 2010, may have created anaerobic conditions

TABLE 4.18
Comparison of Seasonal Nitrogen Concentrations at ICW Influent and Effluent Points Between 2008 and 2010

		NH_3-N (mg/L)					NO_3-N (mg/L)				
			Influent		Effluent			Influent		Effluent	
Season	**Months**	***n***	**Mean**	**SD**	**Mean**	**SD**	***n***	**Mean**	**SD**	**Mean**	**SD**
Spring	1 Feb–30 Apr	22	41	11.9	1	1.9	13	8	6.3	0.2	0.1
Summer	1 May–31 Jul	47	42	10.1	0.3	0.2	45	5	2.1	0.3	0.2
Autumn	1 Aug–31 Oct	34	40	17.3	0.3	0.2	28	4	1.5	0.4	0.2
Winter	1 Nov–31 Jan	17	31	11.5	3	3.1	15	2	1.6	0.3	0.1

n, Sample number; SD, standard deviation.

and decreased biodegradation (Boutilier et al., 2010), and may also partly account for the increased effluent NH_3-N concentrations.

4.5.3.3 Nitrogen Loadings and Removal

Generally, the average (± SD) areal NH_3-N loading rate (245 ± 321.9 mg/m²/d) was higher compared to that of NO_3-N (38 ± 58.3 mg/m²/d). Nevertheless, the areal removal rates for the two N species were consistently high, with an average (± SD) of 240 ± 317.8 mg/m²/d for NH_3-N and 35 ± 54.9 mg/m²/d for NO_3-N. There was a significant linear relationship between the areal loading and removal rates for NH_3-N ($R^2 = 0.99$, $p < 0.01$, $n = 120$) and NO_3-N ($R^2 = 0.99$, $p < 0.01$, $n = 101$) (Figure 4.14), indicating a near complete areal removal rate. The close fit of the points to the regression line also indicates a remarkably constant areal removal rate for both N species.

In general, average annual mass removal efficiencies were high for the ICW. Approximately 92.7% removal was recorded for NH_3-N and 84.4% for NO_3-N. Over the two-year study period, surface inflows carried a total load of 2,802 kg NH_3-N into the ICW system and 98.0% were retained. Similarly, a total load of 441-kg NO_3-N had been received by the ICW and 96.9% retention had been recorded. Hence, nitrogen was effectively removed from the influent wastewater throughout the study period, except during winter (Tables 4.19 and 4.20), where slightly lower N removal was recorded.

The increased HLR, owing to excessive rainfall recorded during this season, might have reduced the HRT in the ICW and contributed to the reduced N removal performance during that period. This phenomenon has been observed in previous studies, which have indicated that pollutant removal efficiencies in CW decreased significantly with HLR (Huang et al., 2000; Tanner et al., 1995; Trang et al., 2010). Usually, the HRT is high at lower HLR. However, at higher HLR, the wastewater passes rapidly through the wetland, reducing the time available for degradation processes to occur effectively. Also, the ICW surface was frozen from late December 2009 through early January 2010, which may have created anaerobic conditions and decreased biodegradation (Boutilier et al., 2010) (Figure 4.15).

TABLE 4.19
Comparison of Seasonal Ammonia Loading and Removal Rates Within Integrated Constructed Wetlands Between 2008 and 2010

			Total Inputs (mg/m²/d)		Total Outputs (mg/m²/d)		Removal Rate	
Season	**Months**	***n***	**Mean**	**SD**	**Mean**	**SD**	**(mg/m²/d)**	**%**
Spring	1 Feb–30 Apr	22	181	242.8	3.0	6.65	187	96.3
Summer	1 May–31 Jul	47	278	347.7	0.5	0.79	275	99.5
Autumn	1 Aug–31 Oct	34	255	395.3	1.2	2.44	253	98.4
Winter	1 Nov–31 Jan	17	204	108.4	25.2	33.60	187	57.6

n, Sample number; SD, standard deviation.

TABLE 4.20
Comparison of Seasonal Nitrate Loading and Removal Rates Within Integrated Constructed Wetlands Between 2008 and 2010

			Total Inputs (mg/m²/d)		Total Outputs (mg/m²/d)		Removal Rate	
Season	**Months**	***n***	**Mean**	**SD**	**Mean**	**SD**	**(mg/m²/d)**	**%**
Spring	1 Feb–30 Apr	13	44	48.8	0.6	0.62	43	94.5
Summer	1 May–31 Jul	45	44	70.3	0.5	0.55	41	96.2
Autumn	1 Aug–31 Oct	28	35	54.9	1.7	3.39	32	88.6
Winter	1 Nov–31 Jan	15	19	16.1	2.5	2.24	16	60.8

n, Sample number; SD, standard deviation.

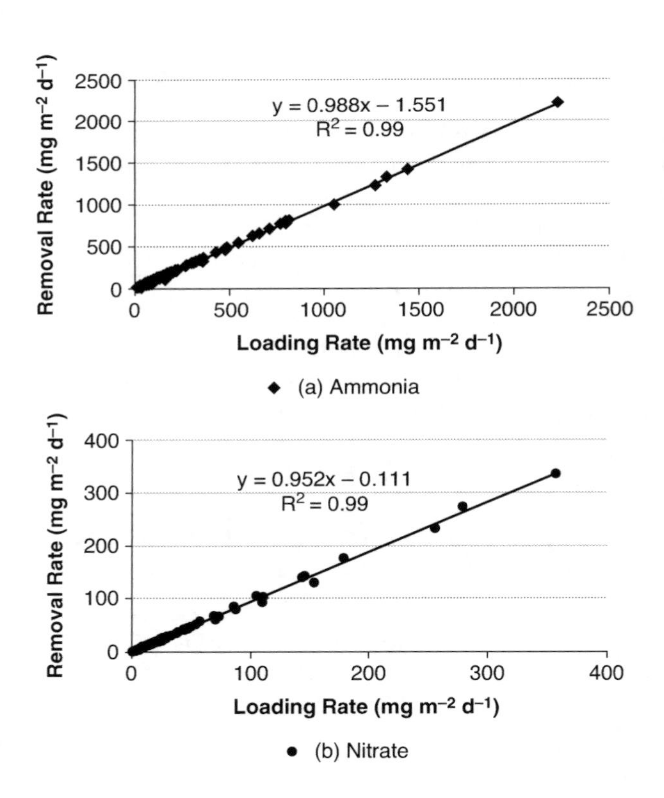

FIGURE 4.15 Areal loading and removal rates for (a) ammonia and (b) nitrate in integrated constructed wetlands between 2008 and 2010.

TABLE 4.21
Area-Based First-Order Removal Rate Constants for Ammonia and Nitrate

Parameter	K (m/yr)			K_{20} (m/yr)			θ
	Mean	SD	*n*	Mean	SD	*n*	
Ammonia	14	16.5	120	15	17.3	101	1.005
Nitrate	11	12.5	101	10	11.3	101	0.984

n, Sample number; SD, standard deviation; K, area-based rate constant; K_{20}, normalized rate constant.

Area-based first-order N removal rate constants (*K*) calculated for NH_3-N and NO_3-N reduction in the ICW were 14±16.5 and 11±12.5 m/yr, respectively (Table 4.21). Average water temperatures ranged between 4°C and 22°C. The average effects of temperature (θ) on N removal rate constants were estimated to be 1.005 for NH_3-N and 0.984 for NO_3-N. This yielded normalized N removal rate constants at 20°C (K_{20}) of 15±17.3 and 10±11.3 m/yr, respectively. The N removal rate constants estimated for the ICW were similar to typical values reported for FWS CW (Kadlec and Knight, 1996; Kadlec and Wallace, 2009). When normalized to 20°C, the K_{20} for N removal increased for NH_3-N and decreased slightly for NO_3-N, indicating that temperature only marginally influenced N removal. Moreover, the estimated θ for the ICW were found to be slightly lower than θ for the reduction of NH_3-N reported by Kadlec and Reddy (2001) and similar to values reported by Kadlec and Wallace (2009), whereas θ values for NO_3-N reduction were generally lower than values reported by Kadlec and Wallace (2009) and Kadlec and Reddy (2001). The lower θ values indicated that the N removal rate constants were independent of temperature and suggested that little of the variability in N removal by the ICW may be attributed to temperature. Also, there was no correlation observed between water temperature and the kinetic rate constants for both NH_3-N and NO_3-N (Figure 4.16). Nevertheless, relatively high N removal rates have been recorded at all times of the year, where the water temperature within the studied ICW ranged only between 4°C and 22°C, further confirming the low influence of temperature on N removal in the ICW. This contrasts with earlier reports that N removal in CW is influenced by temperature (Kadlec and Reddy, 2001).

Previous studies have shown that the biological N removal processes, which are responsible for nitrification and denitrification in CW and accounts for the major N removal pathway, is most efficient at temperature ranges between 25°C and 30°C (Hammer and Knight, 1994; Mitsch et al., 2000; Vymazal, 1999). The temperature range recorded in the ICW barely reached this optimum range. Nevertheless, N removal by the ICW was influenced by seasonality, with slightly higher removal recorded during the warmer months. It is possible that plant nutrient uptake may have influenced this seasonality. According to Kadlec (2003) plants take-up nutrients in the spring and then release them back to the water column during autumn senescence and these seasonal effects may mask the influence of temperature. Also, DO levels

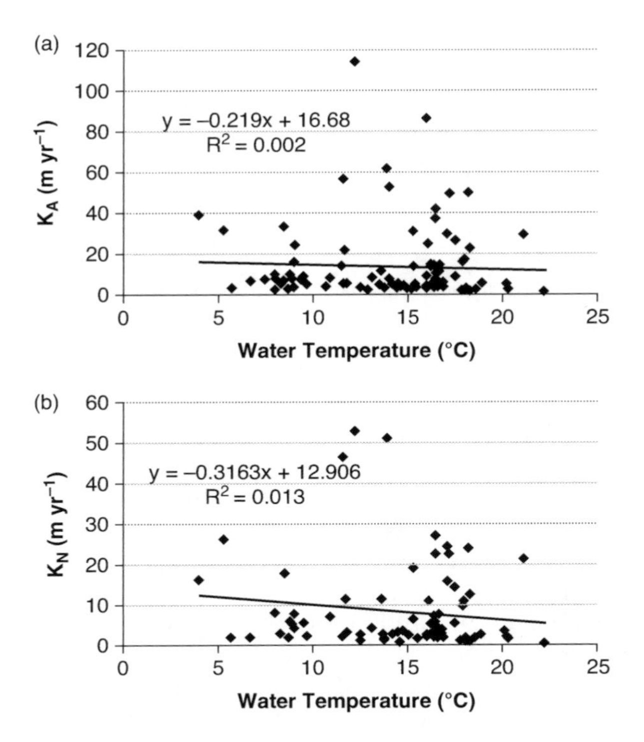

FIGURE 4.16 Water temperature and reaction rate constants for (a) ammonia and (b) nitrate in integrated constructed wetlands between 2008 and 2010.

within the ICW were low. The low DO would contribute to slow biological degradation. This suggests that N removal by the ICW may be due to physical processes. Physical treatment processes are less influenced by temperature. A significant linear relationship was observed between the kinetic rate constants and the loading rates for both NH_3-N and NO_3-N (Figure 4.17), indicating that physical processes indeed may have played a significant role in the N removal performance of the ICW.

4.5.4 Conclusions

Following a detailed evaluation of a two-year (April 2008–May 2010) data set comprising influent and effluent loadings of two nitrogen (N) species, namely, ammonia-nitrogen (NH_3-N) and nitrate-nitrogen (NO_3-N), together with total water budgets, ICW can be effective at removing N pollution from domestic wastewater, with comparatively high areal removal rates at all times of the year. Annual mass removal efficiencies were consistently high for the two N species with average of 92.7% removal for NH_3-N and 84.4% for NO_3-N. Overall, during the two-year operation, the ICW received a total load of 2,802-kg NH_3-N and 441-kg NO_3-N and recorded 98.0% and 96.9% removal, respectively. Average areal removal rates for NH_3-N and NO_3-N

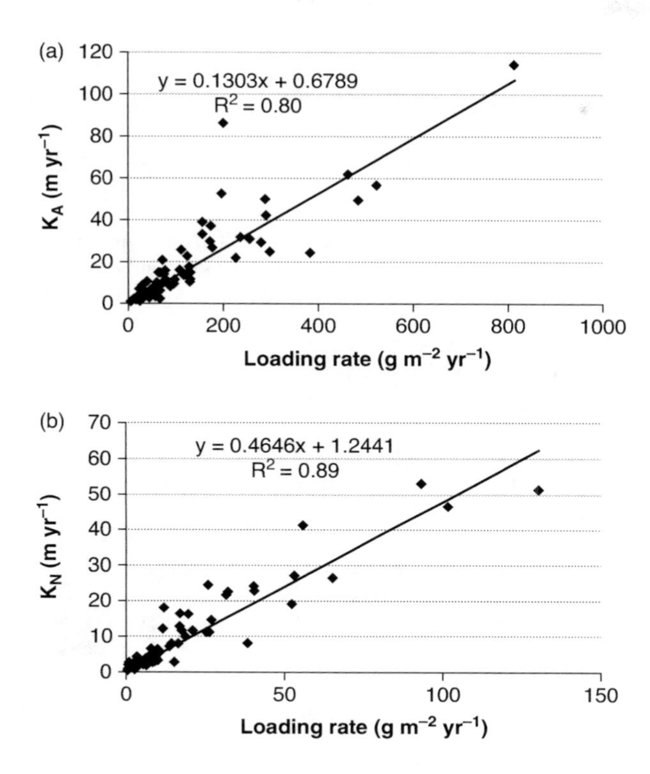

FIGURE 4.17 Nitrogen loading rate and reaction rate constants for (a) ammonia and (b) nitrate in integrated constructed wetlands between 2008 and 2010.

were 240 ± 317.8 and 35 ± 54.9 mg/m²/d, respectively, and showed significant linear correlations with areal loading rates.

Nitrogen removal exhibited some seasonal trends. Removal rates in the summer months were slightly higher. Lowest rates were observed in winter. Areal first-order N removal rate constants in the ICW averaged 14 m/yr for NH_3-N and 11 m/yr for NO_3-N. The normalized areal removal rate constants suggested that N removal in the ICW was marginally affected by temperature. The temperature coefficients (θ), estimated using the modified Arrhenius equation, were low and further validated the low influence of temperature on N removal in the ICW.

REFERENCES

Abed, S.N. and Scholz, M., 2016. Chemical simulation of greywater, *Environmental Technology*, 37(13), 1631–1646.

Ahn, C., Mitsch, W.J. and Wolfe, W.E., 2001. Effects of recycled FGD liner material on water quality and macrophytes of constructed wetlands: A mesocosm experiment, *Water Research*, 35(3), 633–642 (2001).

Al-Hamaiedeh, H. and Bino, M., 2010. Effect of treated grey water reuse in irrigation on soil and plants, *Desalination*, 256(1–3), 115–119.

Al-Jayyousi, O.R., 2003. Greywater reuse: Towards sustainable water management, *Desalination*, 156(1–3), 181–192.

Altundogan, H.S., Altundogan, S., Tumen, F. and Bildik, M., 2002. Arsenic adsorption from aqueous solutions by activated red mud, *Waste Management*, 22(3), 357–363.

American Public Health Association, 1995. Standard Methods for the Examination of Water and Wastewater, 20th edition (Washington, DC: American Public Health Association, American Water Works Association and Water Environmental Federation).

Ann, Y., Reddy, K.R. and Delfino, J.J., 1999a. Influence of chemical amendments on phosphorus immobilization in soils from a constructed wetland, *Ecological Engineering*, 14(1–2), 157–167.

Ann, Y., Reddy, K.R. and Delfino, J.J., 1999b. Influence of redox potential on phosphorus solubility in chemically amended wetland organic soils, *Ecological Engineering*, 14(1–2), 169–180.

APHA,1998. Standard Methods for the Examination of Water and Wastewater (Washington, DC: American Public Health Association (APHA)/American Water Works Association/ Water Environment Federation).

Argun, M.E., Dursun, S., Ozdemir, C. and Karatas, M., 2007. Heavy metal adsorption by modified oak sawdust: thermodynamics and kinetics, *Journal of Hazardous Materials*, 141(1), 77–85.

Arroyo, P., Ansola, G. and Miera, L.E.S., 2013. Effects of substrate, vegetation and flow on arsenic and zinc removal efficiency and microbial diversity in constructed wetlands, *Ecological Engineering*, 51, 95–103.

Asher, W.E. and Pankow, J.F., 1991. Prediction of gas/water mass transport coefficients by a surface renewal model, *Environmental Science and Technology*, 25, 1294–1300.

Ault, T., Velzeboer, R. and Zammit, R., 2000. Influence of nutrient availability on phytoplankton growth and community structure in the Port Adelaide River, Australia: Bioassay assessment of potential nutrient limitation, *Hydrobiologia*, 429(1), 89–103.

Babatunde, A.O., Zhao, Y.Q., O'Neill, M. and O'Sullivan, B., 2008. Constructed wetlands for environmental pollution control: A review of developments, research and practice in Ireland, *Environment International*, 34(1), 116–126.

Bachand, P.A.M. and Horne, A.J., 1999. Denitrification in constructed free-water surface wetlands: I. Very high nitrate removal rates in a macrocosm study, *Ecological Engineering*, 14(1–2), 9–15.

Backstrom, M., Karlsson, S., Backman, L., Folkeson, L. and Lind, B., 2004. Mobilisation of heavy metal by deicing salts in a roadside environment, *Water Research*, 38(3), 720–732.

Bannister, A.F., 1997. Lagoon and reed-bed treatment of colliery shale tip water at Dodworth, South Yorkshire, In: P.L. Younger (Ed.) Mine water Treatment using Wetlands. Proceedings of the CIWEM National Conference (Newcastle: CIWEM), pp. 105–131.

Baskan, M.B. and Pala, A., 2011. Removal of arsenic from drinking water using modified natural zeolite, *Desalination*, 281, 396–403.

Berlitz, B. and Kögler, H., 1997. Water well regeneration with ultra-sonic waves, *BBR, Wasser und Rohrbau*, 48(2), 19–23.

Beutel, M.W., 2006. Inhibition of ammonia release from anoxic profundal sediments in lakes using hypolimnetic oxygenation, *Ecological Engineering*, 28(3), 271–279.

Black and Veatch Corporation, 2010. White's Handbook of Chlorination and Alternative Disinfectants, 5th edn. (Hoboken, NJ: John Wiley & Sons, Inc.).

Boar, R.R., Crook, C.E. and Moss, B, 1989. Regression of *Phragmites australis* reedswamps and recent changes of water chemistry in the Norfolk Broadland, England, *Aquatic Botany*, 35(1), 41–55.

Bocca, B, Pino, A, Alimonti, A. and Forte, G., 2014. Toxic metals contained in cosmetics: A status report, *Regulatory Toxicology and Pharmacology*, 68(3), 447–467.

Boon, P.I. and Mitchell, A., 1995. Methanogenesis in the sediments of an Australian freshwater wetland: Comparison with aerobic decay and factors controlling methanogenesis, *FEMS Microbiology Ecology*, 18(3), 174–190.

Boutilier, L., Jamieson, R., Gordon, R., Lake, C. and Hart, W., 2010. Performance of surface-flow domestic wastewater treatment wetlands, *Wetlands*, 30, 795–804.

Brix, H., 1989. Gas exchange through dead culms of reed, *Phragmites australis* (Cav.) Trin. ex Steud, *Aquatic Botany*, 35(1), 81–98.

Brix, H., 1997. Do macrophytes play a role in constructed treatment wetlands? *Water Science and Techology*, 35(5), 11–17.

Brix, H., 1999. Functions of macrophytes in constructed wetlands, *Water Science and Technology*, 29(4), 71–78.

Brix, H., Arias, C.A. and del Bubba, M., 2001. Media selection for sustainable phosphorus reduction in subsurface flow constructed wetlands, *Water Science and Technology*, 44(11–12), 47–54.

Buddhawong, S., Kuschk, P., Mattusch, J., Wiessner, A. and Stottmeister, U. 2005. Removal of arsenic and zinc using different laboratory model wetland systems, *Engineering in Life Science*, 5(3), 247–252.

Carignan, R. and Kaill, J., 1980. Phosphorus sources for aquatic weeds: Water or sediments? *Science*, 207(4434), 987–989.

Carty, A., Scholz, M., Heal, K., Gouriveau, F., Mustafa, A., 2008. The universal design, operation and maintenance guidelines for farm constructed wetlands (FCW) in temperate climates, *Bioresource Technology*, 99(15), 6780–6792.

Chakravarty, S., Dureja, V., Bhattacharyya, G., Maity, S. and Bhattacharjee, S., 2002. Removal of arsenic from groundwater using low cost ferruginous manganese ore, *Water Research*, 36(3), 625–632.

Chauhan, A.S., Bhadauria, R., Singh, A.K., Sharad, S.S., Lodhi, D.K. and Chaturvedi V.T., 2010. Determination of lead and cadmium in cosmetic products, *Journal of Chemical and Pharmaceutical Research*, 2, 92–97.

Cheng, S.P., Grosse, W., Karrenbrock, F. and Thoennessen, M., 2002. Efficiency of constructed wetlands in decontamination of water polluted by heavy metals, *Ecological Engineering*, 18(3), 317–325.

Chjnacka, K., Saeid, A., Michalak, I. and Mikulewicz, M., 2012. Effects of local industry on heavy metals contents in human hair, *Polish Journal of Environmental Studies*, 21(6), 1563–1570.

Christova-Boal, D., Eden, R.E. and McFarlane, S., 1996. An investigation into greywater reuse for urban residential properties, *Desalination*, 106(1–3), 391–397.

Comino, E, Riggio, V. and Rosso, M., 2013. Grey water treated by a hybrid constructed wetland pilot plant under several stress conditions, *Ecological Engineering*, 53, 120–125.

Convention on Wetlands of International Importance Especially as Waterfowl Habitat, 1971. Ramsar, Iran.

Cooper, P.F., Job, G.D., Green, M.B. and Shutes, R.B.E., 1996. *Reed Beds and Constructed Wetlands for Wastewater Treatment* (Swindon: WRc).

Cullimore, D.R., 2007. *Practical Manual of Groundwater Microbiology*, 2nd edition (Boca Raton, FL: CRC Press).

Decamp, O. and Warren, A., 1998. Bacterivory in ciliates isolated from constructed wetlands (reed beds) used for wastewater treatment, *Water Resources and Technology*, 32(7), 1989–1996.

Decamp, O., Warren, A. and Sanchez, R., 1999. The role of ciliated protozoa in subsurface flow wetlands and their potential as bioindicators, *Water Science and Technology*, 40(3), 91–98.

Diaper, C., Toifl, M. and Storey, M., 2008. Greywater technology testing protocol. CSIRO: Water for a Healthy Country National Research Flagship.
Dixon, A., Butler, D., Fewkes, A. and Robinson, M., 1999. Measurement and modelling of quality changes in stored untreated grey water, *Urban Water*, 1(4), 293–306.
Domenech, L. and Sauri, D., 2010. Socio–technical transitions in water scarcity contests: Public acceptance of greywater reuse technologies in the Metropolitan Area of Barcelona, *Resources, Conservation and Recycling*, 55(1), 53–62.
Donkor, V.A. and Häder, D.-P., 1996. Effects of ultraviolet radiation on photosynthetic pigments in some filamentous cyanobacteria, *Aquatic Microbial Ecology*, 11(2), 143–149.
Dunne, E.J., Culleton, N., O'Donovan, G., Harrington, R. and Olsen, A.E., 2005. An integrated constructed wetland to treat contaminants and nutrients from dairy farmyard dirty water, *Ecological Engineering*, 24(3), 221–234.
Dunne, T. and Leopold, L.B., 1978. *Water in Environmental Planning* (New York: W. H. Freeman and Company).
Dzakpasu, M., Hofmann, O., Scholz, M., Harrington, R., Jordan, S.N. and McCarthy, V., 2011. Nitrogen removal in an integrated constructed wetland treating domestic wastewater, *Journal of Environmental Science and Health, Part A: Toxic/Hazardous Substances and Environmental Engineering*, 7(7), 742–750.
Ehrlich, H.L., 2002. *Geomicrobiology*, 4th edition (New York: Marcel Dekker).
Emmerson, G., 1998. *Every Drop Is Precious: Greywater as an Alternative Water Source* (Brisbane: Queensland Parliamentary Library).
Eriksson, E, Auffarth, K, Henze, M. and Ledin, A., 2002. Characteristics of grey wastewater, *Urban Water*, 4(1), 85–104.
Eriksson, E., Srigirisetty, S. and Eilersen, A.M., 2010. Organic matter and heavy metals in grey-water sludge, *Water SA*, 36(1), 139–142.
Essington, M.E., 2004. *Soil and Water Chemistry: An Integrative Approach* (Boca Raton, FL: CRC Press).
Etherington, J.R., 1983. *Wetland Ecology* (London: Edward Arnold).
FAO, Food and Agriculture Organization of the United Nations, 2007. Coping with water scarcity: Challenge of the twenty first century (Cairo, Egypt: FAO Regional Office for the Near East).
Faulkner, S.P. and Richardson, C.J., 1989. Physical and chemical characteristics of freshwater wetland soil, In: D. A. Hammer (Ed.) *Constructed Wetlands for Wastewater Treatment* (Chelsea: Lewis Publishers) pp. 41–72.
Felföldi, T., Tarnóczai, T. and Homonnay, Z.G., 2010. Presence of potential bacterial pathogens in a municipal drinking water supply system, *Acta microbiologica et immunolica Hungarica*, 57(3), 165–179.
Fenner, R.A. and Komvuschara, K., 2005. A new kinetic model for ultraviolet disinfection of greywater, *Journal of Environmental Engineering – ASCE*, 131(6), 850–864.
Fennessy, M.S., Brueske, C.C. and Mitsch, W.J., 1994. Sediment deposition patterns in restored freshwater wetlands using sediment traps, *Ecological Engineering*, 3(4), 409–428.
Fitria, D., Scholz, M., Swift, G.M. and Hutchinson, S.M., 2014. Impact of sludge floc size and water composition on dewaterability, *Chemical Engineering Technology*, 37(3), 471–477.
Fredrickson, L.H. and Reid, F.A., 1990. Impacts of hydrologic alteration on management of freshwater wetlands, In: J. M. Sweeney (Ed.) *Management of Dynamic Ecosystems* (West Lafayette: North Central Section, Wildlife Society) pp. 71–90.
Friedler, E., Kovalio, R. and Ben-Zvi, A., 2006. Comparative study of the microbial quality of greywater treated by three on-site treatment systems, *Environmental Technology*, 27(6), 653–663.
Friedler, E., Kovalio, R. and Galil, N.I., 2005. On-site greywater treatment and reuse in multi-storey buildings, *Water Science and Technology*, 51(10), 187–194.
Gächter, R. and Meyer, J.S., 1993. The role of micro-organisms in mobilization and fixation of phosphorus in sediments, *Hydrobiologia*, 253(6), 103–121.

Gächter, R., Meyer, J.S. and Mares, A., 1988. Contribution of bacteria to release and fixation of phosphorus in lake sediments, *Limnology and Oceanography*, 33(6, part 2), 1542–1558.

Gambrell, R.P. and Patrick Jr., W. H., 1978. Chemical and microbiological properties of anaerobic soils and sediments, In: D. D. Hook and R. M. M. Crawford (Eds.) *Plant Life in Anaerobic Environments* (Ann Arbor: Ann Arbor Science) pp. 375–423.

Gervin, L. and Brix, H., 2001. Reduction of nutrients from combined sewer overflows and lake water in a vertical-flow constructed wetland system, *Water Science and Technology*, 44(11–12), 171–176.

Gessner, M.O., 2000. Breakdown and nutrient dynamics of submerged *Phragmites* shoots in the littoral zone of a temperate hardwater lake, *Aquatic Botany*, 66(1), 9–20.

Ghosh, D. and Gupta, A., 2012. Economic justification and eco-friendly approach for regeneration of spent activated alumina for arsenic contaminated groundwater treatment, *Resources, Conservation and Recycling*, 61, 118–124.

Green, M., Friedler, E. and Safrai, I., 1998. Enhancing nitrification in vertical-flow constructed wetlands utilizing a passive air pump, *Water Research*, 32(12), 3513–3520.

Green, M.B., Martin, J.R. and Griffin, P., 1999. Treatment of combined sewer overflow at small wastewater treatment works by constructed reed beds, *Water Science and Technology*, 40(3), 357–364.

Greene, A.C., Ogg, C.D., Lynch, K.M., Pope, P.B. and Patel, B.K.C., 2003. Iron-reducing bacteria - ecology, significance and potential uses. In: *Bac-Min Conference Proceedings* (Carlton: Australasian Institute of Mining and Metallurgy).

Gregory, S.V., Swanson, F.J., McKee, W.A. and Cummins, K.W., 1991. An ecosystem perspective of riparian zones, *BioSciences*, 41(8), 540–551.

Groffman, P.M., Gold, A.J. and Addy, K., 2000. Nitrous oxide production in riparian zones and its importance to national emission inventories, *Chemosphere – Global Change Science*, 2(3–4), 291–299.

Gross, A., Kaplan, D. and Baker, K., 2006. Removal of microorganisms from domestic greywater using a recycling vertical flow constructed wetland (RVFCW), *Proceedings of the Water Environment Federation*, 6(6), 6133–6141.

Gross, A., Kaplan, D. and Baker, K., 2007a. Removal of chemical and microbiological contaminants from domestic greywater using a recycled vertical flow bioreactor (RVFB), *Ecological Engineering*, 31(2), 107–114.

Gross, A., Shmueli, O., Ronen, Z. and Raveh, E., 2007b. Recycled vertical flow constructed wetland (RVFCW) – a novel method of recycling greywater for irrigation in small communities and households, *Chemosphere*, 66(5), 916–923.

Gulz, P.A., Gupta, S.K. and Schulin, R., 2005. Arsenic accumulation of common plants from contaminated soils, *Plant Soil*, 272(1–2), 337–347.

Guo, J.X., Hu, L. and Yand, P.Z., 2007. Chronic arsenic poisoning in drinking water in Inner Mongolia and its associated health effects, *Journal of Environmental Science and Health Part A, Toxic/Hazardous Substances and Environmental Engineering*, 42(12), 1853–1858.

Gupta, V.K., Saini, V.K. and Jain, N., 2005. Adsorption of As(III) from aqueous solutions by iron oxide-coated sand, *Journal of Colloid and Interface Science*, 288(1), 55–60.

HACH Company, 2000. *Procedures Manual, Spectrophotometer DR 2010 49300-22*, 7th edn. (Loveland: HACH Company).

Halbach, M., Koschinsky, A. and Halbach, P., 2001. Report on the discovery of *Gallionella ferruginea* from an active hydrothermal field in the deep sea, *InterRidge News*, 10(1), 18–20.

Hallbeck, L. and Pedersen, K., 1991. Autotrophic and mixotrophic growth of *Gallionella ferruginea*, *Microbiology*, 137(11), 2657–2661.

Hallbeck, L. and Pedersen, K., 1995. Benefits associated with the stalk of *Gallionella ferruginea*, evaluated by comparison of a stalk-forming and non-stalk-forming strain and biofilm studies *in situ*, *Microbial Ecology*, 30(3), 257–268.

Hammer, D.A., 1989. *Constructed Wetlands for Wastewater Treatment – Municipal, Industrial and Agricultural* (Chelsea, MA: Lewis Publishers).

Hammer, D.A. and Knight, R.L., 1994. Designing constructed wetlands for nitrogen removal, *Water Science and Technology*, 29(4), 15–27.

Harrington, R. and McInnes, R., 2009. Integrated constructed wetlands (ICW) for livestock wastewater management, *Bioresource Technology*, 100(22), 5498–5505.

Harrington, R., Carroll, P., Carty, A., Keohane, J. and Ryder, C., 2007. Integrated constructed wetlands: Concept, design, site evaluation and performance, *International Journal of Water*, 3(3), 243–256.

Harrington, R., Dunne, E., Carroll, P., Keohane, J. and Ryder, C., 2004. The anne valley project: The use of integrated constructed wetlands (ICWs) in farmyard and rural domestic wastewater management. In: D. Lewis and L. P. Gairns (Eds.) *Agriculture and the Environment, Water Framework Directive and Agriculture. Proceedings of the SAC and SEPA Biennial Conference*, Edinburgh, 24–25 March 2004, pp. 51–58.

Harrington, R. and Ryder, C., 2002. The use of integrated constructed wetlands in the management of farmyard runoff and waste water. In: *Proceedings of the National Hydrology Seminar on Water Resource Management: Sustainable Supply and Demand*, Tullamore, Offaly, 19th November 2002 (Tullamore: The Irish National Committees of the IHP and ICID) pp. 55–63.

Hawke, C.J. and José, P.V., 1996. *Reed Bed Management for Commercial and Wildlife Interests* (Sandy, The Royal Society for the Protection of Birds).

Hedrich, S., Schlömann, M. and Johnson, D.B., 2011. The iron-oxidizing proteobacteria, *Microbiology*, 157(66), 1551–1564.

Heimburg, K., 1984. Hydrology of north-central Florida cypress domes, In: K. C. Ewel and H. T. Odum (Eds.) *Cypress Swamps* (Gainesville, FL: University Presses of Florida) pp. 72–82.

Henkel, S., Weidner, C., Roger, S., Schüttrumpf, H, Rüde, T.R., Klauder, W. and Vinzelberg, G., 2012. Untersuchung der Verockerungsneigung von Vertikalfilterbrunnen im Modellversuch, *Grundwasser*, 17(3), 157–169 (in German).

Henken, K.R. (ed.), 2009. Arsenic in natural environments, In: *Arsenic Environmental Chemistry, Health Threats and Waste Treatment* (Chichester: John Wiley &Sons Ltd.) pp. 69–236.

Ho, Y.S. and McKay, G., 1999. Pseudo-second order model for sorption processes, *Process Biochemistry*, 34(5), 451–465.

Holdren, G.C. and Armstrong, D.E., 1980. Factors affecting phosphorus release from intact lake sediment cores, *Environmental Science and Technology*, 14(1), 79–87.

Houben, G.J., 2006. The influence of well hydraulics on the spatial distribution of well incrustations, *Groundwater*, 44(5), 668–675.

Houben, G. and Treskatis, C., 2019. *Regenerierung und Sanierung von Brunnen – Technische und naturwissenschaftliche Grundlagen der Brunnenalterung und möglicher Gegenmaßnahmen*, 3rd edition (Essen: Vulkan Verlag) (in German).

Houben, G.J. and Weihe, U., 2010. Spatial distribution of incrustations around a water well after 38 years of use, *Groundwater*, 48(1), 53–58.

Hourlier, F., Masse, A., Jaouen, P., Lakel, A., Gerente, C., Faur, C. and Le Cloire, P., 2010. Formulation of synthetic greywater as an evaluation tool for wastewater recycling technologies, *Environment Technology*, 31(2), 215–223.

Houshia, O.J., Abueid, M., Daghlas, A., Zaid, M.O., Al Ammor, J., Souqia, N., Alary, R. and Sholi, N., 2012. Characterization of grey water from country-side decentralized water treatment stations in northern Palestine, *Journal of Environment and Earth Science*, 2(2), 1–8.

Huang, J., Reneau Jr., R.B. and Hagedorn, C., 2000. Nitrogen removal in constructed wetlands employed to treat domestic wastewater, *Water Research*, 34(9), 2582–2588.

Jefferson, B., Burgess, J.E., Pichon, A., Harkness, J. and Judd S.J., 2001. Nutrient addition to enhance biological treatment of greywater, *Water Research*, 35(11), 2702–2710.

Jefferson, B., Laine, A., Parsons, S., Stephenson, T. and Judd, S., 2000. Technologies for domestic wastewater recycling, *Urban Water*, 1(4), 285–292.

Jeppesen, B., 1996. Domestic greywater re-use: Australia's challenge for the future, *Desalination*, 106(1–3), 311–315.

Jimenez-Cedillo, M.J., Olguin, M.T. and Fall, C., 2009. Adsorption kinetic of arsenates as water pollutant on iron, manganese and iron-manganese-modified clinoptilolite-rich tuffs, *Journal of Hazardous Materials*, 163(2–3), 939–945.

Junk, W.J., Bayley, P.B. and Sparks, R.E., 1989. The flood pulse concept in river-floodplain systems, In: D. P. Dodge (ed.) *Proceedings of the International Large River Symposium, Canadian Journal of Fisheries and Aquatic Sciences*, 106(special issue) pp. 11–127.

Jurdo, G.B., Johnson, J., Feeley, H., Harrington, R. and Kelly-Quinn, M., 2010. The potential of integrated constructed wetlands (ICWs) to enhance macroinvertebrate diversity in agricultural landscapes, *Wetlands*, 30(3), 393–404.

Kadlec, R., Knight, R.L., Vymazal, J., Brix, H., Cooper, P.F. and Haberl, R., 2000. *Constructed Wetlands for Pollution Control*, International Water Association (IWA) Specialist Group 'Use of Macrophytes for Water Pollution Control', Scientific and Technical Report Number 8 (London: IWA Publishing).

Kadlec, R.H., 1999. Chemical, physical and biological cycles in treatment wetlands, *Water Science and Technology*, 40(2), 37–44.

Kadlec, R.H., 2002. *Effects of Pollutant Speciation in Treatment Wetlands Design* (Chelsea: Wetland Management Services).

Kadlec, R.H., 2003. Pond and wetland treatment, *Water Science and Technology*, 48(5), 1–8.

Kadlec, R.H., 2009. Comparison of free water and horizontal subsurface treatment wetlands, *Ecological Engineering*, 35(2), 159–174.

Kadlec, R.H. and Knight, R.L., 1996. *Treatment Wetlands* (Boca Raton, FL: CRC Press).

Kadlec, R.H. and Reddy, K.R., 2001. Temperature effects in treatment wetlands, *Water Environment Research*, 73(5), 543–557.

Kadlec, R.H. and Wallace, S.D., 2009. *Treatment Wetlands*, 2nd edition (Boca Raton, FL: CRC Press).

Kamp-Nielson, L., 1974. Mud-water exchange of phosphate and other ions in undisturbed sediment cores and factors affecting exchange rates, *Archiv für Hydrobiologie*, 73(2), 218–237.

Kanel, S.R., Choi, H., Kim, J.Y., Vigneswaran, S. and Wang, G.S., 2006. Removal of Arsenic(III) from Groundwater using Low-Cost Industrial By-products-Blast Furnace Slag, *Water Quality Research Journal of Canada*, 41(2), 130–139.

Karathanasis, A.D., Potter, C.L. and Coyne, M.S., 2003. Vegetation effects on fecal bacteria, biochemical oxygen demand and suspended solids removal in constructed wetland treating domestic wastewater, *Ecological Engineering*, 20(2), 157–169.

Kariuki, F.W., Ngàng, V.G. and Kotut, K., 2012. Hydrochemical characteristics, plant nutrients and metals in household greywater and soils in Homa Bay town, *The Open Environmental Engineering Journal*, 5(1), 103–109.

Kiely, G., 1997. *Environmental Engineering* (Maidenhead: McGraw-Hill International (UK) Limited).

Kneebone, P.E., O'Day, P.A., Jones, N. and Hering, J.G., 2002. Deposition and fate of arsenic in iron- and arsenic-enriched reservoir sediments, *Environmental Science and Technology*, 36(3), 381–386.

Knight, R.L. Kadlec, R.H. and Ohlendorf, H.M., 1999. The use of treatment wetland for petroleum industry effluents, *Environmental Science and Technology*, 33(7), 973–980.

Kuehn, E. and Moore, J.A., 1995. Variability of treatment performance in constructed wetlands, *Water Science and Technology*, 32(3), 241–250.

Kuehn, K.A., Gessner, M.O., Wetzel, R.G. and Suberkropp, K., 1999. Standing litter decomposition of the emergent macrophyte *Erianthus giganteus*, *Microbial Ecology*, 38(1), 50–57.

Kundu, S. and Gupta, A.K., 2006. Adsorptive removal of As(III) from aqueous solution using iron oxide coated cement (IOCC): Evaluation of kinetic, equilibrium and thermodynamic models, *Separation and Purification Technology*, 51(2), 165–172.

Kuschk, P., Wiebner, A., Kappelmeyer, U., Weibbrodt, E., Kastner, M. and Stottmeister, U., 2003. Annual cycle of nitrogen removal by a pilot-scale subsurface horizontal flow in a constructed wetland under moderate climate, *Water Research*, 37(17), 4236–4242.

Kvet, J. and Husak, S., 1978. Primary data on biomass and production estimates in typical stands of fishpond littoral plant communities, In: D. Dykyjová and J. Kvet (eds.) *Pond Littoral Ecosystems* (Berlin: Springer Verlag) pp. 211–216.

Leal, L.H., Soeter, A.M., Kools, S.E., Kraak, M.H.S., Parsons, J.R., Temmink, H., Zeeman, G. and Buisman, C.J.N., 2012. Ecotoxicological assessment of greywater treatment systems with *Daphnia magna* and *Chironomus riparius*, *Water Research*, 46(4), 1038–1044.

Lee, C., Fletcher, T.D. and Sun, G., 2009. Nitrogen removal in constructed wetland systems, *Engineering in Life Sciences*, 9(1), 11–22.

Li, F., Wichmann, K. and Otterpohl, R., 2009. Review of the technological approaches for grey water treatment and reuses, *The Science of the Total Environment*, 407(11), 3439–3449.

Li, Q., Xu, X.T., Cui, H., Pang, J.F., Wei, Z.B., Sun, Z.Q. and Zhai, J.P., 2012. Comparison of two adsorbents for the removal of pentavalent arsenic from aqueous solutions, *Journal of Environmental Management*, 98, 98–106.

Liao, Y., Liang, J. and Zhou, L., 2011. Adsorptive removal of As(III) by biogenic schwertmannite from simulated As-contaminated groundwate, *Chemosphere*, 83(3), 295–301.

Lim, J.W., Chang, Y.Y., Yang, J.K. and Lee, S.M., 2009. Adsorption of arsenic on the reused sanding wastes calcined at different temperatures, *Colloids and Surfaces A: Physicochemical and Engineering Aspects*, 345(1–3), 65–70.

Litter, M.I., Morgada, M.E. and Bundschuh, J., 2010. Possible treatments for arsenic removal in Latin American waters for human consumption, *Environmental Pollution*, 158(5), 1105–1118.

Liu, S., Butler, D., Memon, F.A., Makropoulos, C., Avery, L., Jefferson, B., 2010. Impact of residence time during storage on potential of water saving for grey water recycling system, *Water Research*, 44(1), 267–277.

Lizama, A.K., Fletcher, T.D. and Sun, G., 2012. The effect of substrate media on the removal of arsenic, boron and iron from an acidic wastewater in planted column reactors, *Chemical Engineering Journal*, 179, 119–130.

Lowrance, R., Altier, L.S., Newbold, J.D., Schnabel, R.R., Groffman, P.M., Denver, J.M., Correll, D.L., Gilliam, J.W., Robinson, J.L., Brinsfield, R.B., Staver, K.W., Lucas, W. and Todd, A.H., 1979. Water quality functions of riparian forest buffers in Chesapeake Bay watersheds, *Environmental Management*, 21(5), 687–712.

Luederitz, V., Eckert, E., Lange-Weber, M., Lange, A. and Gersberg, R.M., 2001. Nutrient removal efficiency and resource economics of vertical flow and horizontal flow constructed wetlands, *Ecological Engineering*, 18(2), 157–171.

Lund, L.J., Horne, A.J. and Williams, A.E., 2000. Estimating denitrification in a large constructed wetland using stable nitrogen isotope ratios, *Ecological Engineering*, 14(1–2), 67–76.

Ma, L.Q., Komar, K.M., Tu, C., Zhang, W., Cai, Y. and Kennelley, E.D., 2001. A fern that hyperaccumulates arsenic, *Nature*, 409(6820), 579.

Macedo-Miranda, M.G. and Olguin, M.T., 2007. Arsenic sorption by modified clinoptilolite–heulandite rich tuffs, *Journal of Inclusion Phenomena and Macrocyclic Chemistry*, 59, 131–142.

Machate, T., Noll, B.H.H. and Kettrup, A., 1997. Degradation of phenanthrene and hydraulic characteristics in a constructed wetland, *Water Research*, 31(3), 554–560.

Maiga, Y., Moyenga, D., Nikiema, B.C., Ushijima, K., Maiga, A.H. and Funamizu, N., 2014. Designing slanted soil system for greywater treatment for irrigation purposes in rural area of arid regions, *Environmental Technology*, 35(21–24), 3020–3027.

Maji, S.K., Kao, Y.H. and Liu, C.W., 2011. Arsenic removal from real arsenic-bearing groundwater by adsorption on iron-oxide-coated natural rock (IOCNR), *Desalination*, 280(1–3), 72–79.

Maji, S.K., Pal, A. and Pal, T., 2008. Arsenic removal from real-life groundwater by adsorption on laterite soil, *Journal of Hazardous Materials*, 151(2–3), 811–820.

Mander, Ü., Kuusemets, V. and Ivask, M., 1995. Nutrient dynamics of riparian ecotones: a case study from the Porijõgi River catchment, *Estonia, Landscape and Urban Planning*, 31(1–3), 333–348.

Mandernack, K.W., Lynch, L., Krouse, H.R. and Morgan, M.D., 2000. Sulfur cycling in wetland peat of the New Jersey Pinelands and its effect on stream water chemistry, *Geochimica et Cosmochimica Acta*, 22(5), 3949–3964.

Manning, B.A., Hunt, M., Amrhein, C. and Yarmoff, J.A., 2002. Arsenic(III) and Arsenic(V) reactions with zerovalent iron corrosion products, *Environmental Science and Technology*, 36(24), 5455–5461.

March, J.G., Gual, M. and Orozco, F., 2004. Experiences on greywater re-use for toilet flushing in a hotel (Mallorca Island, Spain), *Desalination*, 164(3), 241–247.

Marchand, L., Mench, M., Jacob, D.L. and Otte, M.L., 2010. Metal and metalloid removal in constructed wetlands, with emphasis on the importance of plants and standardized measurements: A review, *Environmental Pollution*, 158(12), 3447–3461.

Marfil-Vega, R., Suidan, M.T. and Mills, M.A., 2010. Abiotic transformation of oestrogens in synthetic municipal wastewater: an alternative for treatment, *Environmental Pollution*, 158(11), 3372–3377.

Mayer, T. and Kramer, J.R., 1986. Effect of lake acidification on the adsorption of phosphorus by sediments, *Water Air and Soil Pollution*, 31(3–4), 949–958.

Mays, P.A. and Edwards, G.S., 2001. Comparison of heavy metal accumulation in a natural wetland and constructed wetlands receiving acid mine drainage, *Ecological Engineering*, 16(4), 487–500.

McBain, W.J., Cornish, V.E. and Bowden, C.R., 1912. CCXV–Studies of the constitution of soap in solution: Sodium myristate and sodium laurate, *Journal of the Chemical Society, Transactions*, 101, 2042–2056.

McNaughton, S.J., 1966. Ecotype function in the *Typha* community-type, *Ecological Monographs*, 36(4), 297–325.

McNeill, A. and Olley, S., 1998, The effects of motorway runoff on watercourses in south-west Scotland, *Water and Environment Journal*, 12(6) 433–439.

Menz, C., 2016. Oxygen delivering processes in groundwater and their relevance for iron-related well clogging processes– a case study on the quaternary aquifers of Berlin. PhD Dissertation (Berlin: Freie Universität Berlin).

Met Éireann, 2010. Climate of Ireland. https://www.met.ie/climate/climate-of-ireland

Michelson, L.D., Gideon, P.G., Pace, E.G. and Katal, L.H., 1975. US Department of Industry, Office of Water Research and Technology, 74.

Mikkelsen, H., Sivaneson, M. and Filloux, A., 2011. Key two-component systems that control biofilm formation in *Pseudomonas aeruginosa*, *Environmental Microbiology*, 13(7), 1666–1681.

Ministry of Environmental Protection, 1996. *Integrated Wastewater Discharge Standard (GB8978-1996)* (Ministry of Environmental Protection).

Ministry of Health, 2006. *Sanitary Standard for Drinking Water (GB5749-2006)* (Ministry of Health).

Ministry of Water Resource, 2005. *Water Quality-Determination of Arsenic-Atomic Fluorescence Spectrometric Method (SL 327.1–2005)* (Ministry of Water Resource).

Mitsch, W.J. and Gosselink, J.G., 2000. *Wetlands*, 3rd edition (New York: John Wiley & Sons Inc.).

Mitsch, W.J., Horne, A.J. and Narin, W.R., 2000. Nitrogen and phosphorus retention in wetlands – Ecological approaches to solving excess nutrient problems, *Ecological Engineering*, 14, 1–7.

Mohan, D. and Pittman, C.U., 2007. Arsenic removal from water/wastewater using adsorbents — A critical review, *Journal of Hazardous Materials*, 142(1–2), 1–53.

Moore, P.A. and Reddy, K.R., 1994. Role of Eh and pH on phosphorus geochemistry in sediments of Lake Okeechobee, Florida, *Journal of Environmental Quality*., 23(5), 955–964.

Morgan, M.D. and Mandernack, K.W., 1996. Biogeochemistry of sulfur in wetland peat following 3.5 Y of artificial acidification (HUMEX), *Environment International*, 22(5), 605–610.

Moshiri, G.A., 1993. *Constructed Wetlands for Water Quality Improvement* (Boca Raton, FL: CRC Press).

Moss, B., 1998. *Ecology of Fresh Waters*, 3rd edition (Oxford: Blackwell Science Ltd.).

Mungur, S., Shutes, R.B.E., Revitt, D.M. and House, M.A., 1995. An assessment of metal removal from highway runoff by a natural wetland, *Water Science and Technology*, 32(3), 169–175.

Mungur, S., Shutes, R.B.E., Revitt, D.M. and House, M.A., 1997. An assessment of metal reduction by laboratory scale wetland, *Water Science and Technology*, 35(5), 125–133.

Mustafa, A., Scholz, M., Harrington, R. and Carroll, P., 2009. Long-term performance of a representative integrated constructed wetland treating farmyard runoff, *Ecological Engineering*, 35(5), 779–790.

Myers, D., 1988. *Surfactant Science and Technology* (New York: VCH Publishers).

Nairn, R.W. and Mitsch, W.J., 2000. Phosphorus removal in created wetland ponds receiving river overflow, *Ecological Engineering,* 14(1–2), 107–126.

Nazim, F. and Meera, V., 2013. Treatment of synthetic greywater using 5% and 10% garbage enzyme solution, *Bonfring International Journal of Industrial Engineering and Management Science*, 3, 111–117.

Nghiem, L.D., Oschmann, N. and Schäfer, A.I., 2006. Fouling in greywater recycling by direct ultrafiltration, *Desalination*, 187, 283–290.

Nguyen, T.V., Vigneswaran, S., Ngo, H.H. and Kandasamy, J., 2010. Arsenic removal by iron oxide coated sponge: Experimental performance and mathematical models, *Journal of Hazardous Materials*, 182(1–3), 723–729.

Nolde, E., 2000. Greywater reuse systems for toilet flushing in multi-storey buildings – Over ten years experience in Berlin, *Urban Water*, 1(4), 275–284.

Norrstrom, A.C. and Jacks, G., 1998. Concentration and fractionation of heavy metals in roadside soils receiving de-icing salts, *The Science of the Total Environment*, 218(2–3), 161–174.

Nygaard, B. and Ejrnæs, R., 2009. The impact of hydrology and nutrients on species composition and richness: Evidence from a microcosm experiment, *Wetlands*, 29(1), 187–195.

Obarska-Pempkowiak, H. and Klimkowska, K., 1999. Distribution of nutrients and heavy metals in a constructed wetland system, *Chemoshere*, 39(2), 303–312.

Ostendorp, W., 1989. 'Die-back' of reeds in Europe – A critical review of literature, *Aquatic Botany*, 35(1), 5–26.

Overbeck, J., 1988. Qualitative and quantitative assessment of the problem, In: S. E. Jørgensen and R. A. Vollenweider (eds.) *Guidelines for Lake Management: Principles of Lake Management, Vol. 1*, (International Lake Environment Committee, United Nations Environment Programme), Shiga, Japan. pp. 19–36.

Parr, T.W., 1990. Factors affecting reed (*Phragmites australis*) growth in U.K. reed bed treatment systems, In: P. F. Cooper and B. C. Findlater (eds.) *Constructed Wetlands in Water Pollution Control* (Oxford: Pergamon Press) pp. 67–76.

Perkins, J. and Hunter, C., 2000. Removal of enteric bacteria in a surface flow constructed wetland in Yorkshire, England, *Water Research*, 34(6), 1941–1947.

Pidou, M., Avery, L., Stephenson, T., Jeffrey, P., Parsons, S., Liu, S., Memon, F. and Jefferson, B. 2008. Chemical solutions for greywater recycling, *Chemosphere*, 71(1), 147–155.

Pinney, M.L., Westerhoff, P.K. and Baker, L., 2000. Transformations in dissolved organic carbon through constructed wetlands, *Water Research*, 34(6), 1897–1911.

Pinto, U., Maheshwari, B.L. and Grewal, H.S., 2010. Effects of greywater irrigation on plant growth, water use and soil properties, *Resources Conservation and Recycling*, 54(7), 429–435.

Preston, C.D. and Croft, J.M., 1997. *Aquatic Plants in Britain and Ireland* (Colchester: Harley Books).

Rahman, K.Z., Wiessner, A., Kuschk, P., van Afferden, M., Mattusch, J. and Müller, R.A., 2011, Fate and distribution of arsenic in laboratory-scale subsurface horizontal-flow constructed wetlands treating an artificial wastewater, *Ecological Engineering*, 37(8), 1214–1224.

Ramona, G., Green, M., Semiat R. and Dosoretz, C., 2004. Low strength greywater characterization and treatment by direct membrane filtration, *Desalination*, 170(3), 241–250.

Ran, N., Agami, M. and Oron, G., 2004. A pilot study of constructed wetlands using duckweed (*Lemna gibba* L.) for treatment of domestic primary effluent in Israel, *Water Research*, 38(9), 2241–2248.

Ranjan, D., Talat, M. and Hasan, S.H., 2009. Biosorption of arsenic from aqueous solution using agricultural residue 'rice polish', *Journal of Hazardous Materials*, 166(2–3), 1050–1059.

Reddy, K.R., Kadlec, R.H., Flaig, E. and Gale, P.M., 1999. Phosphorus retention in streams and wetlands: A review, *Critical Reviews in Environmental Science and Technology*, 29(1), 83–146.

Rivera, F., Warren, A., Ramirez, E., Decamp, O., Bonilla, P., Gallegos, E., Calderón, A. and Sánchez, J. T., 1995. Removal of pathogens from wastewater by the root zone method (RZM), *Water Science and Technology*, 32(3), 211–218.

Roberts, L.C., Hug, S.J., Ruettimann, T., Khan, A.W. and Rahman, M.T., 2004. Arsenic removal with iron(II) and iron(III) in waters with high silicate and phosphate concentrations, *Environmental Science and Technology*, 38(1), 307–315.

Robinson, H., Harris, G., Carville, M., Barr, M. and Last, S., 1999. The use of engineered reed bed system to treat leachate at Monument Hill Landfill Site, Southern England, In: G. Mulamoottil, E. A. McBean and F. Rovers (eds.) *Constructed Wetlands for the Treatment of Landfill Leachates* (Boca Raton, FL: Lewis Publishers), Chapter 6

Rudd, J.W.M., Kelly, C.A., St. Louis, V., Hesslein, R.H., Furutani, A. and Holoka, M.H., 1986. The role of sulfate reduction in long term accumulation of organic and inorganic sulfur in lake sediments, *Limnology Oceanography*, 31(6), 1281–1291.

Sakadevan, K. and Bavor, H.J., 1999. Nutrient removal mechanisms in constructed wetlands and sustainable water management, *Water Science and Technology*, 40(2), 121–128.

Sansalone, J.J., 1999. Adsorptive infiltration of metals in urban drainage: Media characteristics, *The Science of the Total Environment*, 235(1–3), 179–188.

Sasaki, K., Ogino, T., Endo, Y. and Kurosawa, K., 2003. Field studies on heavy metal accumulation in a natural wetland receiving acid mine drainage. *Materials Transactions*, 44(9), 1877–1884.

Schäfer, A.I., Nghiem, L.D. and Oschmann, N., 2006. Bisphenol A retention in the direct ultrafiltration of greywater, *Journal of Membrane Science*, 283(1–2), 233–243.

Schlesinger, W.H., 1991. *Biogeochemistry: An Analysis of Global Change* (San Diego, CA: Academic Press).

Scholes, L.N.L., Shutes, R.B.E., Revitt, D.M., Purchase, D. and Forshaw, M., 1999. The removal of urban pollutants by constructed wetlands during wet weather, *Water Science and Technology*, 40(3), 333–340.

Scholz, M., 2003. Case study: Design, operation, maintenance and water quality management of sustainable storm water ponds for roof runoff, *Bioresource Technology*, 95(3), 269–279.

Scholz, M., 2004a. Treatment of gully pot effluent containing nickel and copper with constructed wetlands in a cold climate, *Journal of Chemical Technology and Biotechnology*, 79(2), 153–162.

Scholz, M., 2004b. Stormwater quality associated with a silt trap (empty and full) discharging into an urban watercourse in Scotland, *International Journal of Environmental Studies*, 61(4), 471–483.

Scholz, M., 2006. *Wetland Systems to Control Urban Runoff* (Amsterdam: Elsevier).

Scholz, M., 2010. *Wetland Systems – Storm Water Management Control* (Berlin: Springer).

Scholz, M., 2015. *Wetlands for Water Pollution Control*, 2nd edn. (Amsterdam: Elsevier).

Scholz, M. and Lee, B.-H., 2005. Constructed wetlands: A review, *International Journal of Environmental Studies*, 62(4), 421–447.

Scholz, M. and Martin, R.J., 1998. Control of bio-regenerated granular activated carbon by spreadsheet modelling, *Journal of Chemical Technology and Biotechnology*, 71(3), 253–261.

Scholz, M. and Trepel, M., 2004. Water quality characteristics of vegetated groundwater-fed ditches in a riparian peatland, *The Science of the Total Environment*, 332(1–3), 109–122.

Scholz, M. and Xu, J., 2001. Comparison of vertical-flow constructed wetland treatment of wastewater containing lead and copper, *Water and Environment Journal*, 15(4), 287–293.

Scholz, M. and Xu, J., 2002. Performance comparison of experimental constructed wetlands with different filter media and macrophytes treating industrial wastewater contaminated with lead and copper, *Bioresource Technology*, 83(2), 71–79.

Scholz, M., Harrington, R., Carroll, P. and Mustafa, A., 2007. The integrated constructed wetlands (ICW) concept, *Wetlands*, 27(2), 337–354.

Scholz, M., Höhn, P. and Minall, R., 2002. Mature experimental constructed wetlands treating urban water receiving high metal loads, *Biotechnology Progress*, 18(6), 1257–1264.

Singhakant, C., Koottatep, T. and Satayavivad, J., 2009a. Enhanced arsenic removals through plant interactions in subsurface-flow constructed wetlands, *Journal of Environmental Science and Health. Part A: Toxic/Hazardous Substances and Environmental Engineering*, 44(2), 163–169.

Singhakant, C., Koottatep, T. and Satayavivad, J., 2009b. Fractional analysis of arsenic in subsurface-flow constructed wetlands with different length to depth ratios, *Water Science and Technology*, 60(7), 1771–1778.

Smith, C.S. and Adams, M.S., 1986. Phosphorus transfer from sediments to *Myriophyllum spicatum*, *Limnology and Oceanography*, 31(6), 1312–1321.

Smith, R.L., 1980. *Ecology and Field Biology*, 3rd edn. (New York: Harper and Row).

Spieles, D.J., Mitsch, W.J., 2000. The effects of season and hydrologic and chemical loading on nitrate retention in constructed wetlands: A comparison of low- and high nutrient riverine systems, *Ecological Engineering*, 14(1), 77–91.

Spratt, H. G. and Morgan, M.D., 1990. Sulfur cycling in a cedar dominated, freshwater wetland, *Limnology and Oceanography*, 35(7), 1586–1593.

SPSS, 2003. *Analytical Software* (Statistical Package for the Social Sciences (SPSS) Headquarters). Chicago, Illinois, USA.

Srivastava, S., Shrivastava, M., Suprasanna, P. and D'Souza, S.F., 2011. Phytofiltration of arsenic from simulated contaminated water using *Hydrilla verticillata* in field conditions, *Ecological Engineering*, 37(11), 1937–1941.

Stams, A.J.M. and Plugge, C.M., 2009. Electron transfer in syntrophic communities of anaerobic bacteria and archaea, *Nature Reviews Microbiology*, 7(8), 568–577.

Standardization Administration of the People's Republic of China, 2008. Soil Quality Analysis of Total Mercury, Arsenic and Lead Contents in Soils. *Atomic Fluorescence Spectrometry - Part 2: Analysis of Total Arsenic Contents in Soils (GB/T 22105.2–2008)* (Standardization Administration of the People's Republic of China).

Stefanakis, A.I. and Tsihrintzis, V.A., 2012. Heavy metal fate in pilot-scale sludge drying reed beds under various design and operation conditions, *Journal of Hazardous Materials*, 213–214, 393–405.

Stumm, W. and Morgan, J.J., 1996. *Aquatic Chemistry: Chemical Equilibria and Rates in Natural Waters*, 3rd edition (New York: John Wiley & Sons).

Surendran, S. and Wheatley, A., 1998. Grey-water reclamation for non-potable re-use, *Water and Environment Journal*, 12(6), 406–413.

Sutherland, I.W., 2001. Biofilm exopolysaccharides: A strong and sticky framework, *Microbiology*, 147(1), 3–9.

Syafalni, S., Abustan, I., Dahlan, I. and Wah, C.K. 2012. Treatment of dye wastewater using granular activated carbon and zeolite filter, *Modern Applied Science*, 6(2), 37–51.

Tang, X., Huang, S., Scholz, M. and Li, J., 2009. Nutrient removal in pilot-scale constructed wetlands treating eutrophic river water: Assessment of plants, intermittent artificial aeration and polyhedron hollow polypropylene balls, *Water, Air and Soil Pollution*, 197(1–4), 61–73.

Tang, X., Huang, S., Scholz, M. and Li, J., 2011. Nutrient removal in vertical subsurface flow constructed wetlands treating eutrophic river water, *International Journal of Environmental Analytical Chemistry*, 91(7–8), 727–739.

Tanner, C.C., 1996. Plants for constructed wetland treatment systems – A comparison of the growth and nutrient uptake of eight emergent species, *Ecological Engineering*, 7(1), 59–83.

Tanner, C.C., Clayton, J.S. and Upsdell, M.P., 1995. Effect of loading rate and planting on treatment of dairy farm wastewaters in constructed wetland – I. Removal of oxygen demand, suspended solids and faecal coliforms, *Water Research*, 29(1), 17–26.

Tchobanoglous, G., Burton, F.L. and Stensel, H.D., 2003. *Wastewater Engineering: Treatment and Reuse*, 4th edition (New York: Metcalf & Eddy Inc.).

Thirunavukkarasu, O.S., Viraraghavan, T. and Subramanian, K.S., 2003. Arsenic removal from drinking water using granular ferric hydroxide, *Water SA*, 29(2), 161–170.

Thronicker, O. and Szewzyk, U., 2013. Verockerungsanalytik für eine längere Brunnenlebensdauer, *BBR Leitungsbau, Brunnenbau, Geothermie*, 64(10), 70–73 (in German).

Tophof, L., Kreyenschulte, M., Schüttrumpf, H. and Heimbecher, F., 2022. Verockerung wasserbaulicher Filteranlagen: Stand der Wissenschaft und notwendige Untersuchungen, *Grundwasser*, 27, 295–308 (in German).

Trang, N.T.D., Konnerup, D., Schierup, H.H., Chiem, N.H., Tuan, L.A. and Brix, H., 2010. Kinetics of pollutant removal from domestic wastewater in a tropical horizontal subsurface flow constructed wetland system: Effects of hydraulic loading rate, *Ecological Engineering*, 36(4), 527–535.

Travis, M.J., Wiel-Shafran, A., Weisbrod, N., Adar, E. and Gross, A. 2010. Greywater reuse for irrigation: Effect on soil properties, *the Science of the Total Environment*, 408(12), 2501–2508.

Tsihrintzis, V.A. and Gikas, G.D., 2010. Constructed wetlands for wastewater and activated sludge treatment in north Greece: A review, *Water Science and Technology*, 61(10), 2653–2672.

Turner, M.K., 1995. Engineered reed-bed systems for wastewater treatment, *Trends in Biotechnology*, 13(7), 248–252.

U.S. Army Corps of Engineers, 2000. *Wetlands Engineering Handbook - ERDC/EL TR-WRP-RE-21* (United States Army Engineer Research and Development Center Cataloging-in-Publication Data).

U.S. EPA, 1995. *Handbook of Constructed Wetlands: A Guide to Creating Wetlands for Agricultural Wastewater, Domestic Wastewater, Coal Mine Drainage, and Stormwater in the Mid-Atlantic Region. Volume 2: Domestic Wastewater* (Washington, DC: Environment Protection Agency).

U.S. EPA, 2002. *Drinking Water from Household Wells*, EPA 816-K-02-003 (Washington, DC: United States Environmental Protection Agency).

van Beek, C.G.E.M., Breedveld, R.J.M., Juhász-Holterman, M., Oosterhof, A., Stuyfzand, P.J., 2009. Cause and prevention of well bore clogging by particles, *Hydrogeology Journal*, 17, 1877–1886.

van der Valk, A.G. and Davis, C.B., 1978. Primary production of prairie glacial marshes, In: R. E. Good, D. F. Whigham and R. L. Simpson (Eds.) *Freshwater Wetlands: Ecological Processes and Management Potential* (New York: Academic Press) pp. 21–37.

Van Halem, D., Heilman, S.G.L., Amy, G.L. and van Dijk, J.C., 2009. Subsurface arsenic removal for small-scale application in developing countries, *Desalination*, 248(1–3), 241–248.

Verhoeven, J.T.A. and Meuleman, A.F.M., 1999. Wetlands for wastewater treatment: Opportunities and limitations, *Ecological Engineering*, 12(1–2), 5–12.

Vymazal, J., 1999. Nitrogen removal in constructed wetlands with horizontal sub-surface flow – Can we determine the key process? In: J. Vymazal (Ed.) *Nutrient Cycling and Retention in Natural and Constructed Wetlands* (Leiden: Backhuys Publishers) pp. 1–17.

Vymazal, J., 2001. Types of constructed wetlands for wastewater treatment: Their potential for nutrient removal. In: J. Vymazal (Ed.) *Transformations of Nutrients in Natural and Constructed Wetlands* (Leiden: Backhuys Publishers) pp. 1–93.

Vymazal, J., 2007. Removal of nutrients in various types of constructed wetlands, *The Science of the Total Environment*, 380, 48–65.

Vymazal, J. and Krasa, P., 2003. Distribution of Mn, Al, Cu and Zn in a constructed wetland receiving municipal sewage, *Water Science and Technology*, 48(5), 299–305.

Wall, N.A. and Choppin, G.R., 2003. Humic acids coagulation: Influence of divalent cations, *Applied Geochemistry*, 18(10), 1573–1582.

Wang, N. and Mitsch, W.J., 2000. A detailed ecosystem model of phosphorus dynamics in created riparian wetlands, *Ecological Modelling*, 126(2–3), 101–130.

Watson, T.J., Reed, S.C., Kadlec, R.H., Knight, R.L. and Whitehouse, A.E., 1989. Performance expectations and loading rates for constructed wetlands. In: D. A. Hammer (Ed.) *Constructed Wetlands for Wastewater Treatment: Municipal, Industrial and Agricultural* (Chelsea: Lewis Publishers, Inc.) pp. 319–351.

Weihe, U., 2000. Brunnenalterung – Vergleich unterschiedlicher Regenerierverfahren an einem Förderbrunnen im norddeutschen Raum. *BBR, Wasser und Rohrbau*, 51(10), 43–49 (in German).

Wieβer, A., Kuschk, P. and Stotmeister, U. 2002. Oxygen release by roots of *Typha latifolia* and *Juncus effusus* in laboratory hydroponic systems, *Acta Biotechnologica*, 22(1–2), 209–216.

Winward, G.P., Avery, L.M., Frazer-Williams, R., Pidou, M., Jeffrey, P., Stephenson, T. and Jefferson, B., 2008. A study of the microbial quality of grey water and an evaluation of treatment technologies for reuse, *Ecological Engineering*, 32(2), 187–197.

Wood, T.S. and Shelley, M.L., 1999. A dynamic model of bioavailability of metals in constructed wetland sediments, *Ecological Engineering*, 12(3–4), 231–252.

World Health Organization, 2004. Guidelines for Drinking Water Quality – Recommendations (Geneva: World Health Organization).

World Health Organization, 2006. Overview of greywater management health considerations (Amman: Regional Office for the Eastern Mediterranean Centre for Environmental Health Activities). https://iris.who.int/handle/10665/116516

Wrubleski, D.A., Murkin, H.R., van der Valk, A.G. and Nelson, J.W., 1997. Decomposition of emergent macrophyte roots and rhizomes in a northern prairie marsh, *Aquatic Botany*, 58(2), 121–134.

Wu, M., Li, Q., Tang, X., Huang, Z., Lin, L. and Scholz, M., 2014. Arsenic(V) removal in wetland filters treating drinking water with different substrates and plants, *International Journal of Environmental Analytical Chemistry*, 94(6), 618–638.

Ye, Z.H., Lin, Z.Q., Whiting, S.N., De Souza, M.P. and Terry, N., 2003. Possible use of constructed wetland to remove selenocyanate, arsenic, and boron from electric utility wastewater, *Chemosphere*, 52(9), 1571–1579.

Zhang, D., Gersberg, R.M. and Keat, T.S., 2009. Constructed wetlands in China, *Ecological Engineering*, 35(10), 1367–1378.

Zhang, G., Ren, Z., Zhang, X. and Chen., J., 2013. Nanostructured iron(III)-copper(II) binary oxide: A novel adsorbent for enhanced arsenic removal from aqueous solutions, *Water Research*, 47(12), 4022–4031.

Zhao, Y.Q., Sun, G. and Allen, S.J., 2004. Anti-sized reed bed system for animal wastewater treatment: A comparative study, *Water Research*, 38(12), 2907–2917.

Zhu, S.N., Wang, C., Yip, A.C.K. and Tsang, D.C.W., 2015. Highly effective degradation of sodium dodecylbenzene sulphonate and synthetic greywater by Fenton-like reaction over zerovalent iron-based catalyst, *Environment Technology*, 36(11), 1423–1432.

Zurita, F., Toro-Sánchez, C.D., Gutierrez-Lomelí, M., Rodriguez-Sahagún, A., Castellanos-Hernandez, O.A., Ramírez-Martínez, G. and White, J.R., 2012. Preliminary study on the potential of arsenic removal by subsurface flow constructed mesocosms, *Ecological Engineering*, 47, 101–104.

5 Industrial Wastewater Treatment and Modelling

5.1 NUTRIENT AND HYDROCARBON REMOVAL WITHIN CONSTRUCTED WETLANDS

5.1.1 INTRODUCTION

Constructed wetlands (CWs) are engineered systems designed and constructed to utilize the natural processes involving wetland vegetation, soils and their associated microbial assemblages to assist in treating wastewater (Scholz, 2006; Vymazal, 2007). These systems take advantage of physical, chemical and biological processes occurring in natural wetlands, but do so within a semi-controlled environment (Hammer and Bastian, 1989). Previous literature claims that organic matter represented by chemical oxygen demand (COD) and biochemical oxygen demand (BOD), suspended solids, nitrogen, phosphorus and bacteria can be treated well by wetland systems (Sakadevan and Bavor, 1998; Scholz, 2006; Scholz et al., 2002; Vymazal, 1999; Wanda et al., 2007).

CWs were successfully studied for water purification purposes in the 1960s (Seidel, 1966); *Scirpus lacustris* L. (common bulrush) removed organic substances such as nutrients and even toxic chemicals such as phenols. After then, an increasing number of CWs were applied for the treatment of municipal, industrial and agriculture wastewater (Braskerud, 2002; Fraser et al., 2004; Scholz et al., 2002). A better understanding of processes, such as adsorption, uptake by plants and living organisms, biodegradation and transformation and burial, led to the construction of more CWs used to remove organic mfatter, nitrogen and phosphorus from eutrophic river and lake water (Li et al., 2008; Vymazal, 2007). Good treatment performances in terms of petroleum hydrocarbon removal (typically more than 60% reduction of the total input) by CWs were recorded previously (Eke and Scholz, 2008; Keefe et al., 2004; Knight et al., 1999; Salmon et al., 1998).

Petroleum hydrocarbon wastewaters also contain pollutants such as nitrogen and phosphorus and may have substantial COD and BOD values (Knight et al., 1999). However, the major focus of the petroleum industry is on assessing the removal efficiency of hydrocarbons. The potential toxicity of hydrocarbon pollutants may restrain or kill microorganisms living in the wetland plant root zone and filter media (Knight et al., 1999). Nevertheless, COD and even BOD removal efficiencies for wetlands treating toxic hydrocarbons are comparable to wetlands treating other types of wastewaters (Ji et al., 2007; Knight et al., 1999). Nitrogen and phosphorus, which are essential nutrients for successful biodegradation of hydrocarbon pollutants (Cooney, 1984), are better removed in wetlands treating petroleum industry wastewater in comparison to wetlands treating other wastewater categories (APHA, 1998; Knight et al., 1999).

DOI: 10.1201/9781003509370-5

The role of aquatic plants in pollutant removal by CWs has been discussed in detail, previously (Fraser et al., 2004; Li et al., 2008; Scholz, 2006; Scholz et al., 2002). There is sufficient evidence to demonstrate the importance of wetland plants in removing nutrients (Vymazal, 2007). However, previous studies indicated that the adverse effects of petroleum hydrocarbon on plants ranged from short-term reductions in photosynthesis to mortality (Pezeshki et al., 2000; Scrimshaw and Lester, 1996). A recent research study, which assessed the plant growth performance in CWs treating heavy oil-produced water, indicates that the heights of planted reeds reduce with an increase in hydrocarbon concentration (Ji et al., 2007). In summary, previous research paid attention to the harmful effect of hydrocarbons on microbial and plant biomass. However, the role of plants including macrophytes concerning nutrient removal has been neglected.

Another major research area is focusing on the assessment of the uncertainty of temperature and humidity on nutrient removal in CWs treating hydrocarbons. Previous research indicates that organic matter and nitrogen removal predominantly depends on the temperature-sensitive microbial activity within the root zone (Ouellet-Plamondon et al., 2006). In comparison, phosphorus removal in different wetland systems is only indirectly affected by the temperature-sensitive oxygen availability, which influences the redox potential impacting on the phosphorus availability (Bachand and Horne, 2000; Kadlec and Knight, 1996; Tsihrintzis et al., 2007; Wittgren and Maehlum, 1997). In wetlands treating hydrocarbons, however, the effect of temperature on nutrient removal is not as pronounced as those in other areas of wastewater treatment. Therefore, further research is necessary to better predict the impact of environmental boundary conditions on nutrient removal in hydrocarbon treatment wetlands.

This study aims to assess nutrient removal as a function of benzene supply within treatment wetlands. In light of the above considerations revealed by the literature study, research was conducted subject to the following objectives: (a) to assess the potential of low-molecular-weight hydrocarbon removal, using benzene as a representative example, in vertical-flow CWs; (b) to compare the nutrient removal performances within vertical-flow CWs, which were different in terms of contamination, plant development and temperature, and (c) to statistically examine the key and interactive effects of zero and high benzene loading, presence or absence of *P. australis* (Cav.) Trin. ex Steud and natural and controlled temperature conditions on ammonia-nitrogen, nitrate-nitrogen and ortho-phosphate-phosphorus concentrations. Section 5.1 is based on a revised version of an original article by Tang et al. (2010).

5.1.2 Materials and Methods

5.1.2.1 Wetland Design and Temperature Control

From April 2005 to October 2007, 12 analogous vertical-flow CWs (Eke and Scholz, 2008) were designed, constructed and operated predominantly to assess nutrient removal as a function of benzene supply. All experimental wetlands were located at The King's Buildings campus at The University of Edinburgh (Scotland, UK) and constructed with polyethylene columns of 100 mm in diameter and 750 mm in height. The outlet valves were located at the centre of the bottom plate of each wetland column and were connected to 12-mm internal diameter vinyl tubing. This arrangement was used for manually adjusting the flow, and collecting outflow sample water at the same time. Furthermore, ventilation pipes with an internal diameter of 13 mm reaching down to 10 mm above the bottom of each wetland were installed to encourage passive aeration.

The 12 experimental CWs were grouped into two groups of six wetlands each: one group was located in a temperature, light and humidity controlled indoor room, while the other group was operated at natural outdoor conditions. The former group allows for the generation of a data set suitable for modelling because data variability is low, considering that weather changes do not impact on pollutant removal. In comparison, the outdoor experiment represents better a real industrial application in temperate climate (e.g., Edinburgh (1971–2000 means; http://www.metoffice.gov.uk/climate/uk/averages/19712000/sites/edinburgh.html): relatively few sunshine hours (mean of 1,406 hours per annum), low precipitation (mean of 676 mm) and mild temperature (mean monthly minimum of 5.1°C and mean monthly maximum of 12.2°C)) despite the relatively small column sizes, which are, however, not uncommon in the scientific literature (Eke and Scholz, 2008; Scholz, 2006; Scholz et al., 2002).

The packing orders of the six indoor and six outdoor wetlands, which were operated in the same way, were the same. Wetlands 1, 2, 3 and 4 were filled with the following five successive layers of aggregates to a depth of 600 mm: stones (37.5–75 mm), large gravel (10–20 mm), medium gravel (5–10 mm), small gravel (1.2–5 mm) and sand (0.6–1.2 mm). The corresponding layer thicknesses were 10, 15, 10, 15 and 10 cm, respectively. Detailed information is outlined by a previous study summarizing findings related to a proof-of-concept experiment (Eke and Scholz, 2008). Wetlands 5 and 6 were designed as controls and not filled with gravel and sand. The role of *P. australis* in removing benzene, its degradation products and nutrients was also assessed. CWs 1 and 2 were planted with nine individual plants of similar biomass and equal strength. *P. australis* was obtained from a local supplier (Alba Trees Public, East Lothian, UK).

The role of benzene supply and temperature on nutrient removal within treatment wetlands was assessed with the chosen system set-up (see above). Concerning the indoor wetlands, a Denco Local Environmental Control Unit supplied by Denco Limited (East Kilbride, Scotland, UK) was used to control the constant temperature of 15°C and humidity of 60%. The simulation of day and night cycles was conducted with the help of three plant growth lights (Sylvania 15,000 h, 36 W, 1,200 mm, T8 Grolux Fluorescent Tube; supplied by Lyco Direct Limited (Bletchely, Milton Keynes, England, UK). In comparison, no control of any environmental boundary conditions was attempted for the corresponding outdoor wetlands.

5.1.2.2 Influent and Nutrient Supply

Of all petroleum hydrocarbons, benzene is the most problematic one due to its high toxicity and relatively high water solubility (Johnson et al., 2003), which was addressed by thorough mixing of the inflow water for several minutes before introduction to the wetland rigs. Benzene (BDH analytical reagent, C_6H_6 (99.7%), supplied by VWR International Limited (Hunter Boulevard, Lutterworth, England, United Kingdom) was chosen as a representative target pollutant to assess the removal of low-molecular-weight hydrocarbons. Two types of influents were used in a batch flow mode: tap water and tap water spiked with dissolved benzene. Twice per week, wetlands 2, 4 and 6 received tap water and nutrients (see below), while wetlands 1, 3 and 5 received tap water contaminated with 1,000 mg/L benzene and nutrients (see below). The selected benzene concentration compares well with those concentrations quoted by the petroleum industry for produced waters (Cooney, 19984; Eke and Scholz, 2008; Ji et al., 2007; Keefe et al., 2004; Knight et al., 1999; Salmon et al., 1998).

The well-balanced slow-releasing nitrogen (24%), phosphorus (8%) and potassium (14%) Miracle-Gro fertilizer (formerly Osmocote, produced by Scot Europe B. V., The Netherlands) was used as the predominant source of nutrient supply to enhance plant and microbial growth, and to improve the benzene treatment efficiency of the wetland systems. Approximately 8 g of the fertilizer was added directly to all wetlands every two weeks until 29 May 2006, when the concentration was increased to 30 g to assess the increase of nutrient concentrations on hydrocarbon removal. From 26 June 2006, the amount was lowered to 15 g, which was seen as a more realistic nutrient concentration for optimum benzene removal.

5.1.2.3 Sampling and Analysis

Since April 2005, all indoor and outdoor wetlands were fully saturated and flooded to a depth of 10 cm above the top level of the packing media. The wetlands were filled with water (i.e. inflow) directly after being fully drained manually within a short period of less than 10 minutes twice per week. A cycle of filling and emptying of each wetland was performed twice per week. Water samples were taken for benzene analysis once per month until January 2007, and then twice per month afterwards as the systems fully matured.

During the entire experimental period, water samples were collected and analysed twice per week for five days at 20°C nitrogen-allythiourea BOD, COD, ammonia-nitrogen, nitrate-nitrogen, ortho-phosphorus-phosphate, pH, dissolved oxygen (DO) and redox potential. Benzene was determined with a Perkin Elmer gas chromatograph FID Model 9700 and a corresponding headspace sampler HS-101 (Beaconsfield, England, UK). American standard methods (APHA, 1998) were used for all analytical work unless stated otherwise.

All statistical tests were performed using the software Statistical Package for the Social Sciences (SPSS, 2003). In all cases, significance was defined as $p < 0.05$, if not stated otherwise. One-way analyses of variance (ANOVA) and Duncan's multiple range tests (Li et al., 2008) were carried out to assess the differences between means of benzene removal efficiency and effluent nutrient concentrations in different CWs. Three-way ANOVA were applied to examine the influences of benzene concentration, plant presence and environmental boundary condition control and their interactions with each other on the nutrient removal efficiencies in the tested wetlands. For all ANOVA, it was checked that the tested variables were normally distributed. Otherwise, the variables were $\log_{10}$-transformed, which was the most suitable transformation function to bring the variance closer to the mean (García et al., 2004).

5.1.3 Results

5.1.3.1 Benzene Removal

The benzene removal for selected indoor and outdoor wetlands, which received 1,000 mg/L benzene in the inflow water, is summarized in Figure 5.1. The mean reductions for the indoor wetlands were approximately 90% (equates to 900 mg/L or 49 g m^2/d), while much lower removal efficiencies of between roughly 72% and 80% (associated to between 720 and 800 mg/L or between 39 and 43 g m^2/d) were observed for the outdoor CWs (Figure 5.1).

The benzene distribution curves overlapped for the indoor wetlands. The outdoor wetlands 1 and 3 had similar distribution patterns (Figure 5.2). Furthermore, between 20% and 40% of the total benzene effluent concentration values were approximately 0 mg/L for both the indoor and outdoor wetlands (Figure 5.2), which indicates that approximately 100% removal of benzene can be achieved at a probability of between 20% and 40%.

5.1.3.2 Organic Matter and Nutrient Removal

Effluent concentrations of organic matter in terms of BOD and COD are shown in Figure 5.3. Mean effluent organic matter concentrations in benzene treatment wetlands were high and varied between 34.0 and 43.8 mg/L for BOD, and between 199.1 and 453.7 mg/L for COD. A three-way ANOVA indicated that benzene treatment is independent of the presence of *P. australis* (Table 5.1). The control of environmental boundary conditions was significant for BOD removal at $p = 0.011$. Furthermore, the relationship between high benzene concentrations and the control

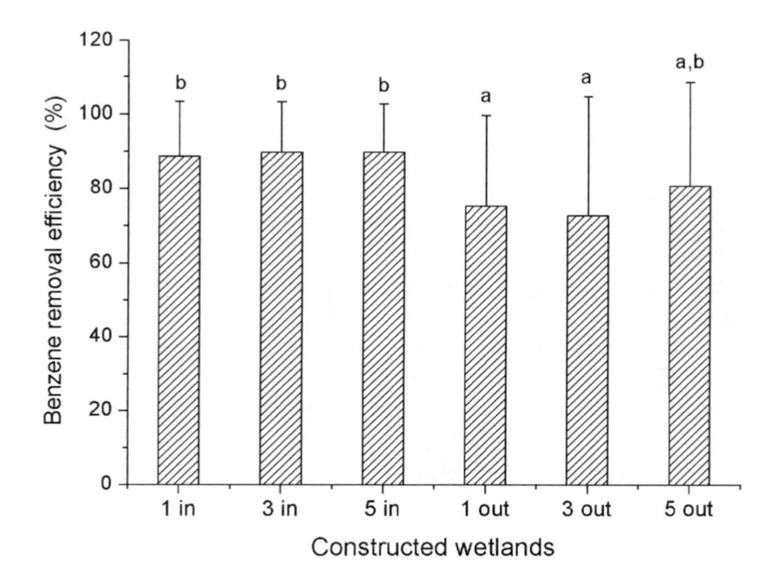

FIGURE 5.1 Mean benzene removal efficiency for experimental constructed wetlands spiked with benzene. Standard deviation bars sharing different letters are significantly different from each other at $p \leq 0.05$ according to Duncan's multiple range tests.

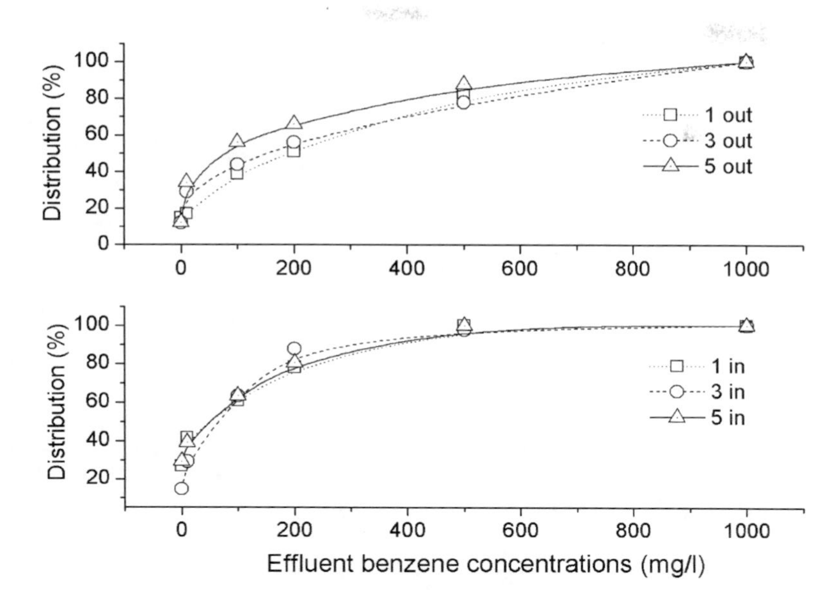

FIGURE 5.2 Distribution of the effluent benzene concentration proportions for outdoor and indoor experimental constructed wetlands spiked with benzene.

of environmental boundary conditions on BOD removal was significant at $p = 0.043$ (Tables 5.1 and 5.2).

The mean effluent nitrogen concentrations were between 26.8 and 48.2 mg/L for ammonia-nitrogen and between 19.1 and 53.8 mg/L for nitrate-nitrogen (Figure 5.4). There was no significant difference in nitrogen removal based on findings from the one-way ANOVA and Duncan's multiply range tests (Figure 5.4). Results obtained by the three-way ANOVA showed no significant effect of benzene, plant presence and environmental boundary condition control as well as their interactions with each other on both ammonia-nitrogen and nitrate-nitrogen removal (Tables 5.3 and 4).

Except for high mean effluent ortho-phosphate-phosphorus concentrations of 23.7 mg/L obtained for the outdoor wetland 2, all other wetlands had similar mean effluent phosphorus concentration ranges between 14.4 and 17.9 mg/L (Figure 5.5). Results from the three-way ANOVA (Table 5.5) indicated that benzene and plant presence, and environmental boundary condition control as well as their interactions with each other had no significant effect on ortho-phosphate-phosphorus removal.

5.1.3.3 Other Water Quality Variables

Mean values of effluent DO, pH and redox potential differed significantly ($p < 0.05$) among indoor and outdoor wetlands (Table 5.6). Multiply comparisons detected significantly higher DO concentrations in outdoor wetlands. Those wetlands treating benzene had significantly lower DO concentrations compared with non-benzene treatment wetland controls (Table 5.6). However, the DO values were obtained from samples taken from the wetland columns and were not directly measured within the

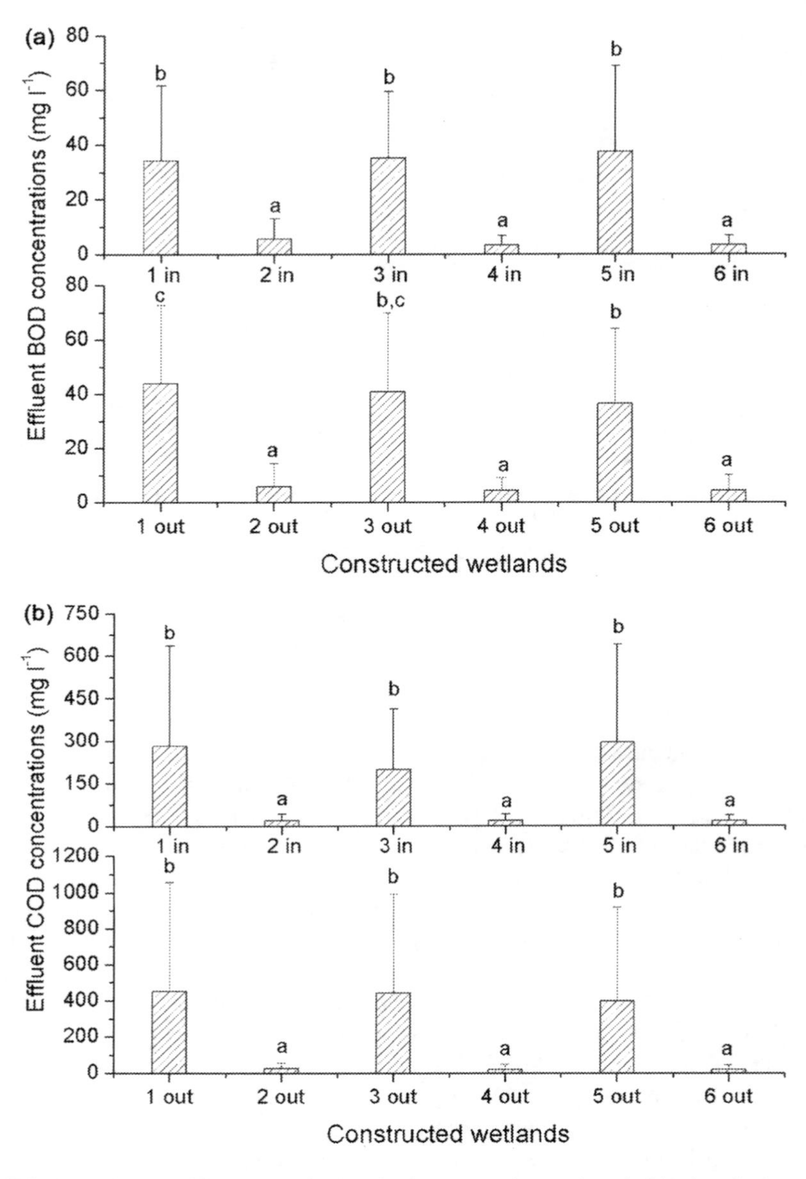

FIGURE 5.3 Mean effluent (a) biochemical oxygen demand and (b) chemical oxygen demand (COD) concentrations for all experimental constructed wetlands. Standard deviation bars sharing different letters are significantly different from each other at $p \leq 0.05$ according to Duncan's multiple range tests.

columns. It is therefore likely that all DO concentrations provided in Table 5.6 are slightly higher due to contact with the atmosphere than the ones that would be present within the actual wetlands.

TABLE 5.1
Results of Three-Way Analyses of Variance Examining the Role of *Phragmites australis* (Cav.) Trin. ex Steud (common reed) Presence (plant), Environmental Boundary Condition (temperature, light and humidity) Control (ENV) and Benzene Concentration on the Effluent Biochemical Oxygen Demand

Source	Sum of Squares	df	F-ratio	*p*
Plant	196.049	1	0.893	0.345
ENV	1,458.880	1	6.649	0.011
Benzene	53,411.130	1	243.422	<0.001
Plant and ENV	11.541	1	0.053	0.819
Plant and benzene	62.501	1	0.285	0.594
ENV and benzene	907.864	1	4.138	0.043
Plant, ENV and benzene	96.251	1	0.439	0.508
Error	52,660.339	240	n/a	n/a

df, degree of freedom; F-ratio, test statistic used to decide whether the sample means are within the sampling variability of each other; *p*, *p*-value in the analysis of variance table, which gives an overall confidence for the fit to be rejected; n/a, not applicable.

TABLE 5.2
Results of Three-Way Analyses of Variance Examining the Role of *Phragmites australis* (Cav.) Trin. ex Steud (common reed) Presence (plant), Environmental Boundary Condition (Temperature, Light and Humidity) Control (ENV) and Benzene Concentration on the Effluent Chemical Oxygen Demand

Source	Sum of Squares	*df*	*F*-ratio	*p*
Plant	0.065	1	0.065	0.694
ENV	0.888	1	0.888	0.145
Benzene	44.625	1	44.625	<0.001
Plant and ENV	0.003	1	0.003	0.937
Plant and benzene	0.068	1	0.068	0.685
ENV and benzene	0.126	1	0.126	0.581
Plant, ENV and benzene	0.053	1	0.053	0.720
Error	99.560	240	n/a	n/a

df, degree of freedom; F-ratio, test statistic used to decide whether the sample means are within the sampling variability of each other; *p*, *p*-value in the analysis of variance table, which gives an overall confidence for the fit to be rejected; n/a, not applicable.

The pH values were significantly higher in wetlands 1, 3 and 5 than those in wetlands 2, 4 and 6. Higher values for the redox potential occurred in outdoor non-benzene treatment wetlands.

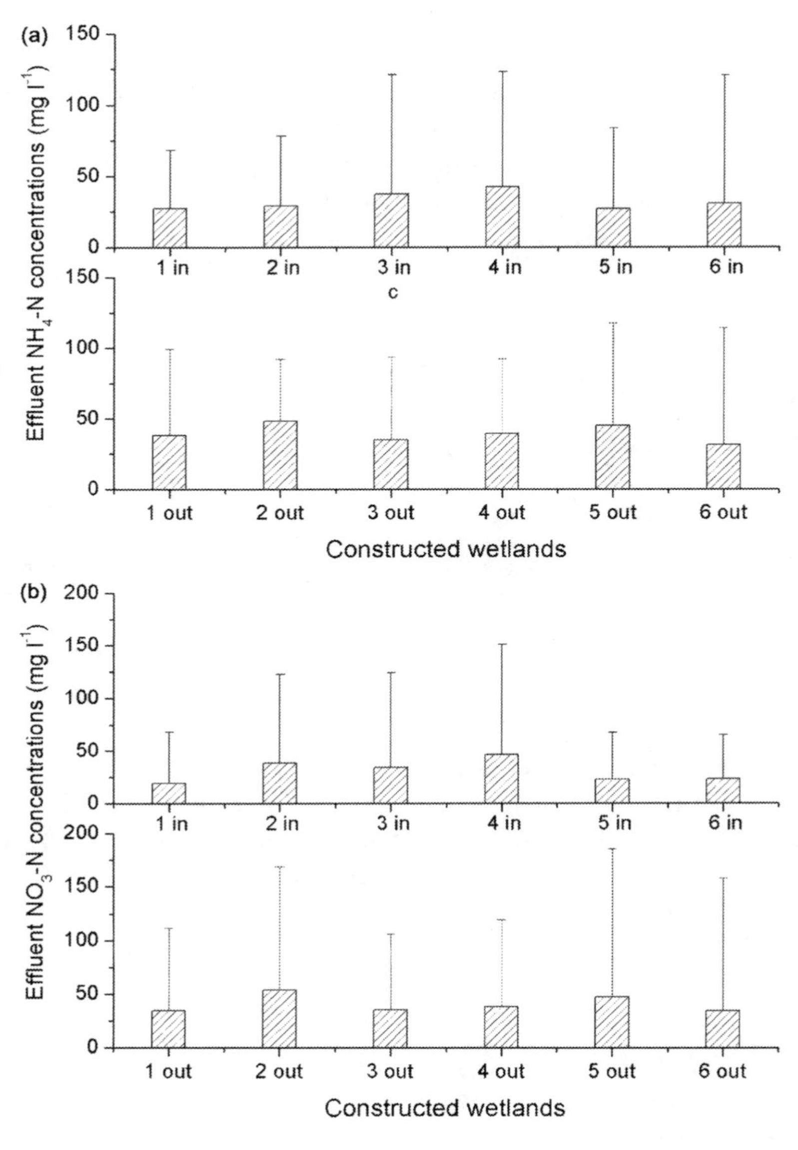

FIGURE 5.4 Mean effluent (a) ammonia-nitrogen (NH_4-N) and (b) nitrate-nitrogen (NO_3-N) concentrations for all experimental constructed wetlands.

5.1.4 Discussion

The results obtained in this study indicated that vertical-flow CWs could remove relatively high benzene concentrations effectively. However, the concentration-based removal efficiencies between 72% and 90% were lower than the range between 90% and 93% reported elsewhere (Ji et al., 2007; Salmon et al., 1998). This might be due

TABLE 5.3
Results of Three-Way Analyses of Variance Examining the Role of *Phragmites australis* (Cav.) Trin. ex Steud (common reed) Presence (plant), Environmental Boundary Condition (Temperature, Light and Humidity) Control (ENV) and Benzene Concentration on the Effluent Ammonia-Nitrogen

Source	Sum of Squares	df	F-ratio	*p*
Plant	0.142	1	0.458	0.499
ENV	0.002	1	0.006	0.936
Benzene	0.143	1	0.460	0.498
Plant and ENV	0.010	1	0.031	0.860
Plant and benzene	0.061	1	0.198	0.657
ENV and benzene	0.036	1	0.115	0.735
Plant, ENV and benzene	0.003	1	0.011	0.918
Error	57.187	240	n/a	n/a

df, degree of freedom; F-ratio, test statistic used to decide whether the sample means are within the sampling variability of each other; *p*, *p*-value in the analysis of variance table, which gives an overall confidence for the fit to be rejected; n/a, not applicable.

TABLE 5.4
Results of Three-Way Analyses of Variance Examining the Role of *Phragmites australis* (Cav.) Trin. ex Steud (common reed) Presence (plant), Environmental Boundary Condition (Temperature, Light and Humidity) Control (ENV) and Benzene Concentration on the Effluent Nitrate-Nitrogen

Source	Sum of Squares	*df*	*F*-ratio	*p*
Plant	0.623	1	1.013	0.316
ENV	0.127	1	0.207	0.650
Benzene	0.195	1	0.317	0.574
Plant and ENV	0.323	1	0.525	0.470
Plant and benzene	0.102	1	0.166	0.316
ENV and benzene	0.006	1	0.010	0.920
Plant, ENV and benzene	0.005	1	0.008	0.927
Error	113.213	240	n/a	n/a

df, degree of freedom; F-ratio, test statistic used to decide whether the sample means are within the sampling variability of each other; *p*, *p*-value in the analysis of variance table, which gives an overall confidence for the fit to be rejected; n/a, not applicable.

to the relatively high influent benzene concentration of 1,000 mg/L in this evaluation compared with ranges between 20 and 60 mg/L reported in the referenced literature sources. Nevertheless, a direct comparison is not possible because information on loading rates, climatic conditions and operational details are partly missing.

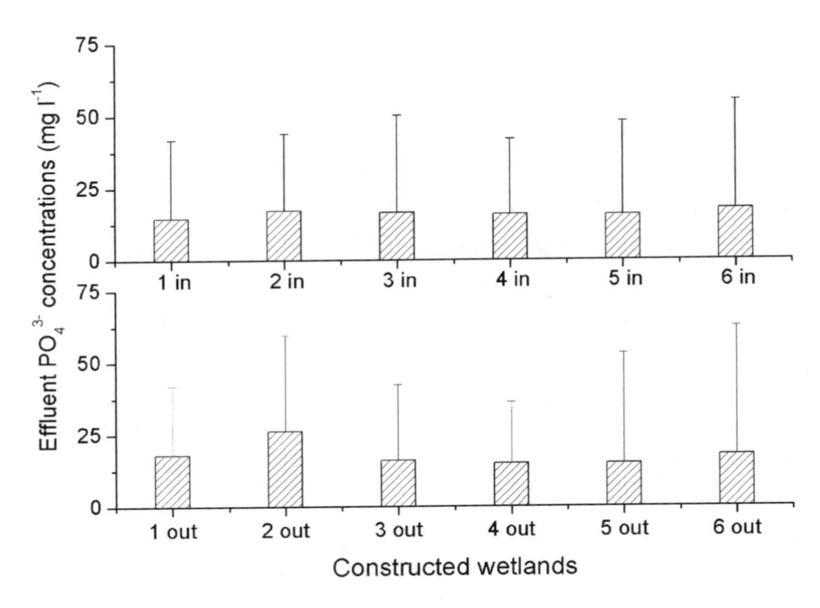

FIGURE 5.5 Mean effluent ortho-phosphate-phosphorus (PO_4^{3-}-P) concentrations of all experimental constructed wetlands.

TABLE 5.5
Results of Three-Way Analyses of Variance Examining the Role of *Phragmites australis* (Cav.) Trin. ex Steud (Common Reed) Presence (Plant), Environmental Boundary Condition (Temperature, Light and Humidity) Control (ENV) and Benzene Concentration on the Effluent Ortho-Phosphate-Phosphorus

Source	Sum of Squares	*df*	*F*-ratio	*p*
Plant	0.446	1	3.653	0.058
ENV	0.039	1	0.322	0.571
Benzene	0.018	1	0.146	0.703
Plant and ENV	0.072	1	0.593	0.442
Plant and benzene	0.052	1	0.424	0.516
ENV and benzene	<0.001	1	0.000	0.985
Plant, ENV and benzene	0.008	1	0.069	0.794
Error	22.459	240	n/a	n/a

df, degree of freedom; F-ratio, test statistic used to decide whether the sample means are within the sampling variability of each other; *p*, *p*-value in the analysis of variance table, which gives an overall confidence for the fit to be rejected; n/a, not applicable.

Results obtained from multiple comparisons implied that the control of temperature significantly influenced hydrocarbon removal (Figure 5.1). Similar observations were also reported elsewhere. For example, the hydrocarbon treatment efficiency is significantly lower in winter than in spring and summer (Salmon et al., 1998).

TABLE 5.6

Mean ± Standard Deviation for Effluent Dissolved Oxygen (DO), pH and Redox Potential (Redox) Concerning the Indoor (in) and Outdoor (out) Experimental Constructed Wetlands

Variable	1 in	2 in	3 in	4 in	5 in	6 in
DO (mg/L)	$2.7^{a} \pm 1.64$	$3.1^{b} \pm 1.78$	$2.7^{a} \pm 1.64$	$3.3^{b} \pm 1.54$	$3.2^{b} \pm 1.39$	$4.9^{f} \pm 1.57$
pH (-)	$6.7^{d} \pm 0.43$	$6.0^{b} \pm 0.57$	$6.4^{c} \pm 0.49$	$5.6^{a} \pm 1.02$	$6.4^{c} \pm 0.58$	$6.4^{c} \pm 0.59$
Redox (mV)	$165.4^{b,c} \pm 58.59$	$166.3^{b,c} \pm 52.08$	$143.0^{a} \pm 56.21$	$177.1^{c,d} \pm 55.72$	$164.6^{b} \pm 53.92$	$175.8^{b,c,d} \pm 51.92$
Variable	1 out	2 out	3 out	4 out	5 out	6 out
DO (mg/L)	$3.9^{c} \pm 2.00$	$4.5^{d,e} \pm 2.03$	$4.3^{c,d} \pm 1.80$	$4.7^{e,f} \pm 2.01$	$6.2^{g} \pm 2.44$	$6.3^{g} \pm 2.36$
pH (-)	$6.7^{d} \pm 0.42$	$5.6^{a} \pm 1.18$	$6.7^{d} \pm 0.43$	$6.0^{b} \pm 0.85$	$6.8^{d} \pm 0.47$	$6.6^{d} \pm 0.65$
Redox (mV)	$172.3^{b,c,d} \pm 59.60$	$191.1^{e} \pm 51.58$	$168.6^{b,c} \pm 45.87$	$181.3^{d,e} \pm 48.35$	$163.5^{b} \pm 51.48$	$164.5^{b} \pm 53.20$

Means marked with different letters are significantly different from each other at $p \leq 0.05$ according to the Duncan's multiple range test. The sample number for all variables and filters was 193.

Approximately 15°C is the lower temperature threshold for effective organic compound removal (Scholz, 2006).

This study determined the factors affecting nutrient removal in benzene treatment wetlands. The application of a slow-releasing fertilizer (see above) contributed to detectable small additional amounts of effluent BOD and COD in wetlands 2, 4 and 6. In contrast, benzene residual and its biodegradation products were the most important source of carbon contributing to high concentrations of organic matter in the treatment wetlands (Figure 5.2), considering that the tap water was virtually free of BOD.

The effluent DO and BOD concentrations in benzene treatment wetlands were significantly lower and higher, respectively, than those in non-benzene treatment wetland controls (Figure 5.1). Consumption of DO and a corresponding increase in BOD effluent concentrations compared to non-benzene treatment wetlands indicated indirectly biodegradation of benzene. This finding was confirmed by the observation made previously (Salmon et al., 1998), showing that biodegradation was responsible for nearly 80% of hydrocarbon reduction, if the corresponding influent concentration was <100 mg/L.

Unlike organic matter removal, the effect of benzene treatment on nitrogen and phosphorus removal was not significant in the wetlands governed by anaerobic and aerobic processes (Tables 5.6 and 5.7). Aerobic conditions are likely to occur in the top water zone, and during and directly after draining of the wetlands. This finding is in good agreement with previous studies (Knight et al., 1999), but does conflict with biodegradation research (Hu et al., 2007) showing that nitrate-nitrogen was a favorable electron acceptor for benzene reduction.

An inhibitory effect of nitrate on hydrocarbon degradation might have occurred. Adequate fertilizer concentrations increase the biodegradation rate, whereas excessive fertilization has often a negative effect (Kadlec and Knight, 1996; Scholz, 2006). However, the nitrogen compounds within the wetland columns cannot be qualified and quantified because they are directly related to the fertilizer used. The release rate of nitrogen from the fertilizer is also unknown because it is a function of the specific environmental conditions within each column.

The relative unimportance of benzene on nitrogen and phosphorus removal in this study could mean the following: the effective removal of hydrocarbon might reduce its adverse effect on nutrient removal. Furthermore, negligible nitrogen and phosphorus concentrations were consumed during the process of benzene biodegradation.

The role of aquatic plants and their effect on nutrient removal in hydrocarbon treatment wetlands need to be discussed. One-way and three-way ANOVA found that *P. australis* had no significant effect on organic matter removal (Figure 5.2; Tables 5.1 and 5.2). Similar removal figures for BOD and COD were reported for planted and unplanted wetlands treating other wastewater types (Ciria et al., 2005; Hench et al., 2003).

A statistical analysis indicated that *P. australis* had no great importance on nitrogen and phosphorus removal in benzene treatment wetlands (Figures 5.4 and 5.5; Tables 5.3–5.5). This observation was in agreement with other findings (Brix, 1997), but in contrast to results reported elsewhere (Ciria et al., 2005; Vymazal, 2007), which indicated that wetland plants played a significant role in nitrogen and phosphorus removal. Studies verified that nutrient uptake by plants is only significant under low nutrient loading conditions (Brix, 1997). Plant biomass accumulation can account between 5.1% and 26.2% of total nitrogen, and between 40.5% and 80.9% of the total phosphorus removal with influent mean concentrations of 4.8 mg/L for total nitrogen and 0.2 mg/L for total phosphorus (Li et al., 2008).

The nutrient uptake efficiency for both aboveground and belowground tissues decreases with greater nutrient availability (Shaver and Melillo, 1984). In this study, high nitrogen and phosphorus loadings (Figures 5.4 and 5.5) were responsible for the relative unimportance of the role of plants played in nutrient removal. It is likely that the higher the inflow load the lower percentage of nutrients can be found within the mature plant biomass.

Environmental boundary condition control was unimportant for COD removal (Figure 5.3). The removal of COD within CWs can occur via aerobic and anaerobic biological processes, as well as by a variety of physical processes including adsorption and filtration (Scholz, 2006). No significant ($p=0.145$) environmental (temperature, light and humidity) sensitivity might imply that physical rather than biological processes were responsible for COD removal in benzene treatment wetlands. However, statistical analyses indicated significant effects of environmental boundary condition control on BOD removal (Figure 5.3).

Microbial degradation plays the dominant role in BOD removal for benzene treatment wetlands and other types of wastewater treatment systems (Ciria et al., 2005; Lim et al., 2001; Scholz, 2006). Reducing the variability of temperature, light and humidity resulted in improved BOD treatment performances in this study (Figure 5.2).

Conflicting opinions concerning the temperature dependency of BOD removal within treatment wetlands can be found in literature (Scholz, 2006). The general temperature dependency of BOD with respect to various biological treatment processes has been reported elsewhere (Metcalf & Eddy Inc., 1996). Similar temperature-dependent relationships have also been outlined for wetlands (Ouellet-Plamondon et al., 2006; Reed and Brown, 1995). In contrast, several other studies suggested the negligible effect of temperature on BOD removal in wetlands (Gerke et al., 2001; Kadlec and Knight, 1996; Knight et al., 1999).

The redox potential has been closely linked to light conditions, and sufficient light leads to higher corresponding redox values (Wießner et al., 2005), because a higher photosynthetic rate leads to more oxygen being transported by the macrophytes into the rhizosphere. Compared to the indoor wetlands, higher values for the redox potential in the outdoor wetlands (Table 5.6 indicated that wetlands operated in natural environments experience much better light conditions than those operated under plant growth lights. Lower effluent BOD concentrations and redox potential values were recorded for indoor benzene treatment wetlands, especially for wetlands 1 and 3, if compared to the corresponding outdoor wetlands (Figure 5.2 and Table 5.6). Low effluent BOD concentrations correlated well with low redox potential values. This confirmed previous findings showing that high removal efficiencies for BOD are related to less reducing conditions (García et al., 2004).

For nitrogen, results from a three-way ANOVA indicated no significant dependence on environmental boundary condition control (Tables 5.4 and 5.5). Nevertheless, many researchers (Bachand and Horne, 2000; Vymazal, 2007; Wießner et al., 2005) reported on the sensitivity of environmental boundary condition control with respect to nitrogen removal within CWs. Processes such as ammonification, nitrification and denitrification are characterized as temperature-dependent in wetlands

(Werker et al., 2002). Nitrification rates are reported to become inhibited at water temperatures of around 10°C, and rates drop rapidly to zero below approximately 6°C (Herskowitz et al., 1987). Denitrification processes have been observed at temperatures as low as 5°C (Brodrick et al., 1988). Compared to the temperature dependency, however, previous studies indicate that oxygen availability is the main variable regulating the processes of nitrogen removal (Bezbaruah and Zhang, 2003; Werker et al., 2002). Low and insufficient oxygen transport by diffusion and by plants through their aerenchyma (airy tissue allowing exchange of gases between shoots and roots) may result in the absence of environmental boundary condition dependency for nitrogen removal according to Bezbaruah and Zhang (2003); see also Table 5.6.

No significant effect of environmental boundary condition control on phosphorus removal was noted (see also above). These findings disagree with those observations made elsewhere (Bachand and Horne, 2000; Kadlec and Knight, 1996), indicating that phosphorus removal was indirectly affected by temperature through its effects on redox potential levels. Under high phosphorus loading conditions (Figure 5.5), similar redox potential ranges (143.0–177.1 mV and 163.5–191.1 mV for indoor and outdoor wetlands, respectively) could be responsible for the lack of relationship between environmental boundary condition control and phosphorus removal (Luo et al., 2002).

The interactions between benzene, *P. australis* and environmental boundary conditions on nutrient removal were also assessed in this study. The combination between high benzene concentrations and the presence of environmental boundary condition control had only a significant effect on BOD removal. Previous work by Knight et al. (1999) showed that BOD removal ranges are between 48% and 73% in hydrocarbon treatment wetlands, and many hydrocarbons are not toxic to microorganisms except at high doses. In this study, low BOD effluent concentrations indicate that the high benzene concentration of 1,000 mg/L did not hinder the overall microbial community in degrading BOD.

5.1.5 Conclusions and Recommendations

Experimental vertical-flow CWs can be an effective option for the treatment of low molecular weight hydrocarbons such as benzene. The control of environmental boundary conditions such as temperature, light and humidity resulted in an additional 15% benzene reduction.

No significant effect of benzene treatment on nitrogen and phosphorus removal was observed. However, this does not mean that these nutrients do not contribute to microbial benzene degradation.

The standard wetland plant *P. australis* (Cav.) Trin. ex Steud (common reed) played a negligible role in organic matter, nitrogen and phosphorus removal. The presence of *P. australis* and its interactions with benzene and environmental boundary conditions (see above) were not significant in terms of nutrient removal. It follows that a filter filled with aggregates is likely to be the preferred unplanted wetland option for benzene removal, if sufficient nutrients are present.

Concerning the presence of environmental boundary condition control, nitrogen and phosphorus removal was not significantly affected by the presence of benzene.

Benzene is generally toxic to microbial biomass responsible for respiration. The BOD is likely to be representative for a mature system, where microbes adjusted to the presence of benzene dominate, as it was the case in Section 5.1.

The research highlights the great potential for wetlands to be used in removing wastewaters from the oil industry. This is particularly relevant for countries where sufficient land is available, and where a low-cost and low-technology treatment solution is preferred over potentially unsustainable traditional technologies. Developing countries such as Nigeria may also choose CWs to clean up contaminated river and lake waters.

Further research in nitrogen and phosphorus sequestration in plant biomass is encouraged. It would also be valuable to assess the impact of different benzene concentrations on plants, and their corresponding efficiencies in treating benzene.

5.2 MESO-SCALE INTEGRATED CONSTRUCTED WETLAND SYSTEM OPERATIONS

5.2.1 Introduction

The examination of experimental CW has previously been conducted using bench-, pilot- and full-scale systems (Carty et al., 2008). Each of these approaches has its advantages and disadvantages. However, the financial construction and operating costs as well as the land area required for siting increases with the size of the model system. In some instances, different designs have been combined to examine the possibility of benefiting from their specific strengths and possibly removing some of the limitations of each. Vertical-flow wetlands combined with horizontal-flow (surface and sub-surface) have been examined in parts of Thailand and in China where available land is limited (Al-Rekabi et al., 2007; Kantawanichkul and Somprasert, 2005; Molle et al., 2008; Vymazal et al., 2004). Vertical-flow systems are often examined in a bench-scale setup as they are often dependent upon the filter media that are being used (Brix et al., 2001; Hua et al., 2010; Platzer and Mauch, 1997; Tong et al., 2007).

These relatively small-scale systems allow the operator greater control and give the possibility to fine-tune. However, operators may have difficulties to scale these systems up. The recycling of the treated wastewater back into the system has shown improvements in the denitrification rates and removal of ammonia from the influent (He et al., 2006; Humenik et al., 1999). This has the knock-on effect of also reducing the outflow from a CW, adding an additional measure of protection to any receiving water body.

Horizontal-flow systems are similarly tested in bench and pilot-scale setups. In North America, there are prominent pilot-sale CWs that have been in operation for several decades (Hunt and Poach, 2001; Hunt et al., 2002; Knight et al., 2000) examining various design parameters including dilution rates (Hill et al., 1997; Hunt et al., 2002; Knight et al., 2000), hydraulic loading (Lee et al., 2004), and planted and unplanted combinations (Poach et al., 2004a, 2004b). These pilot-scale systems are advantageous as they can mimic full-scale systems relatively accurately but can incur a large cost due to the land area required and any potential maintenance that will be necessary during their operation. Meso-scale CW systems have not been commonly used for the examination of interactions and operations within differently designed CW. These systems fit midway between bench- and pilot-scale systems. They are relatively cheap to construct, operate and maintain due to their scale. They are also effective in terms of land use while allowing for examination of specific parameters. Utilizing these factors, a study was performed in the south-east of Ireland to examine their performance in the treatment of diluted anaerobically digested swine wastewater under various treatment operations.

The aim of Section 5.2 is to examine the performance of a standardized meso-scale design. This approach utilizes a large amount of replication which improves statistical confidence in the recorded data. Implementing a standard design template allows for the examination of performance of varying treatment operations with set and controlled parameters. Varying treatments are easily configured to mimic other large-scale CWs that are already in use or to examine experimental operations within a pre-defined environment. Within this experiment, key operations were examined to determine their effects on the treatment efficiency of this meso-scale design. Section 5.2 is revisiting an original article published by Harrington et al. (2011).

5.2.2 Materials and Methods

Sixteen meso-scale CW systems (four operations with four replicates each) were operated at Teagasc, Pig Development Department, Animal & Grassland Research & Innovation Centre, Moorepark, Fermoy, County Cork, Ireland (Figure 5.6) between November 2008 and June 2010. These systems were used to examine key operations identified in the literature including hydraulic loading rates (HLRs), nutrient loading rates and nutrient recycling modes. Each system received the separated liquid from anaerobically digested pig manure. This liquid was diluted using tap water to a known concentration and stored in 1-m^3 containers from where it was pumped to the receiving systems. The parameters of these systems are outlined below:

1. Normal: 37 m^3/ha/day loading rate at 100 mg NH_4/L.
2. Recycling: 37 m^3/ha/day loading rate at 100 mg NH_4/L with 100% effluent recycled through the system weekly.

FIGURE 5.6 Meso-scale system after eight months of operation (note that the overflow containers are sealed).

3. High nutrient loading (HNL): 37 m^3/ha/day loading rate at 200 mg NH_4/L.
4. High flow rate (HFR): 74 m^3/ha/day loading rate at 100 mg NH_4/L.

These systems were constructed using readily available materials, requiring little or no specialist equipment or knowledge. The meso-scale systems were designed to mimic industrial-scale integrated constructed wetland (ICW) design parameters (Carty et al., 2008; Dunne et al., 2005; Harrington et al., 2005; Scholz et al., 2007) and were operated correspondingly. These designs allow for systems to be adapted to accommodate varying influents at set loading rates and concentrations. This design set-up allowed for the examination of a wide range of operations to be subsequently compared with CW designs elsewhere. Their use in outdoor settings maintains them as open biological systems, similar to a pilot-scale system, but also allows for the adaptability and ease of use that is inherent in bench-scale systems.

The meso-scale experiment was explicitly designed to utilize readily available materials and components to keep construction, as well as operational costs to a minimum. The individual cells were made using polypropylene containers (external dimensions: L 600 × W 400 × H 319 mm; internal dimensions: L 555 × W 355 × H 300 mm). The containers gave internal aspect ratios of close to 1.5:1.0 (L:W).

This ratio has been suggested by Scholz et al. (2007) with respect to the construction of ICW in the Anne Valley Project near Waterford, Ireland. Each container was fitted with an overflow pipe of 60-mm diameter to allow effluent to flow to the next cell in the system. The first four cells were filled with a 1-cm diameter aggregate to a depth of 15 cm. The aggregate was thoroughly washed prior to use to remove any dust and sediment, thus ensuring a homogenous substrate throughout all 16 systems. The fourth cell was fitted with a flexible overflow pipe having an internal diameter of 12 mm that fed any effluent from the final cell into an overflow container, allowing the effluent volume, if any, to be recorded. This final cell was covered with a fitted lid. The material being fed into each system was pumped from two containers of 1-m^3 capacity that were housed in a garden shed. These containers contained separated anaerobic digestate liquid, which was diluted to 100- and 200-mg ammonia-nitrogen/L, respectively. The digestate liquid was analysed at the Waterford County Council, Water Research Laboratory, Adamstown, County Waterford, Ireland prior to the filling of each tank. This was done to ensure that the dilutions of the digestate were relatively stable.

The material was added by hand and the dilutions were prepared in situ using tap water. In-line submersible pumps were used to pump the material from these containers to the individual systems. Each pump could deliver 11.75 L/min at a 1-m head. The pumps were connected to electrical timers to regulate the flow and to ensure reliable and consistent flows. These timers allowed for accurate changes to be made to the hydraulic loading of each of the systems. With a set pumping volume, this allowed for the change in loading rates based on a set time. All wetland cells were densely planted (254 individual plants per m^2) with an approximately equal number and amount of mature and healthy stands of *Glyceria maxima* (Hartm.); Holmb. (syn. G. aquatica L.) Wahlenb.; *Glyceria spectabilis* Mert. & W.D.J. Koch; *Molinia maxima* (Hartm.); *Poa aquatica* L.; Reed Mannagrass Reed Sweet grass. The plants were about two years old and were obtained from an existing semi-natural wetland near Kilmeaden, County Waterford, Ireland. These plants were chosen for the design due to their rapid and dense growth rates because they are readily available throughout Ireland and Great Britain and because of their relatively high tolerance of ammonia (Clarke and Baldwin, 2002; Cronk, 1996; Harrington and McInnes, 2009; Hubbard et al., 1999).

5.2.3 Results and Discussion

The ammonia removal rates for the meso-scale ICW systems were like other full-scale ICW systems (Babatunde et al., 2008; Scholz et al., 2007) that have been in operation in the south-east of Ireland for the past decade (Table 5.7). Throughout the 18 months of operation, removal rates from the meso-scale systems remained relatively stable regardless of seasonal changes. This stability was also recorded between replicates themselves (Figure 5.7). Ammonia concentrations in the final effluent of

TABLE 5.7
Mean Removal Rates of Key Parameters for the Meso-Scale Systems

	Operation			
Parameter	**Normal (%)**	**Recycling (%)**	**HNL[a] (%)**	**HFR[b] (%)**
Ammonia-nitrogen	99.7	99.9	99.8	99.1
Molybdate reactive phosphorus	97.7	96.0	94.6	89.1
Nitrite-nitrogen	95.2	96.2	98.4	83.8
Nitrate-nitrogen	93.0	92.2	81.7	75.7
Total organic nitrogen	93.4	93.1	92.9	77.4

[a] HNL, high nutrient loading.
[b] HFR, high flow rate.

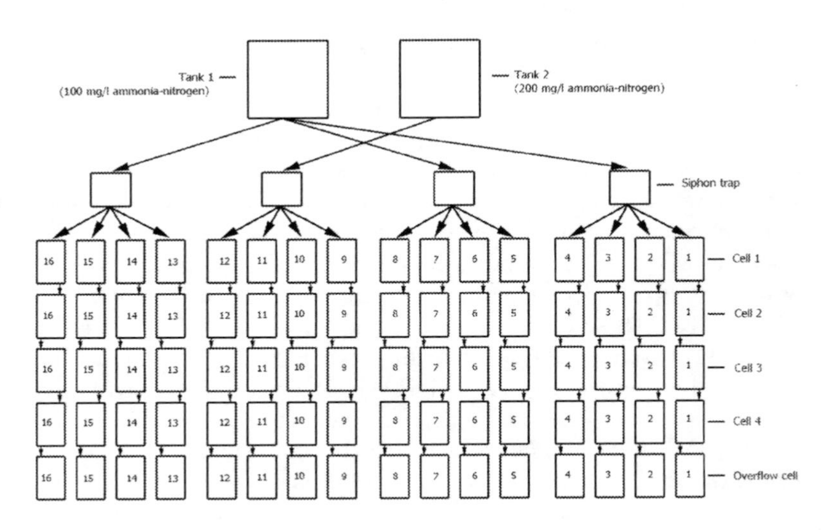

FIGURE 5.7 Schematic layout of meso-scale systems showing flow direction and splitting of inflow from storage tanks.

each treatment remained low between June 2009 and June 2010. Normal treatment showed an average concentration of 0.53-mg ammonia per litre with a standard deviation of ±0.6 mg. Concerning the recycling mode, HNL and HFR treatments showed similar concentration levels during this same time period (Table 5.8).

Over the 18 months when the meso-scale systems were in full operation, the recycling treatments showed the lowest ammonia concentrations in the final effluent (Figure 5.8). The improvement in effluent quality when recycling has been implemented has been seen in previous studies (He et al., 2006; Põldvere et al., 2009). The increase in nitrified water promotes greater denitrification and increases the retention time of the system. The HFR system showed significantly higher corresponding concentrations for both molybdate reactive phosphorus (MRP) and ammonia-nitrogen. Summer months showed very low volumes being collected in the overflow containers

TABLE 5.8
Final Effluent Ammonia Concentration (June 2009–June 2010)

Operation	Mean (mg/L)	Standard Deviation
Normal	0.53	0.600
Recycling	0.17	0.118
HNL[a]	0.69	1.299
HFR[b]	2.09	3.537

[a] HNL, high nutrient loading.
[b] HFR, high flow rate.

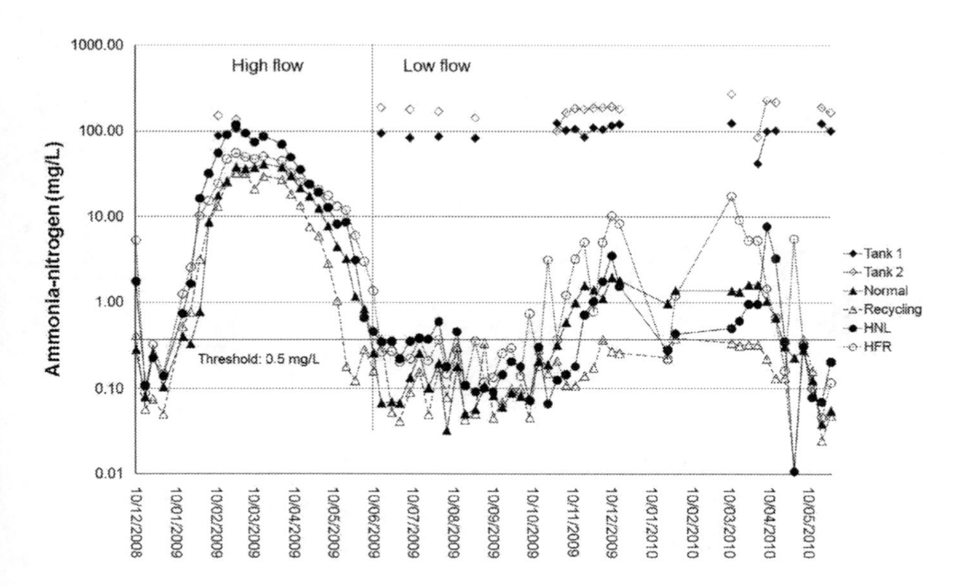

FIGURE 5.8 Ammonia-nitrogen concentrations in the outflow from recycling systems over 18 months of operation.

and high outflow was usually a direct result of heavy rainfall, which gave additional dilution to the effluent material.

Phosphorus removal was also very high in all four treatment systems averaging 97.7, 96.0, 94.6 and 89.1%, respectively (Table 5.8). These removal rates are in line with other full-scale ICW systems that are currently in operation in Ireland treating dairy, cattle, municipal and food production wastewaters (Harrington and McInnes, 2009; Harrington and Ryder, 2002; Scholz et al., 2007; Zhang et al., 2009).

The levels of MRP in the influent were significantly reduced after separation following anaerobic digestion as most had been removed in the solid fraction. The main mechanisms for phosphorus removal are adsorption and sedimentation. While these are also key mechanisms in the nitrogen cycle, they are not the primary removal methods (Vymazal, 2007).

The MRP concentrations in the final effluent maintained provisional discharge limits for Ireland throughout the operational time frame, including the higher hydraulic loading phase (Figure 5.9). Despite the systems only having a washed aggregate substrate, the removal efficiency remained exceptionally high. These rates were maintained during the winter periods, showing that the removal should be maintained year long. Prolonged usage with a limited substrate could potentially yield lower removal rates. However, this has not been seen in ICW systems in operation in Ireland for almost a decade (Harrington and McInnes, 2009).

Considering an exceptionally harsh winter and spring in 2010, the meso-scale systems encountered very few problems regarding maintenance and operation or their treatment capacity. Indeed, the effectiveness of CWs has been shown to work in cold climates such as Finland, Estonia, Canada and Sweden (Bastviken et al., 2009;

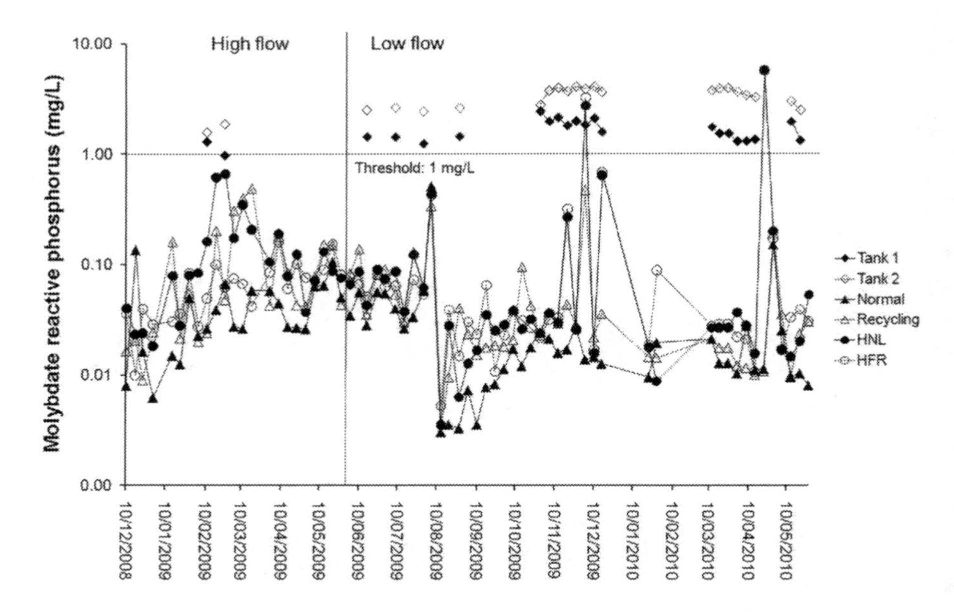

FIGURE 5.9 Molybdate reactive phosphorus concentrations in the outflow from recycling systems over 18 months of operation

Maltais-Landry et al., 2009; Põldvere et al., 2009; Puustinen and Jormola, 2005). Any components that did fail were easily replaced and with minimal, if any, downtime. The high level of replication of the treatments gives great statistical confidence in the data collected. Any significant variability between expected and recorded results was easily identified and the cause could be identified from the replicate itself by comparison with other replicates.

Generally, two replicates are common in most CW studies (Bastviken et al., 2009; Hunt et al., 2002). However, some researchers have used very high replicate counts when examining more biological parameters, such as mosquito populations (Poach et al., 2004a, 2004b). The scaling of these systems when viewed in terms of the materials used is advantageous as it allows for high replication while minimizing the required land area. At the same time, it allows for the sizing of these systems to be relatively large in comparison to bench-scale tests. Throughout the operation of the meso-scale systems, very few problems were encountered and those that did occur were easily corrected with little or no downtime to the systems themselves. The high replication of each treatment allowed for great confidence in the recorded results throughout the sampling period. While this did come at a relatively small cost with respect to required land area, it ensured reliable results and the ability to examine the systems themselves.

5.2.4 Conclusions

The use of this meso-scale 'sandbox' approach for the examination of ICW treatments allows the user a great degree of control over what is essentially an open biological system. The wetlands fully interacted with their surrounding environment, but all key parameters were easily controlled. External factors such as meteorological conditions can be recorded for the immediate area and their effect on the systems can be deduced from the systems themselves as well as the recorded results. The land area required in comparison to bench-scale and indeed some pilot schemes is significant; however, when viewed considering the amount of replication that it allows for, the cost is easily acceptable. The ability to compare a larger number of replicates ensures reliability in the recorded data from each system as well as each treatment operation. Recorded data points that were significantly different to other replicate samples were identified and dealt with. This also has a significant impact on the reliability of any statistical analysis performed on the recorded data set, for which replication is essential.

5.3 HYDRODYNAMIC MODELLING OF CONSTRUCTED WETLANDS

5.3.1 Introduction

5.3.1.1 Storm Water Wetlands

Storm water management includes the use of many devises and strategies with a range of purposes and benefits such as water quality improvement, landscape amenity enhancement, provision of wildlife habitat and flood protection and flow control (Scholz, 2010; Wong et al., 1999). Free-water surface constructed wetlands (FWS CW) and ponds can be used to achieve these benefits. These wetland systems are most useful for storm water containing high concentrations of soluble contaminants, which are hard to remove using other treatment methods. In addition to flood control, FWS CW treat many types of wastewaters and provide an ecological approach to reduce the release of nutrients and toxic materials into the environment through settling and both uptake and filtering by vegetation. The removal rates for FWS CW can reach up to 90% for some pollutants subject to individual case study conditions (Scholz, 2010). The merits of treatment wetlands (compared to conventional wastewater treatment plants) include less cost of construction and operation, simple operation and high removal efficiency of wastewater constituents (New Jersey Department of Environmental Protection, 2004; Walton, 2012).

The removal efficiency for a FWS CW depends upon the hydrodynamic characteristics of passing flow as characterized by the HLR and the hydraulic retention time (HRT). Both characteristics affect the duration of the contact between wastewater and the wetland system. A longer stay of the water in FWS CW could improve the removal capacity of pollutants due to longer interaction times with surfaces and higher sedimentation rates as the result of lower current velocities (Kadlec and Knight, 1996; Reddy and DeBusk, 1987).

The HRT is more prevalent compared to HLR as the guiding parameter in designing most FWS CW. Since the physical and biological CW treatment processes are all a function of time, maximizing the HRT will eventually improve the overall treatment (Conn and Fiedler, 2006; Scholz, 2010). The theoretical HRT can be defined as the water volume contained within the wetland divided by the volumetric flow rate out of the wetland considering steady-state conditions. However, the actual HRT of a CW is less than the theoretical HRT, because of imperfectness of the wetland volume (Persson et al., 1999).

5.3.1.2 Literature Review and Objectives

This section critically assesses the literature concerned with the overall modelling of flows within wetland systems. Several hydrodynamic modelling studies using numerical methods to study various aspects of CW have been undertaken. For example, Moustafa (1997) developed a simple steady-state model to predict phosphorus retention in a wetland treatment system. Somes et al. (1998) used MIKE 21 (two-dimensional depth-averaged model) to simulate the flow hydrodynamics within a wetland and compared the results with data collected from a field-based investigation. Considering flat bottom topography, Walker (1998) used a two-dimensional implicit finite difference model to simulate several rectangular storm water storage ponds with varying length-to-width ratios. Persson et al. (1999) investigated the influence of CW's shape, and inlet/outlet locations and types on the hydrodynamics of 13 hypothetical ponds. They used MIKE-21 (two-dimensional depth integrated hydraulic model) to simulate the progress through the system. Persson et al. (1999) claimed that the hydraulic efficiency coefficient provides a good balance in assessing the hydrodynamic performance of detention systems against the uniformity of flow.

The impact of vegetation on wetland flows has also been studied, previously. Wörman and Kronnas (2005) simulated wetland flow resistance by numerical modelling considering vegetation and the aspect ratio of the wetland. Their study indicated that vegetation contributed to the increase of the variance of the water residence time. Moreover, the aspect ratio had a direct effect on the active water volume and the treatment efficiency. Jenkins and Greenway (2005) applied a two-dimensional numerical model to investigate the effects of emergent fringing and banded vegetation on the hydraulic characteristics of surface water flow CW. The modelling study showed that the wetland shape as well as the vegetation density and spatial distribution has a significant impact on the hydraulic characteristics of the wetland.

The HRT has been identified as a key wetland design parameter. Conn and Fiedler (2006) explored the effects of characteristic bottom topographic features that increase HRT of constructed storm water treatment wetlands using a two-dimensional depth-averaged hydrodynamic model. They applied a modified MacCormack explicit finite difference scheme on a structured computational mesh for solving the governing equations. Numerical simulations of rectangular test wetlands were considered between topographic features and their effects on HRT. They claimed that creating baffled wetlands with multiple vertical-scale topography can markedly increase HRT.

The hydraulic efficiency of a wetland system has been identified as crucial in water quality modelling. Su et al. (2009) proposed the optimal design for different FWS CW according to the hydraulic efficiency index. They chose a horizontal two-dimensional model called TABS-2 for their study. The position and distribution of the outlet retention time distribution (RTD) curve was taken into consideration when calculating the hydraulic efficiency parameter. They also discussed the influence of three factors including the aspect ratio of the wetland, the configuration of the inlet and outlet, and obstruction selection on the hydraulic efficiency. Furthermore, Min and Wise (2010) presented a two-dimensional hydrodynamic and solute transport model of a large-scale and sub-tropical FWS CW using the MIKE21 model.

More recently, complex integrative modelling attempts have been made to better understand wetland processes. Lago et al. (2010) used a numerical model to investigate the effects of water flow, sediment transport and vegetation growth on the spatio-temporal pattern of the ridge and slough landscape of wetlands in the Everglades. The numerical model consisted of governing equations for integrated surface water and groundwater flow, sediment transport and soil accretion, as well as litter production by vegetation growth. Furthermore, Wamsley et al. (2010) applied a numerical storm surge model to assess the sensitivity of surge response to specified wetland loss. They claimed that wetlands have the potential to reduce surges, but the magnitude of attenuation is dependent on the surrounding coastal landscape. Wang et al. (2011) used a three-dimensional numerical model for simulating the total phosphorus transport and removal efficiency in the horizontal sub-surface flow and wavy sub-surface flow CW. Arega (2013) applied a coupled numerical model (depth-integrated hydrodynamic and particle transport model) to study the mixing and transport processes in a tidal wetland located in South Carolina. An integrated two-dimensional depth-average numerical model was developed by Yang et al. (2012) to simulate hydrodynamics and transport of contaminants in a wetland in northern Taiwan.

This study investigates the interactions between the aspect ratio, shape, inflow/outflow configuration, spreading inflow and hydraulic efficiency of FWS CW. A relatively high number (89 in total) of artificial (synthetic) FWS CWs were created and numerically simulated with a two-dimensional depth-averaged hydrodynamic numerical model. The numerical simulation uses a k-ε model for predicting turbulence and the Galerkin finite volume method for unstructured triangular meshes to discretize the governing equations.

The objectives of Section 5.3 are to (a) develop a generalized numerical input-output type model capable of predicting retention time in non-vegetated FWS CW; (b) establish relationships between rectangular CW characteristics and HRT and (c) aid designers of CW in understanding the complex hydraulic effects of wetland geometry and flow hydrodynamics on the overall hydraulic efficiency of the system. Note that this chapter is based on the original article by Zounemat-Kermani et al. (2015), which has been updated.

5.3.2 Methodology

5.3.2.1 Numerical Model and Equations

As the water depth of a CW is shallow, the two-dimensional character of the free surface flow is usually enforced by a horizontal length scale that is much larger than the vertical one, and variation in the vertical direction can therefore be omitted. Under these conditions, three-dimensional equations can be simplified within a depth-averaged two-dimensional model.

In this study, the governing equations of flow and transport were chosen as the two-dimensional depth-averaged equations coupled with a two-equation model of k-ε for turbulence modelling. A FORTRAN code was developed for solving the equations on a triangular unstructured mesh using the Galerkin finite volume method (Sabbagh-Yazdi and Zounemat-Kermani, 2008; Zounemat-Kermani and Sabbagh-Yazdi, 2010).

The integral form of two-dimensional depth-averaged equations can be written as shown in Equation 5.1 (Anastasoiu and Chan, 1997). Equation 5.2 explains the vector of flux function. Moreover, Equation 5.2 can be extended as shown in Equation 5.3 (Ern et al., 2007; Quecedo and Pastor, 2002). The turbulent eddy viscosity (Equation 5.3) is determined from a two-equation depth-average k-ε model (Rodi, 1993), where k is the kinetic energy, ε is the rate of dissipation of kinetic energy and υ_t (Equation 5.4) is the turbulent eddy viscosity.

$$\frac{\partial}{\partial t}\iint_{\Omega} d\Omega + \oint_{S} \mathbf{F}.\mathbf{n}\, ds = \iint_{\Omega} H d\Omega \tag{5.1}$$

where Ω is the domain of interest, $\mathbf{n}$ is the normal vector to S in the outward direction, S is the boundary surrounding Ω, Q is the vector of conserved variables, $\mathbf{F}$ is the vector of flux functions through S and H is the vector of forcing functions.

$$F \cdot n = F^{I} n_x - F^{V} n_y \tag{5.2}$$

where the superscripts I and V denote the inviscous and viscous fluxes, respectively, and n_x and n_y denote the components of the normal vector $\mathbf{n}$.

$$Q=\begin{bmatrix} h \\ hu \\ hv \\ hs \\ hk \\ h\varepsilon \end{bmatrix}, F^I=\begin{bmatrix} hu & hv \\ hu^2 & huv \\ huv & hv^2 \\ hus & hvs \\ huk & hvk \\ hu\varepsilon & hv\varepsilon \end{bmatrix},$$

$$F^V=\begin{bmatrix} 0 & 0 \\ h\upsilon_t 2u_x & h\upsilon_t\left(u_y+v_x\right) \\ h\upsilon_t\left(v_x+u_y\right) & h\upsilon_t 2\left(v_y\right) \\ h\frac{\upsilon_t}{Sc}s_x & h\frac{\upsilon_t}{Sc}s_y \\ h\frac{\upsilon_t}{\sigma_k}k_x & h\frac{\upsilon_t}{\sigma_k}k_y \\ h\frac{\upsilon_t}{\sigma_\varepsilon}\varepsilon_x & h\frac{\upsilon_t}{\sigma_\varepsilon}\varepsilon_y \end{bmatrix}, H=\begin{bmatrix} 0 \\ -gh\frac{\partial\eta}{\partial x}+hvf_c-\frac{\tau_{bx}}{\rho_w}+\frac{\tau_{wx}}{\rho_w} \\ -gh\frac{\partial\eta}{\partial y}+huf_c-\frac{\tau_{by}}{\rho_w}+\frac{\tau_{wy}}{\rho_w} \\ R_s \\ hP_k+hP_{kv}-h\varepsilon \\ hC_{1\varepsilon}\frac{\varepsilon}{k}(P_k)+hP_{\varepsilon v}-hC_{2\varepsilon}\frac{\varepsilon^2}{k} \end{bmatrix} \quad (5.3)$$

where h represents the flow depth, u and v are the horizontal components of velocity, η is the surface water height, Sc is the turbulent Schmidt number, g is the gravitational acceleration, the term $\left(\tau_{bx},\tau_{by}\right)$ represents the bottom friction shear stress (using the Manning's roughness coefficient) and υ_t is the eddy viscosity.

$$\upsilon_t=c_\mu\frac{k^2}{\varepsilon} \quad (5.4)$$

The aforementioned Galerkin finite volume scheme was used to solve the mass balance equation (forth row in Equation 5.3) for the two-dimensional transport of the conservative contaminant. The symbol s is the concentration of the conservative contaminant tracer.

To solve the two-dimensional depth-averaged equations, the Galerkin finite volume method (Zounemat-Kermani and Sabbagh-Yazdi, 2010) was applied. The problem domain is first discretized into a set of triangular cells forming an unstructured computational mesh based on cell vertex overlapping control volumes for inviscid and viscous fluxes (Figure 5.10). By applying the finite volume method, the governing equation is converted into discrete form by integrating over a control volume Ω.

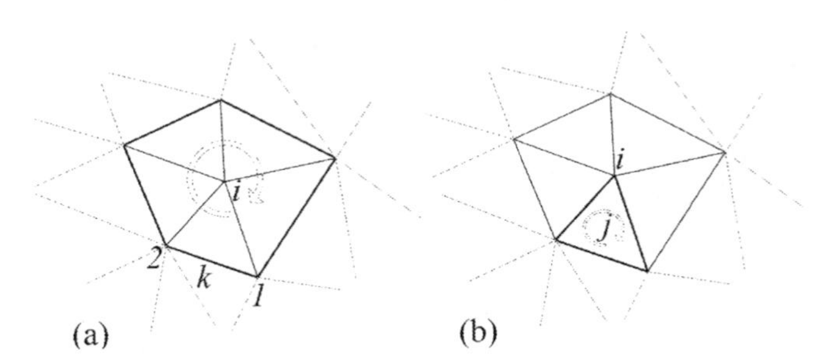

FIGURE 5.10 Illustration of (a) a control volume for inviscid fluxes around a computational node (i) and (b) a secondary control volume for corresponding viscous fluxes.

In Figure 5.10a, the unstructured overlapping control volumes Ωi are formed by gathering triangles meeting a computational node (i). In the final formulation of this method, the shape function vanishes from the formulation via mathematical manipulations. Equation 5.5 for the two-dimensional model can be derived from Zounemat-Kermani and Sabbagh-Yazdi (2010).

$$Q^{n+1} = Q^n + \frac{\Delta t}{\Omega}\left[\sum_{k=1}^{N_{side}}\left(\frac{F_i^c + F_1^c + F_2^c}{2}\right)_k \vec{\Delta l}_k - \frac{3}{2}\sum_{k=1}^{N_{side}} F_j^d \vec{\Delta l}_k + \Omega S^n\right] \tag{5.5}$$

where Ω represents the total area of control volume N_{side} with external edges/faces of $\vec{\Delta l}_k, \vec{\Delta s}_k$ and Δt is the actual time step, which is proportional to the minimum size of the control volumes of the computational domain.

In the present study, a three-stage Runge-Kutta scheme, which damps high frequency errors, is used for stabilizing the explicit time stepping process. In addition, the artificial dissipation terms suitable for the unstructured meshes are applied to stabilize the numerical solution procedure. To dampen unwanted numerical oscillations, a fourth-order artificial dissipation term is added to the above algebraic formula.

5.3.2.2 Hydrodynamics of Free Water-Surface Constructed Wetlands

In FWS CW, the basic design parameters are the storage volumes within its various zones (Scholz et al., 2007). Many wetland management problems and hydraulic malfunctions can be attributed to poor hydrodynamic characteristics within the wetland system. Improving the hydrodynamic performance of FWS CW can be achieved by allowing the flow to pass through the wetland uniformly. As noted by Persson et al. (1999), the hydrodynamics within a wetland are mainly influenced by the wetland's aspect ratio (length-to-width ratio). Also spreading flow at the inlet by using weirs, multiple inlets and islands in front of the inlet, and avoiding dead zones within FWS CW, promotes hydrodynamic activity and consequently improves the hydraulic efficiency.

The retention (residence) time is often used as the principal parameter in establishing the treatment efficiency of a CW. The mean HRT for a CW is the average amount of time the wastewater is being treated within the CW. The HRT has a significant influence on the pollutant removal ratio and is a key factor in designing CW.

Enhancing the HRT will improve the treatment efficiency of CW. The simplest but costly method of increasing HRT is by increasing the CW volume.

Land space limitations make it often difficult to increase CW volume in practice. Therefore, alternative ways to improve and increase the HRT of CW need to be found. The HRT in CW is influenced by hydrologic variables (e.g., rainfall, evapotranspiration and infiltration), geometry (topography, general shape and slope), vegetation (type of vegetation and mass) and hydraulics of CW (configuration of inlet/outlet, flow spreading and inlet rate flow) as discussed by Scholz et al. (2007). This study focuses on the geometry and hydraulic options for increasing the HRT.

To estimate the HRT, a pulse experiment can be undertaken at the inlet by adding a tracer such as chloride with a given concentration within a short time (Scholz and Trepel, 2004). Based on the tracer concentration measurement at the outlet of the wetland, the RTD can be derived. The flow condition inside a CW is revealed by the position and distribution of the RTD curve. The RTD curve is derived either by field measurements or by numerical simulation. The latter has been undertaken in this study.

Figure 5.11 shows the RTD curve of the inflow and outflow developed from the pulse experiment in two different CWs. Under ideal flow conditions (also known as plug flow), the inlet water flows as a single plug. However, the actual HRT of a CW is shorter than the ideal plug flow. A long hydraulic retention time corresponds to long time for the treatment process. The HRT is the ratio of the CW volume to the flow rate of the surface water. The hydraulic retention time considering ideal plug flow conditions can be calculated with Equation 5.6 (Panuvatvanich et al., 2009; Su et al., 2009).

$$T_n = \frac{V}{Q_W} \tag{5.6}$$

where T_n stands for the nominal hydraulic retention time, V is the CW volume and Q_W is the flow rate through the CW.

The hydraulic efficiency λ refers to the extent to which plug flow conditions are approximated and the proportion of the wetland volume utilized in the movement of the inflow through the CW. Many researchers proposed that the hydraulic efficiency coefficient should identify the RTD curve position and distribution (Jenkins and Greenway, 2005; Persson et al., 1999; Su et al., 2009). They defined the hydraulic efficiency of a CW as the ratio of the time taken for a tracer (e.g., chloride as discussed by Scholz and Trepel, 2004) to reach a peak at the CW outlet to the nominal HRT (Equation 5.7).

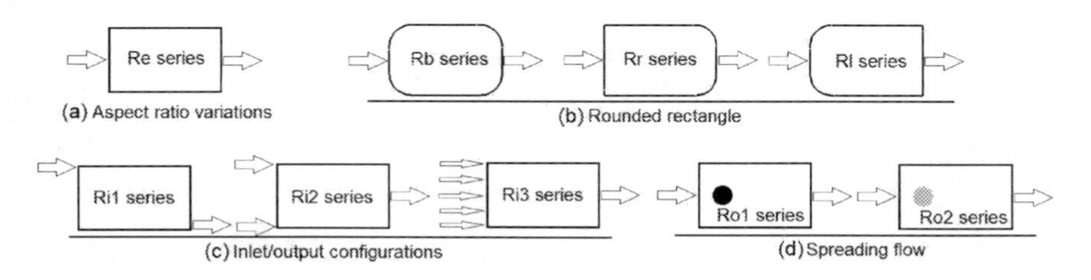

FIGURE 5.11 Illustration of the four main groups (a–d) of rectangular artificial free surface water flow constructed wetlands.

$$\lambda = \frac{T_p}{T_n} \tag{5.7}$$

where T_p (peak time) is the time taken for a conservative tracer to reach a peak at the wetland outlet (Panuvatvanich et al., 2009). Another alternative approach (which is not considered in this study) that can be applied to interpret solute transport and as a result for calculating the hydraulic efficiency is the moments of the breakthrough curve (Luo et al., 2006; Zhu et al., 2009).

A high hydraulic efficiency usually correlates with an efficient FWS CW in terms of water quality improvements. Under ideal flow conditions, T_p equals T_n, which indicates the maximum hydraulic efficiency ($\lambda = 1$ or $\lambda = 100\%$). Nonetheless, ideal flow conditions never occur in semi-natural FWS CW. Therefore, the hydraulic design of a FWS CW should achieve at least near-optimum flow conditions.

The configurations of FWS CW have a direct influence on the hydraulic efficiency. Geometric factors such as the shape and form of a CW and its aspect ratio, the type and location of the inlet/outlet structures and spreading inflow could be considered as potential features, which influence the hydraulic efficiency. In subsequent sections, the effects of several configurations of artificial FWS CW on hydraulic efficiency are considered.

5.3.2.3 Verification of the Numerical Model

To verify the results of the numerical model, comparison between numerical results and experimental measurements of behaviour of the outlet tracer of an experimental pond were considered. The experimental studies were performed in an artificial rectangular free surface pond system in the hydraulic laboratory of Sirjan Azad University for two models (model a: middle/middle inlet/outlet flow without any obstacles; model b: middle/middle inlet/outlet flow with a circular obstacle in the middle of the pond with a diameter of 0.3 m). The system composed of a rectangular pond with an area of $2\,m^2$ ($1\ m \times 2\ m$) and with an average water depth of 0.142 m and a slope of 2.5%. A schematic plan of the experimental set-up is shown in Figure 5.12. The incoming water (discharge of 0.25 L/s) to the pond came from a free surface stilling tank over a rectangular weir. As for the tracer injection, the incoming water

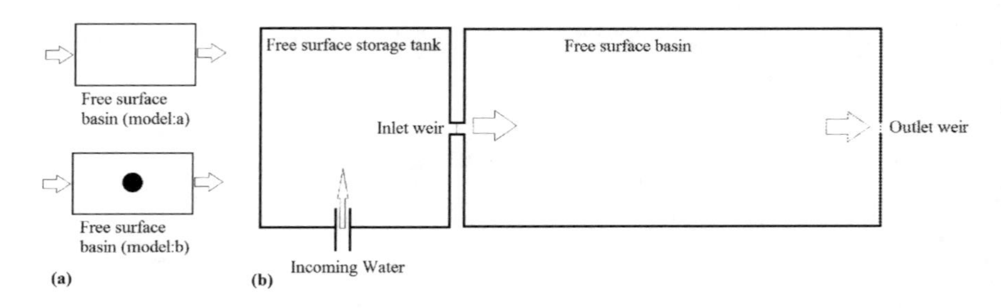

FIGURE 5.12 Schematic illustration of the free water surface pond system; (a) model a: middle/middle inlet/outlet flow without any obstacle and model b: middle/middle inlet/outlet flow with a circular obstacle in the middle of the pond with diameter of 0.3 m and (b) designing details of the experimental set-up.

to the pond was injected abruptly by a total concentration of total dissolved solids (TDS) using 3 L of a sodium chloride solution with a concentration of 100 mg/L. Physical and hydraulic information about the experimental ponds are tabulated in Table 5.9. Measurements for the outlet concentration in terms of TDS were undertaken using a stopwatch, a volumetric flask and a TDS meter.

Figure 5.13 illustrates the results of the RTD curves of experimental measurements and numerical simulation. Table 5.10 compares the results of retention time between numerical results and experimental measurements. As it can be seen from Table 5.10, the numerical model performs well regarding the simulation of the RTD curve (on average 5% absolute relative error with respect to retention time). On the other hand, the simulated HRT are in a good agreement with the observed ones. This confirms the good capability of the numerical model to simulate free surface flow and concentration transport in wetlands.

TABLE 5.9
Physical and Hydraulic Characteristics Concerning the Experimental Free Water Surface Pond Systems

Pond	Height of Inlet/ Outlet Weirs (m)	Slope (%)	Area (m^2)	Volume (L)	Inlet Flow (L/s)
Model a	0.15	2.5	2	284	0.25
Model b	0.15	2.5	1.93	274	0.25

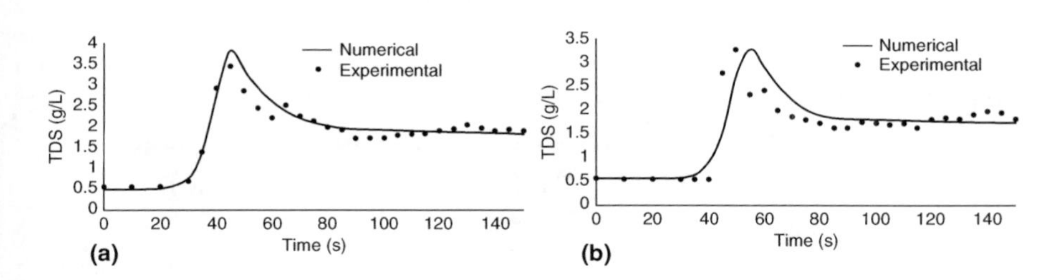

FIGURE 5.13 Simulated and experimental tracer curves (RTD) for free water surface pond; (a) experimental model *a* and (b) experimental model *b*.

TABLE 5.10
Comparison of Retention Time between Experimental and Numerical Results

Pond	Retention Time (s)		Absolute Relative Error (%)
	Experimental	Simulated	
Model a	45	46	2.2
Model b	50	54	8.0

5.3.2.4 Model Configurations of Free Water-Surface Wetlands

The current study analysed 89 different set-ups of rectangular FWS CW to determine the influences of different geometric and hydraulic factors on the hydraulic efficiency. Four main groups of FWS CW were categorized and numerically modelled: CW with different aspect ratios (11 cases), CW with rounded edges (33 cases), inlet/outlet scenarios (33 cases) and spreading flow at the inlet (12 cases). Figure 5.13 shows the four groups of artificial FWS CW. Note that all CWs were numerically modelled considering a depth of 0.5 m and an inflow rate of 50 L/s. General features and characteristics of the above mentioned four groups are described.

Initially, 11 different rectangular artificial FWS CW considering different length-to-width ratios were assessed as part of the Re CW series (Figure 5.11a and Table 5.11). The dimensions of the Re CW with the same areas (3,600 m^2) and volumes (1,800 m^3) are 40 m × 90 m (Re1), 50 m × 72 m (Re2), 60 m × 60 m (Re3), 75 m × 48 m (Re4), 90 m × 40 m (Re5), 120 m × 30 m (Re6), 150 m × 24 m (Re7), 180 m × 20 m (Re8), 240 m × 15 m (Re9), 300 m × 12 m (Re10) and 360 m × 10 m (Re11). The corresponding aspect ratios are 0.44, 0.69, 1.00, 1.56, 2.25, 4.00, 6.25, 9.00, 16.00, 25.00 and 36.00, respectively. For all Re CW, the inflow/outflow configurations were modelled as a centrally located point source (Figure 5.11a). Furthermore, Table 5.12 compares the typical characteristics of CW based on related previous studies with the current one.

The 11 Re CWs were geometrically modified by cutting the rectangles' vertices using a quadrant in three categories of rounded rectangle CW as shown by the Rb, Rr and Rl series (Figure 5.11b). The dimensions of the CW were unchanged (see Re

TABLE 5.11
Re Series of Free Surface Water Flow Constructed Wetland Characteristics and Numerical Results for Different Configurations of Aspect Ratios

Case	Length (m)	Width (m)	Aspect Ratio	Volume (m^3)	T_n[a] (hour)	T_p[b] (hour)	λ[c]	Description
Re(1)	40	90	0.44	1,800	10.00	0.83	8.3	Poor
Re(2)	50	72	0.69	1,800	10.00	1.21	12.1	Poor
Re(3)	60	60	1.00	1,800	10.00	2.22	22.2	Poor
Re(4)	75	48	1.56	1,800	10.00	4.00	40.0	Poor
Re(5)	90	40	2.25	1,800	10.00	4.44	44.4	Poor
Re(6)	120	30	4.00	1,800	10.00	6.39	63.9	Satisfactory
Re(7)	150	24	6.25	1,800	10.00	7.26	72.6	Satisfactory
Re(8)	180	20	9.00	1,800	10.00	8.10	81.0	Good
Re(9)	240	15	16.00	1,800	10.00	8.97	89.7	Good
Re(10)	300	12	25.00	1,800	10.00	9.31	93.1	Good
Re(11)	360	10	36.00	1,800	10.00	9.42	94.2	Good

[a] Nominal hydraulic retention time; [b] peak time; [c] hydraulic efficiency.

TABLE 5.12
Comparison between Various Constructed Wetland Characteristics

Authors	Aspect Ratio Range	Number of Wetlands	Manning's Coefficient	Water Depth (m)	Flow Rate (L/s)	Volume (m^3)	T_n[a] (hour)
Persson et al. (1999)	1–12	4	–	1.5	40	2,700	18.8
Jenkins and Greenway (2005)	0.36–36.00	9	0.035	0.5	200	3,500	4.9
Su et al. (2009)	0.3–30.0	9	0.035	0.7	40	2,100	14.6
This study	0.44–36.00	11	0.035	0.5	50	1,800	10.0

CW series), but the areas and volumes were different (Tables 5.13–5.15). In all of the 33 Rb, Rr and Rl case studies, the inflow was modelled as a point source and both the inflow/outflow were centrally located as midpoints.

Section 5.3 examines the impact of different inlet/outlet configurations by conducting simulations on 33 artificial FWS CW based on different aspect ratios of Re CW (Figure 5.11c). Corner–corner (Ri1 series), corners-midpoint (Ri2 series) and uniform-midpoint (Ri3 series) FWS CW with the same areas (3,600 m^2) and volumes (1,800 m^3) were designed and modelled (Tables 5.16–5.18).

The influence of the surface and submerged circular island type obstructions on the hydraulic efficiency performance was examined for the Ro1 and Ro2 CW series (Table 5.19). Case Re4 (75 m × 48 m) was set up as the benchmark to compare the hydraulic efficiency performance with the Ro1 and Ro2 CW series. Six surface obstructions and six submerged obstructions with a radius of 16 m were designed with 5, 10, 20, 30, 40 and 50 m distances from the midpoint inlet. The areas and volumes of the surface obstructions (Ro1 series) were 3,398.9 m^2 and 1,699.5 m^3, respectively. Although the areas of the submerged obstructions (Ro2) series are equal to the benchmark case (3,600 m^2), the volumes were 1,729.6 m^3 because of the height of the obstructions, which were set to 0.35 m.

5.3.2.5 Application and Modelling

To solve the governing equations (Equations 5.1–5.4) numerically, the flow domain was discretized using an unstructured mesh generated by the Deluaney triangulation technique. Figure 5.14a and b shows the two-dimensional view of the constructed triangular meshes within the computational domain of Rb(4) and Ro1(2) CW, respectively. In comparison, Figure 5.14c shows the three-dimensional view of the generated mesh of the submerged obstacle case (Ro2(3)). Note that local mesh refinement has been imposed near the Ro1 and Ro2 CW series.

In total, 89 different CW configurations were modelled using the developed two-dimensional depth-averaged hydrodynamic model. Each CW had a depth fixed at 50 cm and a steady inflow of 50 L/s. The Manning's coefficient of all non-vegetated wetland beds was 0.035, reflecting the nature of the wetland beds, which mostly composed of created irregular soil with some minor obstructions (Su et al., 2009). For each CW, the conservative tracer (chloride) was injected at the inlet as a plug

TABLE 5.13
Rb Series of Free Surface Flow Constructed Wetland Characteristics and Numerical Results for Different Aspect Ratio Configurations

Case	Length (m)	Width (m)	Rounded Part Radius (m)	Aspect Ratio	Volume (m^3)	T_n[a] (hour)	T_p[b] (hour)	λ[c] (%)	Description	λ Improvement (%)
Rb(1)	40	90	10.0	0.44	1642.90	9.13	0.83	9.1	Poor	9.6
Rb(2)	50	72	10.0	0.69	1642.90	9.13	1.21	13.2	Poor	9.6
Rb(3)	60	60	10.0	1.00	1642.90	9.13	2.24	24.5	Poor	10.2
Rb(4)	75	48	10.0	1.56	1642.90	9.13	4.44	48.7	Poor	21.7
Rb(5)	90	40	10.0	2.25	1642.90	9.13	5.00	54.8	Satisfactory	23.3
Rb(6)	120	30	10.0	4.00	1642.90	9.13	6.85	75.1	Good	17.4
Rb(7)	150	24	10.0	6.25	1642.90	9.13	7.71	84.5	Good	16.3
Rb(8)	180	20	10.0	9.00	1642.90	9.13	8.61	94.3	Good	16.5
Rb(9)	240	15	7.5	16.00	1711.65	9.51	9.25	97.3	Good	8.4
Rb(10)	300	12	6.0	25.00	1743.45	9.69	9.43	97.4	Good	4.6
Rb(11)	360	10	5.0	36.00	1760.75	9.78	9.54	97.5	Good	3.6

[a] Nominal hydraulic retention time; [b] peak time; [c] hydraulic efficiency.

TABLE 5.14
RI Series of Free Surface Flow Constructed Wetland Characteristics and Numerical Results for Different Aspect Ratio Configurations

Case	Length (m)	Width (m)	Rounded Part Radius (m)	Aspect Ratio	Volume (m^3)	T_n[a] (hour)	T_p[b] (hour)	λ[c] (%)	Description	λ Improvement (%)
RI(1)	40	90	10.0	0.44	1721.45	9.56	0.83	8.7	Poor	4.6
RI(2)	50	72	10.0	0.69	1721.45	9.56	1.19	12.5	Poor	3.4
RI(3)	60	60	10.0	1.00	1721.45	9.56	2.22	23.2	Poor	4.6
RI(4)	75	48	10.0	1.56	1721.45	9.56	4.43	46.3	Poor	15.8
RI(5)	90	40	10.0	2.25	1721.45	9.56	4.99	52.1	Satisfactory	17.3
RI(6)	120	30	10.0	4.00	1721.45	9.56	6.83	71.5	Satisfactory	11.8
RI(7)	150	24	10.0	6.25	1721.45	9.56	7.64	79.9	Good	10.0
RI(8)	180	20	10.0	9.00	1721.45	9.56	8.56	89.5	Good	10.5
RI(9)	240	15	7.5	16.00	1755.80	9.75	9.22	94.5	Good	5.4
RI(10)	300	12	6.0	25.00	1771.75	9.84	9.33	94.8	Good	1.9
RI(11)	360	10	5.0	36.00	1780.35	9.89	9.47	95.8	Good	1.7

[a] Nominal hydraulic retention time; [b] peak time; [c]hydraulic efficiency.

TABLE 5.15

Rr Series of Free Surface Flow Constructed Wetland Characteristics and Numerical Results for Different Aspect Ratio Configurations

Case	Length (m)	I. Width (m)	Rounded Part Radius (m)	Aspect Ratio	Volume (m^3)	T_n[a] (hour)	T_p[b] (hour)	λ[c] (%)	Description	λ Improvement (%)
Rr(1)	40	90	10.0	0.44	1721.45	9.56	0.83	8.7	Poor	4.6
Rr(2)	50	72	10.0	0.69	1721.45	9.56	1.21	12.6	Poor	4.6
Rr(3)	60	60	10.0	1.00	1721.45	9.56	2.22	23.2	Poor	4.6
Rr(4)	75	48	10.0	1.56	1721.45	9.56	4.00	41.8	Poor	4.6
Rr(5)	90	40	10.0	2.25	1721.45	9.56	4.47	46.8	Poor	5.2
Rr(6)	120	30	10.0	4.00	1721.45	9.56	6.44	67.4	Satisfactory	5.5
Rr(7)	150	24	10.0	6.25	1721.45	9.56	7.29	76.2	Good	5.0
Rr(8)	180	20	10.0	9.00	1721.45	9.56	8.11	84.8	Good	4.7
Rr(9)	240	15	7.5	16.00	1755.80	9.75	8.99	92.1	Good	2.7
Rr(10)	300	12	6.0	25.00	1771.75	9.84	9.32	94.7	Good	1.7
Rr(11)	360	10	5.0	36.00	1780.35	9.89	9.43	95.3	Good	1.3

[a] Nominal hydraulic retention time; [b] peak time; [c] hydraulic efficiency.

TABLE 5.16

Ri1 Series of Free Surface Flow Constructed Wetland Characteristics and Numerical Results for Different Aspect Ratio Configurations

Case	Length (m)	II. Width (m)	Aspect Ratio	Volume (m^3)	T_n[a] (hour)	T_p[b] (hour)	λ[c] (%)	Description	λ Improvement (%)
Ri1(1)	40	90	0.44	1,800	10.00	2.78	27.8	Poor	233.3
Ri1(2)	50	72	0.69	1,800	10.00	3.06	30.6	Poor	152.9
Ri1(3)	60	60	1.00	1,800	10.00	3.47	34.7	Poor	56.3
Ri1(4)	75	48	1.56	1,800	10.00	4.44	44.4	Poor	11.1
Ri1(5)	90	40	2.25	1,800	10.00	5.00	50.0	Satisfactory	12.5
Ri1(6)	120	30	4.00	1,800	10.00	5.83	58.3	Satisfactory	–8.7
Ri1(7)	150	24	6.25	1,800	10.00	6.67	66.7	Satisfactory	–8.2
Ri1(8)	180	20	9.00	1,800	10.00	7.44	74.4	Good	–8.1
Ri1(9)	240	15	16.00	1,800	10.00	8.33	83.3	Good	–7.1
Ri1(10)	300	12	25.00	1,800	10.00	8.94	89.4	Good	–3.9
Ri1(11)	360	10	36.00	1,800	10.00	9.33	93.3	Good	–0.9

[a] Nominal hydraulic retention time; [b] peak time; [c] hydraulic efficiency.

TABLE 5.17
Ri2 Series of Free Surface Flow Constructed Wetland Characteristics and Numerical Results for Different Aspect Ratio Configurations

Case	Length (m)	Width (m)	Aspect Ratio	Volume (m^3)	T_n[a] (hour)	T_p[b] (hour)	λ[c] (%)	Description	λ Improvement (%)
Ri2(1)	40	90	0.44	1,800	10.00	3.33	33.3	Poor	300.0
Ri2(2)	50	72	0.69	1,800	10.00	3.47	34.7	Poor	187.4
Ri2(3)	60	60	1.00	1,800	10.00	3.61	36.1	Poor	62.5
Ri2(4)	75	48	1.56	1,800	10.00	4.72	47.2	Poor	18.1
Ri2(5)	90	40	2.25	1,800	10.00	5.28	52.8	Satisfactory	18.8
Ri2(6)	120	30	4.00	1,800	10.00	5.97	59.7	Satisfactory	−6.5
Ri2(7)	150	24	6.25	1,800	10.00	6.94	69.4	Satisfactory	−4.4
Ri2(8)	180	20	9.00	1,800	10.00	7.64	76.4	Good	−5.7
Ri2(9)	240	15	16.00	1,800	10.00	8.61	86.1	Good	−4.0
Ri2(10)	300	12	25.00	1,800	10.00	9.03	90.3	Good	−3.0
Ri2(11)	360	10	36.00	1,800	10.00	9.33	93.3	Good	−0.9

[a] Nominal hydraulic retention time; [b] peak time; [c]hydraulic efficiency.

TABLE 5.18
Ri3 Series of Free Surface Flow Constructed Wetland Characteristics and Numerical Results for Different Aspect Ratio Configurations

Case	Length (m)	Width (m)	Aspect Ratio	Volume (m^3)	T_n[a] (hour)	T_p[b] (hour)	λ[c] (%)	Description	λ Improvement (%)
Ri3(1)	40	90	0.44	1,800	10.00	5.00	50.0	Satisfactory	500.0
Ri3(2)	50	72	0.69	1,800	10.00	5.56	55.6	Satisfactory	359.8
Ri3(3)	60	60	1.00	1,800	10.00	6.11	61.1	Satisfactory	175.0
Ri3(4)	75	48	1.56	1,800	10.00	6.67	66.7	Satisfactory	66.7
Ri3(5)	90	40	2.25	1,800	10.00	7.22	72.2	Satisfactory	62.5
Ri3(6)	120	30	4.00	1,800	10.00	8.06	80.6	Good	26.1
Ri3(7)	150	24	6.25	1,800	10.00	8.61	86.1	Good	18.5
Ri3(8)	180	20	9.00	1,800	10.00	9.17	91.7	Good	13.2
Ri3(9)	240	15	16.00	1,800	10.00	9.44	94.4	Good	5.3
Ri3(10)	300	12	25.00	1,800	10.00	9.58	95.8	Good	3.0
Ri3(11)	360	10	36.00	1,800	10.00	9.72	97.2	Good	3.2

[a] Nominal hydraulic retention time; [b] peak time; [c] hydraulic efficiency.

TABLE 5.19
Ro1 and Ro2 Series of Free Surface Flow Constructed Wetland (CW) Characteristics and Numerical Results for Different Aspect Ratio Configurations

Case	Status	Relative Distance from Inlet (m/m)	Volume (m^3)	T_n[a] (hour)	T_p[b] (hour)	λ[c] (%)	Description	λ Improvement (%)
Re(4)	Referred CW	–	1,800	10	4	40	Poor	–
Ro1(1)	Surface	0.07	1,699.5	9.44	6.14	65.0	Satisfactory	62.5
Ro1(2)	Surface	0.13	1,699.5	9.44	5.56	58.8	Satisfactory	47.1
Ro1(3)	Surface	0.27	1,699.5	9.44	4.50	47.7	Poor	19.2
Ro1(4)	Surface	0.40	1,699.5	9.44	4.11	43.5	Poor	8.9
Ro1(5)	Surface	0.53	1,699.5	9.44	4.03	42.7	Poor	6.6
Ro1(6)	Surface	0.67	1,699.5	9.44	4.00	42.4	Poor	5.9
Ro2(1)	Submerged	0.07	1,729.6	9.61	4.44	46.3	Poor	15.6
Ro2(2)	Submerged	0.13	1,729.6	9.61	4.19	43.7	Poor	9.1
Ro2(3)	Submerged	0.27	1,729.6	9.61	4.08	42.5	Poor	6.2
Ro2(4)	Submerged	0.40	1,729.6	9.61	4.06	42.2	Poor	5.5

[a] Nominal hydraulic retention time; [b] peak time; [c] hydraulic efficiency.

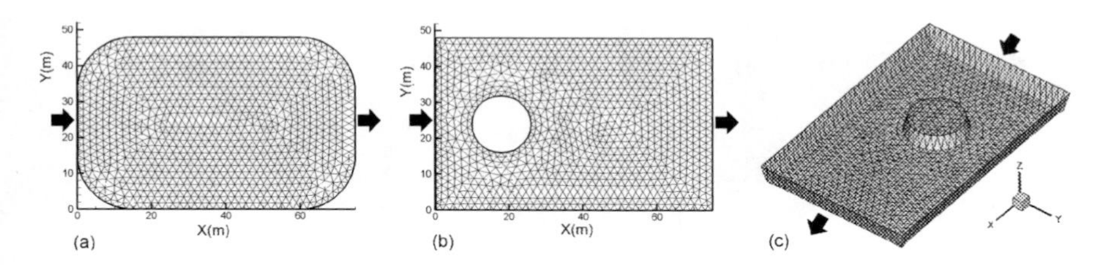

FIGURE 5.14 Two-dimensional triangular unstructured triangular mesh for: (a) Rb(4) case, (b) Rol(2) case and (c) three-dimensional bottom surface formed by triangular mesh for the Ro2(3) case.

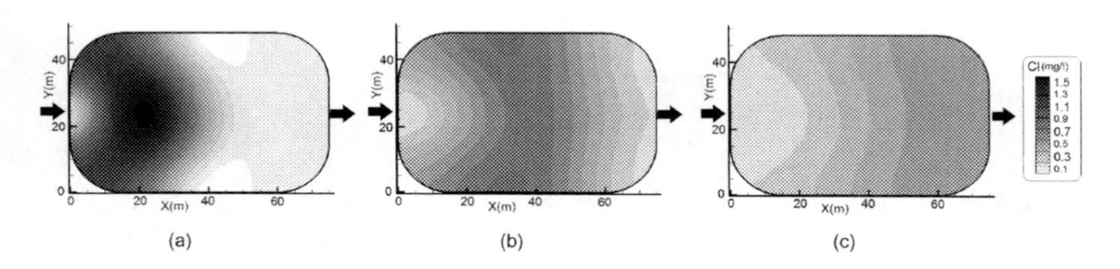

FIGURE 5.15 Chloride (Cl) concentration distributed in constructed wetland Rb(4) after (a) 30 minutes, (b) 135 minutes and (c) 300 minutes.

over a 30-second period. Figure 5.15 illustrates the chloride concentration distribution in CW Rb(4) after 30, 135 and 300 minutes. The chloride concentration got diluted over time.

Typical tracer responses are shown in Figure 5.15 with respect to RTD curves for Re(4) and Re(7) CW as well as the schematic behaviour of ideal plug flow. The model simulations were run for durations of 30 hours to ensure a falling limb shape of the corresponding RTD curves.

As can be seen in Figure 5.16, the RTD curves of Re(4) and Re(6) CW are characterized by steeper rising limbs compared to the receding limbs. This phenomenon is more evident for CW Re(4) with an aspect ratio of 1.56 compared to Re(6) with an aspect ratio of 4. Moreover, the outlet concentration peaks for CW with high aspect ratios resemble that of a plug flow configuration. This can be explained by the presence of dead or at least low circulation zones in CW with lower aspect ratios.

5.3.3 Results and Discussion

5.3.3.1 Wetlands by Different Aspect Ratio Simulations

A steady flow rate of 50 L/s was applied to each CW with a nominal HRT of T_n equal to ten hours to assess the effects of aspect ratio on the distribution of flow through rectangular wetlands. A summary of the wetland configurations modelled is shown in Table 5.19. This study used 11 different aspect ratios ranging from 0.44 to 36.00 to assess the relationship between aspect ratio and hydraulic efficiency. Results indicate that by increasing the aspect ratio, the hydraulic efficiency increases.

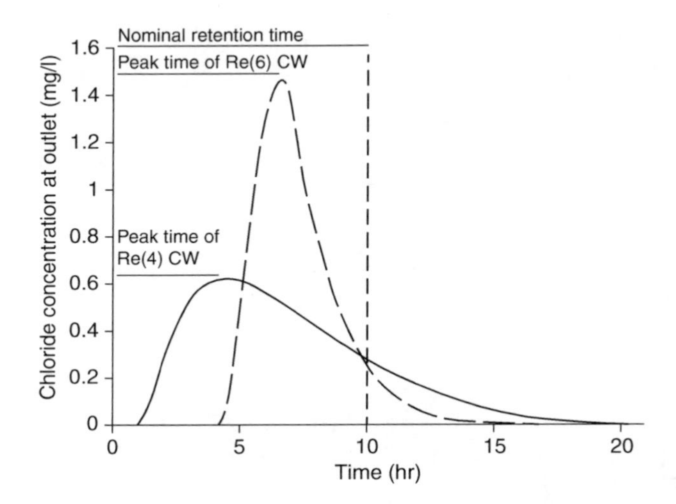

FIGURE 5.16 Retention time distribution curve of the tracer (chloride) based on the tracer concentration at the outlet of the Re(4) and Re(6) constructed wetlands (CW) using numerical simulation.

Jenkins and Greenway (2005), Persson et al. (1999) and Su et al. (2009) investigated the effect of wetland aspect ratio on the hydraulic efficiency and found that wetlands with high aspect ratios are usually associated with higher values for λ. Table 5.12 compares the general characteristics of CW based on these studies with the current one. It should be noted that the range of HRT in these studies and the study of interest are short compared to most realistic wetlands (e.g., 4–14 days according to Avelar et al. (2014) and Tsihrintzis et al. (2007)). However, findings linked to the geometric characteristics support general CW design. Figure 5.18a shows similar curves describing the relationship between aspect ratio and hydraulic efficiency with respect to the four studies. However, the results of this study are closer to those of Jenkins and Greenway (2005). From Table 5.19 and Figure 5.17a, an aspect ratio of more than 4 is beneficial in terms of hydraulic efficiency. Nevertheless, an aspect ratio of at least 8 is recommended for good hydraulic performance.

Figure 5.17b and c indicates the flow distributions and velocity fields in Re(7) CW ($\lambda = 73\%$) and Re(1) CW ($\lambda = 8\%$) using stream traces. The major improvement in terms of λ for Re(7) compared to Re(1) is due to hydraulic characteristics. The recirculation zones (dead zones) of Re(1) CW are much wider than those of Re(7) CW. The low velocity areas in Re(1) CW are more visible than those in Re(7) CW. Note that low velocity areas are relatively white and have light contours (Figure 5.17). Short-circuiting in recirculation zones is responsible for flow characteristics that differ from those typically seen in ideal plug flow conditions. Short peak flow times within CW are often linked to low hydraulic efficiencies. Figure 5.17b shows that the flow uniformly passed through the wetland, which is linked to a high hydraulic efficiency because of the lack of recirculation zones.

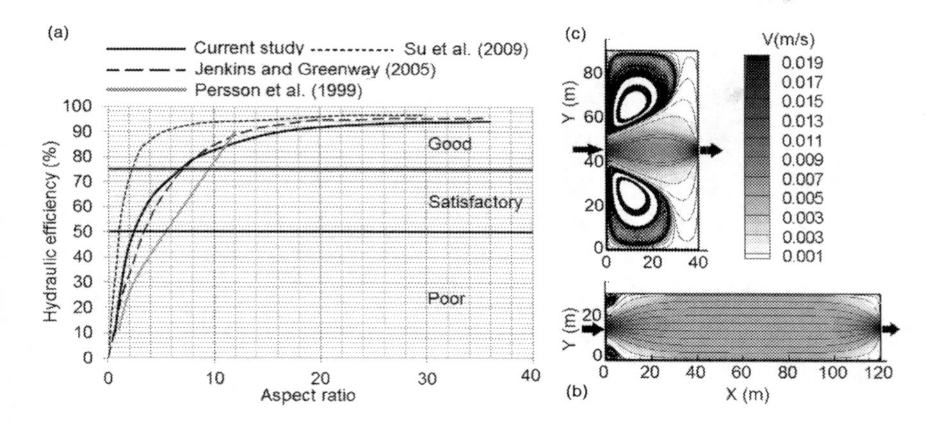

FIGURE 5.17 (a) Relationship between the hydraulic efficiency and the aspect ratio of the Re series of constructed wetlands (CW) for this and related studies (Su et al. 2009; Jenkins and Greenway 2005; and Persson et al., 1999); (b) schematic view of a simulated velocity (V) field and stream traces in Re(6) CW and (c) Re(4).

5.3.3.2 Modified Constructed Wetland Shape Simulations

Generally, two low or dead recirculation zones at the inlet and two low velocity zones at the outlet are formed in rectangular CW with midpoint inlet and outlet compositions (Figure 5.17b; recirculation zones and low velocity zones). The effective volume is therefore not optimal. One simple remedy could be to reduce areas with not optimal hydrodynamic behaviour such as recirculation and low velocity zones.

In this study, each rectangular CW (Re series) was modified by cutting dead recirculation and low velocity zones. Three CW series comprising 33 artificial wetlands were subsequently defined: rounded vertices of CW (Rb series), rounded vertices of the inlet part of CW (Rl series) and rounded vertices of the outlet part of CW (Rr series). In geometry, a vertex is defined as a point where two or more curves, lines or edges meet or intersect.

Tables 5.25–5.27 summarize the characteristics and simulation results for Rb, Rl and Rr CW. Findings show that reducing dead zones and areas of low velocity should improve the hydraulic efficiency. The area and nominal retention time of each modified CW (Rb, Rl and Rr series) is less than the corresponding original rectangular CW (Re series) because of cutting recirculation and low velocity areas. Figure 5.18 compares the velocity fields and stream traces for Re(4) and Rb(4) CW.

Figure 5.18a shows the flow velocity and stream traces for the Re(4) CW. Two large recirculation zones at the inlet and low velocity zones at the outlet appeared. In Figure 5.18b, the geometry is modified by cutting the four corners (rounded vertices), which makes the recirculation and low velocity zones smaller (increasing λ from 0.44% to 48.00%). Reducing dead zones and low velocity areas improves the hydraulic efficiency (Tables 5.25–5.27). Figure 5.18a compares the effect of aspect ratio on the modified shapes CW series (Rb, Rl and Rr) with that on rectangular CW series (Re) using semi-logarithmic graphs. Figure 5.18b illustrates the impact of rounded vertices in terms of hydraulic efficiency as a function of aspect ratio variations for rectangular CW.

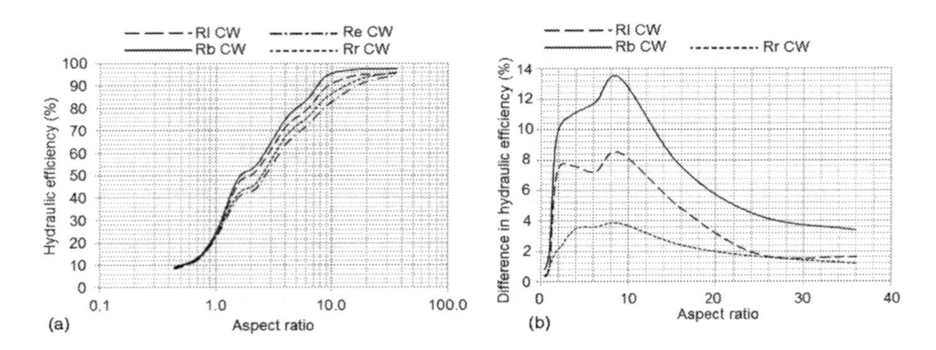

FIGURE 5.18 Simulated velocity (V) field and stream traces in (a) Re(4) and (b) Rb(4) constructed wetlands.

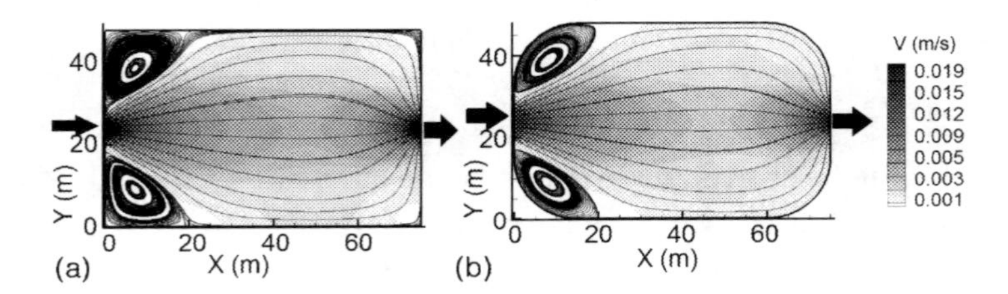

FIGURE 5.19 (a) Illustration of the hydraulic efficiency versus the aspect ratio for Re, Rb, Rl and Rr series of free surface water flow constructed wetlands (CW) and (b) difference hydraulic efficiencies of Rb, Rl and Rr CW versus rectangular CW (Re series) based on the aspect ratio variations.

Results indicate that rounded vertices have the greatest impact on CW with aspect ratios between 2 and 10 (Figure 5.19b). For instance, the Re(5) CW with an aspect ratio of 2.25 had a poor hydraulic performance ($\lambda = 44\%$), but corresponding CW with modified shapes had a satisfactory performance; for example, Rb(5) ($\lambda = 55\%$) and Rr(5) ($\lambda = 52\%$). Furthermore, the Re(7) CW with an aspect ratio of 6.25 had a satisfactory hydraulic performance ($\lambda = 73\%$), but a corresponding modification of the shape led to a good performance as indicated by Rb(7) ($\lambda = 85\%$), Rr(7) ($\lambda = 80\%$) and Rl(7) ($\lambda = 76\%$). From the hydrodynamic point of view, CW with modified shapes performed better than those with a rectangular geometry (Re series). Moreover, the Rb series was superior to the Rl and Rr series.

5.3.3.3 Inlet and Outlet Configuration Simulations

Three types of inlet and outlet configurations (corner/corner as applied in the Ri1 series; both corners/midpoint as shown by the Ri2 series; and uniform/midpoint as applied in the Ri3 series) were designed and modelled using 11 different aspect ratios (Figure 5.11). Tables 5.16–5.18 show the characteristics and simulation results for these three cases. All 33 CW have a volume of 18,000 m^3 and a T_n of 10 hours. Figure 5.19a indicates the results for the inlet and outlet configurations (Ri1, Ri2 and Ri3) as well as those for the midpoint/midpoint CW (Re series) using a semi-logarithmic plot.

Figure 5.19b and c shows the graphical results of the flow field and stream traces for Ri(1) and Ri(3), respectively.

Two large recirculation zones at the inlet and low velocity zones at the outlet appeared at the opposite corners of the Ri1(1) CW (Figure 5.20b). Uniform water flow avoided recirculation zone formation in the inlet area. This led to an increase of the hydraulic efficiency, especially for CW with low aspect ratios (up to 500% for Ri1(3) CW; Figure 5.20c and Table 5.9). Tables 5.13–5.15 and Figure 5.21 indicate that uniform-midpoint cases (Ri(3) series) were superior to other configurations for all tested aspect ratios. Furthermore, for low aspect ratios (<4) using the proposed corners-midpoint (Ri2 series) and corner-corner (Ri1 series) inlet and outlet configuration can improve the hydrodynamics of CW. However, considering aspect ratios of more than 4, the Ri2 and Ri3 forms of inlet and outlet arrangements worsen the treatment efficiency of the CW, and midpoint-midpoint (Re series) arrangements are more appropriate.

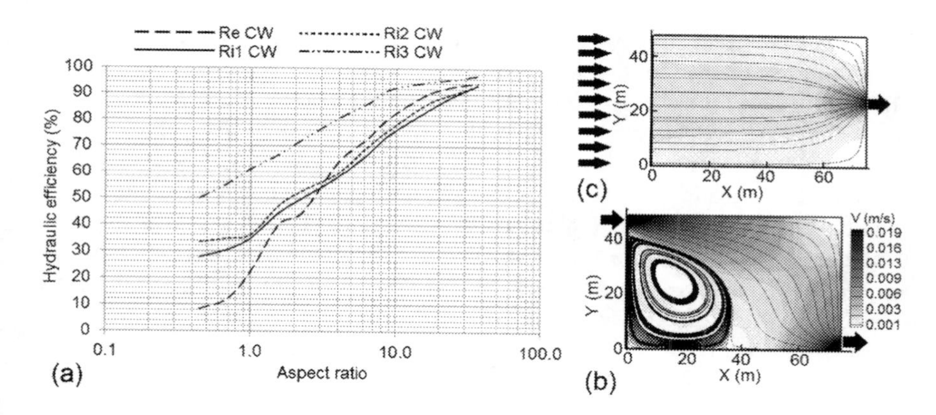

FIGURE 5.20 (a) Simulation results of constructed wetland (CW) hydraulic efficiencies against aspect ratios considering diverse types of inlet and outlet configurations; (b) velocity (V) field and stream traces of Re1(4) CW and (c) velocity field and stream traces of Re3(4) CW.

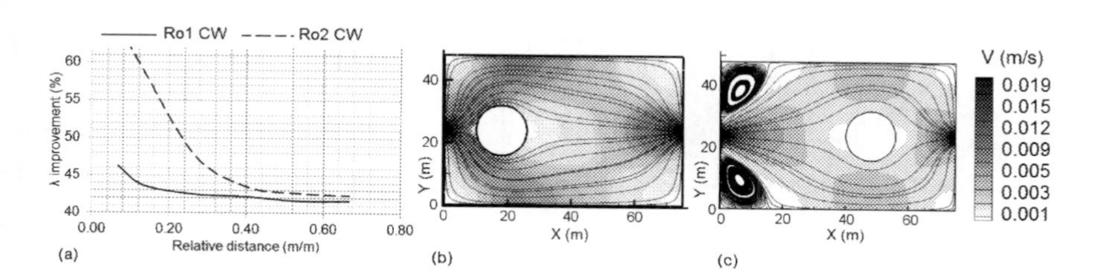

FIGURE 5.21 (a) Rate of hydraulic efficiency improvement versus the relative distance of the obstruction from the inlet part, and distribution of flow field and velocity (V) magnitude for different cases of various obstruction designations: (b) Ro1(2) constructed wetland (CW) and (c) Ro1(5) CW.

5.3.3.4 Obstruction Simulations

This section evaluates the hydraulic efficiency of the Re(4) CW in the presence of an island (obstruction) in front of the inlet part. Six surface (Rol series) and six submerged obstruction states (Ro2 series) were considered. The aspect ratio was set to 1.56, and the wetland area was fixed at 3,600 m^2.

Simulation results for obstructions are listed in Table 5.19. Figure 5.21 compares the hydraulic efficiency of the Rol series CW (surface islands) with those of the Ro2 series (submerged cases). Results indicate that using an emergent obstacle in front of and near the inlet part of the CW is likely to improve the treatment efficiency. For instance, placing the island near the inlet part increased the hydraulic efficiency by up to 63%. This resulted in a shift of the status for the CW Re(4) from poor to satisfactory. The maximum relative distance from the inlet was 0.2 m/m (Table 5.19 and Figure 5.21a). Figure 5.21 illustrates that the surface and submerged obstructions enhance λ by more than 10% for the maximum relative distance of 0.35/0.12, which is the ratio between the distance of the obstacle from the inlet and the length of the CW. It follows that locating an emergent (surface) obstacle far away from the inlet will not improve the hydraulic efficiency.

Table 5.19 and Figure 5.22a show that surface obstacles were superior to submerged obstacles in enhancing the hydraulic efficiency of CW. Submerged islands were not associated with a marked improvement (about 16% only) of λ even if they were located near the inlet of the CW. However, it should be noted that all results are based on the absence of vegetation in the CW. Otherwise, a considerable presence of vegetation such as emergent macrophytes on top of the submerged obstruction would change the characteristics of the obstacle and would make it more similar to an island. Figure 5.22b and c shows the flow conditions with velocity magnitudes and directions for Rol(2) and Rol(5) CW. There are no recirculation zones in Rol(2) CW because of using the surface obstacle in front of the inflow (spreading inflow; Figure 5.22b).

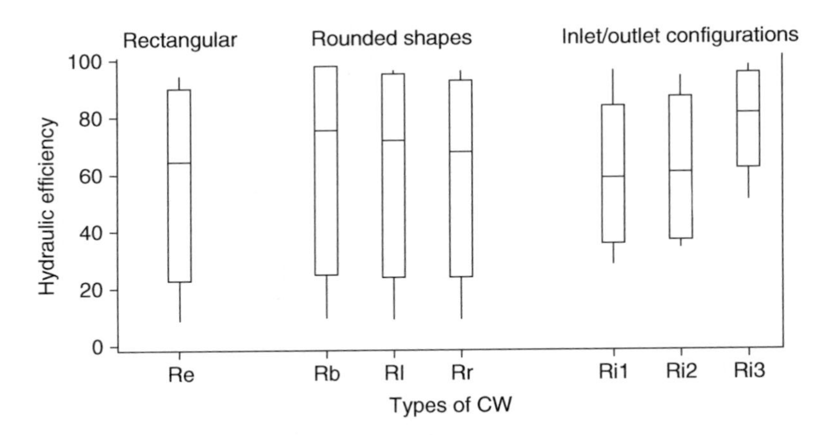

FIGURE 5.22 Graphical illustration of constructed wetland (CW) hydraulic treatment based on the aspect ratio criteria using box plot.

However, by replacing the obstacle near the outlet part, recirculation zones emerge at the corner of the inlet (Figure 5.21c).

5.3.3.5 Discussion of Wetland Designs

Overall findings indicate that the hydraulic efficiency increases with the aspect ratio, which compares well with the literature reviewed by Scholz et al. (2007). As for the basic models of the Re series (rectangular CW), the hydraulic efficiency of the CW improves significantly when the aspect ratio is smaller than four (Table 5.19). Similar findings can be seen for modified rounded CW (Rb, Rl and Rr series; see Tables 5.25–5.27). However, a logarithmic increase of λ for a decreasing aspect ratio can be seen if different inlet and outlet configurations are being tested (Tables 5.16–5.18). The related curves shown in Figure 5.21 for Ri1, Ri2 and Ri3 seem to be linear on logarithmic scale. Considering midpoint-midpoint rectangular CW cases (Re series), an increase of the aspect ratio does not have an obvious influence on λ when the aspect ratio is higher than four. Therefore, if there are no space limitations on constructing CW, the aspect ratio should be larger than nine because λ can reach at least 80%. On the contrary, if there are space restrictions, modified rounded CW (Rb series) should be considered in the design phase, because an aspect ratio of six can reach λ values of 80%.

When designing FWS CW with low aspect ratio (not recommended from a hydraulic point of view), the design engineer should modify the configurations of the inlet and outlet to improve the flow uniformity and to increase λ (Ri1, Ri2 and Ri3 series). Moreover, designing CW with an obstruction (e.g., island) in front of the inlet should improve λ. To evaluate the hydraulic efficiency of different FWS CW cases, Table 5.20 lists λ values for CW with a low aspect ratio of 1.56. Considering that the Re(4) CW has a λ value of 40%, the best way to improve hydraulic treatment in wetlands is to promote a uniform flow entry, which could improve λ by 66.7% (Table 5.20; rank

TABLE 5.20
Comparison of the Hydraulic Efficiency for Different Types of Free Surface Flow Constructed Wetlands (CW) with an Aspect Ratio of 1.56

Case	Definition	λ^a (%)	λ^a Improvement (%)	Rank
Re(4)	Rectangular CW	40.0	–	–
Rb(4)	Modified rounded shape (all corners)	48.7	21.7	3
Rl(4)	Modified rounded shape (left corners)	46.3	15.8	4
Rr(4)	Modified rounded shape (right corners)	41.8	4.6	8
Ri1(4)	Corner/corner inlet/outlet	44.4	11.1	7
Ri2(4)	Corners/midpoint inlet/outlet	47.2	18.1	6
Ri3(4)	Uniform/midpoint inlet/outlet	66.7	66.7	1
Ro1(1)	Surface obstruction	65.0	62.5	2
Ro2(1)	Submerged obstruction	46.3	15.6	5

[a] Hydraulic efficiency.

of Ri3(4) is 1). This design option is followed by locating a surface obstruction in front of the inlet, which results in an improvement of λ by 62.5% (Table 5.20; rank of Ro1(1) is 2).

Figure 5.22 shows the treatment of several types of CW based on the aspect ratio variations using box plots. The lower whisker represents the first CW (aspect ratio of 0.44). The IQR stands for the third CW (aspect ratio of 1.00). The median represents the sixth CW with an aspect ratio of 4.00. The upper IQR stands for the ninth CW (aspect ratio of 16.00), and the upper whisker indicates the eleventh wetland with an aspect ratio of 36.00.

Figure 5.22 indicates that for FWS CW with very low aspect ratios (<1), the main design criterion are the inlet and outlet configurations. The lower whiskers and the IQR of Ri1, Ri2 and Ri3 are noticeably higher than the others. Although the most reasonable type of CW is Ri3, which has a uniform/midpoint inlet and outlet form, this wetland type is unrealistic due to its wide inlet part (e.g., 80 m) and an extremely low corresponding inflow velocity. In that case, the Ri1 and Ri2 series are more applicable. However, using midpoint/midpoint inlet and outlet flows is not recommended.

For FWS CW, low aspect ratios (<4), rounded corners (Rb series with four rounded corners and Rl with left rounded corners) and uniform/mid-point flow are generally advantageous (Figure 5.23; see median line for box plots). Rb (modified shape with four rounded corners) and Ri3 (uniform/mid-point) series of FWS CW with aspect ratios within 4 and 16 are highly recommended (Figure 5.23; see upper IQR of box plots). As for high aspect ratios (>16), designers can choose any type of CW except for the Ri1 (corner/corner) and Ri2 (corners/mid-point) series, which have the lowest upper IQR (Figure 5.23). The best design option is the Rb CW (modified rounded shape for each corner) for which the upper IQR is associated with the upper whisker.

5.3.4 Conclusions and Recommendations

FWS CW can be used to detain storm water and treat a wide range of pollutants from urban runoff and municipal wastewater. For economic and environmental reasons, an optimal design in terms of hydraulic efficiency is important. Numerical simulations can be used to optimize the wetland geometry hydrodynamically. In this study, a two-dimensional depth-averaged, free surface flow solver was developed and applied for 89 artificial (simulated) CW with different aspect ratios and shapes. The solver was implemented on unstructured triangular meshes and the solution methodology was based upon a central Galerkin finite volume formulation.

Results indicate that the aspect ratio is one of the main design criteria, which influence the hydraulic efficiency. For rectangular CW with central point inlet and outlet arrangements, the aspect ratio should preferably be greater than nine. If space restrictions in practice do not allow for the desired aspect ratio, modified rounded rectangular CW should be used. In that case, the aspect ratio has to be at least six to maintain uniform flow. In special conditions with a low aspect ratio, changes to the inlet and outlet configuration and the design of an obstruction close to the inlet of the CW can be considered to increase the hydraulic efficiency. For a low aspect ratio (<6), the hydraulic efficiency associated with the inlet and outlet configuration is as follows: uniform-midpoint > corners-midpoint > corner-corner > midpoint-midpoint.

This study had a limited scope as outlined above. It has to be noted that all findings are based on the absence of significant vegetation in the simulated CW. However, the effect of wetland shape on hydraulic efficiency has a reduced impact as the density and spatial distribution of fringing vegetation increases. The influence of plants in the CW was positive, increasing the removal efficiency especially for nitrogen, phosphorus and potassium. Mass removal of TKN in the CW was 33% greater than that obtained in the unplanted CW. Therefore, the author recommends undertaking of further research to assess the influence of different types of wetland vegetation during different growing seasons in terms of hydraulic efficiency and water quality improvements for real wastewaters.

The effects of evapotranspiration were neglected in this study. However, evapotranspiration can decrease the volumetric flow, thereby increasing the hydraulic retention time and concentrations of dissolved constituents. Also investigating the effects of wind on the mixing factors on the solute tracer and hydraulic efficiency is suggested for further studies. Moreover, applying the moments of tracer breakthrough curves method is recommended for calculating the hydraulic efficiency for comparative purposes in future research.

5.4 SOFT COMPUTING APPROACHES IN MODELLING AERATION PROCESSES

5.4.1 Introduction

Cavitation is common in hydraulic systems including turbines and hydraulic structures such as dams, especially in very large dams. High velocities associated with high flows through spillways and bottom outlets are likely to occur when the water surface elevation for the reservoir rises above the active conservation (normal) pool elevation.

Cavitation within a hydraulic structure is defined as formation of vapor cavities in water. Bubbles are formed when the static pressure of the liquid drops below its vapor pressure (Kramer et al., 2006) and are swept downstream, where pressure is higher, and the bubbles implode. Collapsing bubbles appear over a very short time and cause pressure fluctuations close to the spillway surface (Volkart and Rutschmann, 1984). This instantaneous pressure can cause fatigue and subversion of spillway material, and continued material removal can lead to significant damage (Kells and Smith, 1991). Figure 5.23 provides an example of a spillway collapse due to the cavitation in the chute spillway of Oroville dam in California in 2017.

Peterka (1953) demonstrated that entering of 2% air to the flow greatly reduced the damage of cavitation to concrete dams. Jain and Chao (2011) pointed out that aeration is an effective and low-cost method for reducing cavitation damage. Aerators are used in spillways for artificial air entering to the flow. Aerators are embedded at the upstream end of the overflow, which could be the first place where cavitation damage may occur. The aerators separate the flow from the chute bottom to generate a free jet (Pfister, 2011). Generally, aerators consist of two parts: a discontinuity of the chute bottom and the air duct. The air entrained into the flow is usually provided by lateral

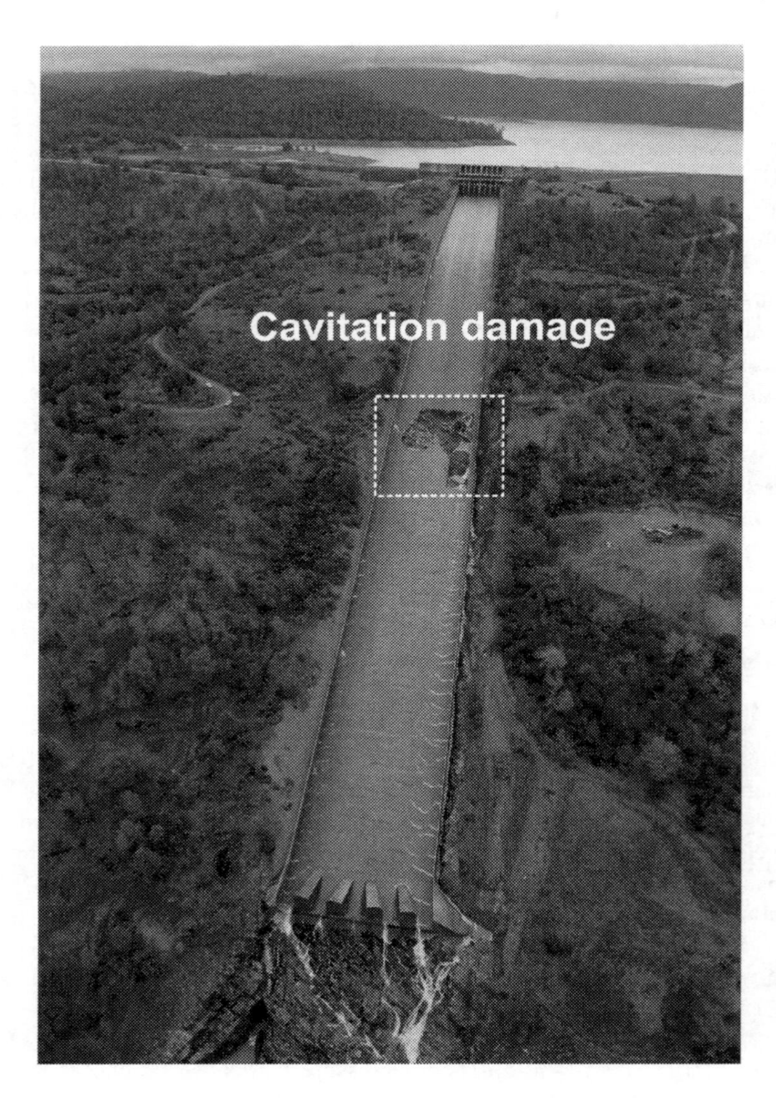

FIGURE 5.23 Spillway collapse due to cavitation phenomenon in the Oroville spillway in 2017 (adapted after unpublished photo received from the California Department of Water Resources).

ducts connected with the atmosphere (Pfister et al., 2011). A schematic spillway aerator placement and aeration is shown in Figure 5.24a.

Aerator shapes include groove, offset and deflector types. Ramps can be use individually or with other shapes of aerators. For example, the use of groove and ramp aerators caused high sub-pressure behind the ramp compared to groove without ramp, increasing the amount of entrained air (Ruan et al., 2007).

The required aerator air flow is one of the most important elements in designing hydraulic structures. The amount of air needed for cavitation protection was

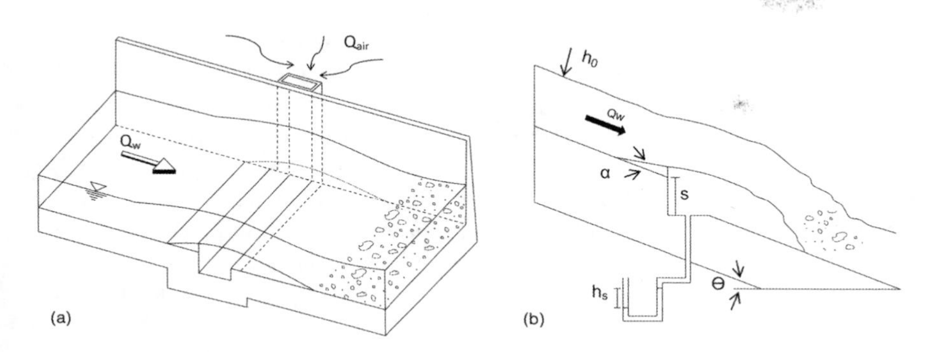

FIGURE 5.24 Schematic plots of (a) spillway aerator placement and (b) effective parameters on an aerator performance.

questioned since about 1960. However, researchers agreed that a small amount of air near to the chutes could significantly reduce the risk of cavitation damage (Kramer et al., 2006). The required aerator air flow can be estimated by using laboratory, experimental and soft computing methods. Although measurements of air demand by building hydraulic models are the most promising method, yet they are costly and time-consuming, thus empirical and soft computing methods could be considered as good alternatives.

Examples of research on empirical methods for tunnel outlets are documented in the literature: U.S. Army Corps of Engineers (1964), Campbell and Guyton (1953), Wisner (1965) and Sharma (1976). Schwarz and Nutt (1963) studied the jet length of flow aerator systems and presented a theoretical equation. Pan et al. (1980) and Pinto et al. (1982) showed empirical relations for the aerator air demand based on jet length. Pfister et al. (2011) referred to research on empirical methods for spillways.

Soft computing methods, including techniques such as neural networks, fuzzy inference systems, genetic algorithms (Najafi et al., 2012) and neuro-fuzzy (ANFIS) (Zounemat-Kermani et al., 2017), have been used to predict tunnel aerator air demands with acceptable prediction accuracy. Zounemat-Kermani and Scholz (2013) pointed out that the neuro-fuzzy method has acceptable accuracy in predicting the required tunnel aerator air demand with low head. Recently, researchers used optimization algorithms to optimize soft computing methods such as ANFIS, neural networks and machine learning. Zaji et al. (2016) predicted the discharge coefficient of a labyrinth side weir using the firefly algorithm (FA) combined with support vector regression. Results showed that the FA increased the accuracy by about 10%. Gholami et al. (2018) predicted the threshold bank profile shape using the adaptive neuro-fuzzy inference system-particle swarm optimization/genetic algorithm (ANFIS-PSO/GA) and confirmed the high accuracy of this method. Salimi et al. (2020) used artificial intelligence (AI) to design sewage transfer system resistant to water hammers and showed that both ANFIS and ANFIS-PSO have acceptable performance figures. Ahmadlou et al. (2019) used the adaptive neuro-fuzzy inference system-biogeography-based optimization (ANFIS-BBO) method to flood susceptibility assessment and showed BBO algorithm increasing the accuracy of ANFIS method.

Generally, the main purpose is to accomplish a comprehensive evaluation of the use of several standard (multi-layer perceptron neural network (MLPNN), radial basis neural network (RBNN), adaptive neuro-fuzzy inference system (ANFIS-FCM) and Wavenet) and hybrid (ANFIS-PSO, ANFIS-GA, ANFIS-FA, ANFIS-BBO) AI-based soft computing models as well as multiple linear regression (MLR) and five empirical models for aerator air demand prediction in dams.

Section 5.4 assesses the usability of both soft computing techniques and empirically based methods for estimating the required air demand of spillway aerators using data from three related experiments. Different statistical approaches were employed to measure the performance of the models. Moreover, comparisons were carried out between the heuristic models and several empirical relations. This chapter is based on an updated original article by Mahdavi-Meymand et al. (2019).

5.4.2 Materials and Methods

5.4.2.1 Empirical Relationships and Linear Regressions

AI-based models including the standard ANFIS-FCM, Wavenet, MLPNN and RBNN are applied for the estimation of the aerator air demand. In the next phase, to challenge the performance of meta-heuristic methods, four heuristic algorithms namely GA, PSO, FA and BBO are combined with ANFIS. In addition to AI-based models, the MLR and some empirical relations are also used for this purpose.

Campbell and Guyton (1953), U.S. Army Corps of Engineers (1964), Wisner (1965) and Sharma (1976) estimated the aeration coefficient (β) of tunnels according to Equations 5.8, 5.9, 5.10 and 5.11, respectively.

$$\beta = 0.04 \times (Fr_c - 1)^{0.85} \tag{5.8}$$

$$\beta = 0.03 \times (Fr_c - 1)^{1.06} \tag{5.9}$$

$$\beta = 0.24 \times \left(Fr_c - 1\right)^{1.4} \tag{5.10}$$

$$\beta = 0.09 \times Fr_c \tag{5.11}$$

where β is the estimated the aeration coefficient and Fr_c is the critical Froude number. Pfister et al. (2011) proposed Equation 5.8 to estimate the required spillway aerator air demand.

$$\beta = 0.0028F_0^{\,2}\left(1 + F_0^{\,2}\tan\alpha\right) - 0.1, \qquad 0 < \alpha < 11.3 \tag{5.12}$$

where F_0 is the Fr number in the aerator location, β is the aeration coefficient (proportion of air discharge to water discharge ($\beta = Q_a/Q_w$)) and α is the ramp angle.

The MLR is denoted by Equation 5.13. The coefficients of the equation were calculated such that the sum-of-square of differences of observed and predicted values is minimized (Zounemat-Kermani and Scholz, 2013).

$$Y = \gamma_0 + \gamma_1 X_1 + \gamma_2 X_2 + \gamma_3 X_3 + \cdots \gamma_n X_n + \varepsilon \tag{5.13}$$

where γ_i are constant regression coefficients, X_i are the elements of the input vector, ε is the random error and Y (Q_{air}) is the output.

5.4.2.2 Neural Network-Based Models

Neural networks have been successfully applied in various fields including hydraulic systems. The MLP is one of the neural network techniques, which has a promising application and great ability to approximate complex functions. An MLPNN includes an input layer, hidden layer(s) and an output layer. According to the pertinent reports on the superiority of a two-layer MLPNN (with one hidden layer and one output layer), a one hidden layer MLPNN was developed (Azizi et al., 2016). Perceptron neural networks can be modelled using Equations 5.14 and 5.15.

$$a_j^1(t) = f\left[\sum_{i=1}^{m} w_{j,i}^1 p_i(t) + b_j^1\right] \qquad 1 \le j \le n \tag{5.14}$$

$$a_k^2(\mathrm{t}) = g\left[\sum_{i=1}^{n} w_{\mathrm{k,j}}^2 a_j^1(\mathrm{t}) + \mathrm{b}_k^2\right] \qquad 1 \le j \le n_o \tag{5.15}$$

where m is the number of inputs, p is the input parameter, w^1 and w^2 are the input and output layer weight matrices, n is the number of neurons in the hidden layer, n_o is the number of outputs, b_1 and b_2 are bias matrixes of hidden and output layers, a^1 and a^2 are the outputs from hidden and output layers and f and g are output transfer functions for hidden and output layers (Asadollahfardi et al., 2012).

In the Levenberg-Marquardt (LM) method, the processor parameters of the neural network are changed by using Equation 5.16 (Costa et al., 2007).

$$\chi_{k+1} = \chi_k - \left[H + \eta I\right]^{-1} J^T e(t) \tag{5.16}$$

where k indicates counter repeated learning, χ is a vector of weights and biases, g_k is the gradient current rate and η is the learning rate, H is the Hessian matrix and J is the Jacobian matrix.

The LM algorithm uses second derivatives without direct use of the Hessian matrix. The second derivatives and the gradient matrix can be estimated using Equation 5.17.

$$H = J^T J, \quad g = J^T e \tag{5.17}$$

The RBNN network is a feed-forward type of artificial neural network (ANN). This neural network was introduced by Bromhead and Lowe (1988) and learns by measuring the Euclidean distance of data (Trajkovic et al., 2000). The RBNN network has a simple structure comprising one input layer, one hidden layer and an output layer. This network uses radial basis functions as excitation of hidden neurons.

The Gaussian function is most popular owing to its flexibility of adjusting the function position and shape via the spread parameter (Ha et al., 2015). Therefore, the Gaussian function was used in Section 5.4 to train the RBNN network. For the RBNN network output of the jth hidden layer, X_p input is calculated by using Equation 5.18.

$$\phi_j = f\left(\frac{\|X_P - U_j\|}{2\sigma_j^2}\right) \tag{5.18}$$

where $\| \ \|$ is the Euclidian norm, U_j is the centre of the jth radial basis function f and σ is the spread of the RBNN that is indicative of the radial distance from the RBNN centre within which the function value is significantly different from zero (Kumar et al., 2012). The outputs of the network are calculated by applying Equation 5.19.

$$y_k = \sum_{j=1}^{L} \phi_j w_{jk} + b_k \tag{5.19}$$

where ϕ_j is the jth output of the hidden layer, w_{jk} is the weight between the jth hidden and the kth output, b_k is biasing of the output node and L is the number of neurons in the hidden layer.

Wavenet belongs to a relatively new class of neural networks with unique capabilities in system identification and classification (Zhou et al., 2004). The wavelet theory provides effective instructions for the construction and initialization of the networks that significantly reduce the training time (Postalcıoğlu et al., 2005). Wavenet is associated with new parameters and can provide a better performance of approximation than ordinary basis function networks such as RBNN (He et al., 2002).

Wavenet is a type of feed-forward ANN with one hidden layer of nodes, whose basis transfer functions are drawn from a family of orthonormal wavelets (Bakshi and Stephanopoulos, 2004). In these networks, only weight training functions do not change. Contrary to ANN, the Wavenet network structure is determined according to the desired signal and with special algorithms, whereas in neural networks trial and error is used to determine the structure of the network. In the Wavenet network structure, various wavelet functions such as Haar, Gaussian, Symlet and DavChyz can be used.

In this study, the Gaussian wavelet function was used for modelling. In general, a Wavenet network is made up of three layers: input, hidden and output. In the input layer, the explanatory variables are introduced to the Wavenet. In the hidden layer, the input variables are transformed to the dilated and translated version of the mother wavelet. In the output layer, the approximation of the target values is estimated (Alexandridis and Zapranis, 2013). The wavelet of the hidden layer in the wavenet (Ψ_j) has been replaced by the sigmoid functions in the neural networks. This layer has been obtained from the mother wavelet (Ψ_j) by applying Equation 5.20.

$$\psi_j(x) = \prod_{k=1}^{n} \psi\left(t_{jk}\right), \qquad t_{jk} = \frac{x_k - bt_{jk}}{a_{jk}} \tag{5.20}$$

where bt is the translation parameter and a is the dilation parameter. The network output can be obtained as shown by Equation 5.21.

$$y = \sum_{j=1}^{m} W2_j \psi_j(x) + b + \sum_{K=1}^{n} W1_k x_k \tag{5.21}$$

where $W1$ and $W2$ are weights, and b represents bias.

An ANFIS is a hybrid ANN benefiting from fuzzy logic for constructing the inference system (Jang, 1993). Mamdani and Sugeno are two types of fuzzy inference systems that can be implemented (Mamdani and Assilian, 1975; Takagi and Sugeno, 1985). The Sugeno system is both more compact and computationally efficient than the Mamdani system (Abbasimehr and Tarokh, 2015). Thus, the Sugeno inference system is used as an adaptive technique for constructing the fuzzy models.

In the basic ANFIS, the gradient descent and the least square method caused the learning algorithm for the ANFIS (Jang et al., 1997). To illustrate the structure and mechanism of the ANFIS model, a fuzzy model based on the Sugeno system is presented using Equation 5.22.

$$\text{if}\left(x_1 \text{ is } A_1\right)\text{and}\left(x_2 \text{ is } A_2\right)\ldots\text{and}\left(x_n \text{ is } A_n\right)\text{then}\left(f_i = p_1x_1 + p_2x_2 + \cdots + p_nx_n + r_1\right) \tag{5.22}$$

where A_1, A_2 and A_n are the fuzzy sets in the antecedents with membership functions $\mu_A(x)$; x_1, x_2 and x_n are input variables; P_1, P_2, P_n and r_1 are the parameters of output functions and f_i is output variable.

Figure 5.25 shows the ANFIS structure with five layers and two inputs. The attachment level of each layer to each fuzzy set is identified according to the relations shown in Equation 5.23.

$$\begin{aligned} O_1^i &= \mu_{Ai}(x) \qquad \text{for } i = 1,2 \\ O_1^i &= \mu_{Bi-2}(y) \qquad \text{for } i = 3,4 \end{aligned} \tag{5.23}$$

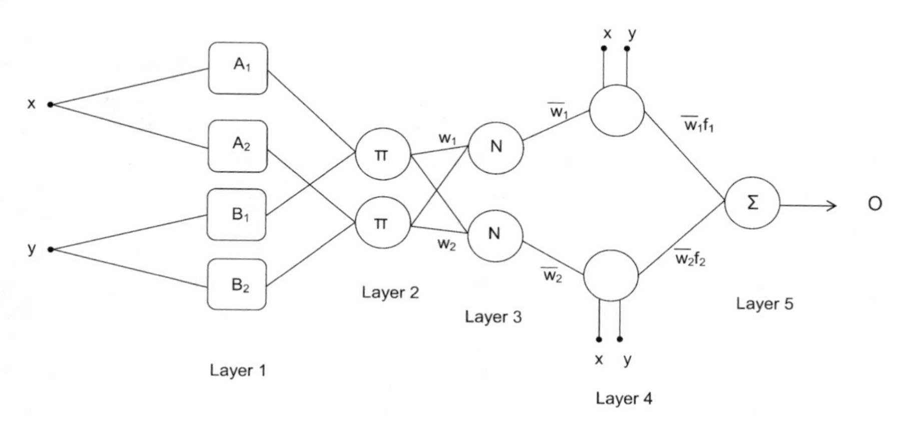

FIGURE 5.25 A schematic diagram example of an adaptive neuro-fuzzy inference system structure with five layers and two inputs.

where x and y are the outputs to node i, and A_i and B_i are the fuzzy sets related to node i.

In the second layer, weights of the rules (operator II) are calculated by multiplying the inputs of this layer by each other (Equation 5.24).

$$O_2^i w_i = \mu_{Ai}(x)\mu_{Bi}(y) \qquad i = 1,2 \tag{5.24}$$

In the third layer, relative weights of the rules are calculated and the effect of the ith rule with respect to other rules (Lin et al., 2012) becomes clear when applying Equation 5.25.

$$O_3^i \ \overline{w_i} = \frac{w_i}{w_1 + w_2} \qquad i = 1,2 \tag{5.25}$$

The fourth layer is the rule layer, which is calculated using the inputs of node I (Equation 5.26).

$$O_4^i = \overline{w_i} \ f_i \tag{5.26}$$

The last layer is the output layer. The Rhe network's output layer is calculated using Equation 5.27.

$$O_5^i = \sum_i \ \overline{w_i} \ f_i \tag{5.27}$$

In this study, the fuzzy c-means (FCM) clustering type of ANFIS has been used.

5.4.2.3 Meta-Heuristic Algorithms

In this chapter, the four meta-heuristic methods of GA, PSO, FA and BBO have been conjugated with the basic ANFIS as four alternative hybrid methods: ANFIS-GA, ANFIS-PSO, ANFIS-FA and ANFIS-BBO. In fact, ANFIS provides the search space and employs optimization methods for finding the best solution.

To begin the modelling process, the initial population (for GA and BBP), particle (for PSO) and firefly population (for FA) are determined randomly. Then ANFIS is trained for each particle or firefly using training data. Fitness values (or cost values) must be calculated by using fitness functions. In this study, the fitness is defined as the root mean square error (RMSE) between the actual output and the prediction output. After the optimization method optimized the ANFIS parameters, these parameters returned to the ANFIS structure, and then the model was ready to use.

The PSO algorithm was proposed by Kennedy and Eberhart (1995) and is inspired by the sociological behaviour of food searching criteria such as birds flying simultaneously, fish swimming behaviour, ant interactions and other animals. The PSO is an effective optimization algorithm, which quickly finds the optimal solution for a non-linear problem (Ali et al., 2017). The PSO is a population that is based on a stochastic optimization algorithm. In populations, individuals interact with each other and learn from their experiences. When an agent finds a better solution (current global best), the overall population of this individual gradually move to a better solution. More details of PSO have been presented by Kim et al. (2017).

The FA is based on the flashing behaviour and characteristics of fireflies. This algorithm was developed by Yang in 2007 (Yang, 2009; Kumar et al., 2017). Yang (2009) presented three rules for the flashing characteristics and formulated the FA based on these rules (Kamarian et al., 2017):

1. All fireflies are unisex, so that any individual firefly will be attracted to other fireflies irrespective of their sex.
2. The attractiveness is proportional to their flashing brightness, since the air absorbs light from the other firefly, causing decreases in attractiveness. Considering only two fireflies, the brighter firefly will be more attractive than the less bright one. In case of no firefly being brighter than an other, the flies move randomly.
3. The brightness of the firefly can be considered as an optimized objective function of problems. For more details of FA, refer to previous publications (Sayadi et al., 2013; Mohanty, 2015; Banerjee and Mandal, 2017).

The GA is known as a robust meta-heuristic method, which is commonly applied for solving optimization problems. This algorithm was first introduced in 1975 by Holland (1975) and was inspired by natural selection and evolutionary theories (Goldberg, 1989). The GA uses several populations of randomly generated individuals (phenotypes) in the solution space for seeking the best optimum. Everyone has specified parameters (genes into each chromosome), which makes the genotype of each individual.

The evolution process is an iterative process through a generation procedure by using operators such as crossover and mutation. The generation procedure is known for having better performance in the subsequent populations based on the fitness function.

In this study, the error criteria of the training set of data are considered as the fitness function of the algorithm. Moreover, the roulette wheel selection is used to determine the best individuals. For attaining more information about the GA, readers are recommended to refer to Wang (1991) as well as Ghamisi and Benediktsson (2015).

The BBO is a meta-heuristic algorithm, which is naturally inspired and extended based on the mathematics of biogeography science. The idea of this algorithm is based on the migration strategy of animal species (Simon, 2008). In comparison, in GA, solution features are called genes, but within the BBO algorithm, they are replaced by a parameter called suitability index variable. The BBO is a population-based optimization algorithm, which instead of using fitness values, uses islands (or habitats). Each island has its habitat suitability index (HSI), which reflects the algorithm's performance. A habitat with a high HSI shows better fitness than a habitat with a low HSI. Concerning the BBO algorithm, emigration and immigration are operators, which are used to improve the solution of optimization challenges. Habitats with higher values of HSI have a low immigrating rate, whereas low values of HSI for a habitat denote a higher immigrating rate of that habitat. More information about BBO can be found elsewhere (Zandieh and Roumani, 2017; Bozorg Haddad et al., 2016; Zhang et al., 2017).

5.4.2.4 Experimental Data

In this study, a total of 1,305 experimental datasets from five scientific reports were used. Exactly 914 experimental sets were retrieved from the reports on the Clyde Dam hydraulic model (Chanson, 1988). This spillway model was in the fluid mechanics laboratory of the Civil Engineering Department at the University of Canterbury, New Zealand. The model scale was 1:15. The flume geometry is characterized by 0.25-m width, 3.6-m length and a slope of 51.33°. The model provided a flow velocity between 3 m/s and 14 m/s, and a *Fr* number in the range of 3–25. Two aerator geometries were used. The first aerator was offset with 0.30-m height and a ramp of 5.7°. The second one was offset with 0.30-m height and no ramp (0°).

The second reference provided 365 datasets from an experiment carried out in the River Engineering Department of the Hydraulics Research Centre at Wallingford (May and Deamer, 1989). The dimensions of the model could be used for testing the model of the prototype aerators at scales between 1:10 and 1:20. The model could be used for prototypes with discharge values of up to 100 m^3/s/m. The flume width and slope angle were 0.3 m and 45.3°, respectively. The aerator construction includes ramps with angles between 4.6° and 9.1°.

The three remaining sources related to experiments conducted by the Tehran Water Research Center in Iran comprised 12 and 4 datasets, which were obtained from the Azad dam and Siyazakh dam hydraulic models, respectively. The remaining 10 datasets were gathered from experiments conducted by Javanbarg et al. (2003).

The Azad dam impacts on the Cham-Gore River in the region of Marivan in Iran. A model was created (scale of 1:33) with a flume geometry including 0.9-m width and an angle of 20°. The aerator geometry can be considered as groove with 0.1896-m height and a ramp of 7°. The flow discharge varied between 0.78 and 0.375 m^3/s, and the *Fr* number ranged between 4.4 and 8.36.

The scale of the Siyazakh model was 1:40. The flume width and angle were 0.5 m and 14.04° in this order. The aerator included an offset with 0.026-m height and a ramp of 5.09°. The flow characteristics comprised *Fr* number ranges between 2.53 and 5.09, and flow discharges between 0.0494 and 0.281 m^3/s.

For the last source (10 datasets), Javanbarg et al. (2003) used a 0.2-m flume with an angle of 14.5°. The flow discharge range was between 0.0215 and 0.0376 m^3/s, and the *Fr* numbers were between 7.7 and 8.19. The aerator geometry consists of a groove with 0.1-m height and three ramps with angles of 7°, 8° and 12°.

Each dataset consists of a dependent value (spillway aerator air demand, Q_{air}) and seven effective parameters as the input vector. The input vector includes flow discharge (Q_W), flow depth at the beginning of the aerator (h_0), the difference between the atmospheric pressure and the pressure under the aerator jet flow (h_s), aerator step height (s), shut angle (Θ), shut width (B) and ramp angle (a) (Figure 5.26b). Table 5.21 shows a summary of the gathered data in this study. The h_s parameter has the highest (inverse) correlation with Q_{air} and Q_w is the least correlated factor.

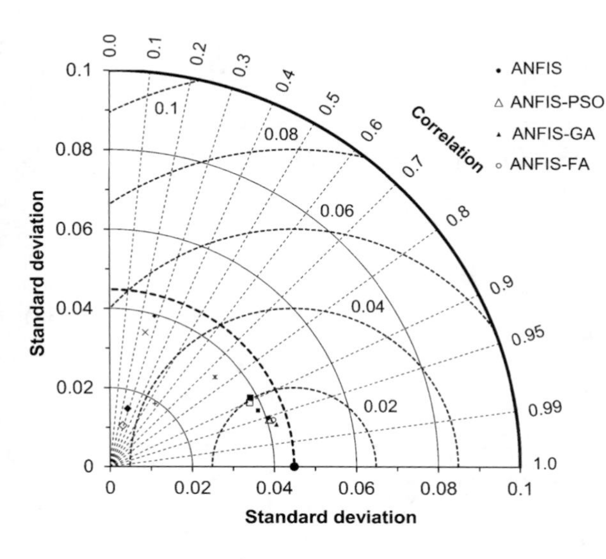

FIGURE 5.26 Taylor's diagram of used methods (Test stage).

TABLE 5.21
Statistical Characteristics of the Data Used in this Study

Parameter	Maximum	Minimum	Mean	Standard Deviation	r vs. Q_{air}
Q_{air} (m^3/s)	0.1806	0	0.035	0.040	1
h_0 (m)	0.405	0.023	0.062	0.029	−0.193
Q_w (m^3/s)	1.312	0.013	0.0996	0.055	−0.008
h_s (pa)	1761.78	−48.45	117.44	145.957	−0.283
s (m)	0.1896	0	0.024	0.0220	0.153
α (°)	12	0	4.914	2.898	0.075
Θ (°)	52.33	14.04	44.978	13.795	0.264
B (m)	0.9	0.2	0.270	0.066	−0.118

Note: Q_{air}, Air discharge; h_0, flow depth at the beginning of the aerator; Q_w, water discharge; h_S, pressure difference between the atmosphere and under the aerator jet flow; s, aerator step height; α, ramp angle; Θ, shut angle; B, shut width; and r is the correlation coefficient.

5.4.2.5 Evaluation of the Model Performances

To compare the performances of the different applied methods, the standard root mean square error (SRMSE), the coefficient of determination (R^2) and the index of agreement (IA) were used. Equations 5.28–5.30 denote the measures applied for the performance comparison.

$$\text{SRMSE} = \frac{\text{RMSE}}{\overline{Qa}} = \frac{\left[\frac{\sum_{i=1}^{N}\left(Qa_i^o - Qa_i^s\right)^2}{N}\right]^{0.5}}{\overline{Qa}} \tag{5.28}$$

$$R^2 = \frac{\left(\sum_{i=1}^{N}\left(Qa_i^o - \overline{Qa^o}\right)\left(Qa_i^s - \overline{Qa^s}\right)\right)^2}{\sum_{i=1}^{N}\left(Qa_i^o - \overline{Qa^o}\right)^2 \sum_{i=1}^{N}\left(Qa_i^s - \overline{Qa^s}\right)^2} \quad (5.29)$$

$$\text{IA} = \frac{\sum_{i=1}^{N}\left(\text{Q}a_i^o - \text{Q}a_i^s\right)^2}{\sum_{i=1}^{N}\left(\left|\text{Q}a_i^s - \overline{Qa^o}\right| + \left|\text{Q}a_i^o - \overline{Qa^o}\right|\right)^2} \quad (5.30)$$

where Qa^o is the measured aerator air flow, Qa^s is the simulated aerator air flow, N is the number of data and $\overline{Qa}$ is the average of air flow.

To set up all models, data were randomly divided into three sets: training data (70% of data), validation data (15% of data) and testing data (15% of data). Training datasets were used for building the models. The validation datasets were utilized for evaluating the models' fitness on the training dataset and avoiding over-training. Finally, the test datasets were applied for model evaluations.

5.4.3 Results

Before starting the modelling process, standardization ((x-min(x))/(max(x)-min(x)) was used to bring all values into the range [0,1]. The architecture of ANN, Wavenet and ANFIS required further investigations to determine their appropriateness. In this study, based on the hold-out method, various MLPNN, RBNN and ANFIS models were established, and the best ones were chosen according to results indicating the least SRMSE of the testing sets.

Results revealed a declining tendency for both MLPNN and RBNN, indicating that an increase of the number of neurons in the hidden layer can result in better performance. The results show a MLPNN with five neurons as well as tangent sigmoid and logarithmic sigmoid activation functions for the hidden and output layers. The RBNN with 81 neurons in the hidden layers had the best performance. The final structure of the applied AI-based methods and specification of the meta-heuristic algorithms are tabulated in Table 5.22.

Table 5.23 provides the results of the training, validation and testing stages for different methods. Training results indicate all AI-based methods modelling aeration phenomena better than the MLR and empirical methods. The MLP has a high accuracy with SRMSE of 0.294, R^2 of 0.931 and IA of 0.982. Test results clearly indicate that the AI-based methods (e.g., ANFIS-FCM) have better performances than the empirical functions.

Hybrid methods of the ANFIS (ANFIS-FA, ANFIS-GA and ANFIS-PSO) and basic ANFIS-FCM showed better performance figures than other models. Although the general performance of these methods was close, ANFIS-GA was superior to the other models with an SRMSE of 0.309, R^2 of 0.935 and IA of 0.982. The U.S. Army Corps of Engineers' empirical relationship was the best among empirical

TABLE 5.22
Final Structure and Characteristics of the Applied Artificial Intelligence-Based Models and Meta-Heuristic Algorithms in this Study

Model/Algorithm	Specifications and Parameters
Adaptive neuro-fuzzy inference system (ANFIS-FCM)	MF shapes: Gaussian MF in the first, second, third and fourth layer as well as linear MF in the fifth layer. FIS type: Sugeno. Derivation of data: subtractive clustering method.
Adaptive neuro-fuzzy inference system-Genetic algorithm (ANFIS-GA)	Mutation percentage: 0.8. Crossover percentage: 0.7. Selection pressure: 8. Mutation rate: 0.5. Number of fuzzy rules: 10.
Adaptive neuro-fuzzy inference system-firefly algorithm (ANFIS-FA)	Mutation Coefficient: 0.2. Attraction Coefficient: 2. Light Absorption Coefficient: 1. Mutation Coefficient Damping Ratio: 0.98.
Adaptive neuro-fuzzy inference system-particle swarm optimization (ANFIS-PSO)	Initial inertia weight: 1. Inertia Weight Damping Ratio: 0.99. Cognitive acceleration (C_1): 1. Social acceleration (C_2): 2. Number of fuzzy rules: 10.
Adaptive neuro-fuzzy inference system-biogeography-based optimization (ANFIS-BBO)	Keep Rate: 0.2. p_m (mutation probability): 0.1. Max immigration and migration rates: 1. Number of fuzzy rules: 10.
Multi-layer perceptron neural network (MLPNN)	Number of layers: 3. Number of neurons in the hidden layer: 5. Hidden layer transfer function: tangent sigmoid. Output layers transfer function: logarithmic sigmoid. $\alpha = 0.1$; $\eta = 0.9$.
Wavenet	Number of layers: 3. Number of neurons in the hidden layer: 5. Wavelet name: Gaussian.
Radial basis neural network (RBNN)	Number of layers: 3. Spread of radial basis functions: 24. Number of neurons: 81.

TABLE 5.23
Summary of the Results of the Applied Methods and Empirical Relations

	Error Criteria								
	Train			Validation			Test		
Model	SRMSE	R^2	IA	SRMSE	R^2	IA	SRMSE	R^2	IA
ANFIS-GA	0.345	0.905	0.975	0.397	0.883	0.969	0.309	0.935	0.982
ANFIS-FA	0.342	0.907	0.975	0.331	0.92	0.979	0.345	0.919	0.977
ANFIS-PSO	0.402	0.872	0.963	0.42	0.874	0.966	0.361	0.912	0.975
ANFIS-FCM	0.45	0.839	0.955	0.433	0.86	0.96	0.374	0.907	0.973
Wavenet	0.51	0.792	0.94	0.547	0.774	0.934	0.452	0.864	0.958
RBNN	0.469	0.824	0.95	0.652	0.688	0.907	0.522	0.815	0.942
ANFIS-BBO	0.476	0.821	0.946	0.529	0.79	0.939	0.549	0.792	0.937
MLPNN	0.294	0.931	0.982	0.885	0.452	0.804	0.799	0.559	0.847
MLR	0.955	0.272	0.628	0.903	0.458	0.825	1.007	0.318	0.603

(Continued)

TABLE 5.23 (*Continued*)
Summary of the Results of the Applied Methods and Empirical Relations

	Error Criteria								
	Train			Validation			Test		
Model	**SRMSE**	R^2	**IA**	**SRMSE**	R^2	**IA**	**SRMSE**	R^2	**IA**
U.S. Army Corps of Engineers (1964)	3.399	0.001	0.08	1.222	0.043	0.459	1.186	0.077	0.404
Campbell and Guyton (1953)	2.088	0.002	0.17	1.199	0.038	0.419	1.216	0.080	0.372
Wisner (1965)	10.985	0	0.016	1.764	0.046	0.446	1.368	0.061	0.496
Sharma (1976)	7.828	0.001	0.047	2.412	0.038	0.387	1.810	0.075	0.489
Pfister-Hager (2011)	13120	0	0	37.24	0.076	0.037	30.184	0.042	0.042

Note: ANFIS, Adaptive neuro-fuzzy inference system; FCM, fuzzy c-means clustering; GA, genetic algorithm; PSO, particle swarm optimization; RBNN, radial basis neural network; BBO, biogeography-based optimization; MLPNN, artificial neural network-multi-layer perceptron neural network; MLR, multiple linear regressions; SRMSE, standard root mean square error; R^2, coefficient of determination; IA, Index of agreement.

relations with an SRMSE of 1.186. The Pfister-Hager equation showed a poor performance with an SRMSE of 30.184. The results indicate that among the four applied meta-heuristic algorithms, GA, FA and PSO improve the performance of the classic method.

The Taylor's diagram is a practical tool to visually assess the performance of different methods. Figure 5.26 shows that the soft computing methods are closer to the observation point (test stage). This indicates that the soft computing methods have better accuracy than empirical methods. In the Taylor's diagram (Figure 5.26), all the ANFIS methods results are very close to each other. However, ANFIS-GA is closer to the observation point, which indicates that this method has a better performance than the other methods.

To compare the performance of different methods, Figure 5.26 represents the results of some applied models via scatter plots for the test stage. For the testing stage (Figure 5.26), to analyse the effect of flow regime on the performance of the applied models, datasets were categorized equivalently into three groups based on the *Fr* number of water flow. In these graphs, the thicker line angles for both verticals and horizontals are 45°. Points close to the 1:1 line indicate superior performance of the corresponding method.

The results of soft computing methods clearly show that the corresponding scatter points are closest to the dash line. ANFIS-GA had the highest coefficient of determination ($R^2 = 0.935$) and best performance between all methods, which were assessed in Section 5.4. The scatter points of this method are closest to the 1:1 line compared to other methods. The results of the ANN methods are more scattered than the other soft computing methods, which implies that the use of a hybrid method can improve the performance of these methods. Although the U.S. Army Corps of Engineers

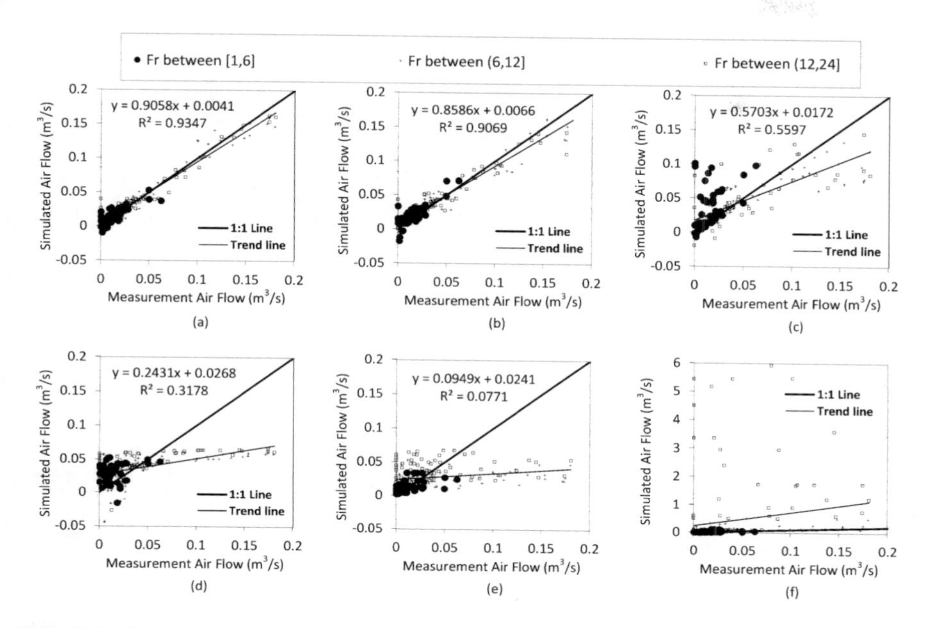

FIGURE 5.27 Scatter plots for various methods at the test stage: (a) adaptive neuro-fuzzy inference system-genetic algorithm; (b) adaptive neuro-fuzzy inference system; (c) multi-layer perceptron neural network; (d) multiple linear regressions; (e) U.S. Army Corps of Engineers and (f) Pfister-Hager.

method has a better performance among the empirical methods, the observed scatter points for this method are too scattered. The scatter points for the Pfister-Hager method are clearly too far away from the 1:1 line, indicating a poor performance linked to this method.

According to the visual analysis of Figure 5.27, most scatter points of the U.S. Army Corps of Engineers method are located below the 45° line. This means that this method is under-simulating the results. In contrast, most scatter points for the Pfister-Hager method are located above the 45° line. This indicates that this method is over-simulating. In general, the AI-based models are neither over- nor under-simulating.

Based on the visual analysis of the scatter plots in Figure 5.27, it can be concluded that the ANFIS models could handle the modelling performance regardless of the value of the *Fr* number. However, in the MLPNN and MLR models, data with *Fr* numbers between 2 and 6 are located far away from the 1:1 line. This indicates that these models show better performances for higher ranges of *Fr* numbers (>6). The U.S. Army Corps of Engineers graph indicates errors and is almost uniform for 3 *Fr* categories. The Pfister-Hager-related scatter plot clearly shows that this method has a better performance for a lower range of *Fr* numbers (<6).

The impact of the *Fr* number is shown in Table 5.24 containing also SRMSE and IA statistics. An assessment of Table 5.24 indicates that in contrary to the empirical

TABLE 5.24
Summary of the Results of the Applied Models According to the Different Ranges of the Froude (*Fr*) Number of the Water Flow

Model	*Fr* Range	SRMSE	IA	Model	*Fr* Range	SRMSE	IA
ANFIS-GA	[1,6]	0.57	0.89	MLPNN	[1,6]	2.256	0.49
	(6,12]	0.316	0.98		(6,12]	0.645	0.88
	(12,24]	0.226	0.985		(12,24]	0.628	0.863
ANFIS-FA	[1,6]	0.524	0.923	MLR	[1,6]	1.632	0.471
	(6,12]	0.379	0.97		(6,12]	0.988	0.573
	(12,24]	0.226	0.986		(12,24]	0.817	0.575
ANFIS-PSO	[1,6]	0.566	0.91	U.S. Army	[1,6]	0.862	0.615
	(6,12]	0.336	0.977	(1964)	(6,12]	1.177	0.389
	(12,24]	0.314	0.971		(12,24]	1.046	0.258
ANFIS-	[1,6]	0.718	0.873	Campbell and	[1,6]	0.858	0.598
FCM	(6,12]	0.349	0.974	Guyton (1953)	(6,12]	1.209	0.393
	(12,24]	0.308	0.972		(12,24]	1.071	0.327
Wavenet	[1,6]	0.82	0.851	Wisner (1965)	[1,6]	1.033	0.619
	(6,12]	0.418	0.961		(6,12]	1.081	0.417
	(12,24]	0.386	0.952		(12,24]	1.46	0.343
RBNN	[1,6]	1.132	0.748	Sharma (1964)	[1,6]	3.267	0.33
	(6,12]	0.488	0.943		(6,12]	1.346	0.54
	(12,24]	0.408	0.947		(12,24]	1.787	0.369
ANFIS-BBO	[1,6]	1.245	0.737	Pfister-Hager	[1,6]	2.041	0.483
	(6,12]	0.541	0.931	(2011)	(6,12]	2.631	0.513
	(12,24]	0.383	0.955		(12,24]	40.513	0.032

Note: ANFIS, Adaptive neuro-fuzzy inference system; GA, genetic algorithm; FA, firefly algorithm; PSO, particle swarm optimization; RBNN, radial basis neural network; BBO, biogeography-based optimization; MLPNN, artificial neural network-multi-layer perceptron neural network; MLR, multiple linear regressions; SRMSE, standard root mean square error; IA, Index of agreement.

relations, all the AI-based models as well as the MLR method have better performance figures for high ranges of the *Fr* number (>12). To explore the potential reason for this observation, the sensitivity analysis of standardized correlation coefficient (SCC) has been applied for the whole dataset. The three categories based on the *Fr* number ranges are shown in Table 5.24.

Table 5.25 also reveals that for the whole dataset, the most effective parameter for estimating Q_{air} is h_s (higher absolute SCC), whereas Q_w is the least effective parameter. Nevertheless, h_s is not the most effective parameter for the lower ranges and higher ranges of the *Fr* number. For $Fr < 6$, Q_w is calculated as the most effective parameter. In comparison, for $Fr > 12$, α is known as the most effective parameter. For low *Fr* numbers (<6), the difference between jet pressure and atmospheric pressure is small (−0.005 Pascal), which makes h_s the least effective input parameter.

TABLE 5.25
Results of the Simple Correlation Coefficient Sensitivity Analysis

	Independent Parameters						
Froude Number Range	h_s **(pa)**	Q_w **(m³/s)**	h_o **(m)**	$\alpha°$	s **(m)**	$\Theta°$	B **(m)**
Whole dataset	−0.283	−0.008	−0.193	0.075	0.153	0.264	−0.118
[1,6]	−0.005	0.256	0.063	0.165	0.113	0.094	0.08
[6,12]	−0.423	0.126	0.108	0.128	0.138	0.303	−0.112
[12,24]	−0.427	−0.05	−0.038	0.495	0.031	0.031	−0.031

h_s, Pressure difference between the atmosphere and under the aerator jet flow; Qw, water discharge; h_0, flow depth at the beginning of the aerator; a, ramp angel; s, aerator step height; Θ, shut angle; B, shut width.

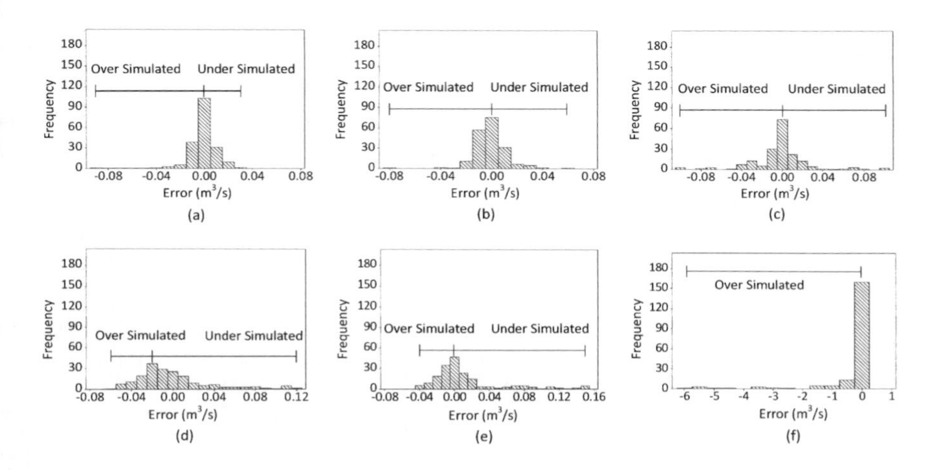

FIGURE 5.28 Histograms of the error frequencies for different methods at the test stage: (a) adaptive neuro-fuzzy inference system-genetic algorithm; (b) adaptive neuro-fuzzy inference system; (c) multi-layer perceptron neural network; (d) multiple linear regressions; (e) U.S. Army Corps of Engineers and (f) Pfister-Hager.

For higher ranges of the Fr number (>6), a strong jet will be formed on the aerator. Moreover, h_S and α have greater effects on the air demand.

Figure 5.28 depicts the histograms of error frequencies for the test stage, which can be used for assessing model performances. In contrast to the other applied models, the ANFIS-GA, ANFIS and MLPNN histograms are close to representing normal distributions. The Pfister-Hager histogram has a more skewed distribution than the corresponding histograms of the other models.

In this study, model results show that the highest score is associated with ANFIS models (especially for ANFIS-GA) followed by the Wavenet model. Figure 5.29 shows the effect of using ANFIS-GA in estimating the required air flow compared to

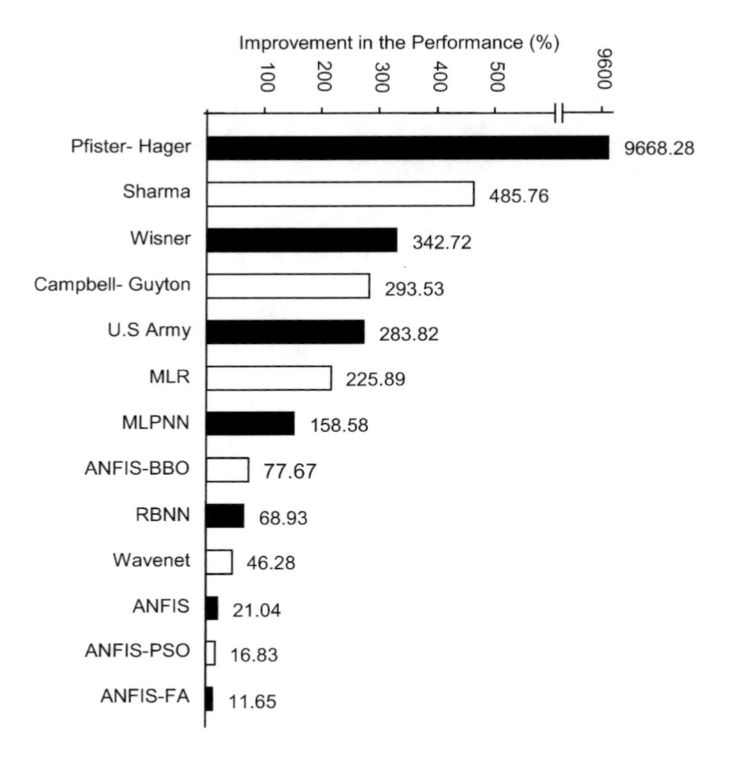

FIGURE 5.29 Improvement performance of the adaptive neuro-fuzzy inference system-genetic algorithm (ANFIS-GA) in comparison to the other applied models using the standard root mean square error statistic (test stage). MLR, multiple linear regressions; MLPNN, multi-layer perceptron neural network; BBO, biogeography-based optimization; RBNN, radial basis neural network; PSO, particle swarm optimization; FA, firefly algorithm.

other selected methods in the test stage. The meta-heuristic GA raises the improvement of the classic ANFIS model by more than 11%.

5.4.4 Discussion

In Section 5.4, several AI-based models and empirical methods were applied to estimate aeration coefficients of spillways. In the previous section, statistical parameters indicated that AI-based models have better results than empirical relations. For $Fr < 6$, ANFIS-FA has better results (SRMSE = 0.379), while for $Fr > 6$, the ANFIS-GA accuracy is higher than for other methods.

The estimated results of the predictive models are statistically compared with the observed data using the Kruskal-Wallis test (Table 5.26). Although there is no significant difference at the 99% confidence level, the test shows that there is a significant difference between the results of the AI-based models versus empirical relations.

An overview of the past reports on the application of AI-based models in simulating air demand in hydraulic structures is presented in Table 5.27. An evaluation of the

TABLE 5.26
Statistical Investigation of Results Obtained via the Kruskal-Wallis Method

Methods	*p*-value	Significantly Different (95%)	Significantly Different (99%)
ANFIS-GA, ANFIS-FA, ANFIS-PSO, ANFIS-BBO, ANFIS-FCM, Wavenet, RBNN, MLPNN, MLR, U.S Army, Campbell-Guyton, Wisner, Sharma, Pfister-Hager	<0.0001	Yes	Yes
ANFIS-GA, ANFIS-FA, ANFIS-PSO, ANFIS-BBO, ANFIS-FCM, Wavenet, RBNN, MLPNN	0.532	NO	NO
U.S Army, Campbell-Guyton, Wisner, Sharma, Pfister-Hager	<0.0001	Yes	Yes

Note: ANFIS, Adaptive neuro-fuzzy inference system; GA, genetic algorithm; FA, firefly algorithm; PSO, particle swarm optimization; BBO, biogeography-based optimization; RBNN, radial basis neural network; MLPNN, artificial neural network-multi-layer perceptron neural network; MLR, multiple linear regressions; SRMSE, standard root mean square error.

TABLE 5.27
Comparison of Present Study and Similar Research Investigations for Simulating Aerators Air Demand

Resource	Method(s)	Purpose	Database	Remarks and Outcomes
Kavianpour and Rajabi (2005)	MLPNN	Predicting tunnel aerator air demand	a. Several data from laboratory experiments performed at the Water Research Institute of Tehran. b. Experiment performed on Folsom model	MLPNN method approximated air demand with good accuracy.
Najafi et al. (2012)	Fuzzy system, ANFIS, Fuzzy-Gas, empirical relationships	Predicting air demand in gated aerator tunnels	a. Several data from laboratory experiments performed at the Water Research Institute of Tehran. b. Experiment performed on Folsom model	Soft computing methods produced best estimation compared to experimental methods. ANFIS performance was better than other methods. Fuzzy Wang–Mendel (WM) was poorest than other soft computing methods.
Zounemat-Kermani and Scholz (2013)	ANFIS, MLPNN, MLR, empirical relationships	Predicting air demand in low-level outlet	Experimental data from Tullis and Larchar work in Utah university	ANFIS and MLPNN had better performance than experimental equation. ANFIS was the best among all methods.

(*Continued*)

TABLE 5.27 (*Continued*)
Comparison of Present Study and Similar Research Investigations for Simulating Aerators Air Demand

Resource	Method(s)	Purpose	Database	Remarks and Outcomes
Zounemat-Kermani et al. (2017)	GEP, Random forest algorithm, Boosted regression trees, empirical relationships	Predicting air demand in bottom outlet	Experimental data from hydraulic models of Karkheh, Jagin, Jareh, Alborz and Kowsar dams in Iran, as well as the Folsom dam in the United State	GEP and classification tree methods performed better than empirical relationships. GEP was better in estimating than classification tree methods.
The present study	ANFIS-FCM, ANFIS-GA, ANFIS-FA, ANFIS-PSO, ANFIS-BBO, Wavenet, RBNN, MLPNN, MLR, empirical relationships	Predicting spillway aerator air demand	a. Experimental sets obtained from the Clyde Dam hydraulic model (Chanson, 1988) b. Experiment carried out in the River Engineering Department of Hydraulics Research, Wallingford (May and Deamer, 1989) c. Several data from laboratory experiments performed at the Water Research Institute of Tehran.	Soft computing methods have better performance than empirical methods. GA, FA and PSO heuristic algorithms improved performance of ANFIS. In soft computing methods, ANFIS-GA has the most accurate results. Based on a sensitivity analysis, the difference between the atmospheric pressure and the pressure under the aerator jet flow (h_s) is the most important parameter in modelling. The Kruskal-Wallis test showed there are no significant differences between soft computing methods for the significance levels of 95 and 99%.

Note: MLPNN, artificial neural network-multi-layer perceptron neural network; ANFIS, adaptive neuro-fuzzy inference system; MLR, multiple linear regressions; GEP, gene expression programming; GA, genetic algorithm; FA, firefly algorithm; PSO, particle swarm optimization; BBO, biogeography-based optimization; RBNN, radial basis neural network; SRMSE, standard root mean square error.

outcomes of the past studies and the present work indicates the superiority of using AI-based models in simulating air demands within hydraulic structures.

5.4.5 Conclusions and Recommendations

This study assessed the performance of various AI-based models including hybrid ANFIS and genetic algorithm, hybrid ANFIS and firefly optimization algorithm, hybrid ANFIS and particle swarm optimization algorithm, hybrid ANFIS and biogeography-based optimization, Wavenet, multi-layer perceptron neural network, radial basis neural network and the MLR method as well as some representative empirical relations (U.S. Army Corps of Engineers, Campbell-Guyton, Wisner, Sharma and Pfister-Hager) in estimating the spillway aerator air flow demand against potential cavitation damage. A close examination of the results indicates that the AI-based models offer better performance than those based on experiments. Results showed that the hybrid models performed better compared to classic models.

The use of GA, FA and PSO meta-heuristic algorithms improved the performance of classic ANFIS by 21.04%, 8.40% and 3.60%, respectively. Among all applied AI-based models, ANFIS-GA has the best performance with an SRMSE of 0.309 and R^2 of 0.935. However, the Kruskal-Wallis test denotes no statistical difference (99%) between all AI-models. Concerning the empirical methods, the U.S. Army Corps of Engineers had the best performance. Results also showed that Pfister-Hager had the lowest score compared to all other assessed models.

However, further investigations using more data sets are recommended. Sensitivity analysis results highlight h_s as the most important parameter in modelling of Fr numbers within the range between 6 and 12. For low Fr numbers, the flow rate is the most important parameter, and for high Fr numbers (>12), the ramp angle is the most effective factor.

REFERENCES

Abbasimehr, H. and Tarokh, M.J., 2015. Trust prediction in online communities employing neurofuzzy approach, *Applied Artificial Intelligence*, **29**(7), 733–751.

Ahmadlou, M., Karimi, M., Alizadeh, S., Shirzadi, A., Parvinnejad, D., Shahabi, H. and Panahi, M., 2019. Flood susceptibility assessment using integration of adaptive network-based fuzzy inference system (ANFIS) and biogeography-based optimization (BBO) and BAT algorithms (BA), *Geocarto International*, **34**, 1252–1272.

Alexandridis, A.K. and Zapranis, A.D., 2013. Wavelet neural networks: A practical guide, *Neural Networks*, **42**, 1–27.

Ali, S.S.A., Kannan, R. and Kumar, M.S., 2017. Exploration of modulation index in multi-level inverter using particle swarm optimization algorithm, *Procedia Computer Science*, **105**, 144–152.

Al-Rekabi, W.S., Qiang, H., Qiang, W.W. and Abbas, A.A., 2007. Utilization of constructed wetlands for wastewater treatment, *European Journal of Scientific Research*, **18**(1), 165–185.

Anastasoiu, K. and Chan, C.T., 1997. Solution of the 2D shallow water equations using the finite volume method on unstructured triangular meshes, *International Journal for Numerical Methods in Fluids*, **24**(11), 1225–1245.

APHA, 1998. *Standard Methods for the Examination of Water and Wastewater*, 20th edition (Washington, DC: American Public Health Association/American Water Works Association/Water Environment Federation).

Arega, F., 2013. Hydrodynamic modelling and characterizing of Lagrangian flows in the West Scott Creek wetlands system, South Carolina, *Journal of Hydro-environment Research*, **7**(1), 50–60.

Asadollahfardi, G., Taklify, A. and Ghanbari, A., 2012. Application of Artificial Neural Network to predict TDS in Talkheh Rud River, *Journal of Irrigation and Drainage Engineering*, **138**(4), 363–370.

Avelar, F.F., de Matos, A.T., de Matos, M.P. and Borges, A.C., 2014. Coliform bacteria removal from sewage in constructed wetlands planted with Mentha aquatic, *Environment Technology*, **35**(16), 2095–2103.

Azizi, S., Ahmadloo, E. and Awad, M.M., 2016. Prediction of void fraction for gas–liquid flow in horizontal, upward and downward inclined pipes using artificial neural network, *International Journal of Multiphysics Flow*, **87**, 35–44.

Babatunde, A.O., Zhao, Y.Q., O'Neill, M. and O'Sullivan, B., 2008. Constructed wetlands for environmental pollution control: A review of the developments, research and practice in Ireland, *Environment International*, **34**(1), 116–126.

Bachand, P.A.M. and Horne, A.J., 2000. Denitrification in constructed free-water surface wetlands II: Effects of vegetation and temperature, *Ecological Engineering*, **14**(1–2), 17–32.

Bakshi, B.R. and Stephanopoulos, G., 2004. Wave-net: A multiresolution, hierarchical neural network with localized learning, *American Institute of Chemical Engineering Journal*, **39**(1), 57–81.

Banerjee, S. and Mandal, D., 2017. Array pattern optimization for a steerable circular isotropic antenna array using the firefly algorithm, *Journal of Computational Electronics*, **16**, 952–976.

Bastviken, S.K., Weisner, S.E.B., Thiere, G., Svensson, J.M., Ehde, P.M. and Tonderski, K., 2009. Effects of vegetation and hydraulic load on seasonal nitrate removal in treatment wetlands, *Ecological Engineering*, **35**(2), 946–952.

Bezbaruah, A.N. and Zhang, T.C., 2003. Performance of a constructed wetland with a sulfur/limestone denitrification section for wastewater nitrogen removal, *Environmental Science and Technology*, **37**(8), 1690–1697.

Bozorg Haddad, O., Hosseini-Moghari, S.M. and Loáiciga, H.A., 2016. Biogeography-based optimization algorithm for optimal operation of reservoir systems, *Journal of Water Resources Planning and Management*, **142**(1), 1–11.

Braskerud, B.C., 2002. Factors affecting phosphorus retention in small constructed wetlands treating agricultural non-point source pollution, *Ecological Engineering*, **19**(1), 41–61.

Brix, H., 1997. Do macrophytes play a role in constructed treatment wetlands? *Water Science and Technology*, **35**(5), 11–17.

Brix, H., Arias, C.A. and del Bubba, M., 2001. Media selection for sustainable phosphorus removal in subsurface flow constructed wetlands, *Water Science and Technology*, **44**(11–12), 47–54.

Brodrick, S.J., Cullen, P. and Maher, W., 1988. Denitrification in a natural wetland receiving secondary treated effluent, *Water Research*, **22**(4), 431–439.

Bromhead, D.S. and Lowe, D., 1988. Multivariate functional interpolation and adaptive networks, *Complex Systems*, **2**, 321–355.

Campbell, F.B. and Guyton, B., 1953. Air demand in gated conduits, *IAHR Symposium*, Minneapolis, USA.

Carty, A., Scholz, M., Heal, K., Gouriveau, F. and Mustafa, A., 2008. The universal design, operation and maintenance guidelines for Farm Constructed Wetlands (FCW) in temperate climates, *Bioresource Technology*, **99**(15), 6780–6792.

Chanson, H., 1988. *Study of air entrainment and aeration devices on spillway model*, Graduate Theses and Dissertations (Canterbury: University of Canterbury).

Ciria, M.P., Solano, M.L. and Soriano, P., 2005. Role of macrophyte *Typha latifolia* in a constructed wetland for wastewater treatment and assessment of its potential as a biomass fuel, *Biosystems Engineering*, **92**(4), 535–544.

Clarke, E. and Baldwin, A.H., 2002. Responses of wetland plants to ammonia and water level. *Ecological Engineering*, **18**(3), 257–264.

Conn, R.M. and Fiedler, F.R., 2006. Increasing hydraulic residence time in constructed stormwater treatment wetlands with designed bottom topography, *Water Environment Research*, **78**(13), 2514–2523.

Cooney, J., 1984. The fate of petroleum pollutants in freshwater ecosystems. In: R.M. Atlas (Ed.) *Petroleum Microbiology* (New York: Macmillan) pp. 399–433.

Costa, M.A., de Pádua Braga, A. and de Menezes, B.R., 2007. Improving generalization of MLPs with sliding mode control and the Levenberg–Marquardt algorithm, *Neurocomputing*, **70**(7–9), 1342–1347.

Cronk, J.K., 1996. Constructed wetlands to treat wastewater from dairy and swine operations: A review, *Agriculture, Ecosystems and Environment*, **58**(2–3), 97–114.

Dunne, E.J., Culleton, N., Donovan, G.O., Harrington, R. and Olsen, A.E. 2005. An integrated constructed wetland to treat contaminants and nutrients from dairy farmyard dirty water, *Ecological Engineering*, **24**(3), 221–234.

Eke, P.E. and Scholz, M., 2008. Benzene removal with vertical-flow constructed treatment wetlands, *Journal of Chemical Technology and Biotechnology*, **83**(1), 55–63.

Ern, A., Piperno, A.E.S. and Djadel, K., 2007. A well-balanced Runge–Kutta discontinuous Galerkin method for the shallow-water equations with flooding and drying, *International Journal for Numerical Methods in Fluids*, **58**, 1–25.

Fraser, L.H., Carty, S.M. and Steer, D., 2004. A test of four plant species to reduce total nitrogen and total phosphorus from soil leachate in subsurface wetland microcosms, *Bioresource Technology*, **94**(2), 185–192.

García, J., Aguirre, P., Mujeriego, R., Huang, Y.M., Ortiz, L. and Bayona, J.M., 2004. Initial contaminant removal performance factors in horizontal flow reed beds used for treating urban wastewater, *Water Research*, **38**(1), 1669–1678.

Gerke, S., Baker, L. and Xu, Y., 2001. Nitrogen transformations in a wetland receiving lagoon effluent: sequential model and implications for water reuse, *Water Research*, **35**(16), 3857–3866.

Ghamisi, P. and Benediktsson, J.A., 2015. Feature selection based on hybridization of genetic algorithm and particle swarm optimization, *Institute of Electrical and Electronics Engineers Geoscience and Remote Sensing Letters*, **12**(2), 309–313.

Gholami, A., Bonakdari, H., Ebtehaj, I., Mohammadian, M., Gharabaghi, B. and Khodashenas, S.R., 2018. Uncertainty analysis of intelligent model of hybrid genetic algorithm and particle swarm optimization with ANFIS to predict threshold bank profile shape based on digital laser approach sensing, *Measurement*, **121**, 294–303.

Goldberg, D.E., 1989. *Genetic Algorithms in Search, Optimization and Machine Learning* (Reading: Addison Wesley).

Ha, Q.P., Wahid, H., Duc, H. and Azzi, M., 2015. Enhanced radial basis function neural networks for ozone level estimation, *Neurocomputing*, **155**, 62–70.

Hammer, D.A. and Bastian, R.K, 1989. Wetlands ecosystems: natural water purifiers? In: D. A. Hammer (Ed.) *Constructed Wetlands for Wastewater Treatment* (Chelsea: Lewis Publishers) pp. 5–19.

Harrington, R., Dunne, E.J., Carroll, P., Keohane, J. and Ryder, C., 2005. The concept, design and performance of integrated constructed wetlands for the treatment of farmyard dirty water. In: E. L. Dunne, K. R. Reddy and O. T. Carton (Eds.) *Nutrient Management in Agricultural Watersheds: A Wetlands Solution* (Wageningen: Wageningen Academic Publishers), pp. 179–188.

Harrington, R. and McInnes, R., 2009. Integrated Constructed Wetlands (ICW) for livestock wastewater management, *Bioresourse Technology*, **100**(22), 5498–5505.

Harrington, R. and Ryder, C., 2002. The use of integrated constructed wetlands in the management of farmyard runoff and waste water. In: *Proceedings of the National Hydrology Seminar on Water Resource Management: Sustainable Supply and Demand* (Tullamore: The Irish National Committees of the International Hydrological Programme and International Commission on Irrigation and Drainage), pp. 55–63.

Harrington, C., Scholz, M., Culleton, N. and Lawlor, P. D., 2011. Meso-scale systems used for the examination of different integrated constructed wetland operations, *Journal of Environmental Science and Health, Part A: Toxic/Hazardous Substances and Environmental Engineering*, **46**(7), 783–788.

He, L.S., Liu, H.L., Xi, B.D. and Zhu, Y.B., 2006. Enhancing treatment efficiency of swine wastewater by effluent recirculation in vertical-flow constructed wetland, *Journal of Environmental Sciences* – **China**, **18**(2), 221–226.

He, Y., Chu, F. and Zhong, B., 2002. A hierarchical evolutionary algorithm for constructing and training wavelet networks, *Neural Computational Applications*, **10**(4), 357–366.

Hench, K.R., Bissonnette, G.K., Sexstone, A.J., Coleman, J.G., Garbutt, K. and Skousen, J.G., 2003. Fate of physical, chemical, and microbial contaminants in domestic wastewater following treatment by small constructed wetlands, *Water Research*, **37**(4), 921–927.

Herskowitz, J., Black, S. and Sewandowski, W., 1987. Listowel artificial marsh treatment project. In: K. R. Reddy and W. H. Smith (Eds.) *Aquatic Plants for Water Treatment and Resource Recovery* (Orlando: Magnolia Publishing Co) pp. 247–254.

Hill, D.T., Payne, V.W.E., Rogers, J.W. and Kown, S.R., 1997. Ammonia effects on the biomass production of five constructed wetland plant species, *Bioresource Technology*, **62**(3). 109–113.

Holland, J.H., 1975. *Adaptation in Natural and Artificial Systems* (Ann Arbor, MI: University of Michigan Press).

Hu, Z.F., Dou, J.F., Liu, X., Zheng, X.L. and Deng, D., 2007. Anaerobic biodegradation of benzene series compounds by mixed cultures based on optional electronic acceptors, *Journal of Environmental Sciences*, **19**(5), 1049–1054.

Hua, G.F., Zhu, W., Zhao, L.F. and Huang, J.Y., 2010. Clogging pattern in vertical-flow constructed wetlands: Insight from a laboratory study, *Journal of Hazardous Materials*, **180**(1–3), 668–674.

Hubbard, R.K., Ruter, J.M., Newton, G.L. and David, J.G., 1999. Nutrient uptake and growth response of six wetland/riparian plant species receiving swine lagoon effluent, *Transactions of the American Society of Agricultural Engineers*, **42**(5), 1331–1341.

Humenik, F.J., Szogi, A.A., Hunt, P.G., Broome, S. and Rice, M., 1999. Wastewater utilization: A place for managed wetlands – review, *Asian-australasian Journal of Animal Sciences*, **12**(4), 629–632.

Hunt, P.G. and Poach, M.E., 2001. State of the art for animal wastewater treatment in constructed wetlands, *Water Science and Technology*, **44**(11–12), 19–25.

Hunt, P.G., Szogi, A.A., Humenik, F.J., Rice, J.M., Matheny, T.A. and Stone, K.C., 2002. Constructed wetlands for treatment of swine wastewater from an anaerobic lagoon, *Transactions of the American Society of Agricultural Engineers*, **45**(3), 639–647.

Jain, W. and Chao, L., 2011. Effects of entrained air manner on cavitation damage, *Journal of Hydrodynamics*, **23**(3), 333–338.

Jang, J.S.R., (1993). ANFIS: Adaptive-network-based fuzzy inference systems, *Institute of Electrical and Electronics Engineers Transactions on Systems, Man and Cybernetics*, **23**(3), 665–685.

Jang, J.S.R., Sun, C.T. and Mizutani, E., 1997. *Neuro-fuzzy and Soft Computing: a Computational Approach to Learning and Machine Intelligence* (New York: Prentice Hall).

Javanbarg, M.B., Zarrati, A.R. and Jalili, M.R., 2003. Effect of ramp angle on aerator performance, In: *Proceedings of the 30th IAHR Congress*, Theme D, Thessaloniki, pp. 727–734.

Jenkins, G.A. and Greenway, M., 2005. The hydraulic efficiency of fringing versus banded vegetation in constructed wetlands, *Ecological Engineering*, **25**(1), 61–72.

Ji, G.D., Sun, T.H. and Ni, J.R., 2007. Surface flow constructed wetland for heavy oil-produced water treatment, *Bioresource Technology*, **98**(2), 436–441.

Johnson, S.J., Woolhouse, K.J., Prommer, H., Barry, D.A. and Christofi, N., 2003. Contribution of anaerobic microbial activity to natural attenuation of benzene in groundwater, *Engineering Geology*, **70**, 343–349.

Kadlec, R.H. and Knight, R.L., 1996. *Treatment Wetlands* (Baca Raton, FL: Lewis Publishers).

Kamarian, S., Shakeri, M. and Yas, M.H., 2017. Thermal buckling optimisation of composite plates using firefly algorithm, *Journal of Experimental and Theoretical Artificial Intelligence*, **29**(4), 787–794.

Kantawanichkul, S. and Somprasert, S., 2005. Using a compact combined constructed wetland system to treat agricultural wastewater with high nitrogen, *Water Science and Technology*, **51**(9), 47–53.

Keefe, S.H., Barber, L.B., Runkel, R.L. and Ryan, J.N., 2004. Fate of volatile organic Compounds in constructed wastewater treatment wetlands, *Environmental Science and Technology*, **38**(7), 2209–2216.

Kells, J.A. and Smith, C.D., 1991. Reduction of cavitation on spillways by induced air entrainment, *Canadian Journal of Civil Engineering*, **18**(3), 358–377.

Kennedy, J. and Eberhart, R., 1995. Particle swarm optimization, In: *Proceeding of the Institute of Electrical and Electronics Engineers International Conference on Neural Networks 1995*, Vol. **4** (Perth: International Conference on Neural Network) pp. 1942–1948.

Kim, H.H., Kim, D.G., Choi, J. Y. and Park, S.C., 2017. Tire mixing process scheduling using particle swarm optimization, *Computational and Industrial Engineering*, **110**, 333–343.

Knight, R.L., Kadlec, R.H. and Ohlendorf, H.M., 1999. The use of treatment wetlands for petroleum industry effluents, *Environmental Science and Technology*, **33**, 973–980.

Knight, R.L., Payne, V.W.E., Borer, R.E., Clarke, R.A. and Pries, J.H., 2000. Constructed wetlands for livestock wastewater management, *Ecological Engineering*, **15**(1–2), 41–55.

Kramer, K., Hager, W.H. and Minor, H.E., 2006. Development of air concentration on chute spillways, *Journal of Hydrological Engineering*, **132**(9), 908–915.

Kumar, R., Rajan, A., Talukda, F.T. and Dey, N., 2017. Optimization of 5.5-GHz CMOS LNA parameters using firefly algorithm, *Neural Computing and Applications*, **28**, 3765–3779.

Kumar, S., Ojha, A., Goyal, M., Singh, R. and Swamee, P., 2012. Modeling of suspended sediment concentration at Kasol in India using ANN, Fuzzy Logic, and Decision Tree Algorithms, *Hydrologic Engineering*, **17**(3), 394–404.

Lago, M.E., Miralles-Wilhelm, F., Mahmoudi, M. and Engel, V., 2010. Numerical modelling of the effects of water flow, sediment transport and vegetation growth on the spatiotemporal patterning of the ridge and slough landscape of the Everglades wetland, *Advances in Water Resources*, **33**(10), 1268–1278.

Lee, C.Y., Lee, C.C., Lee, F.Y., Tseng, S.K. and Laio, C.J., 2004. Performance of subsurface flow constructed wetland taking pre-treated swine effluent under heavy loads, *Bioresource Technology*, **92**(2), 173–179.

Li, L.F., Li, Y.H., Biswas, D.K., Nian, Y.G. and Jiang, G.M., 2008. Potential of constructed wetlands in treating the eutrophic water: Evidence from Taihu Lake of China, *Bioresource Technology*, **99**(6), 1656–1663.

Lim, P.E., Wong, T.F. and Lim, D.V., 2001. Oxygen demand, nitrogen and copper removal by free-water-surface and subsurface-flow constructed wetlands under tropical conditions, *Environment International*, **26**(5–6), 425–431.

Lin, X., Sun, J., Palade, V., Fang, W., Wu, X. and Xu, W., 2012. Training ANFIS parameters with a quantumbehaved particle swarm optimization algorithm, In: Y. Tan, Y. Shi and Z. Ji (Eds.) *Advances in Swarm Intelligence, ICSI 2012*, vol. **7331**. Lecture Notes in Computer Science (Berlin: Springer).

Luo, A., Zhu, J. and Ndegwa, P.M., 2002. SE-structures and environment removal of carbon, nitrogen, and phosphorus in pig manure by continuous and intermittent aeration at low redox potentials, *Biosystems Engineering*, **82**(2), 209–215.

Luo, J., Cirpka, O. A. and Kitanidis, P. K., 2006. Temporal-moment matching for truncated breakthrough curves for step or step-pulse injection, *Advances in Water Resources*, **29**(9), 1306–1313.

Mahdavi-Meymand, A., Scholz, M. and Zounemat-Kermani, M., 2019. Challenging soft computing approaches in modelling complex hydraulic phenomenon of aeration process. *ISH Journal of Hydraulic Engineering*, **27**, 58–69.

Maltais-Landry, G., Maranger, R. and Brisson, J., 2009. Effect of artificial aeration and macrophyte species on nitrogen cycling and gas flux in constructed wetlands, *Ecological Engineering*, **35**(2), 221–229.

Mamdani, E.H. and Assilian, S., 1975. An experiment in linguistic synthesis with a fuzzy logic controller, *International Journal of Man-machine Studies*, **7**(1), 1–13.

May, R.W.P. and Deamer, A.P., 1989. Performance of aerators for dam spillways. HR Report SR 198.

Metcalf & Eddy Inc., 1991. *Wastewater Engineering: Treatment, Disposal, and Reuse*, 3rd edn. (New York: McGraw-Hill, Inc.).

Min, J.H. and Wise, W.R., 2010. Depth-averaged, spatially distributed flow dynamic and solute transport modelling of a large-scaled, subtropical constructed wetland, *Hydrological Processes*, **24**, 2724–2737.

Mohanty, D.K., 2015. Application of firefly algorithm for design optimization of a shell and tube heat exchanger from economic point of view, *International Journal of Thermal Science*, **102**, 228–238.

Molle, P., Prost-Boucle, S. and Lienard, A. 2008. Potential for total nitrogen removal by combining vertical flow and horizontal flow constructed wetlands: A full-scale experiment study, *Ecological Engineering*, **34**(1), 23–29.

Moustafa, M.Z., 1997. Graphical representation of nutrient removal in constructed wetlands, *Wetlands*, **17**, 493–501.

Najafi, M.R., Kavianpour, Z., Najafi, B., Kavianpour, M.R. and Moradkhani, H., 2012. Air demand in gated tunnels – a Bayesian approach to merge various predictions, *Hydroinformatics*, **14**(1), 152–166.

New Jersey Department of Environmental Protection, 2004. *New Jersey Storm Water Best Management Practices Manual* (Trenton: Division of Watershed Management, New Jersey Department of Environmental Protection).

Ouellet-Plamondon, C., Chazarenc, F., Comeau, Y. and Brisson, J., 2006. Artificial aeration to increase pollutant removal efficiency of constructed wetlands in cold climate, *Ecological Engineering*, **27**(3), 258–264.

Pan, S., Shao, Y., Shi, Q. and Dong, X., 1980. The selfaeration capacity of the water jet over the aeration ramp, *Shuili Xuebao*, **5**, 13–22 (in Mandarin).

Panuvatvanich, A., Koottatep, T. and Koné, D. 2009. Hydraulic behaviour of vertical-flow constructed wetland under different operating conditions, *Environment Technology*, **30**(10), 1031–1040.

Persson, J., Somes, N.L.G. and Wong, T.H.F., 1999. Hydraulics efficiency of constructed wetlands and ponds, *Water Science and Technology*, **40**(3), 291–299.

Peterka, A.J., 1953. The effect of entrained air on cavitation pitting. In: *Proceedings of the Minnesota International Hydraulics Convention, St. Anthony Falls Hydraulic Laboratory* (Minneapolis: University of Minnesota).

Pezeshki, S.R., Hester, M.W., Lin, Q. and Nyman, J.A., 2000. The effects of oil spill and clean-up on dominant US Gulf coast marsh macrophytes: A review. *Environmental Pollution*, **108**(2), 129–139.

Pfister, M., 2011. Chute aerators: Steep deflectors and cavity subpressure, *Journal of Hydraulics Engineering*, **137**(10), 1208–1215.

Pfister, M., Lucas, J. and Hager, W.H., 2011. Chute aerators: Preaerated approach flow, *Journal of Hydraulics Engineering*, **137**(11), 1452–1461.

Pinto, N.L. de S., Neidert, S.H., and Ota, J.J., 1982. Aeration at high velocity flows, *Water Power and Dam Construction*, **34**(2), 34–38.

Platzer, C.H.R. and Mauch, K., 1997. Soil clogging in vertical-flow reed beds - mechanisms, parameters, consequences and ... solutions? *Water Science and Technology*, **35**(5), 175–181.

Poach, M.E., Hunt, P.G., Reddy, G.B., Stone, K.C., Johnson, M.H. and Grubbs, A., 2004a. Swine wastewater treatment by marsh-pond-marsh constructed wetlands under varying nitrogen loads, *Ecological Engineering*, **23**(3), 165–176.

Poach, M.E., Hunt, P.G., Reddy, G.B., Stone, K.C., Matheny, T.A., Johnson, M.H. and Sadler, E.J., 2004b. Ammonia volatilization from marsh-pond-marsh constructed wetland treating swine wastewater, *Journal of Environmental Quality*, **33**(3), 844–851.

Põldvere, E., Karabelnik, K., Noorvee, A., Maddison, M., Nurk, K., Zaytsev, I. and Mander, Ü., 2009. Improving wastewater effluent filtration by changing flow regimes – investigations in two cold climate pilot scale systems, *Ecological Engineering*, **35**(2), 193–203.

Postalcıoğlu, S., Erkan, K. and Doğru Bolat, E., 2005. Comparison of Wavenet and Neuralnet for system modelling. In: *Knowledge-Based Intelligent Information and Engineering Systems*, Lecture Notes in Computational Science, vol. **3682**, pp. 100–107. https://doi.org/10.1007/11552451_14.

Puustinen, M. and Jormola, J., 2005. Constructed wetlands for nutrient retention and landscape diversity. In: *21st European Regional Conference 2005* (Frankfurt an der Oder: International Commission on Irrigation and Drainage).

Quecedo, M. and Pastor, M., 2002. A reappraisal of Taylor–Galerkin algorithm for drying–wetting areas in shallow water computations, *International Journal for Numerical Methods in Fluids*, **38**(6), 515–531.

Reddy, K.R. and DeBusk, T.A., 1987. State-of-the-art utilization of aquatic plants in water pollution control, *Water Science and Technology*, **19**(10), 61–79.

Reed, S.C. and Brown, D.S., 1995. Subsurface flow wetlands-a performance evaluation, *Water Environment Research*, **67**(2), 244–248.

Rodi, W., 1993. *Turbulence Models and Their Application in Hydraulics – A State of the Art Review*, 3rd edn. (Delft: International Association for Hydraulic Research).

Ruan, S.P., Wu, J.H., Wu, W.W. and Xi, R.Z., 2007. Hydraulic research of aerators on tunnel spillways, *Journal of Hydrodynamics*, **19**(3), 330–334.

Sabbagh-Yazdi, S.R. and Zounemat-Kermani, M., 2008. Vertex base unstructured finite volume solution on depth averaged tidal currents on 3D bed, *Iranian Journal of Science and Technology Transaction B: Engineering*, **32**(5), 563–570.

Sakadevan, K. and Bavor, H.J., 1998. Phosphate adsorption characteristics of soils, slags and zeolite to be used as substrates in constructed wetland systems, *Water Research*, **32**(2), 393–399.

Salimi, S., Karami, H., Farzin, S., Hassanvand, M., Azad, A. and Kisi, O., 2020. Design of water supply system from rivers using artificial intelligence to model water hammer, *ISH Journal of Hydraulic Engineering*, **26**(2), 153–162.

Salmon, C., Crabos, J.L., Sambuco, J.P., Bessiere, J.M., Brasseres, A., Caumette, P. and Baccou, J.C., 1998. Artificial wetland performances in the purification efficiency of hydrocarbon wastewater, *Water Air and Soil Pollution*, **104**, 313–329.

Sayadi, M.K., Hafezalkotob, A. and Jalali Naini, S.G., 2013. Firefly-inspired algorithm for discrete optimization problems: An application to manufacturing cell formation, *Journal of Manufacturing Systems*, **32**(1), 78–84.

Scholz, M., 2006. *Wetland Systems to Control Urban Runoff* (Amsterdam: Elsevier).

Scholz, M., 2010. *Wetland Systems – Storm Water Management Control* (Berlin: Springer Verlag).

Scholz, M., Harrington, R., Carroll, P. and Mustafa, A., 2007. The Integrated Constructed Wetlands (ICW) concept, *Wetlands*, **27**(2), 337–354.

Scholz, M., Höhn, P. and Minall, R., 2002. Mature experimental constructed wetlands treating urban water receiving high metal loads, *Biotechnology Progress*, **18**(6), 1257–1264.

Scholz, M. and Trepel, M., 2004. Hydraulic characteristics of groundwater-fed open ditches in a peatland, *Ecological Engineering*, **23**(1), 29–45.

Schwarz, I. and Nutt, L.P., 1963. Projected nappes subjected to transverse pressure, *Journal of Hydraulic Engineering*, **89**, 97–104.

Scrimshaw, M.D. and Lester, J.N., 1996. The occurrence and effects of organic micropollutants in salt marsh sediments with reference to the UK Essex coast: A review, *Environment Technology*, **17**(9), 951–964.

Seidel, K., 1966. Reinigung von Gerwässern durch höhere Pflanzen, *Naturwissenschaften*, **53**, 289–297 (in German).

Sharma, H.R., 1976. Air-entrainment in high-head gated conduits, In: *Proceedings of the American Society of Civil Engineers Journal of Hydraulics Division*, HY11.

Shaver, G.R. and Melillo, J.M., 1984. Nutrient budgets of marsh plants: Efficiency concepts and relation to availability, *Ecology*, **65**(5), 1491–1510.

Simon, D., 2008. Biogeography-based optimization. *Institute of Electrical and Electronics Engineers Transactions on Evolutionary Computing*, **12**(6), 702–713.

Somes, N.L.G., Persson, J. and Wong, T.H.F., 1998. Influence of wetland design parameters on the hydrodynamics of storm water wetlands, In: *Proceedings of HydraStorm, 3rd International Conference on Stormwater Mangement*, Adelaide, pp. 123–128.

SPSS, 2003. *Analytical Software*, Statistical Package for the Social Sciences (SPSS) (Chicago: SPSS Headquarters).

Su, T.-M., Yang, S.-C., Shih, S.-S. and Lee, H.-Y., 2009. Optimal design for hydraulic efficiency performance of free-water-surface constructed wetlands, *Ecological Engineering*, **35**(8), 1200–1207.

Takagi, T. and Sugeno, M., 1985. Fuzzy identification of systems and its applications to modelling and control, *Institute of Electrical and Electronics Engineers Transactions on Systems, Man and Cybernetics*, **15**(1), 116–132.

Tang, X., Scholz, M., Eke, P.E. and Huang, S., 2010. Nutrient removal as a function of benzene supply within vertical-flow constructed wetlands, *Environmental Technology*, **31**(6), 681–691.

Tong, W., Zhu, W. and Ruan, A., 2007. Primary study of clogging mechanisms of substrates in vertical flow constructed wetland. *Journal of Lake Science*, **19**(1), 25–31.

Trajkovic, C., Stankovic, M. and Todorovic, B., 2000. Estimation of FAO Blaney-Criddleb Factor by RBF Network. *Journal of Irrigation and Drainage Engineering*, **126**(4), 268–270.

Tsihrintzis, V.A., Akratos, C.S., Gikas, G.D., Karamouz, D. and Angelakis, A.N., 2007. Performance, and cost comparison of a fws and a vsf constructed wetland system, *Environmental Technology*, **28**(6), 621–628.

U.S. Army Corps of Engineers, 1964. Hydraulic design criteria, air demand regulated outlet works. Sheet 050-1.

Volkart, P., and Rutschmann, P., 1984. Air entrainment devices (air slots). Mitteilung 72, Versuchsanstalt fur Wasserbau, Hydrologie und Glaziologie (Zurich: ETH Zurich).

Vymazal, J., 1999. Removal of BOD5 in constructed wetlands with horizontal subsurface flow: Czech experience, *Water Science and Technology*, **40**(3), 133–138.

Vymazal, J., 2007. Removal of nutrients in constructed wetlands, *The Science of the Total Environment*, **380**(1–3), 48–65.

Vymazal, J., Greenway, M., Tonderski, K., Brix, H. and Mander, Ü., 2004. Constructed wetlands for wastewater treatment. In: J. T. A. Verheven, B. Beltman, R. Bobbink and D. F. Whigham (Eds.) *7th INTECOL International Wetland Conference*, Utrecht, The Netherlands (Berlin: Springer-Verlag) pp. 69–96.

Walker, D.J., 1998. Modeling residence time in storm water ponds, *Ecological Engineering*, **10**(3), 247–262.

Walton, W.E., 2012. Design and management of free water surface constructed wetlands to minimize mosquito production, *Wetlands Ecology and Management*, **20**, 173–195.

Wamsley, T.V., Cialone, M.A., Smith, J.M., Atkinson, J.H. and Rosati, J.D., 2010. The potential of wetlands in reducing storm surge, *Ocean Engineering*, **37**(1), 59–68.

Wanda, H., Vaccab, G., Kuschkb, P., Krügerc, M. and Kästnerb, M., 2007. Removal of bacteria by filtration in planted and non-planted sand columns, *Water Research*, **41**(1), 159–167.

Wang, J., Huang, S.-L., He, C.-D. and Ng, C.-O., 2011. Numerical analysis of the performance of horizontal and wavy subsurface flow constructed wetlands, *Journal of Hydrodynamics, Series B*, **23**(3), 339–347.

Wang, Q.J., 1991. The genetic algorithm and its application to calibrating conceptual rainfall-runoff models, *Water Resources Research*, **27**(9), 2467–2471.

Werker, A.G., Dougherty, J.M., McHenry, J.L. and Van Loon, W.A., 2002. Treatment variability for wetland wastewater treatment design in cold climates, *Ecological Engineering*, **19**(1), 1–11.

Wießner, A., Kappelmeyer, U., Kuschk, P. and Kästner, M., 2005. Influence of the redox condition dynamics on the removal efficiency of a laboratory-scale constructed wetland. *Water Research*, **39**(1), 248–256.

Wisner, P., 1965. On the role of the Froude criterion for the study of air entrainment in high velocity flows. In: *Proceedings of the 11th International Association for Hydraulic Research (IAHR Congress)*, Madrid.

Wittgren, H.B. and Maehlum, T., 1997. Wastewater treatment wetlands in cold climates, *Water Science and Technology*, **35**(5), 45–53.

Wong, T.H.F., Breen, P.F., Somes, N.L.G. and Lloyd, S.D., 1999. *Managing Urban Stormwater Using Constructed Wetlands*, 2nd edn. (Clayton: Monash University).

Wörman, A. and Kronnas, V., 2005. Effect of pond shape and vegetation heterogeneity on flow and treatment performance of constructed wetlands, *Journal of Hydrology*, **301**(1–4), 123–128.

Yang, C.-P., Lung, W.-S., Kuo, J.-T., Lai, J.-S., Wang, Y.-M. and Hsu, C.-H., 2012. Using an integrated model to track the fate and transport of suspended solids and heavy metals in the tidal wetlands, *International Journal of Sediment Research*, **27**(2), 201–212.

Yang, X.S., 2009. Firefly algorithms for multimodal optimization, In: O. Watanabe, and T. Zeugmann (Eds.) *Proceedngs of the 5th International Conference on Stochastic Algorithms: Foundations and Applications*, SAGA, Sapporo, **5792**, pp. 169–178.

Zaji, A.O., Bonakdari, H., Khodashenas, S.R. and Shamshirband, S., 2016. Firefly optimization algorithm effect on support vector regression prediction improvement of a modified labyrinth side weir's discharge coefficient, *Applied Mathematics and Computation*, **274**, 14–19.

Zandieh, M. and Roumani, M., 2017. A biogeography-based optimization algorithm for order acceptance and scheduling, *Journal of Industrial Production Engineering*, **34**(4), 312–321.

Zhang, L., Scholz, M., Mustafa, A. and Harrington, R., 2009. Application of the self-organising map as a tool for an integrated constructed wetland agroecosystem treating agricultural runoff, *Bioresource Technology*, **100**(2), 559–565.

Zhang, M., Jiang, W., Zhou, X., Xue, Y., and Chen, S., 2017. A hybrid biogeography-based optimization and fuzzy C-means algorithm for image segmentation, *Soft Computing*, **23**, 2033–2046.

Zhou, B., Shi, A., Cai, F. and Zhang, Y., 2004. Wavelet neural networks for nonlinear time series analysis, *Lecture Notes in Computer Science*, **3174**, 430–435.

Zhu, J., Cai, X. and Yeh, T-C.J., 2009. Analysis of tracer tomography using temporal moments of tracer breakthrough curves, *Advances in Water Resources*, **32**, 391–400.

Zounemat-Kermani, M., Rajaee, T., Ramezani-Charmahineh, A. and Adamowski, A., 2017. Estimating the aeration coefficient and air demand in bottom outlet conduits of dams using GEP and decision tree methods, *Flow Measurement and Instrumentation*, **54**, 9–19.

Zounemat-Kermani, M. and Sabbagh-Yazdi, S.R., 2010. Coupling of two- and three-dimensional hydrodynamic numerical models for simulating wind-induced currents in deep ponds, *Computer and Fluids*, **39**(6), 994–1011.

Zounemat-Kermani, M. and Scholz, M., 2013. Computing air demand using the Takagi-Sugeno Model for dam outlets, *Water*, **5**(3), 1441–14566.

Zounemat-Kermani, M., Scholz, M. and Tondar, M.-M., 2015. Hydrodynamic modelling of free water-surface constructed storm water wetlands using a finite volume technique, *Environmental Technology*, **36**(20), 2532–2547.

6 Sludge Dewatering Tests

6.1 ASSESSMENT OF THE CAPILLARY SUCTION TIME TEST

6.1.1 Introduction

Dewatering tests are indispensable to quantify the easiness of removing moisture from slurry and sludge with high moisture content of >90% (Scholz, 2005a). Dewaterability tests can also be used to assess the viscosity of slurry, which is needed in the design of industrial dewatering facilities and equipment. Volume reduction of sludge is important to reduce the environmental and financial burden of disposal, and to alleviate land capacity constraints, which are mainly a function of weight and volume. The reduction of volume and weight of sludge is one of the costliest processes in wastewater treatment and is achieved by reducing the moisture content. Dewatering is also an essential step before incineration can be undertaken efficiently (Scholz, 2005a).

Apart from the capillary suction time (CST) test, several other methods to estimate dewaterability are available: specific resistance to filtration (SRF), conditioned filtrate, filtrate total solids and streaming current. Nevertheless, the most used method to measure the dewaterability of sludge is the CST test, which has been proven to be cost-effective, rapid and simple to execute (Dentel and Abuorf, 1995; Scholz, 2005a, 2005b). The CST test can also be used to determine the required amount of a specific conditioning chemical to achieve optimal properties of dewaterability (Jin et al., 2004). The CST method is therefore a handy tool for testing dewaterability in numerous industrial, chemical and environmental applications. A review of dewaterability testing and current trends in CST research has been published in 2005 (Scholz, 2005a).

The CST apparatus was developed by Gale and Baskerville in 1967, and since then it has been used worldwide in various applications and disciplines (Lee and Chen, 2004). However, the standard CST test has a major drawback of inconsistency of the results and relatively high consumable costs associated with the use of the Whatman No. 17 chromatographic paper. Therefore, there is a great industrial demand for a revised CST apparatus and testing procedure based on statistically tested methodologies. There is also a need for comprehensive understanding of the underlying theory of dewaterability testing and the models that allow more precise prediction of the dewaterability of different types of sludge in a relevant industrial and chemical engineering context.

The purpose of this study was to introduce novel CST testing methodologies and materials to reduce the variation of CST data, and operational consumable costs. This was done by assessing the applicability of various alternative papers with improved characteristics (i.e. less expensive, isotropic and smaller pore diameter), the usefulness of different funnel geometries and the influence of stirring on CST measurements with respect to heavy and light sludge types. Several mixtures of synthetic sludge were also formulated to simulate the physical and CST properties

DOI: 10.1201/9781003509370-6

of natural heavy sludge types. Section 6.1 is based on the paper by Sawalha and Scholz (2008), which has been revisited and updated.

6.1.2 Materials and Methods

6.1.2.1 Apparatus and Test Procedures

A patented CST device (Model 304B CST) provided by Triton Electronics Ltd. was used to conduct the experiments. The device consists of a cylindrical steel funnel (filtrand reservoir) resting on the filter paper fitted between two Perspex plates with electrode sensors across the top plate (Lee and Hsu, 1993). The electrodes are placed at a standard interval of 3.7 mm, and at distances of 18.6 and 22.3 mm from the centre of the funnel. The electrodes are connected to a timer. The CST values are automatically displayed on a screen in seconds. The recorded CST value is a measure of the time required for the waterfront to move through a stretch of paper positioned between the two electrodes (Triton Electronics Ltd., product information on www.tritonel.com).

An adequate and representative amount of suspension is poured into the funnel of the CST device until the liquid level reaches the top rim of the funnel. The pressure difference between the filtrand and the paper is typically 5–10 kPa and originates from the capillary pressure difference across the liquid-air meniscus of the wetting front in the paper (Meeten and Smeulders, 1995). The capillary suction pressure forces the filtrate to be sucked from the suspension into the porous media, and a cake on top of the paper is subsequently leftover. The capillary suction pressure of the porous paper is about twice as large as the hydrostatic pressure head within the funnel. Therefore, it can be assumed that the CST value is independent of the quantity of the liquid in the funnel if there is sufficient liquid to generate the suction pressure (Lee and Chen, 2004; Meeten and Smeulders, 1995; Scholz, 2005a). The rate at which filtrate permeates through the paper varies depending on the condition of the sludge and the filterability of the cake formed on the paper. A large CST value indicates a high SRF.

6.1.2.2 Papers

The standard paper commonly used with the CST apparatus is the Whatman No. 17 chromatographic paper (Whatman Plc., Brentford, UK), which was considered as the benchmark paper for the purpose of Section 6.1. The Whatman No. 17 chromatographic paper has been used since the invention of the CST apparatus. It is a chromatographic paper made of cellulose with a high flow rate of 190 mm per 30 minutes and with a mean pore diameter of 8 μm. However, the paper has some disadvantages in the context of the CST test including its anisotropic (i.e. having properties that differ to the direction of measurement) properties and relatively oversized pores besides its relatively high cost which is considered a consumable running cost.

Therefore, alternative cheaper filter and chromatographic papers from different sources (Fisher 200 chr manufactured by Fisher Scientific Ltd., Loughborough, UK, and distributed by Camlab in the UK; HOVO TO w/s from Hollingsworth and Vose Ltd., Kentmere, UK; Carlson EE1.OH from Carlson Filtration Ltd., Barnoldswick,

TABLE 6.1
Summary of Key Paper Properties and Costs (Volunteered by the Manufacturer and/or Supplier)

Paper	Type	FR[a]	BW[b]	Thk[c]	PD[d]	Isotropic	Cost[e]
Whatman No. 17 chr[f]	chr[f]	190	413	920	8	No	0.14
Fisher 200 chr[f]	chr[f]	180–220	440	900		No	0.07
HOVO TO w/s			280		6.7	Yes	0.01
Carlson			425–475	800–1,200	5–6	Yes	0.02
SS1107		121	140	280	4–7	Yes	0.02
SS 3205		136	95	200	6–12	Yes	0.01
SS 3324 chr[f]	chr[f]	180	280	730		No	
MN 440B	chr[f]	130–145	400	1,000		No	
FN 30	chr[f]	240	320	900		No	0.09
BF 4	BP[g]		550	1,300		No	0.15
BF 3	BP[g]	170	280	500		No	0.09
FN 8	chr[f]		330	760		No	0.07

[a] Flow rate (mm/30 minutes); [b] basis weight (g/m^2); [c] thickness (μm), [d] pore diameter (μm); [e] cost per CST paper (£); [f] chromatographic paper; [g] blotting paper.

UK; SS 1107, SS 3205 and SS 3324 chr from Schleicher and Schüll Microscience GmbH (part of the Whatman Group since 2004), Dassel, Germany; MN 440B and MN 280 from Macherey and Nagel GmbH, Düren, Germany, and FN 30, BF 4, BF 3 and FN 8 from Munktell and Filtrak GmbH, Germany) were also tested, and subsequently compared with the Whatman No. 17 chromatographic paper (Table 6.1).

6.1.2.3 Funnel Geometry and Additional Stirrer

Two different sizes of cylindrical funnels were tested: a 10-mm-diameter funnel with a volume of 5.62 cm^3 for light and fast filtering sludges, and an 18-mm-diameter funnel with a volume of 6.36 cm^3 for heavy and slow filtering sludges. A rectangular funnel was used to overcome the negative effect (i.e. uneven water distribution) of the anisotropic (see above) property of many papers including the standard Whatman No. 17 chromatographic paper. The rectangular funnel was designed to have a similar volume in comparison to the 18-mm cylindrical funnel. Dimensions of 18-mm width, 18-mm length and 20-mm height were therefore chosen. The rectangular funnel was tested with both light and heavy sludges, and corresponding results were compared with data obtained with both types of cylindrical funnels. The standard contact area between liquid and paper was enlarged to fit the rectangular funnel, but the probes were placed at the same positions as with the circular funnel.

The CST device was altered by the incorporation of a stirrer into the design. The purpose was to evaluate the possible influence of sedimentation on the CST values. The key objective was to quantify the corresponding effect on the mean and variability of CST results. The stirrer was adapted from an inexpensive toy (manufactured by K'nex) designed for kids. The toy motor is battery-operated and drives an L-shaped stirring shaft. The horizontal part of the shaft is 6-mm long and 3 mm in diameter. The maximum number of rotations per minute is only 35, which is considerably lower than the corresponding number of the stirrer used by the research team in the past (Scholz, 2005a, 2005b). Experience has shown that the reduction in stirrer speed helps to avoid the destruction of weak sludge flocs. The deviation of the shaft from the centre during the rotation was observed to be <1 mm. The stirrer was only used with heavy and slow filtering sludge types. Therefore, the stirrer was only applied with the 18-mm-diameter circular and the rectangular funnels. The stirrer was fixed to the rack positioned 2 mm above and at a right-hand angle (i.e. 90°) to the paper.

6.1.2.4 Different Types of Sludge and Other Suspensions

Samples of primary sludge were obtained from Scottish Water (wastewater treatment plant in North Queensferry, near Edinburgh, UK). The samples were drawn from holding tanks, which collect sludge from primary settlement tanks and septic tanks. Typically, primary sludge contains readily settleable matter from wastewater and usually has a high organic content originating mainly from faecal matter and food scraps. Therefore, primary sludge is a typical representative of heavy sludges. The samples were stored at 4°C in a refrigerator, constantly aerated and used for testing within 4 days.

Samples of surplus activated sludge (i.e. sludge going to waste) were collected from the same wastewater treatment plant as for the primary sludge (see above). However, the samples were taken from the final sedimentation tanks of the activated sludge process. The surplus activated sludge usually contains light flocculent biological solids and is a typical representative of light sludge from secondary sedimentation processes. Corresponding samples usually have higher moisture content in comparison with other sludge types and therefore are associated with greater dewaterability problems. The CST tests with surplus activated sludge were performed within 2 days of sampling. Meanwhile, the samples were stored at 4°C and constantly aerated in a refrigerator. The aeration system was designed with the purpose to avoid destruction of the fragile sludge flocs.

Samples of gully pot sludge (Lee et al., 2005) were gathered from a series of randomly selected gully pots located on roads ranging from representative 'A' class roads to the streets of minor traffic importance in Edinburgh. The gully pot sludge contains both organic and inorganic matter, and its composition varies greatly in space and time (Lee et al., 2005). The samples from different sources were mixed and subsequently strained with a sieve of 250-μm-diameter pore size to simulate preliminary treated wastewater. Thereafter, they were stored at 4°C, continuously aerated and used for testing within 5 days. Physical properties of the different types of natural sludge used for testing, and a comparison with the properties taken from the literature are presented in Table 6.2.

TABLE 6.2
Typical Properties of Tested Real Sludges

Liquid	DMC[a]	WC[b]	SS[c]	pH
Preliminary treated gully plot sludge	416	94	400–450	6–7.5
Primary sludge	33.5		3–40	6–7
Surplus activated sludge	8.5	99.2	6–10	7–7.5
Primary sludge	50–100	95		5–7
Secondary sludge	5–30	99.3		6–7
Digested sludge	40–120			7.2–7.5

[a] Dry matter content (g/L); [b] water content (%) and [c] suspended solids (g/L).

6.1.2.5 Surrogates for Real Sludges

Different recipes for synthetic sludge to be used as a surrogate for real industrial sludge were considered and tested to overcome the problem of time-varying characteristics and properties of the real sludges. The purpose was to produce synthetic sludges with CST values of similar magnitude in comparison to natural sludges. A non-empirical qualitative and quantitative assessment of the modifications of the CST test procedure can only be made with stable reference sludge. The synthetic sludge can be designed, for example, to resemble the physical properties (i.e. rheology (science of the deformation and flow of matter) and dewaterability) of typical sludge flocs and the behaviour of activated sludge, if conditioned with a cationic synthetic polymer (Örmeci and Vesilind, 2000). The dispersion of flocs determines the density, particle size and distribution and porosity of activated sludge flocs. These properties determine the distribution of water in sludge and, in turn, properties such as filterability, easiness to settle and rheology (Sanin and Vesilind, 1999).

The selected synthetic sludges were therefore based on the mixtures of clay slurries, which are the most common constituents to produce stable and reproducible synthetic sludge. Different concentrations of mainly two types of clay (i.e. kaolin and bentonite) were tested for their CST values. The cohesive property of kaolin clay particles is primarily due to their edge-to-face electrostatic alignments. Bentonite clay undergoes osmotic swelling when exposed to moisture, and its shear strength is mainly due to van der Waals attraction forces and partly due to interparticle friction. Kaolin is known to be adequate to simulate individual bacteria (Chu and Lee, 2005). In addition to kaolin and bentonite clay, which are supposed to simulate bacteria, cellulose fibres are used as substitute for filamentous microorganisms, sodium alginate to simulate microbial extracellular polymers and calcium ions are surrogates for bridging cations. Calcium ions have a high affinity for extracellular polymer structure, and calcium is involved in bio-flocculation (Sanin and Vesilind, 1999).

The first mixture contained 100 mL of distilled water and 10% w/w kaolin. The second mixture contained 100 mL of distilled water and 2% w/w kaolin. The third mixture was created by mixing 100 mL of distilled water with 1% w/w bentonite.

The fourth mixture contained 100 mL of distilled water, 7% w/w kaolin and 3% w/w bentonite. The fifth mixture was made up of 100 mL of 85-mM sodium chloride solution, 0.1% w/w kaolin clay, 10 mg/mL of sodium alginate, 60 mg/mL of cellulose fibre and 548-mg/mL $CaCl_2$ $6H_2O$. The sixth mixture contained 100 mL of 85-mM sodium chloride solution, 5% w/w kaolin clay, 10-mg/mL sodium alginate, 60-mg/mL cellulose fibre and 548-mg/mL $CaCl_2$ $6H_2O$. Finally, the seventh mixture was made up of 100 mL of 85-mM sodium chloride solution, 3.33% w/w kaolin clay, 1.67% w/w bentonite clay, 10 mg/mL of sodium alginate, 60-mg/mL cellulose fibre and 548 mg/mL of $CaCl_2$ $6H_2O$.

The first five mixtures were prepared by suspending kaolin and/or bentonite in distilled water and stirring the ingredients until they became homogeneously mixed. Mixtures 6 and 7 were prepared as described in the work of Örmeci and Vesilind (2000). Kaolin and bentonite were mixed and then suspended in 85 mM of sodium chloride solution. Cellulose fibres and sodium alginate were added, and the suspension was stirred for 3 hours. During this time, alginate was adsorbed. Afterwards, calcium was added to the mixture, and the suspension was considered fit for purpose when the flocculation process was finished. The sludge was thoroughly mixed before each use to make sure the suspension could be regarded as homogenous to all intense and purposes.

6.1.2.6 Statistical Analysis

About five to ten sampling replicates were tested on each occasion. All tested treatment combinations were independent of each other and assigned randomly to the experimental units at times. Therefore, a completely random design of the experiment was assumed for all tests. However, it should be noted that not every theoretically possible combination of variables was tested. For example, the 10-mm funnel was only used with light sludge and the 18-mm funnel was only applied for slow filtering sludge as recommended by the manufacturer and independent researchers (Lee and Hsu, 1993). In total, 1279 CST tests were carried out. This figure excludes preliminary tests.

Individual analysis of variance tests were conducted to statistically quantify the differences between the mean CST values for all key variables. Moreover, an overall analysis of variance test was performed to test if the CST values were statistically significantly different in terms of the type of paper, use of stirrer and funnel size and geometry for different types of sludge. The data were then classified according to the presence or absence of stirring action. Descriptive statistics for CST data sets resulting from the use of different filter papers were calculated. Similarly, mean CST values and other descriptive statistics were calculated for different types of funnels when stirring action was applied. Additionally, the mean CST values for each type of paper used were compared with the Whatman benchmark paper using multiple range t-tests with Bonferroni adjustment. Similar multiple range tests were carried out comparing the mean CST values of the two circular funnels with the rectangular funnel. Homogeneity of variance tests were also conducted to quantify the differences in consistency for different papers when measuring CST values. Statistical analyses were performed using the Statistical Analysis System (SAS) Program 9.1 (SAS Institute Inc., Cary, New York, U.S.).

6.1.3 Results and Discussion

6.1.3.1 Testing of Alternative Filter Papers

The overall analysis of variance test showed that the CST values were significantly influenced ($p < 0.01$) by the type of filter or chromatographic paper, funnel geometry and type of sludge, but not by stirring action. It follows that mean CST values were not significantly different with and without stirring. However, the overall analysis of variance test does not allow the user to judge which variable causes the overall test result to be statistically significant. Therefore, multiple comparison t-tests between the mean CST values for the Whatman No. 17 chromatographic paper and CST values for other types of paper were applied.

Tables 6.3 and 6.4 show the least square mean of CST values and other statistics for different papers used for testing the filterability of gully pot, primary and surplus activated sludges and synthetic sludge, respectively, without the use of a stirrer. The mean CST values and corresponding variances with respect to the Fisher 200 chr paper were very similar to the values obtained with the Whatman No 17 chromatographic benchmark paper. There were no significant differences between the mean CST values of the two papers, except when using the surplus activated sludge where the difference was only below two seconds. However, this relatively small difference was statistically significant. There was also a significant reduction in testing time for heavy and slow filtering sludge and improvement in test repeatability (i.e. reduced variability) when using the SS 1107 and SS 3205 papers compared with the Whatman benchmark paper.

On the other hand, the use of the Carlson paper resulted in a large significant increase in testing time compared with the control paper. However, results for the other paper types (e.g., MN 440B and BF 4) were not consistent for different sludge types.

TABLE 6.3
Capillary Suction Time Values for Different Papers and Real Sludge

Variable	N[a]	LSM[b]	Median	CV[c]
		Primarily Treated Gully Pot Sludge		
Whatman No. 17 chr[d]	70	130.2	82.5	74.2
Fisher 200 chr[d]	70	107.1	66.4	75.9
HOVO TO w/s	70	102.3	62.2	128.7
Carlson	70	262.1[e]	143.2	92.8
SS 1107	70	82.9[e]	53.9	68.6
SS 3205	70	81.5[e]	57.8	54.2
SS 3324 chr	70	118.3	65.9	74.4
		Primary Sludge		
Whatman No. 17 chr[d]	22	46.9	41.0	46.0
Fisher 200 chr[d]	22	30.2	19.1	68.5
HOVO TO w/s	10	39.3	22.8	83.7
Carlson	10	107.0[e]	90.2	43.5

(Continued)

TABLE 6.3 (*Continued*)
Capillary Suction Time Values for Different Papers and Real Sludge

Variable	N[a]	LSM[b]	Median	CV[c]
SS 1107	10	22.6[e]	20.1	43.5
SS 3205	10	16.8[e]	12.4	62.8
SS 3324 chr	10	39.8	30.9	53.2
MN 440B	12	14.0[e]	13.4	29.3
FN 30	12	11.5[e]	10.4	34.3
BF 4	12	18.4[e]	16.6	37.5
BF 3	12	54.9	45.3	62.7
FN 8	12	25.2[e]	22.2	36.4
Surplus Activated Sludge				
Whatman No. 17 chr[d]	50	11.3	9.9	52.7
Fisher 200 chr[d]	50	9.4[e]	7.5	67.6
HOVO TO w/s	10	7.9[e]	5.6	73.3
Carlson	10	20.7[e]	19.0	45.3
SS 1107	10	19.7[e]	16.2	34.7
SS 3205	10	10.6	4.8	102.0
SS 3324 chr	10	8.1	5.8	67.2
MN 440B	40	9.6	7.5	67.0
FN 30	40	7.5[e]	6.5	65.8
BF 4	40	13.7[e]	10.3	76.6
BF 3	40	13.6[e]	13.1	45.7
MN 280	40	8.7[e]	8.0	45.2

[a] Number of samples; [b] least square mean; [c] coefficient of variation = standard deviation/mean;
[d] chromatographic paper and [e] values are significantly different ($p < 0.05$) from the benchmark Whatman No. 17 chr paper.

TABLE 6.4
Capillary Suction Time Values for Different Papers Applying Synthetic Sludge

Paper	N[a]	LSM[b]	Median	CV[c]
Whatman No. 17 chr[d]	10	36.0	30.1	81.0
Fisher 200 chr[d]	10	50.5	56.4	78.4
MN 440B	10	135.5[e]	116.5	93.3
FN 30	10	54.4	61.4	80.8
BF 4	10	109.8[e]	106.8	82.2
BF 3	10	59.1	65.3	72.6
FN 8	10	27.5	35.8	71.4

[a] Number of samples; [b] least square mean; [c]coefficient of variation = standard deviation/mean and
[d] chromatographic paper; [e]values are significantly different ($p < 0.05$) from the benchmark Whatman No. 17 chr paper.

The heterogeneity of variance test showed that the variances of the corresponding CST values were significantly different ($p < 0.05$) for various papers. This explains the differences that can be observed, if the coefficients of variation of several types of papers are compared with each other (Table 6.3). The isotropic papers HOVO TO w/s, Carlson EE1.OH, SS 1107 and SS 3205 did not show noticeable improvement in testing time or consistency compared with anisotropic papers (Whatman No. 17 chr, Fisher 200 chr and SS 3324 chr).

6.1.3.2 Testing of Stirring and Rectangular Funnel Use

Stirring was applied when using the 18-mm-diameter circular and rectangular funnels to try to alleviate or reduce the unfavourable effect of sedimentation, which is particularly problematic for very heavy and slow filtering types of sludge. The overall effect of stirring was not statistically significant ($p = 0.08$). Estimates of least square means of CST values for different types of paper when a stirrer was in operation are presented in Table 6.4. Generally, differences between mean CST values for different papers are similar with and without stirring, except for the Fisher 200 chr paper, which showed a reduction in testing time, if a stirrer was applied. Although the mean CST values are generally not affected by stirring, there seems to be a reduction in test variability for some papers (e.g., BF 3 and BF 4). However, the overall effect of stirring seems to be insignificant or even counter-productive when comparing individual types of sludge (Tables 6.5 and 6.6).

The use of a rectangular funnel instead of the standard circular funnels clearly affected the mean and variance of CST values for all types of sludge tested in this experiment. The mean test time was significantly shorter using the rectangular funnel compared with both the 10-mm and the 18-mm-diameter circular funnels (Table 6.7).

TABLE 6.5
Capillary Suction Time Values for Different Filter Papers and Stirred Sludges

Paper Type	N[a]	LSM[b]	Median	CV[c]
Whatman No. 17 chr[d]	35	74.1	63.3	84.7
Fisher 200 chr[d]	35	44.7[e]	52.9	46.5
HOVO TO w/s	30	47.1[e]	32.4	87.8
Carlson	20	112.7[e]	133.3	54.4
SS 1107	20	36.6[e]	42.6	44.7
SS 3205	20	30.3[e]	34.8	29.1
SS 3324 chr[d]	20	61.3	73.5	38.7
MN 440B	5	68.6	94.7	20.8
FN 30	5	17.9[e]	47.4	35.5
BF 4	5	170.6[e]	237.6	32.0
BF 3	5	255.4[e]	304.4	20.4
FN 8	5	81.0	119.4	18.5

[a] Number of samples; [b] least square mean; [c] coefficient of variation = standard deviation/mean and [d] chromatographic paper; [e] values are significantly different ($p < 0.05$) from the benchmark Whatman No. 17 chr paper.

TABLE 6.6
Capillary Suction Times for Different Sludge Types

Stirring	N[a]	LSM[b]	Median	CV[c]
	Preliminary Treated Gully Pot Sludge with 18-mm Funnel			
Stirred	100	72.2	62.2	54.8
Unstirred	210	76.7	66.1	53.5
	Preliminary Treated Gully Pot with Rectangular Funnel			
Stirred	10	67.6[d]	69.5	7.3
Unstirred	140	53.6	48.8	47.5
	Primary Sludge with 18-mm Funnel			
Stirred	70	109.3[d]	80.5	78.4
Unstirred	77	43.4	35.1	81.0
	Synthetic Sludge with 18-mm Funnel			
Stirred	35	19.1[d]	18.6	27.0
Unstirred	35	34.2	31.7	36.7
Synthetic Sludge with Rectangular Funnel				
Stirred	5	17.5	10.0	37.0
Unstirred	35	18.6	17.2	30.1

[a] Number of samples; [b]least square mean; [c]coefficient of variation = standard deviation/mean and [d]values are significantly different ($p < 0.05$) from the benchmark Whatman No. 17 chr paper.

TABLE 6.7
Capillary Suction Time Values for Different Types of Funnels and Sludges

Sludge/Funnel	N[a]	LSM[b]	Median	CV[c]
	Preliminarily Treated Gully Pot Sludge			
10-mm-diameter circular funnel	140	248.8[d]	187.7	72.4
18-mm-diameter circular funnel	310	75.8[d]	63.7	54.0
Rectangular funnel	150	54.4	50.5	45.7
	Primary Sludge			
18-mm-diameter circular funnel	147	75.3[d]	56.8	95.4
Rectangular funnel	77	24.0	16.0	81.8
	Surplus Activated Sludge			
10-mm-diameter circular funnel	175	17.6[d]	15.4	34.0
Rectangular funnel	175	5.9	4.2	53.0
Synthetic sludge				
10-mm-diameter circular funnel	35	136.4[d]	116.5	53.0
18-mm-diameter circular funnel	70	35.7	22.7	45.8
Rectangular funnel	40	16.7	17.0	33.3

[a] Number of samples; [b] least square mean; [c] coefficient of variation = standard deviation/mean and [d] values are significantly different ($p < 0.05$) from the benchmark Whatman No. 17 chr paper.

The advantage of using a rectangular funnel was particularly apparent when heavy gully pot and primary sludges were tested. These sludges usually require long testing times and are therefore expected to be influenced more by sedimentation in comparison with the light sludge types such as the surplus activated sludge. Similarly, the consistency of the test was markedly improved by using the rectangular funnel compared to the circular funnels, except for surplus activated sludge. However, the surplus activated sludge was not tested with the 18-mm-diameter circular funnel, which is directly comparable with the rectangular funnel. Additionally, the mean CST values of the surplus activated sludge were small. This resulted in large coefficients of variation, even though the absolute magnitudes of variation (expressed as standard deviation) were not noticeably large.

6.1.3.3 Synthetic Sludge

A further objective of Section 6.1 was to develop a reasonably inexpensive synthetic sludge, which could be used as a stable benchmark suspension for the calibration of equipment and for reference purposes. Several mixtures of chemicals (see above) were tested, but only one showed promising results. Comparing the CST values of different types of real with synthetic sludges showed that the surrogate was effective in simulating the primary sludge, because the mean CST values were not significantly statistically different from each other (Figure 6.1).

However, the mean CST values of the synthetic sludge were significantly greater than the mean CST values of the surplus activated sludge. Similarly, the CST values were significantly different from each other, and smaller than those of the gully pot sludge.

6.1.3.4 Discussion

This study has evaluated the standard CST test and investigated potential improvement through altering the testing methodology and device. The test reliability can

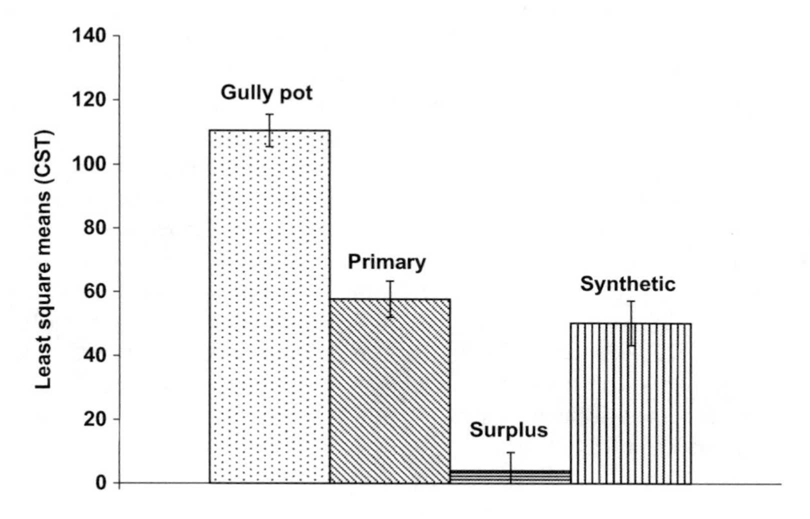

FIGURE 6.1 Least square means (± standard errors) of capillary suction time (CST) test values for different types of sludge.

be improved with increased testing repeatability and reduced variations in testing results. The testing cost and time can also be reduced using less costly papers with at least similar performance compared to the standard Whatman paper. The results obtained here can be considered highly credible because of the use of large number of samples and tests of all practically possible combinations of variables (paper, use of stirrer, funnel size and geometry and sludge type).

A statistical analysis was used to compare means and variances of different alterations of the CST tests. It was found that the mean of the CST values of the standard paper was very close to those obtained with Fisher 200 chr paper. Consequently, this paper can be considered as an adequate surrogate for the Whatman No. 17 chromatographic standard paper which is known to have been arbitrarily chosen as the standard paper (Scholz, 2005a). In addition, the price of the Whatman No. 17 chromatographic paper in comparison to all other papers is relatively high for a consumable item (Lee and Hsu, 1993). It was also shown that the variability and CST testing time can be reduced with other paper types compared with the standard one. This may be attributed to the large pore size (8 μm) of the Whatman paper, which can result in erroneous CST values for some sludge types depending on the floc size. The large pore size of Whatman No 17 paper allows for small-sized particulate invasion, and thereby blocking of the wet front movement. Consequently, a suitable paper should have smaller pore sizes to allow for constant filtrate movement in comparison to the Whatman No 17 paper (Meeten and Smeulders, 1995).

The interactions between different types of paper with various types of sludge were assessed in the analysis of variance and were found to be statistically significant ($p < 0.01$). Therefore, for different types of sludge tested, there could be different papers that perform the best. However, the use of different types of papers for different sludge types may complicate the interpretation of testing results. Different types of sludge with different floc properties can have highly variable CST values. The floc physical properties of flocculating ability, hydrophobicity, surface charge and sludge viscosity are important factors known to significantly affect the water binding ability of the sludge flocs which is a major determinant of the CST test (Jin et al., 2004).

It was hypothesized that the use of stirring would reduce the rate of sludge sedimentation on the top of the paper. This was expected to reduce the possibility of sludge particle invasion of the pores of the filter paper and thereby to reduce testing time and variability among different types of sludge with different particle size and total solid content. However, the introduction of stirring was found to have no significant effect on testing time or variability. The insignificant effect of stirring on CST results may be considered as proof that the sedimentation effect is much smaller than the capillary suction forces of the CST test which themselves seem to be not affected by stirring. This is supported by the finding that the CST value is a function of the free water content of the tested sludge rather than the bound portion (Smollen, 1990). Stirring is not expected to change the proportion or the amount of free water content if the floc matrix is not destroyed.

Similarly, the author found that stirring of most sludge types during a CST test had a trivial impact on reducing sedimentation rates. Other researchers such as Chen et al. (1995) also found that the effect of sedimentation on CST is negligible with papers having small to medium pore sizes. However, a variable speed stirrer may be more helpful to tackle the possible effect of sedimentation on CST test where several rotational speeds can be tested.

Paper water flow can be physically described by the theory of water flow in porous media. This theory is based on the material balance of the water and Darcy's law. The suction and the hydraulic conductivity of both the paper and the suspension are functions of the water content (Scholz, 2005b). The introduction of a rectangular funnel is expected to improve the easiness and accuracy of mathematical modelling of any CST data. This is because of the linear movement of the wet front through all types of paper, if a rectangular funnel is used. The filtrate flow (waterfront movement) with the use of a rectangular funnel is expected to be largely unidirectional, which allows for exact correction of the hydraulic resistance of the filter paper (Lee et al., 2005). In contrast, the wet front movement has an elliptical shape, if a circular funnel is used. Modelling becomes therefore more complicated, because of the non-linear nature of the movement of the wet front. The anisotropic property of some papers is due to the presence of longitudinal grains, which cause the filtrate to move faster along them than across (Lee and Hus, 1992). Therefore, it is more complex to theoretically describe and model the CST values obtained with the circular funnel (Lee and Hsu, 1994; Scholz, 2005a; Scholz et al., 2019). Additionally, the rectangular funnel is expected to reduce the effect of sedimentation, which would otherwise result in the overestimation of the testing time when ignored (Tiller and Li, 2001). Consequently, the CST values are shorter if the rectangular instead of the circular funnel is used as shown in this experiment. Previously, the research team showed for a limited data set that the use of a rectangular funnel instead of the commonly used circular funnel results in more repeatable CST values (Lee and Hsu, 1993).

6.1.4 Conclusions and Recommendations

In comparison to the standard Whatman No. 17 chromatographic paper, alternative cheaper filter or chromatographic papers can be used without significantly changing the range of expected CST values, and the repeatability of most tests. Several alternative papers such as the Fisher 200 chromatographic paper will lead to a faster CST test and improve test consistency as well. The use of a stirrer for heavy sludge did not significantly change the testing time or consistency. However, the introduction of the rectangular funnel was the most promising innovation and can help reducing testing time and unwanted data variability due to instability. A synthetic reference sludge can be prepared from inexpensive ingredients to simulate the primary sludge.

6.2 MIXER IMPACTS ON SLUDGE DEWATERABILITY

6.2.1 Introduction

Water and wastewater treatment plants produce large volumes of sludge, and dewatering and disposal of sludge accounts for approximately 40% of the treatment costs of a treatment plant (Hernando et al., 2010). Coagulation is an essential process in water and wastewater treatment plants (Zhan et al., 2011). Rapid mixing is a process that has a strong influence on all stages of coagulation processes including sludge formation. Not only does rapid mixing disperse the chemical in the water, but it also establishes the formation of coagulant hydrolysis products. Precipitate formation of coagulant hydrolysis products is the agent that has the responsibility not only to destabilize the contaminant, but also to determine sludge production (AWWA, 1999). The effectiveness of coagulation depends on many factors such as rapid mixing, coagulant characteristics, pH, alkalinity, temperature and contaminant characteristics. Among all of these factors, the most important one is rapid mixing (Mhaisalkar et al. 1991).

In water and wastewater treatment plants, rapid mixing can be carried out with a wide range of mixers and reactor configurations, any of which will produce different shearing rates, different flocculent aggregate sizes and hence different rates of flocculent agglomeration. Much work has been published around rapid mixing in relation to the coagulation process (Chul et al., 2006; Yukselen and Gregory, 2004). However, the influence of different shapes and types of mixers on floc formation and stability has been neglected in the scientific literature, possibly because of the associated complexity of the coagulation process. Empirical findings show that the use of different types of mixers produce different floc types and sizes. Amongst other factors, this is due to different shear rates influencing the rate of floc agglomeration. Previous research on the efficiency of different shear-inducing devices on flocculation indicated that different types of mixers produce different sizes of flocs (Serra et al., 2008; Torres et al., 1991. The various types of mixers produced different shear rates; thus, they influenced the rates of floc agglomeration.

The effect of hydraulic turbulence on turbidity removal was assessed for rapid mixing units in the past (Park, 2003). They concluded that for designing and operating rapid mixing units, rapid mixing intensity was defined as the product of the velocity gradient (G) value and the mixing time (t). Rapid mixing technology has the drawback of the formation of dead zones depending on the tank design; for example, dead zones are a greater problem in rectangular compared to circular tanks. For the most effective turbidity removal processes, non-identical paddle rotating speeds and G values for different shapes of jars have been found and listed in tables (Spicer et al., 1996). Furthermore, researchers recommend that mixing intensity and mixer shape should also be determined for different hydraulic conditions and types of water (Park, 2003).

Although some basic studies have been carried out with regard to rapid mixing and sludge dewaterability (Kan et al., 2002, Mhaisalkar et al., 1991; Park, 2003; Rossini et al., 1990; Sawalha, 2010; Torres et al., 1991; Yukselen and Gregory, 2004; Zhan et al., 2011), comprehensive research studies on specific aspects of this relationship are required. Further research may provide insights into the scientific aspects of the floc formation processes and the factors determining the efficiency of sludge dewatering in water and wastewater treatment systems.

The aim of Section 6.2 is to analyse the influence of mixer types and geometries on sludge formation. Recommendations will be made for the use of these types of mixers in water and wastewater treatment plants. A range of parameters that might affect the coagulation process, using CST to measure sludge dewaterability, will be assessed. The objectives of this study are therefore to investigate (a) the influence of different shapes and types of mixers, (b) different rapid mixing velocities and (c) different coagulants on the coagulation process during sludge dewatering simulated by using the CST indicator. Fitria et al. (2013) have provided the basis for this chapter.

A detailed discussion on the advantages and disadvantages of different dewaterability indicators such as CST and the SRF is beyond the scope of Section 6.2. The rapid and cost-effective use of the CST test has led to its wide application in environmental engineering and technology (Houghton et al., 2000; Trias et al., 2004; Thaveemaitree et al., 2003) as reviewed by Scholz (2005). The test limitations including relatively high variability and difficulty to model have been outlined by Sawalha and Scholz (2008, 2009). In comparison, the SRF test is rather expensive and time-consuming. This test is often seen as more scientific than the CST test because of its predominantly hydrodynamic character. However, electrokinetic effects such as a change in the pH value will lead to significant differences in test results (Agerbæk and Keiding, 1993).

6.2.2 Materials and Methodology

Five types of mixers have been used (radial, axial, wheel, 3-blade and magnetic) to investigate their influence on sludge dewaterability tests using the CST apparatus. These types have been chosen based on the information provided by companies producing and/or selling standard mixers used by the water and wastewater industry (Chemineer, Ltd., Cranmer Road, Derby DE21 6XT; Promix Mixing Equipment & Eng. Ltd., Columbus Road, Mississauga L5T 2G9, Canada).

The coagulants, such as aluminium sulphate (alum), ferric sulphate (ferric) and *Moringa oleifera*, have been investigated. Alum and ferric are used because they are the most common coagulants applied in water and wastewater treatment plants, particularly of the developed world (Bektas et al., 2004; Liang et al., 2009; Shi et al., 2007). *M. oleifera* is a natural coagulant (Ahmed et al., 2010), which has been applied as a new alternative coagulant with none of the negative effects of metal-based coagulants (Ndabingengesere et al., 1995).

Real raw wastewaters and sludges have unstable properties; that is, the water quality is constantly changing. Therefore, relatively more stable synthetic raw water has been used to obtain consistent water properties, which is required for research purposes. A kaolin solution has been used in this investigation. The sample preparation was undertaken by adding 1 g of kaolin to 100 mL of distilled water followed by mixing at 1200-rpm velocity using a magnetic stirrer for 2 minutes. This sample was freshly prepared for every experimental run.

A 100-mL kaolin solution sample was poured into a glass beaker, and sulphuric acid (H_2SO_4) or sodium hydroxide (NaOH) was added as a pH adjuster to obtain a pH of around 6.5, which is realistic for the water associated with the simulated treatment units. Thereafter, 5 g of a coagulant (alum, ferric or *M. oleifera*) was added after starting the mixing process. Rapid mixing between 60 and 100 rpm was employed for different rapid mixing velocity parameters for a period of 1 minute. Meanwhile, for different rapid mixing times, between 10- and 90-s mixing time was employed at 100-rpm mixing velocity. Rapid mixing was followed by 15 minutes of slow mixing. Sedimentation was subsequently allowed, and 6 mL of sludge was collected after 15 minutes to measure its dewaterability with a CST apparatus. A Whatman 17 paper was used with the CST apparatus. All experiments were undertaken at constant laboratory temperature of 20 ± 1°C to avoid CST variations due to viscosity changes.

6.2.3 Results and Discussion

6.2.3.1 Different Rapid Mixing Velocities and Rapid Mixing Times

The influence of different rapid mixing velocities and shapes of mixers on sludge dewaterability testing using the CST test was investigated. Figures 6.2–6.4 report on the sludge dewaterability results as a function of different shapes of mixers, different

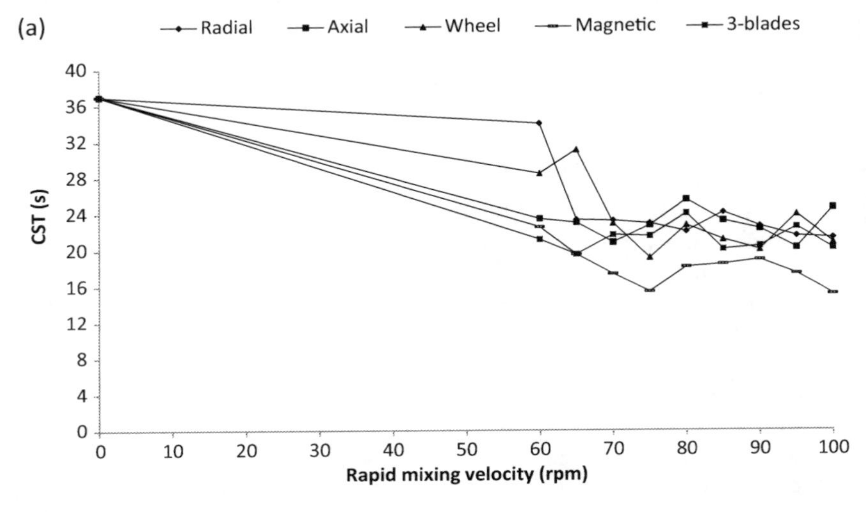

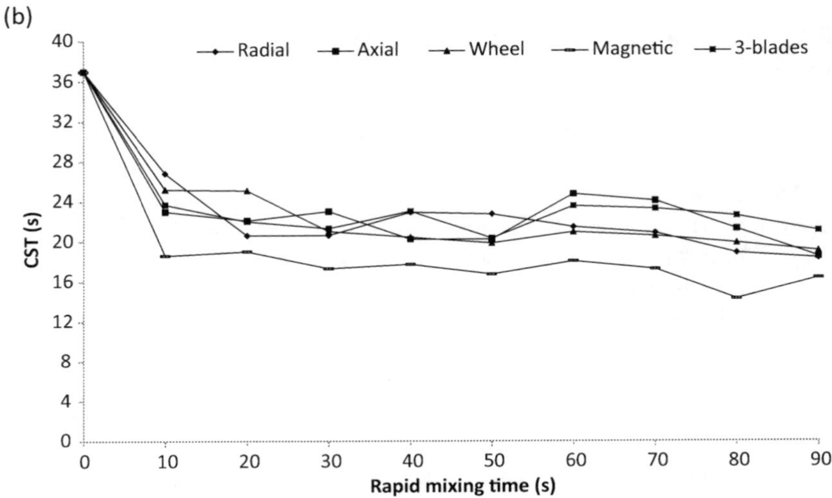

FIGURE 6.2 Influence of different shapes of mixers on sludge dewaterability using aluminium sulphate as a coagulant: (a) velocity versus capillary suction time (CST) and (b) time versus CST.

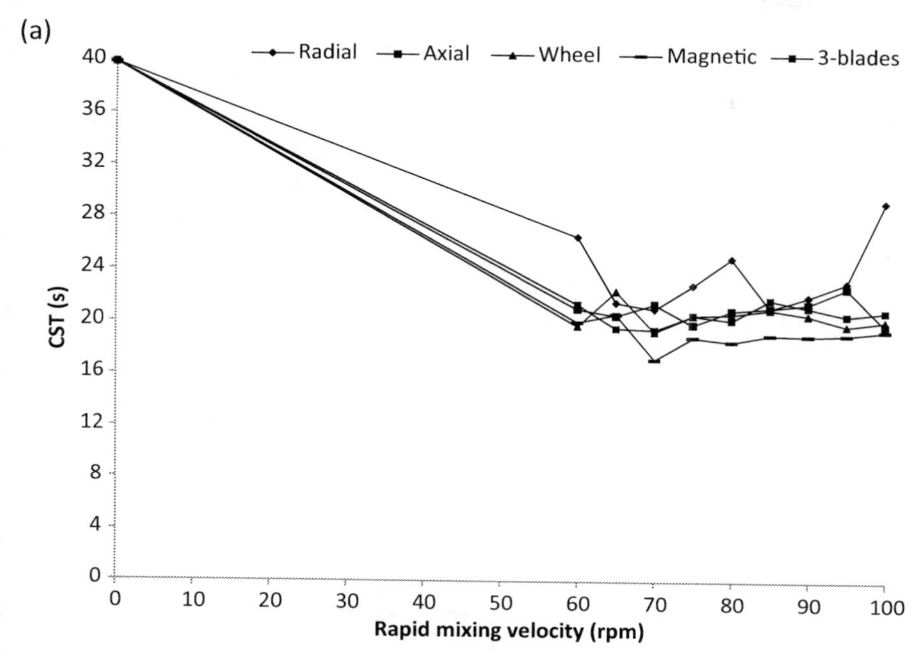

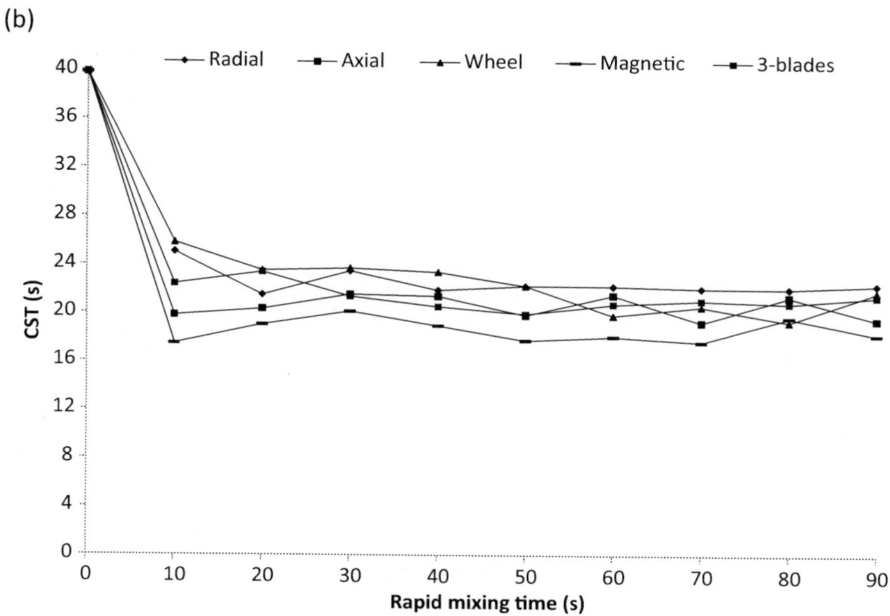

FIGURE 6.3 Influence of different shapes of mixers on sludge dewaterability using ferric chloride as a coagulant: (a) velocity versus capillary suction time (CST) and (b) time versus CST.

(a)

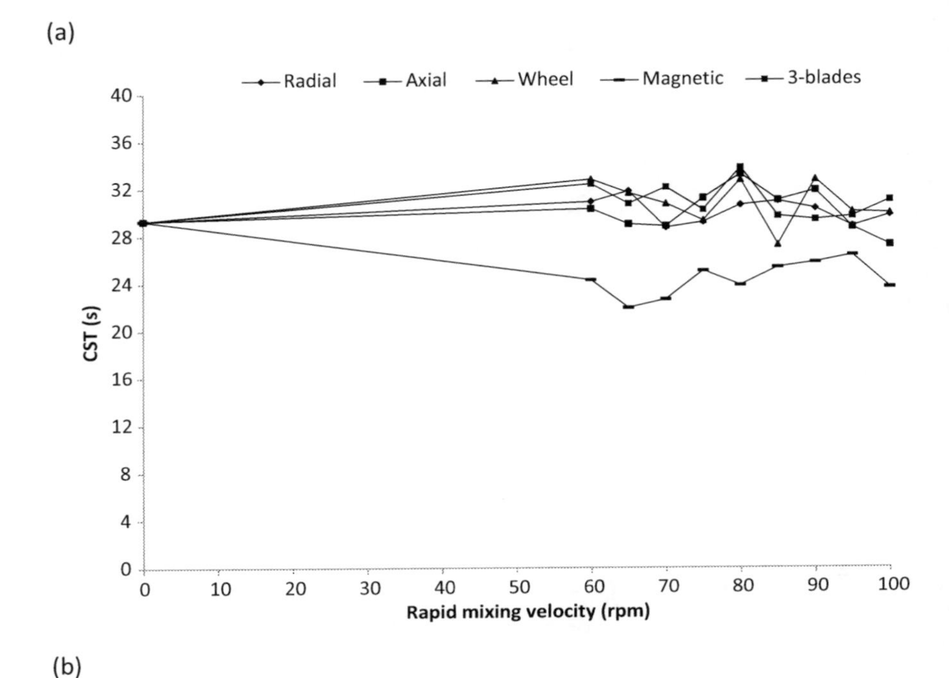

(b)

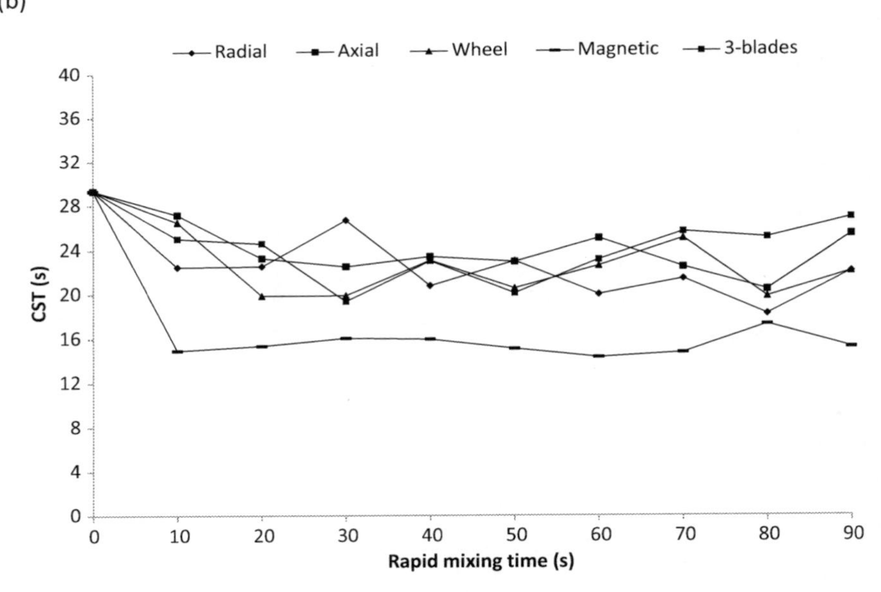

FIGURE 6.4 Influence of different shapes of mixers on sludge dewaterability using *Moringa oleifera* as a coagulant: (a) velocity versus capillary suction time (CST) and (b) time versus CST.

rapid mixing velocities and times and different coagulants. Figures 6.2–6.4 show findings for alum, ferric and *M. oleifera*, respectively. Figures 6.2a, 6.3a and 6.4a show the influence of different shapes of mixers and different rapid mixing velocities on sludge dewaterability. In comparison, Figures 6.2b, 6.3b and 6.4b indicate the impact of different shapes of mixers and different rapid mixing times on sludge dewaterability.

Using different shapes of mixers followed by different rapid mixing velocities, rapid mixing times and different coagulants should produce different results based on the literature review (see Section 6.2.1). In general, different rapid mixing velocities and times did not influence sludge dewaterability. However, different shapes of mixers and different coagulants did show an influence on sludge dewaterability.

For increasing rapid mixing velocities and rapid mixing times, sludge dewaterability, which is indicated by the CST value, is similar for most findings. It follows that different velocities and times do not have an impact on sludge dewaterability. It is likely that even the optimum dose of coagulant has a reduced impact as velocities and times increase. This can be explained by the fact that rapid mixing velocities and times have their optimum values (Mhaisalkar et al., 1991). It seems that relatively low velocities and times are associated with the optimum values.

Extended rapid mixing velocities may break flocs. Highly turbid water does not require a very high mixing intensity compared to low turbid water (Mhaisalkar et al., 1991). Coagulant hydrolysis products formed shortly after coagulant dosing. Prolonged rapid mixing limited floc growth, possibly due to the formation of small flocs during the rapid mixing process. To avoid this type of floc formation, optimum rapid mixing times are required (Yukselen and Gregory, 2004).

6.2.3.2 Different Shapes of Mixers

All results indicate that a magnetic stirrer produces the lowest CST value, while the other mixers offer similar results regarding sludge dewaterability. The greatest difference has been recorded for *M. oleifera* (Figure 6.4) followed by ferric (Figure 6.3) and finally alum (Figure 6.2).

The magnetic stirrer introduces turbulence at the bottom of the beaker glass. In contrast, the other mixers operate higher up in the beakers. Due to its position and how it operates, the magnetic stirrer produces a different vortex from the other products; it leads to a greater vortex and diverse eddy conditions. The magnetic stirrer produces more intense turbulence responsible for mixing the entire water body in the glass beaker. The use of other mixers, however, results in less intense turbulence. Moreover, hydraulic dead zones are created in the outer areas. A larger vortex disperses the coagulant better than a smaller vortex, so an immediate contact between the coagulant and a particle can be reached. Rapid mixing influences floc formation due to its intensity leading to the dispersion of the coagulant into water (Mhaisalkar et al., 1991). Each reaction pathway produces a different coagulant hydrolysis product. This affects the floc formation due to the interaction between the coagulant and the contaminant (Mhaisalkar et al., 1991).

A coagulant is used to optimize sludge dewaterability conditions. In Figure 6.2, it can be seen that by using *M. oleifera* as a coagulant, a magnetic stirrer results in comparably greater CST value differences in contrast to alum and ferric. This is likely due to the difference in sludge volume that is produced by each coagulant (Ndabingengesere et al., 1995). *M. oleifera* produces a smaller volume of sludge compared with a metal-based coagulant, because it only stimulates small contaminants to gather to one another without generating a precipitated coagulant. On the other hand, metal-based coagulants such as alum and ferric are associated with larger volumes of sludge. The application of metal-based coagulants results not only in the agglomeration of contaminants, but also in the precipitation of coagulant hydrolysis products (Ndabingengesere and Narasiah, 1997).

The size of a contaminant is an important factor for the assessment of sludge dewaterability; the bigger the floc size, the less the overall floc water content and the easier the dewatering process. As small contaminants attach to one another, a more condense contaminant agglomeration is being created (Mhaisalkar et al., 1991). The bond water within the agglomeration can be removed relatively easily from the sludge by the dewatering process (Katsiri and Katsiri, 1987).

Tables 6.8 and 6.9 provide statistical information about means and corresponding variability, which is expressed as standard errors and coefficients of variation. Table 6.8 presents sludge dewaterability values for experiments using different mixer shapes, rapid mixing velocities and coagulants. In general, mean CST values for tests using ferric as a coagulant are the lowest compared to alum and *M. oleifera*. About 80% of the lowest CST values from five different shapes of mixers are produced

TABLE 6.8
Capillary Suction Time (CST) Value Based on Different Shapes of Mixers, Different Rapid Mixing Velocities and Different Coagulants

Mixer Shape	Coagulant	Mean of CST ± SE	CV
Radial	Alum	23.6 ± 1.53[a]	20.49
	Ferric	25.2 ± 1.83	22.97
	Moringa	30.0 ± 0.33[b]	3.46
Axial	Alum	24.4 ± 1.49	19.33
	Ferric	22.8 ± 1.9[a]	26.42
	Moringa	30.0 ± 0.56[b]	5.92
Wheel	Alum	24.8 ± 1.80	22.97
	Ferric	22.4 ± 1.96[a]	27.68
	Moringa	30.6 ± 0.58[b]	6.02
Magnetic	Alum	25.8 ± 1.54[b]	18.89
	Ferric	21.1 ± 2.10[a]	31.41
	Moringa	24.8 ± 0.66	8.48
3 Blades	Alum	22.8 ± 1.63	22.61
	Ferric	22.5 ± 1.94[a]	27.26
	Moringa	30.8 ± 0.47[b]	4.88

SE, standard error; CV, coefficient of variation (=100 × (standard deviation/mean)).
[a] Lowest CST value.
[b] Highest CST value.

TABLE 6.9
Capillary Suction Time (CST) Value Based on Different Shapes of Mixers, Different Rapid Mixing Times and Different Coagulants

Mixer Shape	Coagulant	Mean of CST ± SE	CV
Radial	Alum	23.6 ± 1.25	19.13
	Ferric	24.3 ± 1.75[b]	22.83
	Moringa	22.6 ± 1.01[a]	14.18
Axial	Alum	23.0 ± 1.26	19.78
	Ferric	22.2 ± 1.88[a]	26.77
	Moringa	24.2 ± 0.81[b]	10.69
Wheel	Alum	23.4 ± 1.31	20.19
	Ferric	24.0 ± 1.86[b]	24.49
	Moringa	22.9 ± 1.01[a]	14.05
Magnetic	Alum	22.8 ± 1.28[b]	20.31
	Ferric	20.7 ± 2.14	32.77
	Moringa	16.8 ± 1.41[a]	26.47
3 Blades	Alum	23.4 ± 1.19	18.39
	Ferric	23.2 ± 1.87[a]	25.57
	Moringa	24.2 ± 0.94[b]	12.31

SE, standard error; CV, coefficient of variation (=100 × (standard deviation/mean)).
[a] Lowest CST value.
[b] Highest CST value.

by using ferric as a coagulant. *M. oleifera* produces the highest CST value (80%), although while using a magnetic stirrer.

In Table 6.9, the influence of different mixer shapes, rapid mixing times and coagulants on CST can be seen. The data show that the lowest CST values are being produced for tests using *M. oleifera* (60%). In comparison, findings are inconclusive for the highest CST value.

The results show that rapid mixing velocities impact on CST values more than rapid mixing times. This finding is supported by previous research (Mhaisalkar et al., 1991) indicating that rapid mixing velocity is of high importance in the coagulation process. Mixing influences sludge dewaterability, and to ensure high-quality results in sludge dewaterability testing, sufficient mixing is needed to increase sludge dewaterability (Sawalha, 2010; Wang et al., 2009).

6.2.4 Conclusions and Recommendations

Section 6.2 shows that different types of mixers impact on the sludge dewatering properties. The rapid mixing velocity is more important than rapid mixing time, if assessed by the CST test. Different coagulants produced different sludge dewatering conditions.

The magnetic stirrer was linked to the lowest CST values. However, magnetic stirrer systems are not commonly used in industry. Elevated rapid mixing velocities have a significant impact on sludge dewaterability properties. The addition of ferric as a coagulant produced the lowest CST values. In contrast, the highest CST values were recorded when using *M. oleifera*.

More research is required to investigate the effect of different shapes of mixers on sludge dewaterability with different real water samples at various realistic temperatures. The verification of the experimental findings by testing industrial filter presses supported by the CST test should also be undertaken. Furthermore, CST results could be compared with other dewaterability indicators such as the SRF to obtain better interpretations of hydrodynamic processes, if the sludge quality remains constant.

6.3 NOVEL DEWATERABILITY ESTIMATION TEST

6.3.1 Introduction

Sludge contains solids of varying inorganic and organic nature and sizes, but predominantly liquid, which is commonly water. Industries such as water and wastewater utilities, breweries and pulp and paper manufacturers produce large volumes of sludge every day (Yang et al., 2015). Sludge dewatering and disposal accounts for approximately 40% of the treatment costs (Hernando et al., 2010).

The CST test, which essentially measures the advancement of liquid drawn from sludge within a chromatographic paper, is often applied to optimize coagulation processes in industry (Zhou et al., 2015). Coagulation is an essential process in water and wastewater treatment plants (Zhan et al., 2011; Zhou et al., 2015). The influence of different shapes and types of mixers on floc formation and stability has been neglected in the scientific literature due to the complexity of the coagulation process (Kim et al., 2006; Yukselen and Gregory, 2004; Zhou et al., 2015). Empirical findings show that the use of different types of mixers produces different floc types and sizes (Xu et al., 2010), which have different dewaterability characteristics. Furthermore, rapid mixing units also impact on the dewaterability characteristics of sludge. The effect of hydraulic turbulence on turbidity removal has been assessed for rapid mixing units (Park, 2003). Furthermore, Fitria et al. (2014) assessed the impact of sludge floc size and water composition on dewaterability. The reader may wish to refer to Scholz (2005a) for a full review of the subject matter.

Dewatering tests are indispensable to quantify the ease of removing liquids from slurry and sludge with a moisture content of above 90% (Scholz, 2005a, 2005b). An indication of dewaterability can be used to help to characterize the viscosity of slurry, which is needed in the design of industrial dewatering facilities and equipment (Dentel and Abuorf, 1995). Volume reduction of sludge is important to reduce the environmental and financial burden of disposal, and to alleviate land capacity constraints, which are mainly a function of weight and volume. Therefore, dewatering is also an essential step before incineration can be undertaken efficiently (Yang et al., 2015).

Different methods to estimate how easy it is to dewater sludge include CST, SRF (Teng et al., 2018), conditioned filtrate, filtrate total solids and streaming current (Chen et al., 1996). The most commonly used method is currently the CST test, which has been proven to be cost-effective, rapid and simple-to-execute (Dentel and Abuord, 1995; Sawalha and Scholz, 2008, 2009; Scholz, 2005a, 2005b). The standard test provides only one value per experiment. However, the multi-radii CST device provides up to five measurements but is based on the same principles as the standard CST device.

The CST apparatus was developed in 1967, and since then, it has been used worldwide in various applications and disciplines (Lee and Chen, 2004). However, the standard CST test has major drawbacks of inconsistency of results and relatively high consumable costs associated with the use of the Whatman No. 17 chromatographic paper, which is the standard CST paper. The capillary suction pressure generated by this non-homogenous paper is used to suck water and fine solids radially (difficult to treat mathematically) from the sludge, and the time taken for the waterfront to pass between two electrodes constitutes the CST. Fine solids can partly clog the pores within the paper leading to elevated CST values (Sawalha and Scholz, 2008, 2009; Scholz, 2005a).

The novel dewaterability estimation test (DET) apparatus has been developed to address the scientific shortfalls of the traditional CST test (Sawalha and Scholz, 2009) and the complexity of laboratory tests such as resistance to filtration (Teng et al., 2018). The aim of this chapter is to outline the design, operation and performance of the new DET invention. The corresponding objectives are to (a) make a case for the need of the DET apparatus, (b) compare the CST and DET tests with each other and (c) outline the operation of the DET device by using test examples. Note that Section 6.3 is based on an updated version of an original article by Scholz et al. (2019).

6.3.2 Method Developments

6.3.2.1 Capillarity Suction Time

Figure 6.5 shows the standard CST device (Model 304B CST) provided by Triton Electronics Ltd., which was used to conduct the experiments. The equipment consists of a cylindrical steel funnel resting on the filter paper fitted between two Perspex plates with electrode sensors across the top plate (Lee and Hsu, 1993). The electrodes are placed at a standard interval of 3.7 mm, and at distances of 18.6 and 22.3 mm from the centre of the funnel. The electrodes are connected to a timer. The recorded CST value is a measure of the time required for the waterfront to move through a stretch of paper positioned between the two electrodes.

An adequate and representative amount of suspension is poured into the funnel of the CST device until the liquid is level with the top rim of the funnel. The pressure difference between the funnel and the paper is typically 5–10 kPa, and it originates from the capillary pressure difference across the liquid-air interface of the wetting front in the paper (Meeten and Smeulders, 1995). The capillary suction pressure forces the filtrate to be sucked from the suspension into the porous media, and a cake on top of the paper is subsequently leftover.

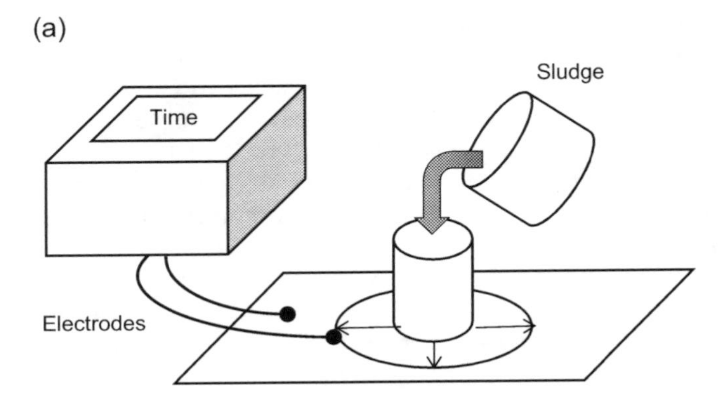

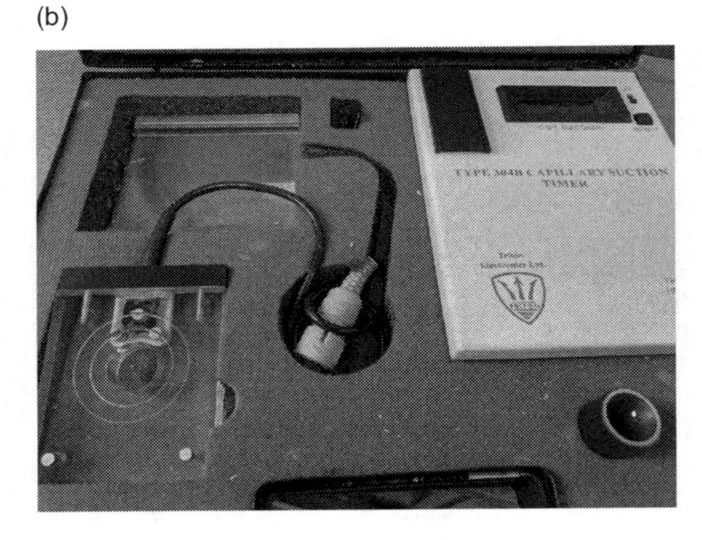

FIGURE 6.5 Standard apparatus (model 304B CST) to measure the capillary suction time (CST): (a) sketch and (b) photograph

The capillary suction pressure of the porous paper is about twice as large as the hydrostatic pressure head within the funnel. Therefore, it can be assumed that the CST value is independent of the quantity of the liquid in the funnel if there is sufficient liquid to generate the suction pressure (Lee and Hsu, 1993; Meeten and Smeulders, 1995; Scholz, 2005a). The rate at which the filtrate permeates through the paper varies depending on the condition of the sludge and the filterability of the cake formed on the paper. A large CST value indicates a high SRF.

6.3.2.2 Dewaterability Estimation Test Device

The DET test relies on both the DET equipment and the corresponding software. The equipment is available as a prototype (Figure 6.6), which is made of stainless steel to allow for high durability and precision. Key components of the device

(a)

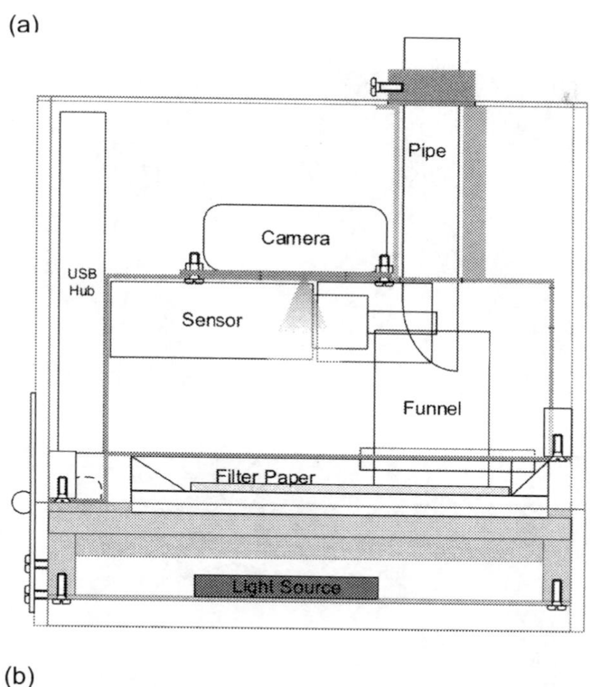

(b)

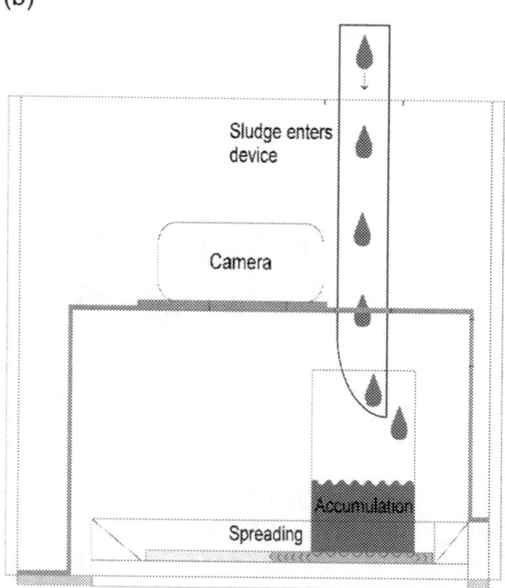

FIGURE 6.6 Components of the dewaterability estimation test (DET) prototype (stainless steel casing): (a) outline sketch; (b) sketch showing the travel stages of sludge; (c) picture of key components; (d) picture of key assembled components and (e) picture of adding sludge process.

(Continued)

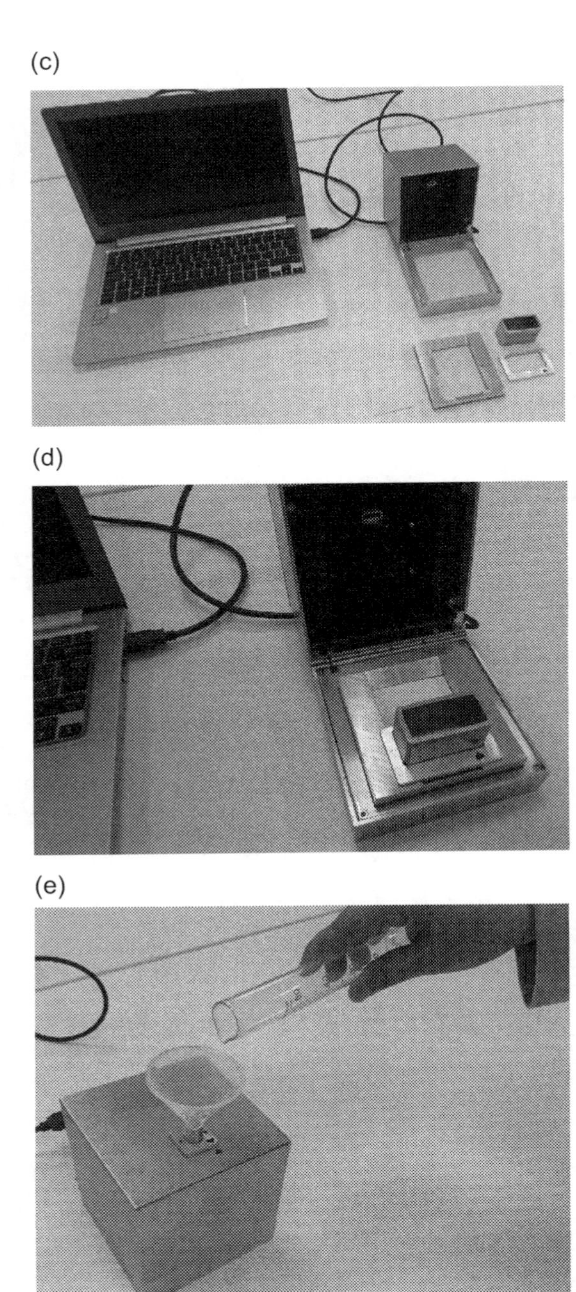

FIGURE 6.6 (*Continued*) Components of the dewaterability estimation test (DET) prototype (stainless steel casing): (a) outline sketch; (b) sketch showing the travel stages of sludge; (c) picture of key components; (d) picture of key assembled components and (e) picture of adding sludge process.

(in no particular order) are the slot funnel, camera, light-emitting diode, light diffuser, sensors to measure temperature and humidity, laptop (hosting the DET software), fan and thermal paste for cooling.

The water distribution during the start of the experiment has been studied carefully to allow for a uniform initial distribution of the sludge onto one side of the paper. This allows for a straight waterfront development, parallel to the side of the filter where the sludge touches the paper for the first time. To achieve a uniform distribution, a slot funnel outlet type (located parallel to the paper) was chosen. Furthermore, to avoid edge effects impacting on the waterfront, the funnel length had to be sufficient, and the funnel had to be located some distance away from the end of the paper to avoid water escaping through gaps between the funnel and the filter paper at the edges (Figure 6.6).

Finding the appropriate weight of the funnel was also an important design consideration. Therefore, experiments with different funnel heights (including 2.5 and 4.55 cm) were performed. The waterfronts of higher (and heavier) funnels performed consistently better in comparison to the shorter funnels. A tall and heavier funnel flattens the paper, subsequently reducing seepage. However, the act of flattening might influence the speed of water spread, because the filter paper is more compacted. Experiments seem to support this theory at least for the standard CST filter paper. Therefore, it was necessary to find empirically the best paper and most suitable funnel using trial and error.

The negative impact of warped paper, particularly, if the paper is very thick, is influenced by the dimensions of the funnel (the wider, the greater the risk) and the funnel weight (the lighter, the less the paper will be flattened). A solution was therefore the selection of a small, but heavy funnel. In order not to increase the local pressure on the filter paper too much, this was achieved by increasing the wall thickness of the funnel to an optimum.

Sawalha and Scholz (2012) have shown that temperature fluctuations lead to high sludge dewatering variability characteristics. Therefore, the fan component in the DET device was seen as essential for the experiment to keep the measurement temperature constant and close to the temperature of the environment in which the sludge will be dewatered. The light source emits heat to raise the temperature by up to 2°C for experiments lasting more than 3 minutes, which alters the viscosity of the sludge, and therefore the time required for the liquid to travel through the paper.

6.3.2.3 Software

The DET software architecture is shown in Figure 6.7. The software uses an area range for checking how the sludge moves, and then analyses liquid characteristics. The software detects the spread across the area of interest. Measurement parameters include average time, maximum time, minimum time, centre location time, standard deviation (expressed in percentage) and the magnitude between the top trace and the side trace.

When auto-stop has been enabled by the user, the software regularly performs the image analysis in the background, while a measurement is being taken. If it is detected that the liquid has spread across the area of interest, the measurement will be stopped automatically.

The dynamic frame rate is also an important feature of the DET device. The user no longer must select a recording frame rate depending on the expected duration of a measurement. Using the automatic setting for the frame rate, the software starts recording with a high frame rate and lowers it, thereafter, logarithmically, the longer the measurements take.

The specific requirements of the DET software (Figure 6.7) are to access the hardware, record data (images, temperature and humidity), detect and follow the spread of the sludge, store results as well as visualize and present relevant data to the user. The software is based on two open-source software libraries (OpenCV and Qt) commonly used for image processing and graphical user interface functionality.

The software was developed with the C++ programming language and follows a modular design. Figure 6.8 shows a binary image analysis example applying the DET software. There are two software tools with graphical user interface: one for end-users (essential functions only) and a second one that provides extended features for calibration and settings. The area between the dotted lines is the area of interest for the DET measurements. Note that the black patches on top of Figure 6.8b is data noise, which is not being considered by the software, because it is outside the area of interest.

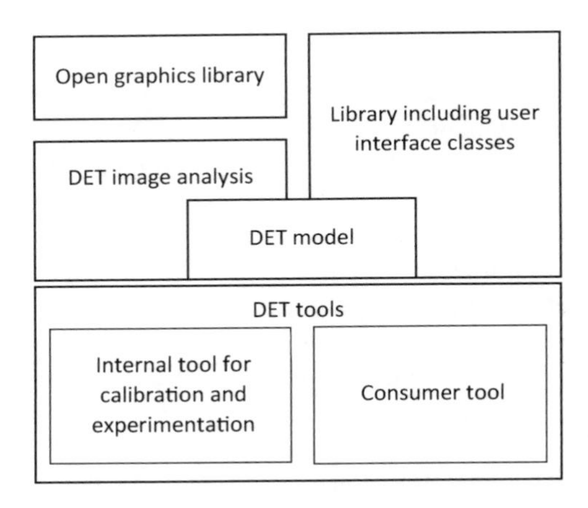

FIGURE 6.7 Dewaterability estimation test (DET) software architecture.

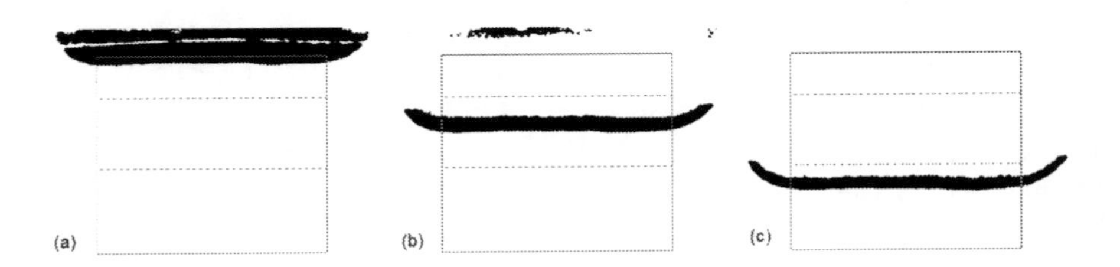

FIGURE 6.8 Binary (black-and-white) image analysis with the dewaterability estimation test software: (a) start of the test and waterfront development; (b) the waterfront has passed the first line of interest and (c) the waterfront has passed the second line of interest.

The method developed for the DET project uses motion tracking to follow the waterfront throughout the recording. The analysis is restricted to an area of interest, a rectangular region within the full-size recording canvas. The processing pipeline for each frame (i.e. picture taken from the webcam) is as follows:

1. Subtract the previous frame: the result reflects the movement of the waterfront between the previous and the current frame.
2. Median filter to reduce noise.
3. Binarization via adaptive threshold, which creates a bitonal black-and-white image.
4. Validation: if the area of the change is smaller than 15% of the area of interest (or if the maximum threshold is reached, continue with step 5. Otherwise, increase the threshold and go back to step 3.
5. Add the result from step 3 to an accumulative image.
6. Measure the water spread for each measurement strip: (a) follow the water flow, pixel row by pixel row, and count the pixels that were classified as 'wet filter paper'. If a certain number of pixels is found, classify the row as 'wet filter paper'; (b) the spread is defined as the maximum number of continuous 'wet filter paper' pixel rows, starting from the beginning of the area of interest, and (c) translate measurements from pixel units into relative spread (percentage of coverage of the area of interest).
7. Clean up the accumulative image: clear all pixels after the current waterfront position to remove noise (e.g., Figure 6.8b).
8. Failsafe: check if there are two or more measurement strips where no spread time could be established. If this is not the case, the process is finished. If there are two or more problematic strips, re-analyse with a different algorithm: (a) always subtract the first frame, (b) apply a fixed threshold and (c) measure water spread.

The software records in a dynamic frame rate, starting with many frames per minute at first, and then slowing down to reduce the amount of recorded data for long measurements. Data will still be produced for each frame using interpolation for frames where no image analysis was applied.

6.3.2.4 Conducted Experiments and Statistical Analysis

About 450 experiments in total were conducted to test different (filter) papers, types of sludge and funnels. Due to instability in real sludge properties (Douziech et al., 2018), synthetic sludge was used to simulate consistent properties, which is required for research purposes (Fitria et al., 2014). The synthetic sludge A recipe was chosen following published guidelines (Hu et al., 2011). The solution was prepared by adding the items displayed in Table 6.10 to 1 L of hot tap water followed by adding 10 g of kaolin. Kaolin was used to simulate the total suspended solids concentrations of 1%, which are like those available in synthetic raw water (Fitria et al., 2014). Moreover, all chemicals shown in Table 6.10 were supplied by Sigma Aldrich Co. Ltd. (Gillingham, UK). After that, the solution was mixed well using a 1,200-rpm

TABLE 6.10
Characteristics of Tested Sludges

Abbreviations	Sludge Name	Location of Origin	Composition/Characteristics
A	Synthetic sludge	Not applicable	Dextrin (150 mg/L); ammonium (130 mg/L); yeast extract (120 mg/L); glucose 100 (mg/L); soluble starch (100 mg/L); sodium carbonate (150 mg/L); detergent commercial (10 mg/L); sodium dihydrogen orthophosphate (100 mg/L); potassium sulphate (8.3 mg/L) and kaolin (10,000 mg/L).
B	Mixed wastewater sludge	Coventry	Real sludge (primary settlement tank) obtained from a sewage treatment plant serving an estate with highly variable domestic but mainly industrial wastewater.
C	Ochre sludge	Bacup	Aluminium (177.8 mg/kg); boron (37.0 mg/kg); calcium (25677.4 mg/kg); cadmium (8.7 mg/kg); chromium (22.5 mg/kg); copper (95.7 mg/kg); iron (4,70458.5 mg/kg); magnesium (286.6 mg/kg); manganese (4,276.3 mg/kg); nickel (15.8 mg/kg); zinc (70.4 mg/kg) and sulphur (15,033.8 mg/kg).
D	Domestic wastewater sludge	Stoke-on-Trent	Highly variable but thick domestic sludge obtained from septic tanks after conditioning.

mixing intensity and a magnetic stirrer for 5 minutes. This synthetic domestic wastewater was prepared fresh and was always stored in the fridge to avoid uncontrolled growth of microorganisms (Fitria et al., 2014).

Test papers such as CST (equivalent to Whatman No. 17), BF3 and EE 2.0H have been explored. The Whatman No. 17 paper (Whatman Plc, Brentford, England, UK) is a chromatographic paper made of cellulose with a high flow rate of 6.33 mm/minute and with a mean pore diameter of 8 μm, basic weight of 413 g/m^2 and thickness of 920 μm. However, the paper has some disadvantages in the context of the CST test including its anisotropic properties and oversized pores besides its high cost. Therefore, alternative cheaper filter and chromatographic papers from different sources were used: BF3 filter paper (Santa Cruz Biotechnology, Inc., Heidelberg, Germany) and EE2.0H filter paper (Carlson Filtration ltd, Barnoldswick, UK) were also tested and subsequently compared with the Whatman No. 17 chromatographic paper. The blotting non-Isotropic filter paper (BF3) has a flow rate of 13 mm per min, while the Isotropic EE 2.0H paper has a flow rate of 2,245 mm/minute with basic weight of 700 g/m^2 and thickness of 3,000 μm.

These funnels have been tested: highly thick with dimensions of 40 mm × 20 mm × 45.5 × 2 mm and medium thick with 40 mm × 20 mm × 35 mm × 2 mm. Ten measurements for each paper, sludge (20 mL) and funnel were taken per experimental run.

The IBM SPSS Statistics Version 20 was applied. Comparisons between two independent variables were performed using the T-Test, when data are normally distributed, while the Mann–Whitney U-test was used instead for not normally distributed data. Moreover, the one-way analysis of variance (ANOVA) was applied to determine whether there are any significant differences between the means of three or more groups, which are normally distributed, while the Kruskal–Wallis test was used instead for non-normally distributed data (Field, 2009).

6.3.3 Results and Discussion

6.3.3.1 Dewaterability Estimation Test Device

Due to variability and instability in real sludge properties (Douziech et al., 2018; Yang et al., 2015), synthetic sludge was used to simulate consistent properties (Fitria et al., 2014). Tables 6.11 and 6.12 show the DET device results for different funnel types when using synthetic domestic wastewater sludge and corresponding statistical

TABLE 6.11
Overview of the Dewaterability Estimation Test Device Results for Different Funnel Types and Synthetic Domestic Wastewater Sludge (10 Samples Were Tested per Each Variable)

Parameter	Unit	Funnel Types	
		High Thick Funnel	Medium Thick Funnel
CST-Filter Paper (Equivalent to Whatman No. 17)			
Average time	S	1,328	1,447
Minimum time	S	1,211	1,310
Maximum time	S	1,550	1,661
Centre time	S	1,215	1,327
Standard deviation	%	7.8	7.3
Magnitude	Mm	2	2
BF3-Filter Paper			
Average time	S	1,437	1,243
Minimum time	S	1,246	1,086
Maximum time	S	1,858	1,513
Centre time	S	1,377	1,195
Standard deviation	%	11.7	10.8
Magnitude	Mm	3	3
EE.2OH- Filter Paper			
Average time	S	2,636	1,400
Minimum time	S	1,529	564
Maximum time	S	3,705	2,802
Centre time	S	2,576	935
Standard deviation	%	32.5	45.0
Magnitude	Mm	8	11

Highly thick with dimensions of 40 mm×20 mm×45.5×2 mm and medium thick with 40 mm×20 mm×35 mm×2 mm.

TABLE 6.12
Overview of Dewaterability Estimation Testing Device Results When Using Different Filter Paper Types (10 Samples Were Tested per Each Variable)

Parameters	Unit	Filter Paper Type		
		CST	BF3	EE.2OH
		Synthetic Sludge (A)		
Average time	S	1,328	1,437	2,636
Minimum time	S	1,211	1,246	1,529
Maximum time	S	1,550	1,858	3,705
Centre time	S	1,215	1,377	2,576
Standard deviation	%	7.8	11.7	32.5
Magnitude	mm	2	3	8
		Coventry Sludge (B)		
Average time	S	191	109	158
Minimum time	S	166	98	129
Maximum time	S	266	126	193
Centre time	S	194	99	154
Standard deviation	%	15.2	8.4	20.1
Magnitude	mm	3	3	3
		Ochre Sludge (C)		
Average time	S	27	21	16
Minimum time	S	24	19	9
Maximum time	S	30	23	28
Centre time	S	25	20	15
Standard deviation	%	6.3	5.3	50.8
Magnitude	mm	2	2	3

analysis. Findings indicated that the required time for synthetic sludge dewatering in terms of average time, minimum time, maximum time and centre time were lower for the high thick funnel than those for the medium thick funnel (Table 6.11) when using CST filter paper type. Concerning Table 6.11, note that no readings were obtained when using Stoke-on-Trent sludge (D), which is very thick, and can therefore not be sucked-up.

Statistical analysis results (Table 6.12) indicated that there are no significant differences ($p > 0.05$) in sludge dewaterability time for the two funnel types when the samples were tested using Whatman No. 17 filter paper. On the other hand, average time, minimum time, maximum time and centre time were recorded to be greater for a high thick funnel than those for a medium thick funnel when using BF3 and EE-2OH filter papers (Table 6.12) to dewater the sludge type A showing some significant ($p < 0.05$) statistical differences (Table 6.12) mainly for the BF3 type indicating the impact of filter paper properties on sludge dewaterability as discussed by others (Lee et al., 1994; Sawalha and Scholz, 2008, 2009, 2012; Tiller and Li, 2001). However, the waterfronts of higher (and heavier) funnels performed consistently better in comparison to the shorter funnels. This can be explained by avoiding warping of the filter paper when the funnel is too light. A tall and heavier funnel flattens the paper, subsequently reducing seepage. Nevertheless, the act of flattening might influence the water spread speed, because the filter paper is more compacted (Scholz, 2005a). The negative impact of a warped paper, particularly if the paper is very thick, is influenced by the dimensions of the funnel (the wider, the greater the risk) and the funnel weight (the lighter, the less the paper will be flattened). Based on that, the author decided to continue the experimental work with the high thick funnel.

Table 6.13 show the DET device results when using different filter paper types per each sludge type and corresponding statistical analysis for the impacts of different filter papers (Table 6.14) and sludge types (Table 6.15). Results highlight that the CST and BF3 filter papers brought about lower and more stable results than the EE.2OH paper (Table 6.14). However, different sludge types might require specific

TABLE 6.13
Statistical Assessment of Different Funnel Type Impacts on the Dewaterability Estimation Test Device Results (10 Samples Were Tested per Each Variable)

Parameter	Shapiro–Wilk Test (*p*-value)[a]	Statistical Test	*p*-value[b] (HTF & MTF)
	CST Paper (Equivalent to Whatman No. 17)		
Average time	0.976	T-test	0.062
Minimum time	0.858	T-test	0.092
Maximum time	0.724	T-test	0.270
Centre time	0.732	T-test	0.050
Standard deviation	0.071	T-test	0.658
Magnitude	0.047	Mann–Whitney	0.961
	BF3 Paper		
Average time	0.153	T-test	0.031
Minimum time	0.316	T-test	0.047
Maximum time	0.000	Mann–Whitney	0.514
Centre time	0.773	T-test	0.017
Standard deviation	0.000	Mann–Whitney	0.231
Magnitude	0.239	T-test	0.055
	EE2.OH Paper		
Average time	0.837	T-test	0.375
Minimum time	0.025	Mann–Whitney	0.182
Maximum time	0.771	T-test	0.606
Centre time	0.179	T-test	0.336
Standard deviation	0.325	T-test	0.451
Magnitude	0.151	T-test	0.138

HTF, highly thick funnel and MTF, medium thick funnel.

[a] Test of normality (if the p-value >0.05, data are normally distributed; if the p-value <0.05, data are not normally distributed; [b] p-value, probability of the statistical test. Results are statistically significantly different only if the p-value <0.05.

TABLE 6.14
Statistical Assessment of Different Filter Paper Impacts on the Dewaterability Testing Device Results (10 Samples Were Tested per Each Variable)

Parameters	Shapiro–Wilk Test (*p*-value)[a]	Statistical Test	*p*-Values[b] for Different Paper Combinations			
			CST&BF3& EE.2OH	CST&BF3	CST& EE.2OH	BF3&EE.2OH
Synthetic Domestic Sludge						
Average time	<0.001	Kruskal–Wallis	0.052	n.a.	n.a.	n.a.
Minimum time	<0.001	Kruskal–Wallis	0.478	n.a.	n.a.	n.a.
Maximum time	<0.001	Kruskal–Wallis	0.073	n.a.	n.a.	n.a.
Centre time	<0.001	Kruskal–Wallis	0.068	n.a.	n.a.	n.a.
Standard deviation	<0.001	Kruskal–Wallis	0.004	0.771	0.005	0.002
Magnitude	<0.001	Kruskal–Wallis	0.001	0.510	<0.001	0.002
Coventry Sludge						
Average time	0.367	ANOVA	<0.001	<0.001	0.233	0.041
Minimum time	0.045	Kruskal–Wallis	<0.001	<0.001	0.169	0.015
Maximum time	<0.001	Kruskal–Wallis	<0.001	<0.001	0.168	0.003
Centre time	<0.001	Kruskal–Wallis	<0.001	<0.001	0.197	0.003
Standard deviation	<0.001	Kruskal–Wallis	0.997	n.a.	n.a.	n.a.
Magnitude	0.178	ANOVA	0.289	n.a.	n.a.	n.a.
Ochre (90%) Sludge						
Average time	0.143	ANOVA	<0.001	<0.001	0.016	0.268
Minimum time	<0.001	Kruskal–Wallis	<0.001	0.001	<0.001	0.177
Maximum time	0.014	Kruskal–Wallis	0.002	<0.001	0.098	0.151
Centre time	0.033	Kruskal–Wallis	<0.001	0.001	<0.001	0.299
Standard deviation	<0.001	Kruskal–Wallis	0.007	0.156	0.053	0.002
Magnitude	0.025	Kruskal–Wallis	0.001	0.004	0.368	0.001

n.a., not applicable as the difference among the variables is not significant.

[a] Test of normality (if *p*-value >0.05, data are normally distributed; if *p*-value <0.05, data are not normally distributed); [b] *p*-value, probability of the statistical test. Results are statistically significantly different only if the *p*-value <0.05.

filter papers to obtain optimal results. In theory, an almost endless amount of filter papers and sludge types could have been tested, which is practically impossible but might be justified and feasible for very specific and stable industrial liquid wastes (Yang et al., 2015).

TABLE 6.15
Statistical Assessment of Different Sludge Impacts on Dewaterability Estimation Test Device Results

Parameter	Shapiro–Wilk Test (*p*-value)[a]	Statistical Test	*p*-values[b] for Different Sludge Combinations			
			A & B & C	A & B	A & C	B & C
CST Paper (Equivalent to Whatman No. 17)						
Average time	<0.001	Kruskal–Wallis	<0.001	0.011	<0.001	0.006
Minimum time	<0.001	Kruskal–Wallis	<0.001	0.034	<0.001	0.019
Maximum time	<0.001	Kruskal–Wallis	<0.001	0.013	<0.001	0.007
Centre time	<0.001	Kruskal–Wallis	<0.001	0.025	<0.001	0.012
Standard deviation	<0.001	Kruskal–Wallis	0.039	0.689	0.057	0.016
Magnitude	0.043	Kruskal–Wallis	0.079	n.a.	n.a.	n.a.
BF3 Paper						
Average time	<0.001	Kruskal–Wallis	<0.001	0.006	<0.001	0.008
Minimum time	<0.001	Kruskal–Wallis	<0.001	0.005	<0.001	0.003
Maximum time	<0.001	Kruskal–Wallis	<0.001	0.002	<0.001	0.006
Centre time	<0.001	Kruskal–Wallis	<0.001	0.008	<0.001	0.033
Standard deviation	<0.001	Kruskal–Wallis	0.017	0.523	0.038	0.006
Magnitude	0.460	ANOVA	<0.001	0.027	<0.001	<0.001
EE.2OH Paper						
Average time	<0.001	Kruskal–Wallis	<0.001	0.026	<0.001	0.014
Minimum time	<0.001	Kruskal–Wallis	0.002	0.042	<0.001	0.055
Maximum time	<0.001	Kruskal–Wallis	<0.001	0.026	<0.001	0.014
Centre time	<0.001	Kruskal–Wallis	<0.001	0.017	<0.001	0.017
Standard deviation	<0.001	Kruskal–Wallis	0.113	n.a.	n.a.	n.a.
Magnitude	0.001	Kruskal–Wallis	0.002	0.019	<0.001	0.362

n.a., not applicable as the difference among the variables is not significant; A, synthetic sludge; B, Coventry sludge; C, ochre sludge; D, Stoke-on-Trent sludge; ANOVA, one-way analysis of variance.

[a] Test of normality (if the p-value > 0.05, data are normally distributed; if the p-value <0.05, data are not normally distributed; [b] p-value, probability of the statistical test. Results are statistically significantly different only if the p-value < 0.05.

6.3.3.2 Capillary Suction Time Device

Table 6.16 shows an overview of CST device results when using different filter paper types. A corresponding statistical assessment of different filter paper and sludge impacts on the CST results can be found in Tables 6.17 and 6.18, respectively. Findings indicate that the CST test only produces results for half of the tested sludge

TABLE 6.16
Overview of Capillary Suction Time (CST) Device Results When Using Different Filter Paper Types (10 Samples Were Tested per Each Variable)

Parameters	Unit	Filter Paper Type		
		CST	BF3	EE.2OH
Coventry Sludge (B)				
Average time	S	734	577	1,128
Minimum time	S	113	378	108
Maximum time	S	575	748	1,781
Centre time	S	-	-	-
Standard deviation	-	951.1	139.8	566.9
Magnitude	mm	-	-	-
Ochre Sludge (C)				
Average time	S	31	49	180
Minimum time	S	11	42	56
Maximum time	S	38	70	416
Centre time	S	-	-	-
Standard deviation	-	9.2	8.5	93.6
Magnitude	mm	-	-	-

TABLE 6.17
Statistical Assessment of Different Filter Paper Impacts on the Capillary Suction Time Device Results (10 Samples Were Tested per Variable)

Parameters	Shapiro–Wilk Test (p-value)[a]	Statistical Test	p-Values[b] for Different Paper Combinations			
			CST&BF3&EE.2OH	CST&BF3	CST&EE.2OH	BF3&EE.2OH
Coventry Sludge						
Average time	0.001	Kruskal–Wallis	0.018	0.014	0.668	0.014
Ochre (90%) Sludge						
Average time	<0.001	Kruskal–Wallis	<0.001	0.031	<0.001	0.038

[a] Test of normality (if the p-value >0.05, data are normally distributed; if the p-value <0.05, data are not normally distributed); [b] p-value, probability of the statistical test. Results are statistically significantly different only if the p-value <0.05.

TABLE 6.18
Statistical Assessment of Different Sludge Impacts on the Capillary Suction Time (CST) Device Results (10 Samples Were Tested per Each Variable)

Parameters	Shapiro–Wilk Test (*p*-value)[a]	Statisti-cal Test	*p*-values[b] (B & C)
	CST Paper (Equivalent to Whatman No. 17)		
Average time	<0.001	Mann–Whitney	<0.001
	BF3 Paper		
Average time	0.001	Mann–Whitney	<0.001
	EE.2OH Paper		
Average time	0.001	Mann–Whitney	0.002

Results are statistically significantly different only if the p-value < 0.05. B, Coventry sludge; C, ochre sludge.
[a] Test of normality (if p-value > 0.05, data are normally distributed; if p-value < 0.05, data are not normally distributed; [b]p-value, probability of the statistical test.

samples. No readings were obtained when using synthetic sludge (A) and Stoke-on-Trent sludge (D), which cannot be sucked-up by the papers, because they are too thick. Centre time and magnitude could not be measured by the CST device. Moreover, the data are highly variable, which indicates low reproducibility. This can be explained by the flaws of the CST test outlined in the methodological development explained above.

6.3.3.3 Test Comparisons

Table 6.19 shows a comparison between DET and CST tests. The filter paper EE 2.0 H is highly anisotropic and very thick. Therefore, sludge spreading appears to be random and takes a long time using the CST apparatus (even no measurements for sludge A could be recorded). Particles within the sludge are not retained fully by paper EE 2.0 H and are taken along with the liquid during the test.

When comparing the performances of both competing devices, the following claims can be made for the tests outlined in Table 6.19: the DET apparatus is usually more reliable (lower standard deviations) than the CST device. The new device gives faster results than the CST apparatus (Table 6.19).

6.3.3.4 Software Performance

The DET apparatus is fully supported by the DET software for testing in the lab and field. The software performs image analysis in the background while measurements are being taken. The new equipment produces considerably more data (multiple points of measurement, deceleration of dewatering throughout measurement and video recording) than the CST test. The DET device is also the only dewaterability test supported by image analysis.

The DET software worked without any complications. However, on occasions, the software did not recognize the waterfront within paper EE 2.0H, which was considered, however, inappropriate for the test as explained above.

TABLE 6.19
Comparison of the Dewaterability Estimation Test (DET) with the Capillary Suction Time (CST) Test

Test	CST A	BF3 A	EE 2.0H A	CST B	BF3 B	EE 2.0H B	CST C	BF3 C	EE 2.0H C
				Number of Measurements					
DET	10	10	10	10	10	10	10	10	10
CST	n.a.	n.a.	n.a.	10	10	10	10	10	10
				Average Measurement Times (s)					
DET	1,328	1,437	2,636	191	109	159	27	21	16
CST	n.a.	n.a.	n.a.	734	577	1,128	31	49	180
				Relative Standard Deviations (%)					
DET	7.8	12	33	15	8	20	6	5	51
CST	n.a.	n.a.	n.a.	15	24	50	30	17	52

A, synthetic sludge; B, Coventry sludge; and C, ochre sludge; n.a., not applicable; DET, dewaterability estimation test; CST, capillary suction time.

6.3.4 Conclusions and Recommendations

The DET apparatus is faster and more reliable than the CST apparatus with respect to the sludge types tested and filter papers used. It can obtain readings that the CST apparatus is not able to provide particularly for heavy sludge. In contrast to the CST, the DET apparatus is supported by image analysis software allowing for the recording and analysis of as many pictures as required.

The commercial potential of the patented DET apparatus is high, since it should replace the CST apparatus, which currently dominates the market due to lack of competition by other portable devices. The author recommends further work on testing the new device in various sludge-producing industries, and to produce tables allowing practitioners to easily transform their previous CST data into DET equivalent ones, wherever scientifically justifiable. Furthermore, quantitative comparisons and qualitative relations to measures of real dewaterability linked to industrial dewatering technologies should be provided.

REFERENCES

Agerbæk, M.L. and Keiding, K., 1993. On the origin of specific resistance to filtration, *Water Science and Technology*, **28**(1), 159–168.

Ahmed, T., Kanwal, R., Hassan, M., Ayub, N. and Scholz, M., 2010. Coagulation and disinfection in water treatment using *Moringa*, *Proceedings of the Institution of Civil Engineers – Water Management*, **163**(8), 381–388.

American Water Works Association (AWWA) 1999. *Coagulation and flocculation; Water Quality and Treatment*, 5th edition (New York: Mc Graw Hill).

Bektas, N., Akbulut, H., Inan, H. and Dimaglo, A., 2004. Removal of phosphate from aqueous solutions by electro coagulation, *Journal of Hazardous Materials*, **106**(2–3), 101–105.

Chen, G.W., Lin, W.W. and Lee, D.J., 1995. Effects of particle sedimentation on capillary suction time (CST), *Chinese Institute of Chemical Engineers*, **26**(6), 371–377.

Chen, G.W., Lin, W.W. and Lee, D.J., 1996. Capillary Suction Time (CST) as a measure of sludge dewaterability, *Water Science Technology*, **34**(3–4), 443–448.

Chu, C.P. and Lee, D.J., 2005. Comparison of dewaterability and floc structure of synthetic sludge with activated sludge, *Environment Technology*, **26**(1), 1–10.

Dentel, S.K. and Abuorf, M.M., 1995. Laboratory and full-scale studies of liquid stream viscosity and streaming current for characterization and monitoring of dewaterability, *Water Research,* **29**(12), 2663–2672.

Douziech, M., Rosique Conesa, I., Benítez-López, A., Franco, A., Huijbregts, M. and van Zelm, R., 2018. Quantifying variability in removal efficiencies of chemicals in activated sludge wastewater treatment plants – a meta-analytical approach. *Environmental Science & Processes and Impacts*, **20**(1), 171–182.

Field, A., 2009. *Discovering Statistics Using SPSS*, 3rd edn. (London: SAGE Publication).

Fitria, D., Swift, G.M. and Scholz, M., 2013. Impact of different shapes and types of mixers on sludge dewaterability, *Environmental Technology*, **34**(7), 931–936.

Fitria, D., Scholz, M., Swift, G.M. and Hutchinson, S., 2014. Impact of sludge floc size and water composition on dewaterability. *Chemical Engineering Technology*, **37**(3), 471–477.

Hernando, M.R., Labanda, J. and Llorens, J., 2010. Effect of ultrasonic waves on the rheological features of secondary sludge. *Biochemical Engineering Journal*, **52**(2–3), 131–136.

Houghton, J.I., Stephenson, T. and Quarmby, J., 2000. The impact of digester retention time on microbial extracellular polymer production and sludge dewaterability, *Environmental Technology*, **21**(10), 1157–1165.

Hu, B., Wheatley, A., Ishtchenko, V. and Huddersman, K., 2011. The effect of shock loads on SAF bioreactors for sewage treatment works. *Chemical Engineering Journal*, **166**(1), 73–80.

Jin, B., Wilén, B.M. and Lant, P., 2004. Impacts of morphological, physical and chemical properties of sludge flocs on dewaterability of activated sludge. *Chemical Engineering Journal*, **98**(1–2), 115–126.

Kan, C., Chuang, C. and Pan, J.R., 2002. Time requirement for rapid mixing in coagulation; colloid and surfaces, *Colloids and Surfaces A: Physicochemical and Engineering Aspects*, **203**(1–3), 1–9.

Katsiri, N. and Katsiri, A.K., 1987. Bound water content of biological sludges in relation to filtration and dewatering, *Water Research*, **21**(11), 1319–1327.

Kim, H.-C., Hong, J.H. and Lee, S., 2006. Fouling of microfiltration membranes by natural organic matter after coagulation treatment: A comparison of different initial mixing conditions, *Journal of Membrane Science*, **283**(1), 266–272.

Lee, D.J. and Chen, G.W., 2004. On some aspects of capillary suction apparatus tests. *Chinese Institute of Chemical Engineers*, **25**(1), 35–44.

Lee, D.J. and Hus, Y.H., 1992. Fluid flow in capillary suction apparatus. *Industrial and Engineering Chemistry Research,* **31**(10), 2379–2385.

Lee, D.J. and Hsu, Y.H., 1993. Cake formation in capillary suction apparatus. *Industrial and Engineering Chemistry Research,* **32**(6), 1180–1185.

Lee, D.J. and Hsu, Y.H., 1994. A rectangular capillary suction apparatus. *Industrial and Engineering Chemistry Research,* **33**(6), 1593–1599.

Lee, D.J., Chen, G.W. and Hsu, Y.H., 1994. On some aspects of Capillary Suction Apparatus tests. *Industrial and Engineering Chemistry Research*, **25**(1), 35–44.

Lee, B.-H., Scholz, M. and Horn, A., 2005. Constructed wetlands for the treatment of concentrated stormwater runoff (Part A). *Environmental Engineering Science*, **23**, 191–202.

Liang, Z., Wang, Y.X., Zhou, Y. and Lu, H. 2009. Coagulation removal of melanoidins from biologically treated molasses waste water using ferric chloride, *Chemical Engineering Journal*, **152**(1), 88–94.

Mhaisalkar, V.A., Parasivam, R. and Bhole, A., 1991. Optimizing physical parameter of rapid mixing design for coagulation-flocculation on turbid waters, *Water Research*, **25**(1), 43–52.

Meeten, G.H. and Smeulders, J.B.A.F., 1995. Interpretation of filterability measured by the capillary suction time method, *Chemical Engineering Science*, **50**(8), 1273–1279.

Ndabingengesere, A. and Narasiah, K.S., 1997. Quality of water treated by coagulation using Moringa oleifera seeds, *Water Research*, **32**(1), 781–791.

Ndabingengesere, A. and Narasiah, K.S. and Talbot, BG., 1995. Active agents and mechanism of coagulation of turbid waters using Moringa oleifera, *Water Research*, **29**(2), 703–710.

Örmeci, B. and Vesilind, P.A., 2000. Development of an improved synthetic sludge: A possible surrogate for studying activated sludge dewatering characteristics, *Water Research*, **34**(4), 1069–1078.

Park, N.S., 2003. Examining the effect of hydraulic turbulence in rapid mixer on turbidity removal with CFD simulation and PIV analysis, *Journal of Water Supply Research and Technology – AQUA*, **52**(2), 95–108.

Rossini, M., Garrido, G. and Galluzo, M., 1990. Optimization of the coagulation-flocculation treatment: Influenced of rapid mixing parameters, *Water Research*, **3**(8), 1817–1826.

Sanin, F.D. and Vesilind, P.A., 1999. A comparison of physical properties of synthetic sludge with activated sludge, *Water and Environmental Research*, **71**(2), 191–196.

Sawalha, O., 2010. CST: Developments in testing methodology and reliability of results. PhD thesis (Edinburgh: The University of Edinburgh).

Sawalha, O. and Scholz, M., 2008. Assessment of Capillary Suction Time (CST) test methodologies, *Environmental Technology*, **28**(12), 1377–1386.

Sawalha, O. and Scholz, M., 2009. Innovative enhancement of the design and precision of the capillary suction time testing device, *Water Environmental Research*, **81**(11), 2344–2352.

Sawalha, O. and Scholz, M., 2012. Impact of temperature on sludge dewatering properties assessed by the Capillary Suction Time. *Industrial and Engineering Chemistry Research*, **51**(6), 2782–2788.

Scholz, M., 2005a. Review of recent trends in capillary suction time (CST) dewaterability testing research, *Industrial and Engineering Chemistry Research*, **44**(22), 8157–8163.

Scholz, M., 2005b. Revised capillary suction time (CST) test to reduce consumable costs and improve dewaterability interpretation, *Journal of Chemical Technology and Biotechnology*, **81**(3), 336–344.

Scholz, M., Almuktar, S., Clausner, C. and Antonacopoulos, A., 2019. Highlights of the novel Dewaterability Estimation Test (DET) device, *Environmental Technology*, **41**(20), 2594–2602.

Serra, T., Colomer, J. and Logan, B.E., 2008. Efficiency of different shear devices on flocculation, *Water Research*, **42**(4–5), 1113–1121.

Shi, B., Li, G., Wang, D., Feng, C.H. and Tang, H. 2007. Removal of direct dyes by coagulation: the performance of preformed polymere aluminium species, *Journal of Hazardous Materials*, **143**(1–2), 567–574.

Smollen, M., 1990. Evaluation of municipal sludge drying and dewatering with respect to sludge volume reduction, *Water Science and Technology* **22**(12), 153–162.

Spicer, P.T., Keller, W. and Pratsinis, S.E., 1996. The effect of impeller type on floc size and structure during shear induced flocculation, *Journal of Colloid and Interface Science*, **184**(1), 112–122.

Teng, J., Shen, L., Yu, G., Wang, F., Li, F., Zhou, X., He, Y. and Lin, H., 2018. Mechanism analyses of high specific filtration resistance of gel and roles of gel elasticity related with membrane fouling in a membrane bioreactor, *Bioresource Technology*, **257**, 39–46.

Thaveemaitree, Y., Polprasert, C. and Seung-Hwan, L., 2003. Application of electrochemical process for landfill leachate treatment with emphasis on heavy metal and organic removal, *Environmental Technology*, **24**(9), 1135–1145.

Tiller, F.M. and Li, W.P. 2001. Modified capillary suction theory with effects of sedimentation for rectangular cells. *Journal of the Chinese Institute of Chemical Engineers*, **32**(5), 391–399.

Torres, F.E., Russel, W.B. and Schowalter, W.R. 1991. Floc structure and growth kinetics for rapid shear coagulation of polystyrene colloids, *Journal of Colloid and Interface Science*, **142**(2), 554–574.

Trias, M., Mortula, M.M., Hu, Z. and Gagnon, G.A., 2004. Optimizing settling conditions for treatment of liquid hog manure, *Environmental Technology*, **25**(8), 657–965.

Wang, Y., Gao, B.Y., Xu, X.M., Xu, W.Y. and Xu, G.Y., 2009. Characterization of floc size, strength and structure in various aluminium coagulants treatment, *Journal of Colloid and Interface Science*, **332**(2), 354–359.

Xu, W., Gao, B., Yue, Q. and Wang, Y., 2010. Effect of shear force and solution pH on flocs breakage and re-growth formed by nano-Al13 polymer, *Water Research*, **44**(6), 1893–1899.

Yang, G., Zhang, G. and Wang, H., 2015. Current state of sludge production, management, treatment and disposal in China, *Water Research*, **78**, 60–73.

Yukselen, M.A. and Gregory, J. 2004. The reversibility of floc breakage. *International Journal of Mineral Processing*, **73**(2–4), 251–259.

Zhan, X., Gao, B., Yue, Q., Wang, Y. and Cao, B., 2011. Influence of velocity gradient on aluminium and iron floc property for NOM removal from low organic matter surface water by coagulation, *Chemical Engineering Journal*, **166**(1), 116–121.

Zhou, Z., Yang, Y. and Li, X., 2015. Effects of ultrasound pretreatment on the characteristic evolutions of drinking water treatment sludge and its impact on coagulation property of sludge recycling process, *Ultrasonics Sonochemistry*, **27**, 62–71.

7 Climate Change, Water Availability and Food

7.1 DIGITAL FILTERING ALGORITHMS SIMULATING CLIMATE VARIABILITY ON RIVER FLOW

7.1.1 Introduction

Separation of the river flow hydrograph into a baseflow and direct runoff is a valuable method to comprehend sub-surface water contribution to the stream, particularly, when dealing with various water resources regulation problems (Lu et al., 2015; Mohammed and Scholz, 2016). These procedures also have been applied to calculate the sub-surface water proportion of hydrological resources and to contribute to the estimate of recharge. The direct runoff element denotes the extra discharge contributed by groundwater and surface runoff. However, the baseflow element indicates constant groundwater flow additions to streamflow, which is often important for water resource assessments (Brodie and Hostetler, 2005).

The baseflow index (BFI) represents the ratio of the baseflow to the total flow that can be calculated by applying a hydrograph separation method. Generally, this index can vary from 0.15 to 0.20 for an impermeable basin and be more than 0.95 for a large capacity basin. The BFI relates to basin characteristics such as soil type, geology, topography, plants and regional climate (WMO, 2009; Price, 2011).

Numerous procedures have been suggested for baseflow separation, which are usually classified into three main groups: graphical methods (Linsley et al., 1988), digital filtering algorithm (DFA) frequency assessment and recession analysis (Welderufael and Woyessa, 2010; Mohammed and Scholz, 2016). Nathan and McMahon (1990) differentiated between techniques for constant separation by various flow proportions (Tallaksen and van Lannen, 2004). Graphical methods separate the flow hydrograph by linking the intersecting point of the baseflow and the runoff upon the bottom point at the rising limb of the flow hydrograph, where it is anticipated that all discharges are considered as baseflow (Linsley et al., 1988; Welderufael and Woyessa, 2010). These methods separate the baseflow by many techniques that differ in their complexity; constant flow rate, constant slope and concave technique (Al-Faraj and Scholz, 2014). Lim et al. (2005) highlighted that the major disadvantage of the graphical procedures is the lack of reliable results even for the same flow data set.

The United Kingdom Institute of Hydrology (UKIH) method is applied for daily mean flow data and is founded on the identification and interpolation of turning discharge values in a measured streamflow time series (Piggott et al., 2005). The turning points represent the time in days and equivalent discharge values, where the recorded streamflow value is anticipated to be the total baseflow. For turning point calculations, the flow values are divided into a series of five-day parts and the lowest values

DOI: 10.1201/9781003509370-7

of flow within each part (an x and y pair where x_i is the day on which the lowest flow value of y_i happened) are chosen and distinct to candidate turning points. Then, the candidate turning point is compared to the preceding and following parts. Turning points are tested using the condition $0.9 \times y_i < \text{minute } (y_{i-1}, y_{i+1})$ (Piggott et al., 2005). The baseflow temporal variation is computed by piecewise linear interpolation connected by sequential sets of turning points.

The DFA is the most utilized approach in streamflow separation that divides the baseflow by a processing method (Welderufael and Woyessa, 2010; Mohammed and Scholz, 2016). This procedure can be automated (Eckhardt, 2005; Mohammed and Scholz, 2016). For Section 7.1, two of the DFAs have been assessed, the Eckhardt (2005) and the frequently applied Chapman (1999) procedures.

The flow duration curve (FDC) can be defined as a curve showing the percentage of time that a streamflow is likely to be equivalent to or is greater than a specific value. This plot describes the capability of the basin to provide flows of various magnitudes. The lower and upper parts of the plot profile are essential in evaluating the channel and basin attributes. The high flow area form indicates the type of flood that the basin is describing, while the form of the low flow region characterizes the basin capacity to sustain low flow during the dry season. Generally, Q_{50} is seen as the median flow. A flow $\geq Q_{50}$ is understood as low flow (Welderufael and Woyessa, 2010; Mohammed and Scholz, 2016). The Q_{90}/Q_{50} (i.e. the recorded flow $\geq 90\%$ and $\geq 50\%$ of the time ratio, respectively) ratio denotes a percentage of the river flow sub-surface water aquifer contribution or the corresponding proportion of the baseflow component (Gordon et al., 2004; Al-Faraj and Scholz, 2014; Stewart, 2015).

The baseflow timing and quantity can be impacted on by many factors, such as climate, human activities (e.g., river damming) and basin characteristics. Many scientists have analysed the baseflow contribution to streamflow (e.g., Partington et al., 2012; Fan et al., 2013; Mei and Anagnostou, 2015; Rumsey et al., 2015; Stewart, 2015; Wan et al., 2015; He et al., 2016; Lott at al., 2016; Miller et al., 2016; Mohammed and Scholz, 2016). However, they have only used and/or compared many traditional techniques without taking into consideration climate change, drought phenomena and/or human-induced activities. To address the knowledge gap, this study will consider such impacts on the baseflow of a stream.

The impact of a combination of anthropogenic intervention and climate change on the streamflow of a basin differs subject to climatic zone arrangement. Consequently, this should be investigated at a local scale (e.g., basin or sub-basin). Recently, due to universal climate variability and local human-induced stresses, many areas have been affected by floods and droughts. Thus, evaluating the impact of such changes is important for understanding the mechanisms of hydrological responses at a basin scale, regional water resource organization and flood and drought protection. The Lower Zab River Basin (LZRB) has witnessed substantial alterations in hydro-climatic variables as well as human activities in recent years. Accordingly, this basin can be considered as an excellent case study for evaluating the hydrological influence of climate variability and anthropogenic intervention.

The aim of Section 7.1 is to apply the methodology by Mohammed and Scholz (2016) to assess the impact of climate change and human-induced drought events on the baseflow contribution to a streamflow. In addition, multiple aspects of basin hydrology are

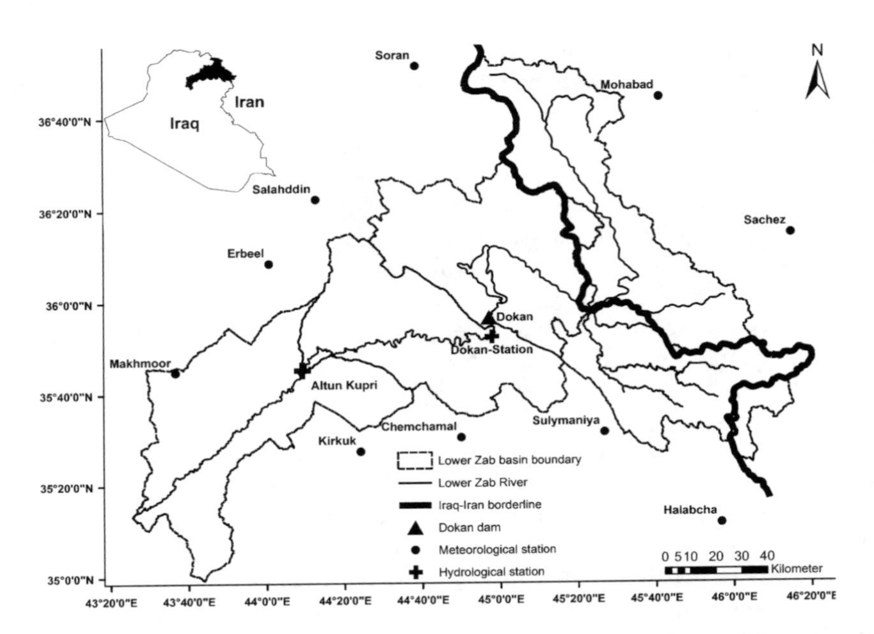

FIGURE 7.1 Overview of the Lower Zab River Basin, between Iraq and Iran, shape file showing the locations of hydro-climatic stations.

addressed. This will be accomplished by addressing the following objectives: (a) to investigate the spatio-temporal characteristics of meteorological data at monthly and annual time periods; (b) to apply the developed approach by Mohammed and Scholz (2016) to study the effects of climate change, drought events and anthropogenic interventions on the baseflow of drought events on the LZRB (Figure 7.1); (c) to quantify the spatio-temporal baseflow contribution to LZRB and (d) to analyse the sensitivity of the baseflow and the reservoir inflow to the collective impact of climate change.

Section 7.1 is based on the original article by Mohammed and Scholz (2018a). The research can be considered as a comparative basin study during which climate change impact, drought events, human intervention, precipitation (*P*), potential evapotranspiration (PET) and streamflow alteration are assessed. By providing a full picture of the prospective climate change and the expected human-induced impacts, the study results should be beneficial for engineers and policymakers, so that they can develop an effective adaptation strategy to cope with and to mitigate the potential impacts of these changes. Therefore, the assessed methodology addresses the challenge whether a simple implementation, although applied to one regime, would easily be replicated to other water resources systems (Figure 7.2).

7.1.2 Case Study and Methodologies

7.1.2.1 Case Study Site

The Lower (Lesser or Little) Zab River is the key stream of the Tigris River located near Erbil (north-eastern part of Iraq). The watercourse system can be found at

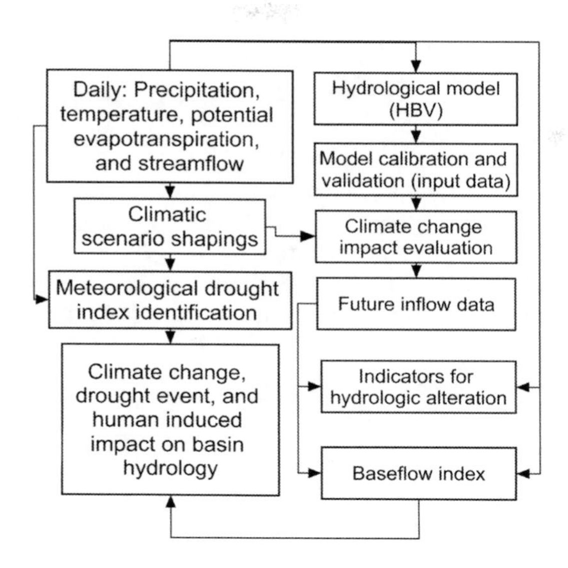

FIGURE 7.2 The proposed methodology for assessing the impacts of climate change, human-induced and drought events on the groundwater contribution to river flow.

latitudes 36°50′ N–35°20′ N and longitudes 43°25′ E–45°50′ E (Mohammed et al., 2018); see Figure 7.1 The associated river basin is subject to a varied range of hydro-climatic conditions as well as upstream and downstream structural improvements (Mohammed and Scholz, 2017a). Accordingly, the basin has been selected as an illustrative case study for both semi-arid and arid areas. The stream initiates from the Zagros Mountains in Iran and flows approximately 370 km southeast and southwest through north-western Iran and northern Iraq before joining the Tigris nearby Fatha city, which is situated around 220 km north of Baghdad (Mohammed and Scholz, 2016), with a total distance of approximately 302 km. The river basin area is approximately 19,254 km^2 with nearly 76% located in Iraq. The average yearly volume of the river at Dokan and Altun Kupri-Goma is about 6 billion cubic meters (BCM) and 7.8 BCM in this order (Mohammed et al., 2017); Figure 7.1 Dokan is the main dam that has been constructed within the upper part of the basin to control the discharge of the river, store water for irrigation purposes and provide hydroelectric power.

7.1.2.2 Meteorological Drought

This study applied the reconnaissance drought index (RDI) for drought identification (Vangelis et al., 2013). This index is founded on the relation between two aggregated quantities of precipitation and PET (Tsakiris and Vangelis, 2005). The RDI can be expressed in alpha ($RDI_{\alpha k}$), normalized (RDI_n) and standard (RDI_{st}) forms. The RDI_{st} can be applied for drought harshness calculations and the $RDI_{\alpha k}$ can be used as an index of aridity. A positive RDI_{st} indicates a wet period, and a negative number indicates a dry period compared to the natural regional conditions. The $RDI_{\alpha k}$ is usually estimated by Equation 7.1:

$$\mathrm{RDI}_{\alpha_0^i} = \frac{\sum_{j=1}^{12} P_{ij}}{\sum_{j=1}^{12} \mathrm{PET}_{ij}} \quad i = 1 \text{ to } N \text{ and } j = 1 \text{ to } 12 \tag{7.1}$$

where P_{ij} and PET_{ij} refer to the precipitation and the PET of the jth month and the ith hydrological year, which in Iraq starts in October, and N is the total number of hydrological years for the equivalent climate.

The values of $\mathrm{RDI}_{\alpha k}$ are equal to both the lognormal and the gamma distributions for many locations at various time periods (Tigkas, 2008). By utilizing the gamma distribution, RDI_{st} can be estimated by Equation 7.2:

$$\mathrm{RDI}_{st}^{i} = \frac{y_i - \overline{y}}{\hat{\sigma}_y} \tag{7.2}$$

where y^i is the $\ln(\alpha_{ki})$, $\ddot{y}$ is its arithmetic average and $\hat{\sigma}_y$ is the associated standard deviation (SD).

Equation 7.3 can be applied for RDI_{st} evaluation for the gamma distribution (Tigkas, 2008):

$$g(x) = \frac{1}{\beta^{\gamma} \times \Gamma(\gamma)} x^{\gamma-1} e^{\frac{x}{\beta}} \text{ for } x > 0 \tag{7.3}$$

where γ, β and $\Gamma(\gamma)$ are the shape, the scale parameter and the gamma function, respectively. The parameters γ and β of the gamma probability distribution can be estimated by Equations 7.4–7.6:

$$\gamma = \frac{1}{4A}\left(1 + \sqrt{1 + \frac{4A}{3}}\right) \tag{7.4}$$

$$\beta = \frac{\overline{x}}{\gamma} \tag{7.5}$$

$$A = \ln(\overline{x}) - \frac{\sum \ln(x)}{N} \tag{7.6}$$

The drought severity rises when RDI_{st} values are minimum. The RDI_{st} values can be categorized into eight classes (Vangelis et al., 2013; Mohammed and Scholz, 2017c) according to Table 7.1. For more details regarding this index, readers may refer to earlier research studies (Tigkas et al., 2012; Mohammed et al., 2017b).

7.1.2.3 Flow Duration Curve and Digital Filtering Algorithms

Daily flows at two key hydrological stations (Dokan and Altun Kupri-Goma Zerdela) have been analysed by applying the methodology that has been suggested by Mohammed and Scholz (2016). The Eckhardt (2005) approach is used to achieve low

TABLE 7.1
Drought Classification Based on Both the Standardized Reconnaissance Drought Index (RDI_{st}) Value

No.	RDI_{st} Range	Drought Classes	No.	RDI_{st} Range	Drought Classes
1	≥2.0	Extremely wet	5	0.00 to −0.99	Near normal
2	1.99–1.50	Very wet	6	−1 to −1.49	Moderately dry
3	1.49–1.00	Moderately wet	7	−1.5 to −1.99	Severely dry
4	0.99–0.00	Normal	8	≤−2.0	Extremely dry

pass filtering of the flow hydrograph to separate baseflow, which is mathematically expressed by Equation 7.7:

$$BF_t = \frac{(1 - BFI_{max}) \times \alpha \times BF_{t-1} + (1 - \alpha) \times BFI_{max} \times TF_t}{1 - \alpha \times BFI_{max}} \quad (7.7)$$

where BF (m^3/s) is the separated baseflow, BFI is the baseflow index, TF (m^3/s) is the total flow and α is the Eckhardt filter parameter, subject to $BF_t \leq TF_t$.

Two parameters are needed for the Eckhardt recursive method identification (Eckhardt, 2005): (a) the recession constant α (Echardt's parameter), which stems from the recession curve of the hydrograph valuation, and (b) the BFI_{max} that cannot be measured but enhanced based on the results of other methods. Eckhardt (2005) introduced three typical BFI_{max} amounts for diverse hydrological and hydrogeological environments: $BFI_{max} = 0.80$ for perennial streams with porous aquifers, $BFI_{max} = 0.50$ for ephemeral streams subjected to permeable aquifers and $BFI_{max} = 0.25$ for perennial streams comprising impermeable aquifers. As a first approximation, this study used $BFI_{max} = 0.25$ (Eckhardt, 2005; Stewart, 2015).

Chapman (1999) discussed the second recursive DFA that can be estimated by Equation 7.8:

$$DF_t = \frac{3 \times \alpha - 1}{3 - \alpha} \times TF_{t-1} + \frac{2}{3 - \alpha} \times (TF_t - TF_{t-1}) \quad (7.8)$$

where DF (m^3/s) is the direct runoff, TF (m^3/s) is the total flow, α is the filter parameter and t is the individual time step.

Based on the suggested approach, the outcomes of the FDC study have been combined with the results from Equation 7.7 to gain an α value after considering $BFI_{max} = 0.25$ (Eckhardt, 2005) for perennial rivers with predominantly porous aquifers as a first estimation. First, the long-term mean annual fraction of the total flow from the baseflow was estimated after obtaining the Q_{90} and Q_{50} values by applying the FDC method, connecting Equation 7.7 with FDC (Mohammed and Scholz, 2016; Al-Faraj and Scholz, 2014). Considering $\alpha = 0.925$ as an initial number (Arnold and Allen, 1999; Smakhtin, 2001), filtering of the daily flow is undertaken for various values of the filter parameter α up to a point when the BFI is equivalent to the ratio Q_{90}/Q_{50}. By applying the filtered α, several baseflow time series can be gained.

Secondly, linear regression models should be performed between the annual BFI and the corresponding runoff for all considered periods at both the upstream and the downstream sub-basins to investigate climate change, drought events and river regulating impacts on the baseflow contribution (Mohammed and Scholz, 2016).

7.1.2.4 Data and Analysis

Meteorological data: Daily meteorological data, which include precipitation as well as minimum and maximum air temperature from ten stations with elevations extending from 319 to 1,536 m, were gathered for the periods from 1979/1980 to 2012/2013 (Table 7.2 and Figure 7.1).

Hydrological data: This study evaluated the daily flow data at two key hydrometric stations, which are the Dokan (Latitude 35° 53′ 00″ N; Longitude 44° 58′ 00″ E) and Altun Kupri-Goma Zerdela (Latitude 35° 45′ 41″ N; Longitude 44° 08′ 52″ E) stations. The drainage area for the former is estimated to be 12,096 km^2 and data are available between 1931 and 2013, whereas the corresponding values for the latter are 8,509 km^2 and data for this sub-basin are available for the period between 1931 and 1993 (Mohammed and Scholz, 2017b).

Geospatial data: Iraqi boundaries and the LZRB shape files have been downloaded from the Global Administrative Areas (GADM, 2012) and the Global and Land Cover Facility (GLCF, 2015) databases, respectively. The GADM is a spatial database of the location of the world's administrative areas (or administrative boundaries) for use in GIS and similar software. The GADM describes where these

TABLE 7.2
Statistical Properties of the Meteorological Variables After Applying a Non-parametric Test for the Decadal Change

		T_m^d (°.C)		P^e (mm)		PET^f (mm)	
Sub- basin	**Station Name**	**M-K[c]**	***p*-value**	**M-K[c]**	***p*-value**	**M-K[c]**	***p*-value**
US[a]	Sulymaniya	0.358[h]	<0.01	−0.301[h]	<0.01	0.201	0.09
	Halabcha	0.572[h]	<0.01	−0.522[h]	<0.01	0.316[h]	<0.01
	Sachez	0.079	0.50	−0.328[h]	0.01	0.193	0.10
	Mahabad	0.603[h]	<0.01	−0.573[h]	<0.01	0.525[h]	<0.01
	Salahddin	0.452[h]	<0.01	−0.472[h]	<0.01	0.220	0.06
	Soran	0.380[h]	<0.01	−0.426[h]	<0.01	0.241[g]	0.05
DS[b]	Kirkuk	0.422[h]	<0.01	−0.553[h]	<0.01	0.420[h]	<0.01
	Makhmoor	0.462[h]	<0.01	−0.536[h]	<0.01	0.243	0.04
	Erbeel	0.351[h]	<0.01	−0.371[h]	<0.01	0.203	0.09
	Chemchamal	0.345[h]	<0.01	−0.412[h]	<0.01	0.139	0.24

Negative (−) and positive values indicate the decreasing and increasing trends, respectively.
[a] Upstream; [b] downstream; [c] Mann-Kendall non-parametric test; [d] mean air temperature; [e] precipitation; [f] potential evapotranspiration. [g] Correlation is significant at the 0.05 level (2-tailed). [h] Correlation is significant at the 0.01 level (2-tailed).

administrative areas are located, and for each area, it provides some attributes such as its name and corresponding variant names. Whereas GLCF is a centre for land cover discipline with a focus on study utilizing remotely sensed satellite data and products to access land cover change for local, national and global systems.

For hydro-climatic stations, site projections, weighted average calculations and stream catchment outlines, ArcGIS 10.3 has been used. Table 7.2 summarizes the location details of the meteorological stations. Statistical analyses including the daily hydro-climatic datasets, comprising trend test, monthly and annual amounts, modifications and the filling of data gaps were completed by the Statistical Program for Social Sciences (ITS, 2016). The estimation of PET (mm) was performed by the Food and Agriculture Organization Penman-Monteith standard method (Allen et al., 1998) that was estimated based on the reference evapotranspiration ET_0 (mm) calculator version 3.2 (FAO, 2012).

The weighing mean method has been used for the rainfall spatial distribution estimation. The process is often seen as essential for engineering praxis. For more information and detailed descriptions of this method, the reader may refer to many previous studies such as Mohammed et al. (2018).

The DrinC 1.5.73 software has been utilized to calculate the RDI_{st} values. The Indicators of Hydrologic Alteration software version (IHA 7.1) (The Nature Conservancy, 2009) was utilized to evaluate the natural flow regime alteration that resulted from climate change linked to human-induced activities. For the FDC estimation and baseflow separation, HydroOffice (2015) for BFI+3.0 and FDC was used (https://hydrooffice.org/Downloads?Items=Software), applying the methodology developed by Mohammed and Scholz (2016).

The RS MINERVE 2.5 (2016) free software was applied for the simulation of free surface runoff flow formation and propagation (Foehn et al., 2016) and used to run the Hydrologiska Byråns Vattenbalansavdelning (HBV) rainfall-runoff model (https://www.crealp.ch/down/rsm/install2/archives.html). To detect the trend in the long-term hydro-climatic data, the distribution-free Mann-Kendall (M-K) method was applied (Tabari and Talaee, 2011; Robaa and AL-Barazanji, 2013). This powerful technique does not assume a specific data distribution (Mohammed et al., 2017). Table 7.2 reveals the M-K test results for the key meteorological variables.

7.1.2.5 Hydrological Model and Climatic Scenario

Rainfall-runoff models have been used widely to obtain streamflow data since such data are not easily obtainable. These models contain a series of equations that attempt to simulate the variety of the interrelated procedures, which participate in hydrological processes. The hydrological models are usually categorized based on many criteria such as process description, solution mechanism and scale. Many types are used in the literature such as lumped, distributed, continuous-time and event-based models as well as conceptual and black-box models (Kaleris and Langousis, 2017).

Based on a set of weather data, Section 7.1 applied the HBV rainfall-runoff model to simulate the basin runoff (Mohammed et al., 2018). HBV is an illustration of a semi-distributed conceptual model estimating daily discharge based on daily precipitation and temperature and monthly estimates of potential evaporation as input data. For runoff simulation, the model was initially calibrated and subsequently validated for normal climate. The performance of the HBV model was evaluated by using the following statistical criterion (Equations 7.9–7.12):

$$\mathrm{RMSE} = \sqrt{\frac{1}{n}\sum_{i=1}^{n}\left[(R_{\mathrm{obs}})_i - (R_{\mathrm{sim}})_i\right]^2} \tag{7.9}$$

$$\mathrm{IoA} = 1 - \frac{\sum_{i=1}^{n}\left[(R_{\mathrm{obs}})_i - (R_{\mathrm{sim}})_i\right]^2}{\sum_{i=1}^{n}\left[\left|(R_{\mathrm{obs}})_i - \bar{R}_{\mathrm{obs}}\right| + \left|(R_{\mathrm{sim}})_i - \bar{R}_{\mathrm{obs}}\right|\right]^2} \tag{7.10}$$

$$r = \sqrt{\frac{\sum_{i}^{n}\left[(R_{\mathrm{obs}})_i - \bar{R}_{\mathrm{obs}}\right]\left[(R_{\mathrm{sim}})_i - \bar{R}_{\mathrm{sim}}\right]}{\left\{\sum_{i=1}^{n}\left[(R_{\mathrm{obs}})_i - \bar{R}_{\mathrm{obs}}\right]\right\}^{0.5}\left\{\sum_{i=1}^{n}\left[(R_{\mathrm{sim}})_i - \bar{R}_{\mathrm{sim}}\right]\right\}^{0.5}}} \tag{7.11}$$

$$\mathrm{MAE} = \frac{1}{n}\sum_{i=1}^{n}\left|(R_{\mathrm{obs}})_i - (R_{\mathrm{sim}})_i\right| \tag{7.12}$$

where RMSE is the root mean square error (without dimension), IoA is the index of agreement (without dimension), r is the correlation coefficient (without dimension), MAE is the mean absolute error, $R_{\mathrm{obs}(i)}$ is the measured streamflow (mm/month) at time step i, $R_{\mathrm{sim}(i)}$ is the predicted streamflow (mm/month) at time step i, $\bar{R}_{\mathrm{obs}}$ is the average amount of the recorded values (mm/month) and n is the number of the data point(s).

For the climate change prediction, the delta perturbation scenario has been applied (Tigkas et al., 2012; Al-Faraj et al., 2014; Mohammed and Scholz, 2017b, 2018b). By using the delta perturbation scenario, the historical climatic variables (mean air temperature and/or precipitation) are perturbed incrementally (and rationally) through arbitrary amounts as shown in Equation 7.13; steps of 2% within the range from 0% to −40% for precipitation and from 0% to +30% for PET. These scenarios comprise all potential basin-wide climate change predictions (336 scenarios).

$$\mathrm{AMV}_t = \mathrm{OMV}_t \mp \mathrm{RA} \times \mathrm{OMV}_t \tag{7.13}$$

where AMV_t is the anticipated meteorological variable (mm) at a specific time step t, OMV_t is the observed meteorological variable (mm) at a specific time step t and RA is the added or subtracted ratio (%).

For simulating different climate change scenarios, hydrological years from 1988 to 2000, which are characterized by a mean number of $\mathrm{RDI}_{\mathrm{st}}$ near to zero, were utilized. Even though the delta perturbation scenario does not address changes in the probability distribution of weather properties and seasonality of, for example, streamflow in the future, it is still a productive technique to identify tipping points when a reservoir is anticipated to fail catastrophically in supplying sufficient water (Mohammed and Scholz, 2017b). The delta perturbation scenarios often represent a realistic set of variations that are physically reasonable. Recently, numerous researchers have adopted this scenario (e.g., Tigkas et al., 2012; Al-Faraj and Scholz, 2014; Soundharajan et al., 2016; Reis et al., 2016; Mohammed and Scholz, 2017b).

7.1.3 Results and Discussion

7.1.3.1 Meteorological Trends

To identify the decadal trend of the main meteorological variables, a distribution-free test has been conducted. Table 7.2 reveals the analysis for ten meteorological stations that are distributed between the upstream and the downstream sub-basins for the time horizon between 1979 and 2013. The analysis shows that there is a positive tendency in yearly mean temperature (T_m) at approximately 90% ($p<0.01$) of the locations. The significant warming trends in yearly T_m changed from 0.37°C to 1.91°C per decade. The findings are largely in agreement with previous studies (Robaa and AL-Barazanji, 2013; UNESCO, 2014; Mulder et al., 2015; Mohammed et al., 2017b).

A declining trend in precipitation with an average degrease of about 162 mm per decade was observed (Figure 7.3a). The basin annual and maximum precipitation is approximately 720 and 1,222 mm, respectively, whereas the equivalent lowest 250 mm was allocated to the time period between 2007 and 2008. The statistical analysis results of the meteorological data revealed that the climate of the study area is getting warmer and drier due to climate change over the past 35 years.

Furthermore, the average yearly precipitation altered spatially from 56 mm at Kirkuk station, which is placed in the lower sub-basin, to 1,369 mm at Sulymaniya station, which is placed in the upper sub-basin. This shows that the upstream sub-basin, which is considered high compared to the downstream sub-basin, had higher precipitation values than the downstream one.

The analysis also displayed that long-term PET has risen significantly ($p<0.01$) by approximately 77% with a decadal growth that extended to its maximum of about 36 mm (Table 7.2). Both variations have contributed significantly ($p<0.01$) to the decrease of the river flow and specified that, during the past 35 years, climate change has considerably affected the meteorological characteristics of the LZRB. Climate variation produced a change in precipitation, mean temperature and PET, extending the gap between the basin water storage capability and the equivalent water need (Figure 7.3a and b).

7.1.3.2 Spatio-Temporal Distribution

Based on the outcomes of Equation 7.7, a 0.982 value of α yielded BFI equivalent to the value resulting from the low flow analysis. After considering this value in the baseflow separation technique, the following information has been gained: firstly, over the hydrological period that spans from 1931 to 2013, the annual baseflow values varied from 0.448 billion cubic meters (bcm) in 2006 to 3.54 bcm in 1968. The

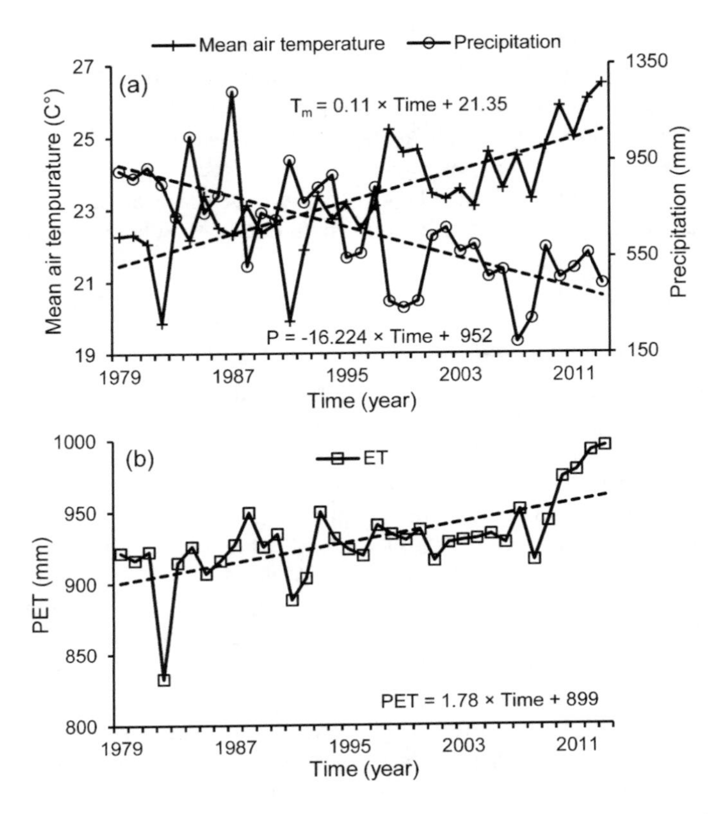

FIGURE 7.3 Long-term trends for the key meteorological variables (a) mean air temperature and precipitation and (b) potential evapotranspiration (PET) over the Lower Zab River Basin for the time span from 1979 to 2013.

equivalent annual baseflow varied from nearly 37 to almost 293 mm. However, the long-term baseflow value and the equivalent SD were 1.43 and 0.60 bcm in this order. Accordingly, the long-term average annual value of baseflow and the associated SD were predicted to be around 118 and about 50 mm, respectively. The results show that the storage of the Dokan sub-basin aquifer decreased by about 87%, which indicates that during the past 35 years climate change has negatively affected the availability of water within the basin.

Figures 7.4a–c as well as Equations 7.1.3.2.2, 7.1.3.2.8 and 7.1.3.2.14 (Table 7.3) show a strong correlation between the separated baseflow and the total flow. Furthermore, Figure 7.4d shows that the annual temporal values of the BFI fluctuated from 0.26 (1958) to 0.42 (1997) and the long-term value of BFI in addition to the equivalent SD were 0.35 and 0.027 in this order. Outcomes expose that around 70% of the yearly BFI values were located within the mean±SD ranges.

Based on Equation 7.8, results expose that a 0.925 value of an α yields BFI that is equivalent to the value gained from FDC. Therefore, the annual values of baseflow changed from 0.84 in 2006 to 7.26 bcm in 1968. The equivalent annual baseflow varied from about 69 to closely 600 mm. The long-term average annual value

TABLE 7.3
The Developed Linear Regression Models for the Upstream and Downstream Sub-basins at the Three Considered Time-Periods Using Both the Eckhardt Algorithm Linked to the Flow Duration Curve and the Chapman Digital Algorithm

Time Period	Sub-basin	Algorithm	Equation Formulation	R^2
Pre-damming	Upstream	Eckhadt	$BF^a = 0.24 \times TF^b + 1.3$	0.99
		Chapman	$BF^a = 0.48 \times TF^b + 2.7$	0.92
		Eckhadt-Chapman	$BF^a_{Eckhadt} = 1.95 \times BF^a_{Chapman} + 0.65$	0.91
	Down-stream	Eckhadt	$BF^a = 0.25 \times TF^b + 0.1$	0.99
		Chapman	$BF^a = 0.51 \times TF^b - 3.6$	0.97
		Eckhadt-Chapman	$BF^a_{Eckhadt} = 2.10 \times BF^a_{Chapman} - 3.73$	0.97
Post-damming	Upstream	Eckhadt	$BF^a = 0.23 \times TF^b + 5.0$	0.80
		Chapman	$BF^a = 0.48 \times TF^b + 1.3$	0.99
		Eckhadt-Chapman	$BF^a_{Eckhadt} = 1.70 \times BF^a_{Chapman} + 11.1$	0.80
	Down-stream	Eckhadt	$BF^a = 0.12 \times TF^b + 0.2$	0.99
		Chapman	$BF^a = 0.52 \times TF^b - 5.2$	0.99
		Eckhadt-Chapman	$BF^a_{Eckhadt} = 4.20 \times BF^a_{Chapman} - 6.2$	0.99
Integrated	Upstream	Eckhadt	$BF^a = 0.24 \times TF^b + 1.8$	0.97
		Chapman	$BF^a = 0.52 \times TF^b - 5.2$	0.99
		Eckhadt-Chapman	$BF^a_{Ekhadt} = 2.10 \times BF^a_{Chapman} - 5.2$	0.99
	Down-stream	Eckhadt	$BF^a = 0.25 \times TF^b + 1.3$	0.99
		Chapman	$BF^a = 0.49 \times TF^b + 1.0$	0.97
		Eckhadt-Chapman	$BF^a_{Ekhadt} = 2.10 \times BF^a_{Chapman-}6.9$	0.98

[a] baseflow; [b] total runoff.

of baseflow and the equivalent SD were 2.86 and 1.23 bcm, correspondingly. The baseflow long-term average yearly value and the corresponding SD were 237 and around 102 mm in this order. Equation 7.8 results also show that the storage of the Dokan sub-basin aquifer decreased by about 88%, which confirms results obtained by applying Equation 7.7, and indicates that, during the past 35 years, climate change has negatively affected the availability of water within the basin.

Figure 7.4a–c shows the separated baseflow linked with the total flow. The developed linear models (Table 7.4 and Equations 7.1.3.2.3, 7.1.3.2.9 and 7.1.3.2.15) also show a strong correlation. The temporal variation of annual BFI ranged from 0.51 (1968) to 0.54 (1972). The average long-term BFI value and the matching SD were 0.53 and 0.01 in that order. Accordingly, roughly 61% of the BFI yearly values were expected to be within the mean±SD range. Furthermore, the results demonstrated a good relationship between Equations 7.7 and 7.8 as shown in Table 7.3 (Equations 7.1.3.2.3, 7.1.3.2.9 and 7.1.3.2.15).

The BFI, which is the ratio Q_{90}/Q_{50}, represents volumes of water that the river might has obtaining from groundwater flow or other delayed shallow sub-surface

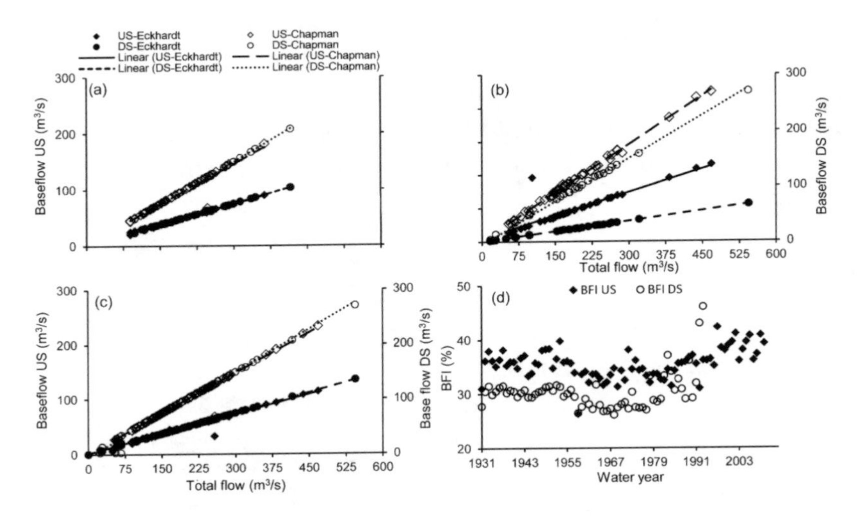

FIGURE 7.4 The linear regression models for the relationships between the separated baseflow using the Eckhardt and Chapman methods and the total runoff at upstream (US) and downstream (DS) sub-basins for (a) pre-damming; (b) post-damming; (c) integrated time periods and (d) the annual baseflow index (BFI) variability as a function of time for the US and the DS sub-basins, estimated by the Eckhardt recursive digital filtering algorithms coupled with the flow duration curve (FDC).

sources regarding the studied periods. A high BFI value shows the basin ability to recharge the stream during a prolonged dry season. The BFI is associated with many river basin characteristics such as soil type and geology, topography, vegetation and weather. The analysis of the results indicates that there is a moderate variation in the BFI values at the upstream sub-basin, which can be attributed to the increase in the groundwater influence on the total flow of the river. Therefore, more consideration should be given to evaluate the aquifer characteristics and comprehend the aspects that might cause such alterations for improving management of sub-surface resources within the basin.

The results from Equation 7.7 show that a 0.925 value for α yields a BFI equating to that obtained from FDC. During the years from 1931 to 1993, the annual baseflow magnitude changed between 0.01 (1993) and 4.20 bcm (1968). The long-term average yearly baseflow value and the equivalent SD were 1.60 and 0.76 bcm, respectively. Figure 7.4a–c shows the relations between the entire flow and the separated baseflow. The estimated linear model represented by Equations 7.1.3.2.4, 7.1.3.2.10 and 7.1.3.2.16 (Table 7.3) shows a good correlation. However, Figure 7.4d displays that the annual temporal values of the BFI varied from 0.26 (1968) to 0.82 (1992) with 0.30 and 0.12 long-term BFI and SD, respectively. Outcomes indicate that nearly 94% of the yearly BFI values are placed within the mean ± SD range.

Based on Equation 7.8, a 0.925 value for α creates a BFI that is equivalent to the value estimated by FDC. During the hydrologic years from 1931 to 1993, the annual baseflow values varied from 0.01 (1993) to 8.42 bcm (1968). The long-term average yearly baseflow value and SD were 3.08 and 1.61 bcm, respectively. The values of the

BFI varied between 0.21 (1989) and 0.82 (1992), while the BFI long-term value was 0.51 ±0.11. Around 94% of the BFI annual values are placed within the mean±SD ranges range, and the developed Equations 7.1.3.2.5, 7.1.3.2.11 and 7.1.3.2.17 (Table 7.3) show a robust coefficient of determination (R^2). The results also demonstrate a strong relationship between Equations 7.7 and 7.8 as shown in Table 7.3 (Equations 7.1.3.2.6, 7.1.3.2.12 and 7.1.3.2.18).

The results of Equations 7.7 and 7.8 are identical at the downstream sub-basin and show that there was a reduction in the storage of the sub-basin aquifer by about 100%, and indicate that, during the past 35 years, climate change has negatively affected the availability of water within the basin. Furthermore, the temporal analysis of the BFI indicates that during the hydrological year 1968, the contribution of the groundwater to the LZRB total hydrograph is larger (about 74%) than that during the hydrologic year 1989, which may attribute to the reduction of the basin water storage availability because of climate change.

The spatial analysis of the BFI shows that the filter parameter values of 0.982 and 0.925, which were obtained from Equations 7.7 and 7.8, respectively, are not the same for the upper and the lower sub-basins, which shows that the aquifers of the two sub-basins have different features. Accordingly, to improve the water resources management of the basin, more attention should be given to assess the characteristics of the aquifer and comprehend the aspects that might cause such alterations.

7.1.3.3 Human-Induced Impact

To explore the potential human-induced impact (i.e. river damming) on the subsurface water contribution to the total flow of streamflow, this study considered three-time periods representing the pre-damming, post-damming and integrated time periods. At the upstream sub-basin, the first period covers the hydrological years 1931 to 1965 (considered as pre-damming), the hydrological years that span from 1966 to 2013 is considered as the post-damming period, and the integrated period covers the hydrological years between 1931 and 2013. The corresponding periods at the downstream sub-basin were 1931–1965, 1966–1993 and 1931–1993, respectively.

At the upstream sub-basin (Dokan site), the Q_{90} and Q_{50} values for the three considered periods were 35 and 101, 31 and 100 as well as 33 and 100 m^3/s, correspondingly. So that the Q_{90}/Q_{50} ratios were 35%, 31% and 33% in that order, representing the water capacities that the stream would obtain from groundwater recharge and/or other delayed low groundwater resources during the considered time periods. At the downstream sub-basin (Altun Kupri-Guma Zerdela sit), the corresponding Q_{90} and Q_{50} values were 40 and 132, 17 and 127 as well as 31 and 129 m^3/s, respectively, so that the Q_{90}/Q_{50} ratios were 30%, 14% and 24% in this order.

The results show that the BFI values during the pre-damming periods were considerably larger compared to those during the post-damming and the integrated time periods, which can be attributed to the decrease in the subsurface water contribution to the LZRB flow due to water yielded from the Dokan reservoir throughout the dry months. Moreover, this, in turn, decreases the BFI. Consequently, and for the purposes of the development of the groundwater flow management, more attention should be given to the evaluation of the groundwater aquifer properties and understand the features potentially causing these variations.

For the pre-damming period, a 0.982 value for α has been produced by Equation 7.7. The annual baseflow volume period varies from 0.632 (1958) to 2.826 bcm (1953). The corresponding annual baseflow values were between 30 and 132 mm. The long-term baseflow average annual value was 1.478 ± 0.512 bcm. However, the annual baseflow magnitudes estimated by Equation 7.8 during the same period varied from 1.38 (1958) to 5.71 bcm (1953). The equivalent annual baseflow was between 65 and 266 mm. The long-term annual mean BF was 2.91 ± 1.05 bcm.

For the post-damming period, the annual baseflow values changed between 0.437 bcm in 2006 and 3.647 bcm in 1968. These values were obtained from Equation 7.7, and the equivalent annual baseflow changed from about 21 to 170 mm. The long-term annual mean baseflow volume was 1.45 ± 0.70 bcm. Using Equation 7.8, the annual baseflow values varied from 0.871 bcm (2006) to 7.077 bcm (1968). The equivalent annual baseflow changed from approximately 40 to 330 mm. The long-term annual mean baseflow figure was 2.88 ± 1.31 bcm. These findings show that due to the impact of climate change during the past 35 years, the basin aquifer storage decreased by about 77% and 88% for pre-regulated and post-regulated time periods, respectively. This can directly affect the availability of water within the basin.

Furthermore, Figure 7.5a and b displays the variations of the total flow, the baseflow, and the BFI values for the three periods within the two sub-basins, which indicate that the derived baseflow values during the studied time periods display similar patterns. Variations have been witnessed for the BFI during the post-damming period at the upstream sub-basin (Figure 7.5a), which could be attributed to the water yield from the Dokan reservoir during the dry season.

7.1.3.4 Drought and Climate Change

To study the impact of climate change and drought phenomena on a basin hydrology, it is important (as a first step) to investigate the impact of such changes on the hydro-climatic variables and the BFI. The dry and wet years are determined by the minimum and the maximum flows, respectively. Yoo (2006) suggested that when the average yearly precipitation of the basin is greater than the average precipitation $(P_{av}) + 0.75 \times SD$, then the hydrological year is defined as wet. However, when precipitation is no more than the characterized $P_{av} - 0.75 \times SD$, then the year is considered as dry. Therefore, years with annual precipitation values within this range can be regarded as normal hydrological years: $P_{av} - 0.75 \times SD \leq P \leq P_{av} + 0.75 \times SD$ (Yang et al., 2008). Figure 7.6 displays wet and dry annual thresholds combined with the time series of the BFI.

Figure 7.6 displays wet and dry year thresholds coupled with the BFI time series. A significant ($p < 0.01$) rise in the sub-basin P_{av} of nearly 44% was noticed for the hydrological year 1987. A noteworthy change in the flow of about 118% resulted from an increase in the amount of precipitation, which in turn reduced the groundwater involvement to TF. In contrast, for the years from 1998 to 2001 and from 2006 to 2008, a steep drop in the sub-basin P_{av} of about 40% and 60%, respectively, was calculated.

Figure 7.6a and b shows wet and dry year thresholds coupled with precipitation and the BFI. A significant ($p < 0.01$) rise in the basin average precipitation of nearly 44% has been noticed for the hydrological year 1987. A noteworthy change in the

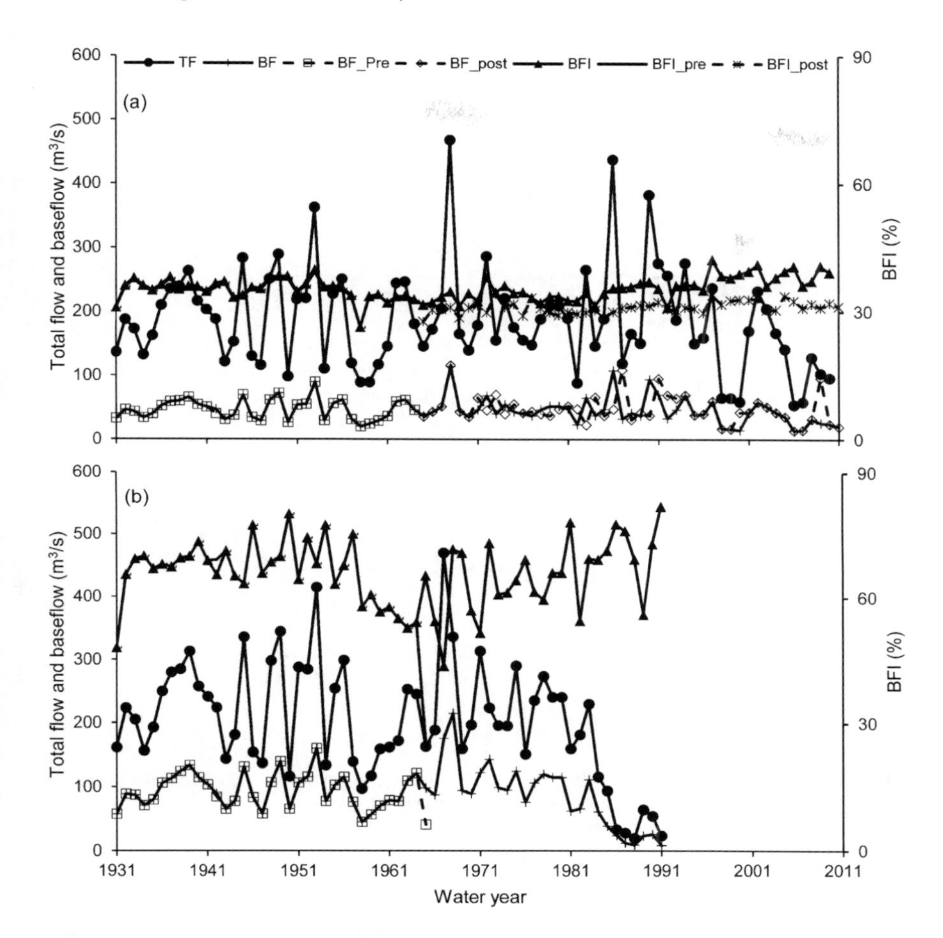

FIGURE 7.5 Long-term monthly variation of the total flow (TF), the baseflow (BF) and the baseflow index (BFI) estimated by the Eckhardt filtering algorithm coupled with the flow duration curve for the pre-alteration and post-alteration time periods at (a) the Dokan hydrometric station and (b) the Altun Kupri-Goma Zerdela station.

flow of about 118% resulted in an increase in the amount of precipitation, which in turn reduced the groundwater involvement to the total flow. However, the hydrological years from 1998 to 2001 as well as from 2006 to 2008 are linked to steep drops in the mean precipitation to approximately 40% and 60%, respectively. A decline of about 66%, 77% and 79% (the equivalent annual average flow values of 0.35×10^9, 0.31×10^9 and $0.34\times10^9 m^3$) for the hydrological years 1998/1999, 1999/2000 and 2000/2001 in river flow resulted from the reduction in the basin average precipitation, which caused an increase in the groundwater contribution (represented by the BFI). Nevertheless, 52%, 80% and 83% streamflow decreases (equivalent to 0.76×10^9, 0.29×10^9 and $0.31\times10^9 m^3$ annual mean flow volumes) were observed during the hydrological years 2006/2007, 2007/2008 and 2008/2009, respectively. Additionally,

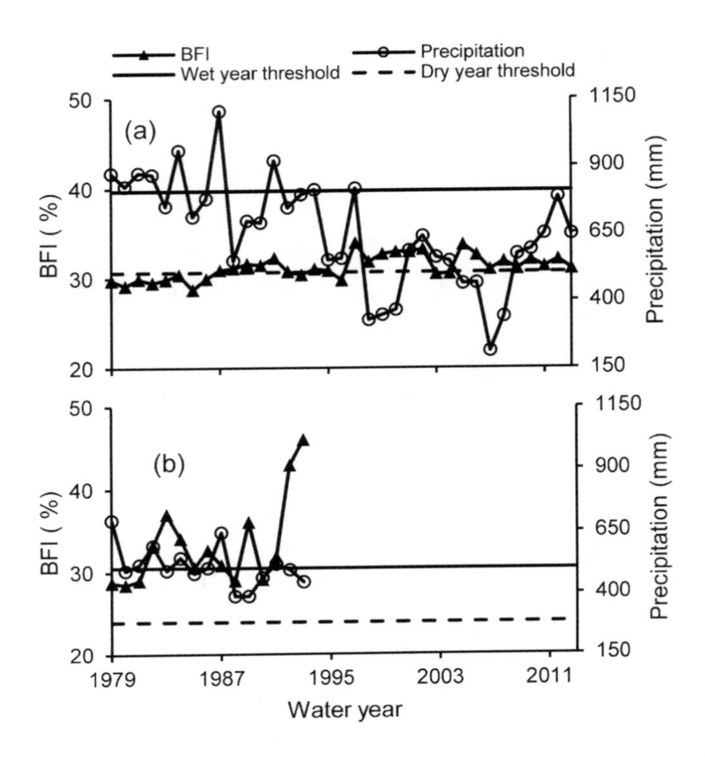

FIGURE 7.6 Long-term baseflow index (BFI) with both wet and dry year thresholds coupled with long-term average precipitation for the time periods between 1979 and 2013 with respect to (a) Dokan and (b) Altun Kupri-Goma Zerdela stations. (For Altun Kupri station, there are no data available for the hydrological years from 1995 to 2013.)

the hydrological years 1991–2013 experienced a steep drop during which the flow change ranged between 75% and 86% (equivalent to 0.31×10^9 and $1.24 \times 10^9 m^3$) indicated the absolute minimum and maximum yearly average storage capacities, consequently, the baseflow raised from approximately 31% to nearly 35%.

Figure 7.7a and b presents RDI_{st} values estimated for the studied sub-basins, showing that a non-symmetric annual drought pattern and wet periods were detected, particularly in the upper sub-basin. These results indicate that the studied area suffers from a severe drought during the last few decades and the severity of drought increased as the number of months with long durations of precipitation shortages and PET growth increases. Droughts were identified on a regular basis at the upstream sub-basin for the hydrological years 1998/1999, 1999/2000, 2000/2001, 2007/2008 and 2008/2009 (see Section 7.1.3.5). The corresponding mean RDI_{st} values were −1.84, −1.67, −1.45, −2.91 and −1.53. These findings, which complement several earlier studies (Fadhil, 2011; UNESCO, 2014), indicate that climate change has increased the severity of droughts within the studied climatic zone. At the upstream sub-basin (Figure 7.7a), the BFI increased from approximately 30% to 35%, while, at the downstream sub-basin (Figure 7.7b), no drought episodes have been recorded.

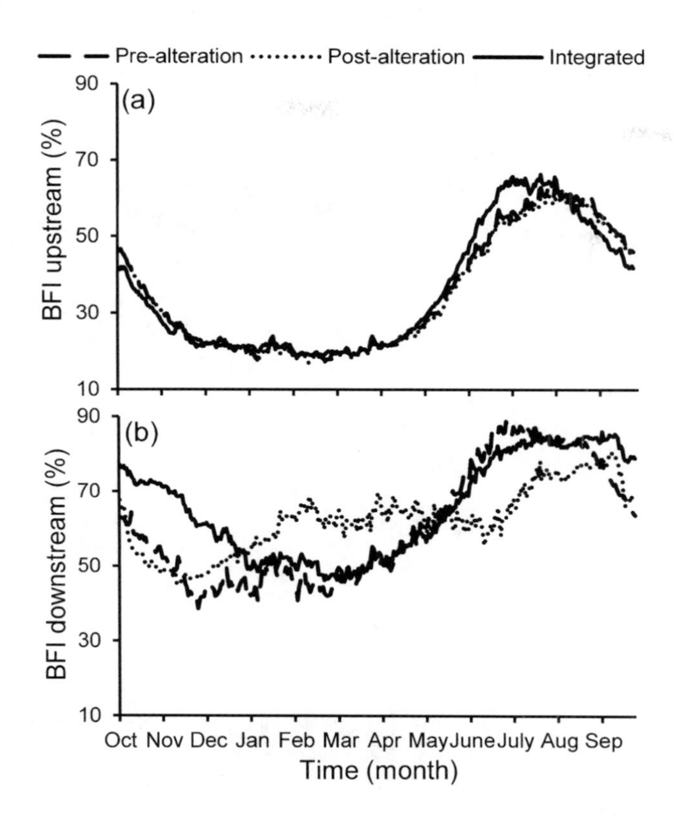

FIGURE 7.7 The standardized reconnaissance drought index (RDI_{st}) coupled with the long-term baseflow index (BFI) for the Lower Zab River Basin between 1979 and 2013 concerning (a) Dokan and (b) Altun Kupri-Goma Zerdela stations in this order.

7.1.3.5 Seasonal Analysis

Furthermore, results of the seasonal BFI variability at the upstream sub-basin (Figure 7.8a) show that this sub-basin is likely to recharge LZRB with groundwater. This contribution started to rise significantly ($p < 0.05$) from April until obtaining an absolute maximum number by the end of June. Then, this level of recharge continued without change until the mid of August followed by a marginal drop in September. In general, the BFI entries usually showed high variations over the dry period.

Figure 7.8b reveals that the long-term BFI at the downstream sub-basin is nearly doubled that of the upstream sub-basin for the months from October to April, whereas they are close to each other for the remaining period. This variability in BFI may result from the difference in the studied time periods, since the period horizon for the Dokan sub-basin spanned from 1931 to 2013 (82 years), whereas for the Altun Kupri-Goma Zerdela sub-basin, ranged from 1931 to 1993 (62 years). Even though the catchment area of the Dokan sub-basin (12,096 km^2) is larger than that of the Altun Kupri-Goma Zerdela (8,509 km^2), the former is characterized by both a higher precipitation rate and elevation (Table 7.4). This would increase the contribution of runoff and the interflow to the total flow of the Dokan sub-basin, which in turn

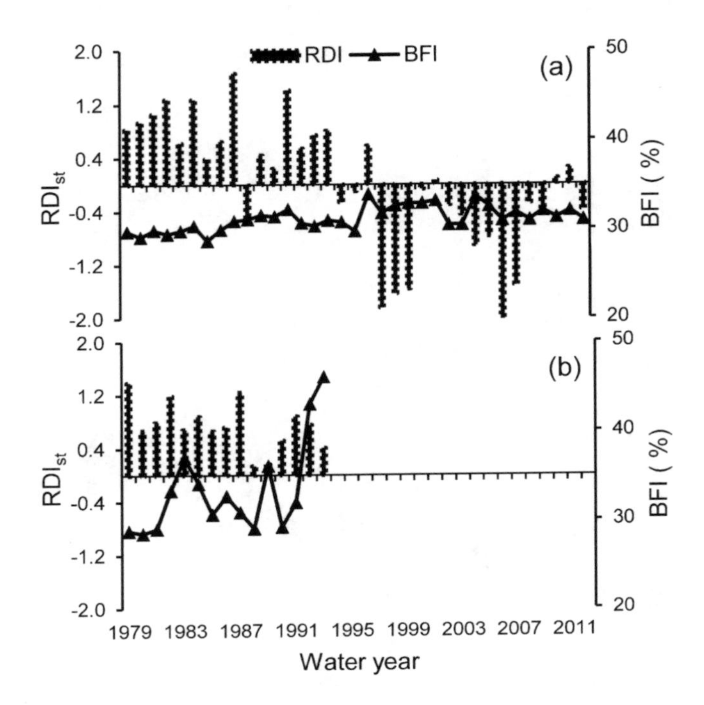

FIGURE 7.8 Seasonal variations of the baseflow index (BFI) estimated by the Eckhardt filtering algorithm coupled with the flow duration curve for the three studied time periods at (a) Dokan and (b) Altun Kupri-Goma Zerdela stations. Note: For Altun Kupri station, there are no data available for the hydrological years from 1995 to 2013. (RDI, reconnaissance drought index)

TABLE 7.4
Station Addresses with Corresponding Average Precipitations and the Sub-area Sizes

Sub-basin	Station ID	Lat[c] (°)	Long (°)	Elev[e] (m)	Sub-area (km^2)	Av[f] P[g] (mm)	Av[f] PET[h] (mm)
US[a]	Sulymaniya	35.53	45.45	885	4,479.57	772	1,989
	Halabcha	35.44	45.94	651	735.60	585	980
	Sachez	36.25	46.26	1,536	1,182.79	462	1,550
	Mohabad	36.75	45.70	1,356	2,593.31	886	920
	Salahddin	36.38	44.20	1,088	1,641.07	652	2,058
	Soran	36.87	44.63	1,132	1,463.30	813	1,433
DS[b]	Kirkuk	35.47	44.40	319	1,693.76	342	897
	Makhmoor	35.75	43.60	306	3,008.41	361	934
	Erbil	36.15	44.00	1,088	979.76	575	935
	Chemchamal	35.52	44.83	701	2827.46	738	2,075

[a] Upstream; [b] downstream; [c] latitude; [d] longitue; [e] elevation; [f] average; [g] precipitation; and [h] potential evapotranspiration.

decreases the contribution of the baseflow to the LZRB discharge. The research findings specify that through the last few years, climate variability and drought events have harmfully affected the water resources availability of LZRB.

Figure 7.9 reveals the anticipated variations of RDI_{st} associated with the potential impact of future reduction in the precipitation under the collective effect of PET increase. The figure indicates that the severity of drought would increase because of climate change.

Furthermore, considering the climate change impacts on the separated baseflow as well as the inflow to the Dokan reservoir, Figures 7.10 and 7.11 show examples of the sensitivity analysis of these two hydrological characteristics under the collective impacts of climate change scenarios.

The HBV model has been utilized to simulate the reservoir inflow, which has been firstly calibrated and validated based on the normal climatic condition. During the model calibration, the statistical performances, root mean square error (RMSE), index of agreement (IoA), correlation coefficient (r) and mean average error (MAE)

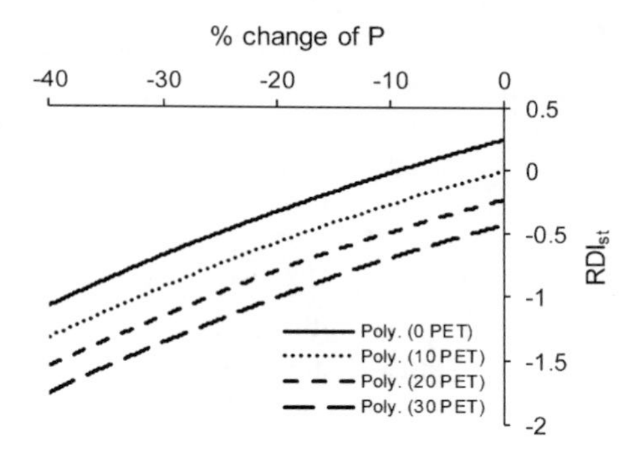

FIGURE 7.9 Anticipated standardized reconnaissance drought index (RDI_{st}) for the Lower Zab River Basin for four climate change scenarios based on the potential evapotranspiration (PET). (P, precipitation.)

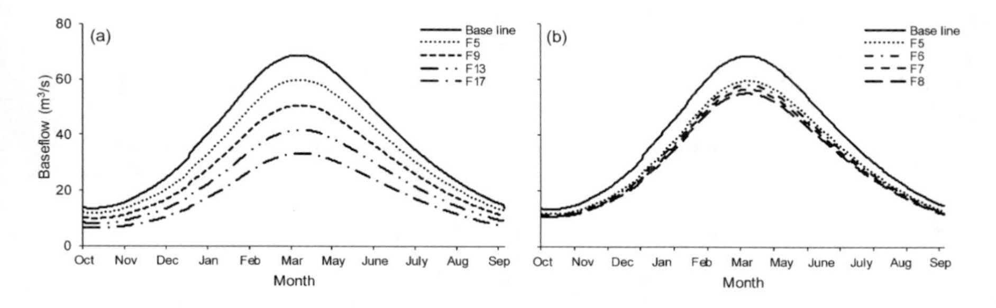

FIGURE 7.10 Sensitivity analysis of the separated baseflow with respect to the impact of (a) precipitation (P) reduction and (b) potential evapotranspiration (PET) increase. (Future scenarios F_5 (10% reduction in P linked with 0% increase in PET); F_6, F_7 and F_8 (10% reduction in P linked with 10, 20 and 30% increase in PET, respectively) and F_9 (20% reduction in P linked with 0% increase in PET))

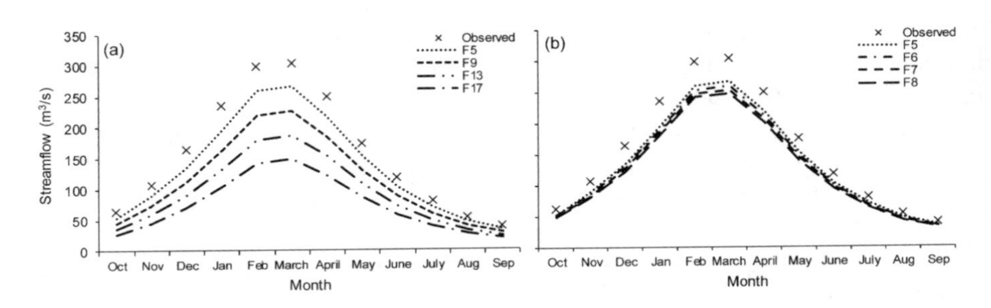

FIGURE 7.11 Sensitivity analysis of the reservoir inflow with respect to the impact of (a) precipitation (P) reduction and (b) potential evapotranspiration (PET) increase. (Future scenarios are F_5 (10% reduction in P linked with 0% increase in PET); F_6, F_7 and F_8 (10% reduction in P linked with 10, 20 and 30% increase in PET, respectively) and F_9 (20% reduction in P linked with 0% increase in PET).)

were 0.73, 0.99, 0.93 and 0.65, respectively, whereas the values were 0.68, 0.99, 0.84 and 0.60 during validation (refer also to Mohammed and Scholz, 2017b). The findings emphasized that the model can confidentially be used for more research such as running the delta perturbation scenarios and calculating the relative change (%) of the mean yearly flow relative to the natural climate condition.

Figures 7.10a and 7.11a show that the baseflow and the reservoir inflow are considered sensitive to seasonal variations of precipitation. However, they can be considered less sensitive to the variations of PET (Figures 7.10b and 7.11b). Additionally, Figures 7.10 and 7.11 demonstrate how climate change would strongly cause drops in the separated baseflow and the reservoir inflow. Compared to the base line and the observed inflow values, the peaks of the anticipated values of these variables would decrease, and there would be a noticeable alteration in their magnitude, which can lead to a dramatic impact on the availability of water resources of the basin.

Figure 7.12a–c shows that there would be a wide variability in the simulated inflow, baseflow and BFI values, respectively. For example, the simulated values of inflow, baseflow and BFI based on the Future 6 (F_6) (10% precipitation reduction linked with 10% PET increase) scenario could be as low as 36%, 34% and 22%, respectively, which indicates the severity of climate change in the study area. The climate change impact on the runoff estimations generally follows the impact on both baseflow and BFI. Coefficient of variation (Cv) for the simulated inflow, separated baseflow and BFI is as high as 0.45, 0.51 and 0.31, respectively. This coefficient can be considered as an indication of the variability or uncertainty of these values.

7.1.4 Conclusions and Recommendations

A generic methodology has been applied to evaluate the impact of climate change, drought phenomena and anthropogenic activities on the baseflow contribution to the total flow of an arid basin. The results indicate that there was a similar pattern in the annual temporal variations of the separated baseflow for the upstream and the downstream of the basin. Very good correlations have been noticed between the separated baseflow and the total flow at both locations as well as between the two considered

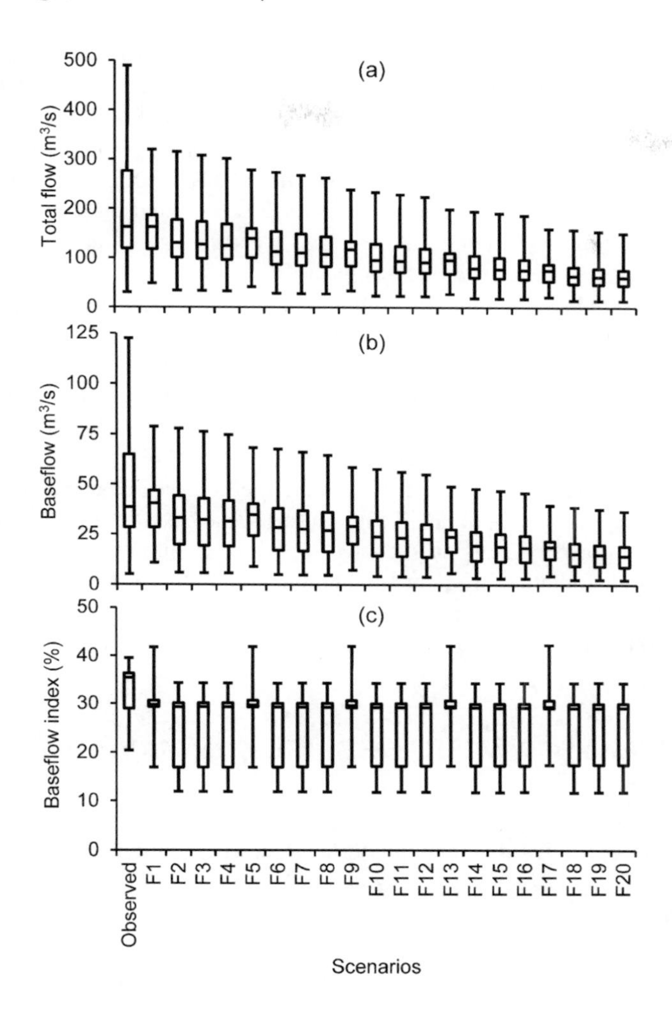

FIGURE 7.12 Box plot indicating (a) the total runoff; (b) the baseflow and (c) the baseflow index at the inlet of the basin. (Future F_1 to F_4 (0% reduction in precipitation (P) linked with 0%, 10%, 20% and 30% increase in potential evapotranspiration (PET), respectively); F_5 to F_8 (10% reduction in P linked with 0%,10%, 20% and 30% increase in PET, respectively); F_9 to F_{12} (20% reduction in P linked with 0%,10%, 20% and 30% increase in PET, respectively); F_{13} to F_{16} (30% reduction in P linked with 0%,10%, 20% and 30% increase in PET, respectively) and F_{17} to F_{20} (40% reduction in P linked with 0%,10%, 20% and 30% increase in PET, respectively).)

algorithms. The average yearly BFI increased from 0.14 to 0.38 for the sub-basins, respectively, which indicates that 14%–38% of the long-standing flow of the basin can be sustained by groundwater flow. There were relatively large differences in the baseflow contributions between the upstream and the downstream sub-basins, which can be attributed to the variations in the studied periods and the drainage area of each sub-basin and the existence of aquifers with greater releases in the downstream location compared to the upstream one.

The alteration of the hydrologic environment can be considered as one of the significant impacts of the projected climate change in the reservoir area. The outputs of the HBV model recommend a serious reduction in the Dokan reservoir inflow due to the precipitation reduction and rise in the PET. As an example, a decrease of approximately 21% in streamflow is likely to result in 10% precipitation decrease and 30% PET increase. The inflow peak value will decrease, and there will be a clear shift in the corresponding magnitude, which can lead to a significant impact on the availability of water resources within the basin.

The suggested approach is useful, simple and supports the preparation for future basin management, particularly when challenging hydrological models cannot be applied appropriately, because of a deficiency in the observed data. This method bridges the gap in the decision-makers' understanding of the climate variability influence, drought phenomena and human activities on the baseflow contribution to a streamflow that is in an arid climatic condition.

Further research is recommended to evaluate the aquifer behaviour and the corresponding data variability. A detailed exploration of the baseflow contribution to the total flow is important, because of shortages in water resources and corresponding conflicts among different stakeholders and countries. Furthermore, the study should be repeated for different climatic regions to assess climate change in addition to both drought phenomena and river regulation impact on groundwater contributions to runoff.

7.2 POTENTIAL FOR WASTEWATER REUSE IN IRRIGATION

7.2.1 Introduction

7.2.1.1 Wastewater Treatment and Water Resources Shortage

In 1970, the first wastewater treatment plant was constructed in Northern Yemen (Figure 7.13) according to Agro Vision Holland (1993). Stormwater ponds (Scholz et al., 2005) and waste stabilization pond (WSP) systems have been constructed at many central locations in Yemen (Almasi and Pescod, 1996). It is notable that WSP systems have been favoured above conventional treatment systems and more advanced technologies such as oxidation ditches, up-flow reactors and trickling filters due to their advantages in cost efficiency and relatively high treatment performance, particularly in arid areas.

The simple design and construction and the reduced costs in operation and maintenance are the main assets of WSP (Mara, 2004). A modern WSP can produce an effluent quality that meets the World Health Organization's (WHO) recommendations for wastewater reuse for crop irrigation (Mara, 2000). A WSP is appropriate for the treatment of wastewater in developing countries, where affordable land availability is normally high (Alcalde et al., 2003; Arthur, 1983), and elevated average temperatures support biological treatment (Mara, 1997; Pearson and Mara, 1998; World Health Organization, 1987). However, WSPs are often regarded as having two major disadvantages; they require large areas of land and are subject to high water losses through evaporation.

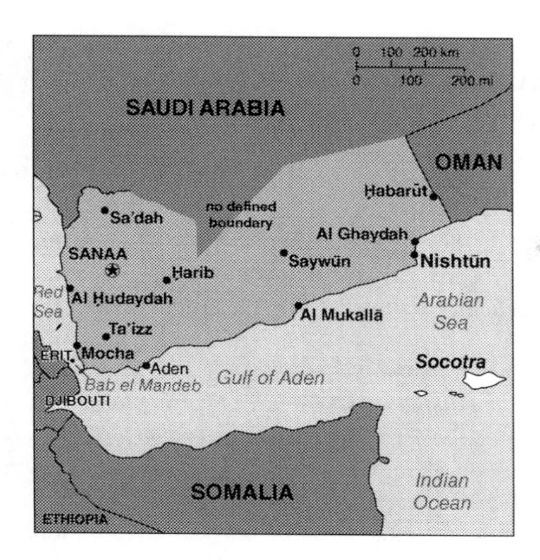

FIGURE 7.13 Map showing Yemen, its capital Aden and the neighbouring countries.

Extreme freshwater shortages are becoming the main concern for many countries in the world, particularly to developing countries in Africa and the Middle East. Sixteen out of 29 states in the Near East are classified as having a deficit in water resources, with less than the threshold of 500 m^3/cap/a of the renewable freshwater resources (Bazza, 2003). For example, Yemen is among these countries, where the estimated water resources availability has even fallen to between 110 and 150 m^3/cap/a in recent years, which is well below the minimum acceptable water threshold (Almas, 2001).

The main factors that contributed to the water crisis in Yemen and other Arabic countries are an increase in population growth, an imbalance between water availability and requirement, limited water resources, contamination of groundwater resources (Foppen, 2003; Metwali, 2003) and a lack of appropriate policies for water allocation. Concerning the latter, the agricultural sector in Yemen, for example, uses at least 90% of the national water resources for irrigation of predominantly qat, a recreational drug.

The water shortage problems in Aden are typical for other major cities of the Arabic Peninsula. The water produced for domestic consumption in the Greater Aden area (Figure 7.13) for the years 2002–2004 are approximately 13.5M, 13.0M and 12.2M m^3/year, respectively. This compares with an actual domestic consumption of 63, 60 and 57 l/cap/d, respectively. The population of Aden for the years 2002–2004 was 574,500, 600,200 and 612,300, respectively (National Water Supply and Sanitation Authority, 2002, 2003, 2004). These figures show that the water production for domestic consumption is decreasing, while the population of Aden is increasing during the same period.

Moreover, the water demand for landscaping and the suburban agriculture sector in Aden has increased. For example, the cultivated land comprised 820, 1,720 and 4,427 ha in 1979, 1989 and 2003, respectively (Ministry of Planning, 2003).

The amount of water required for irrigation is high (Almas, 2001). Consequently, more stress is put on the already decreasing urban groundwater resources. Eventually, the water consumption per capita must decrease because of depleting groundwater resources particularly in urban areas.

The aquifers contain high salinity levels of approximately 2,300 mg/L, which is more than twice of the recommended threshold of 1,000 mg/L (National Water Supply and Sanitation Authority, 2003; World Health Organization, 1993). The salinity is estimated to increase because groundwater levels are frequently lower than the sea level, which is particularly the case for coastal provinces (Figure 7.13).

7.2.1.2 Water Reuse Constraints

The main constrains for reusing wastewater in Islamic countries can be divided into (a) public attitude in terms of social and religious aspects towards water reuse and (b) technical aspects affecting the quality of treated wastewater. The former constraints are the focus of Section 7.2.1.2.

The concept of wastewater reuse must be introduced to all members of the public, which should be involved in the decision-making processes. Particularly farmers and farm workers as well as other members of the public encountering reused water, need to be educated by politicians, Imams and public health officers with respect to the benefits and risks of using treated wastewater (Almas, 2001).

However, the public consultation concerning the reuse of treated wastewater for irrigation purposes is at least in theory completed, because a Fatwa (legal ruling by the religious scholars) made by the Islamic Council of Research and Consultation resolved this challenge in 1979. The Council declared that treated wastewater could be used for all purposes if it meets the required health standards (Ali, 1983; Farooq and Ansari, 1983; Mara, 2000; Uzomah and Scholz, 2002).

On a positive note, several other Islamic countries that are currently facing a water shortage crisis have adopted the reuse of effluent from treated wastewater for irrigation practice as the only available significant resource to recover from over-exploration of groundwater resources. Jordan, the Kingdom of Saudi Arabia and Tunisia have made good progress on reusing treated wastewater. It is Jordan's policy to increase the proportion of treated wastewater to irrigate fields to 30% of the national water demand (Abu-sharar et al., 2003; Almas, 2001; Bahari and Brssaud, 1996; Hussien and Al-Saati, 1999).

Nevertheless, most of the public in Yemen and other more traditional Islamic countries consider the reuse of treated wastewater for irrigation purposes still as Haram (not legitimate or against the Islam). Strong opposition concerning water reuse comes from farmers representing approximately 90% of the water consumers. It is the responsibility of the National Water Supply and Sanitation Authority (NWSA) to address these concerns. The High Council for Justice (Islamic legislation Institute) should cooperate with the NWSA to issue relevant Fatwas. Moreover, the High Council for Justice ought to encourage the Imams of the Mosques to conserve water as the Mosques in Yemen are the heavy users of potable water because their patrons perform a cleansing (Wadu) ritual before prayer. The water that is used for cleansing is of very high quality, because no cleaning agents such as soap are used in this ritual.

An example for public consultation concerning the reuse of treated wastewater for irrigation purposes is at least in theory completed due to a Fatwa stating that polluted water might regain its original characteristics of purity by self-purification or dilution (Mara, 2000). Moreover, a current Fatwa in the Kingdom of Saudi Arabia permits the use of adequately treated wastewater in agriculture. However, the necessary precautions to avoid transmission of diseases must be taken (Turtun, 1999).

Despite the NWSA's concerns over the depletion of local water resources and its initiatives to seek alternative water resources, there is no formal policy or framework of legislation for wastewater reuse in irrigation. Wastewater reuse has not yet been institutionalized in Yemen. In comparison, the Hashemite Kingdom of Jordan and Tunisia face the same water crisis and public perception problem. However, wastewater reuse had been institutionalized in these countries (Ministry of Water and Irrigation, 1998; Shobha Shetty, 2005). Other countries should follow this example.

In the Holy Koran, water is mentioned in more than 80 verses; for example, "We made from water everything" (Holy Koran 21:30). Water receives much attention in both Sunna and Sharia and prevents Muslims from polluting water. The prophet Muhammad said, for example, "Let no one of you urinate in stagnant water".

The NWSA is a key stakeholder and should therefore initiate a national debate on the water crisis with particular emphasis on the concept of wastewater reuse for irrigation. This national debate should involve all stakeholders in the management of water resources; that is, political leaders, central government, parliament, Sheiks, traditional establishments, religious men (Mullahs), decentralized offices, wise men, donors, private sector, press and media and users (predominantly farmers).

7.2.1.3 Aims and Objectives

The first aim of Section 7.2 is to study the design, operation and treatment performance of a typical WSP system located in the city of Aden, the political and economic centre of Yemen (Figure 7.13). The second aim is to discuss the quality of the treated sewage and its potential for recycling in the context of an Islamic society. The objectives are to assess (a) the characteristics of the sewage (influent) received by the WSP; (b) the characteristics of treated sewage (effluent); (c) if the effluent meets the WHO recommendations for wastewater reuse for crop irrigation; (d) the impact of effluent in polluting agricultural land, if reused; (e) alternative options for utilization of effluent in parks, landscaping and Greenbelt projects and (f) if Yemen can be used as a case study for other arid countries. Section 7.2 is based on an updated version of an original article by Almas and Scholz (2006).

7.2.2 Case Study, Materials and Methods

7.2.2.1 Study Area and Pond System

The City of Aden commissioned its first large full-scale sewerage treatment plant (Figures 7.14 and 7.15) in 1985 (Almas, 2001; JICA, 1990). A total population of 123,000 is served by the plant that has a mean flow rate of 29,000 m^3/d. The entire predominantly domestic sewage load is first admitted to a single anaerobic pond. The effluent flow from this is subsequently split into two, feeding two parallel treatment

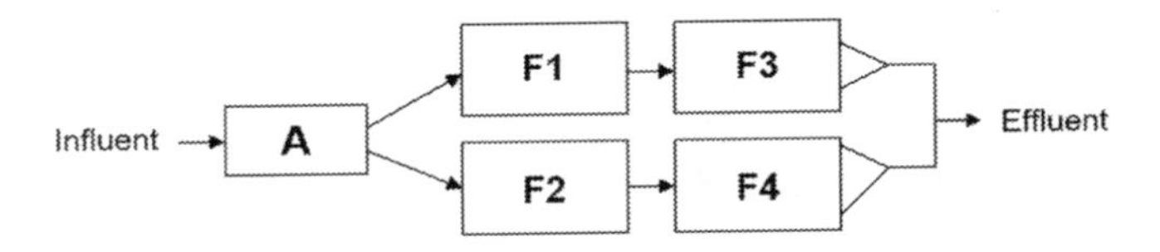

FIGURE 7.14 Layout of the wastewater stabilization pond system (anaerobic (A) and facultative (F) ponds) in Aden.

FIGURE 7.15 Photograph of a representative facultative pond taken on 20 July 2002.

trains of facultative ponds operated in series of two WSPs each (Figures 7.14 and 7.15). The sewage enters the anaerobic pond via a central inlet chamber located at the centre of each pond. The surface of the anaerobic pond is almost square. Effluent from the anaerobic pond is abstracted via a system of discharge pipes and subsequently transferred to the facultative ponds (Figure 7.14).

The facultative ponds also have rectangular surface areas. Pre-treated wastewater enters the facultative ponds via inlets equally distributed along one side of the pond and leaves the pond via two equal outlet pipes along the opposite side of the facultative ponds. Figure 7.15 shows the surface of the WSP having a total area of 22 ha. The hydraulic retention time of the anaerobic pond is 1.4 days. In comparison, the facultative ponds have retention times of 5 days each, so that the retention times of each of the two facultative pond series are 10 days.

The typical average biochemical oxygen demand (BOD) of the influent to ponds is 375 mg/L, and typical numbers of faecal coliforms (FCs) counted in the influent are on average 3.6×10^6 colony forming units (CFU)/100 mL (Almas, 2001; Gitec Consultants and Dorch Consultants, 1997). Typical WSP treats raw sewage of domestic nature usually comprising a mixture of predominantly human excreta (urine and faeces) and grey water (originating from kitchens and bathrooms) according to Almas (2001).

7.2.2.2 Sample Collection and Water Quality

The performance of the WSP was analysed by taking samples from the pond influent (raw sewage) and from the effluent of the ponds (Figures 7.14 and 7.15). The samples were collected twice a month for a period of one year to study the effect of seasonal variation on the performance of WSP. Monthly and seasonal averages were calculated for the sewage samples to assess the pond treatment performances.

All samples were analysed for five-day at 20°C BOD, chemical oxygen demand (COD), FC, ammonium-nitrogen (NH_4-N), total phosphorus (TP) and dissolved oxygen (DO) were analysed in the laboratory located on-site. DO, pH and temperature (*T*) were measured in situ. The former two variables were measured at two different depths in the ponds (Figure 7.14; F3 and F4). All tests including BOD and COD were carried out according to the American standard methods (American Public Health Association, 1995). All inflow and outflow samples were collected between 9 and 10 AM (local time).

The DO was measured with a DO meter Model 15 A (Thermo Electric Corporation) using British Orion electrodes. The pH values were determined with the pH meter Model M240 (Coring, Sudbury, UK) that was fitted with a Ross Combination Electrode 8102 (Orion, Beverly, U.S.). The DO and pH readings were taken at the same time. The suspended solids (SS) were measured by filtering 50 mL of the sample through a pre-weighed glass fibre Millipore filter paper (standard grade FP103; pore diameter of 6–8 μm; thickness of 170 μm). The total dissolved solids were measured by a gravimetric method according to the American Public Health Association (1995).

7.2.3 Results and Discussion

7.2.3.1 Inflow Water Analysis

The characteristics of raw sewage (influent for the anaerobic pond) are summarized in Table 7.5. The raw wastewater contains relatively high organic matter (e.g., BOD)

TABLE 7.5
Key Characteristics of Raw Sewage (Inflow to the Treatment Plant) Based on 24 Samples between July 2001 and June 2001

Variable	Unit	Mean	Standard Deviation
Temperature	°C	24	2.5
pH	–	7.4	0.1
Chemical oxygen demand	mg/L	763	7.5
Biochemical oxygen demand	mg/L	367	9.6
Suspended solids	mg/L	304	88.0
Ammonium-nitrogen	mg/L	38	1.9
Dissolved oxygen	mg/L	0.0	0.00
Faecal coliforms	CFU[a]	3×10^6	0.1×10^6
Total phosphorus	mg/L	15	1.3

[a] colony forming unit/100 mL.

and SS loads. Highly concentrated raw sewage is typical for locations in Yemen and other arid and semi-arid countries. The mean evaporation for Aden during the study period was 9.7 mm/d. Moreover, high concentrations are linked to the low water consummation rates (between 60 and 65 l/cap/d) and high temperatures at the same time (Almas, 2001; Gitec Consultants and Dorch Consultants, 1997; Veenstra et al., 1995).

7.2.3.2 Performance of the Anaerobic Pond

Wastewater stabilization ponds are normally arranged in series rather than in parallel to increase the overall treatment performance (Almas, 2001; World Health Organization, 1987). The arrangements usually comprise anaerobic ponds followed by facultative ponds and sometimes also by maturation and/or polishing ponds (Figure 7.14). Neither maturation nor polishing ponds were applied in this case study. The purpose of this arrangement is for the anaerobic pond to absorb the high organic loading of the wastewater. The associated BOD and SS removal efficiencies of the ponds are usually between 50% and 60%, and sometimes even up to 75%, if they are designed and operated in agreement with international guidelines (Mara, 1997, 2000; Smith and Scott, 2002; World Health Organization, 1987).

However, Table 7.6 indicates that the BOD and SS removal efficiencies are lower than expected for normal anaerobic ponds (see above). Concerning highest and lowest BOD removal efficiencies, the figures for June 2002 and January 2002 were 43% and 25%, respectively. The removal efficiencies for SS were highest (27%) for December 2001, February 2002 and May 2002 and lowest (20%) for September 2001 and June 2002.

TABLE 7.6
Mean Removal Efficiency (RE) of the Anaerobic Waste Stabilization Pond in Aden between July 2001 and June 2002

Month	COD[a] (mg/L)	RE (%)	BOD[b] (mg/L)	RE (%)	SS[c] (mg/L)	RE (%)
July 01	525	30	262	31	258	24
Aug 01	515	31	254	33	261	23
Sept 01	530	29	256	32	259	20
Oct 01	540	28	251	33	247	25
Nov 01	563	24	247	35	250	24
Dec 01	569	24	270	28	240	27
Jan 02	561	25	285	25	244	26
Feb 02	550	26	260	31	239	27
Mar 02	563	26	255	32	253	23
Apr 02	542	27	247	35	251	23
May 02	530	29	236	37	238	27
Jun 02	528	29	250	43	261	20

[a] Chemical oxygen demand; [b] biochemical oxygen demand; [c] suspended solids.

These figures indicate that the performance of the anaerobic pond was inefficient. Considerable sludge accumulation in the ponds was observed in Aden. The typical water depths were between 0.7 and 1.0 m, and the mean retention times were 1.4 days per pond. While the design criteria are usually set for sludge removal every 5 years (Almas, 2001; Mara, 1997; World Health Organization, 1987), sludge was not removed within 5 years of plant commissioning (Almas, 2001).

It follows that the main reason for the lack of removal efficiency and subsequent poor effluent quality is the accumulation of sludge in the ponds. This is consistent with cases reported elsewhere; for example, WSP in Tanzania (Almas and Jefferies, 2001). These WSPs have commonly not been desludged for 15 years. Therefore, large quantities of sludge accumulated at the bottom layers of both the anaerobic and all facultative ponds leading to poor treatment performance.

The absence of a desludging programme has a great impact on the performance of a pond by altering its hydraulic regime due to a decrease in the pond's effective volume, change of the shape of the bottom surface and reduction of the retention time of the ponds (Almas, 2001; Mbwele et al., 2004; Nelson et al., 2004). Therefore, periodic sludge removal is usually required to sustain both long-term self-sustainability and compliance with WHO guidelines that allow for the reuse of treated wastewater for crop irrigation practices.

Furthermore, overloading in terms of the maximum aerial BOD load is a frequent reason for the malfunctioning of anaerobic ponds. The aerial BOD design load was only 0.5535 BOD/kg/m^2, while the observed actual load for the anaerobic pond was three times higher (Almas, 2001). It follows that the anaerobic pond of the WSP system was under-designed for the current load. It is therefore recommended to either enlarge the existing pond or to construct a second anaerobic pond nearby (Figure 7.14).

7.2.3.3 Performance of Facultative Ponds

Tables 7.7–7.9 summarize the wastewater quality effluent of the facultative ponds (equal to the overall plant effluent) between July 2001 and June 2002. The mean BOD, COD and SS concentrations were 52.1, 289.8 and 156.6 mg/L, respectively. Concerning the key nutrients, mean NH_4-N and TP concentrations were 22.3 and 6.7 mg/L, respectively. The DO concentration was 1.4 mg/L. The concentrations of virtually all key water quality variables were remarkably stable indicating a mature treatment system (Table 7.8) according to Almas (2001).

Pathogenic bacteria (Table 7.7) and potentially other microorganisms may impact on the public health of rural communities if treated wastewater is recycled for crop irrigation (Farooq and Ansari, 1983; Hussien and Al-Saati, 1999). Moreover, the bathing water quality could suffer, if the effluent is discharged to rivers or directly to the sea via sea outfall pipes polluting coastal beaches used for recreational activities (Almas, 2001; Uzomah and Scholz, 2002).

FC numbers within the final effluent of the WSP range between 400 and 700 CFU/100 mL for both July 2001 and June 2002 (Table 7.8). This compares well with a WHO guideline threshold of 1,000 CFU/100 mL (Alcalde et al., 2003; Mara, 2000; Mbwele et al., 2004; Smith and Scott, 2002). The final effluent of this treatment plant could therefore be used for restricted irrigation of sand dunes and trees within the Green Belt of Aden (Almas, 2001).

TABLE 7.7
Characteristics of Treated Wastewater (Mean Effluent from a Representative Facultative Pond) between July 2001 and June 2002

Date	pH (–)	COD[a] (mg/L)	BOD[b] (mg/L)	SS[c] (mg/L)	NH_4-N[d] (mg/L)	TP[e] (mg/L)	DO[f] (mg/L)	T[g] (°C)	FC[h] (CFU[i]/100 mL)
08/07/01	8.4	290	54	150	20	7	1.4	25	500
21/07/01	8.3	289	53	152	21	6	1.3	28	700
08/08/01	8.1	291	50	154	23	8	1.1	27	600
21/08/01	8.3	288	51	155	24	6	1.5	25	400
10/09/01	8.0	290	52	153	21	8	1.4	27	500
27/09/01	8.4	288	51	156	22	7	1.5	25	600
10/10/01	8.1	289	49	160	20	6	1.6	26	500
23/10/01	8.2	287	48	155	19	5	1.4	25	600
14/11/01	8.3	292	53	151	24	7	1.5	26	600
20/11/01	8.5	290	55	157	22	6	1.3	26	700
06/12/01	8.2	289	51	154	23	7	1.4	20	700
18/12/01	8.3	288	53	156	21	6	1.6	20	800
05/01/02	8.1	287	51	160	24	7	1.5	20	600
15/01/02	8.0	291	50	162	22	6	1.3	20	500
09/02/02	8.5	292	54	161	23	76	1.3	20	600
26/02/02	8.4	286	55	157	21	5	1.2	20	500
11/03/02	8.3	289	51	158	20	7	1.4	20	600
26/03/02	8.1	286	50	156	24	6	1.3	25	700
10/04/02	8.4	293	53	155	21	8	1.2	25	700
19/04/02	8.4	291	52	159	20	9	1.4	25	600
07/05/02	8.6	294	56	158	24	7	1.7	26	400
23/05/02	8.6	293	55	154	23	7	1.3	26	700
08/06/02	8.1	292	51	168	25	6	1.4	26	500
17/06/02	8.6	291	52	156	26	7	1.1	26	500

[a] Chemical oxygen demand; [b] biochemical oxygen demand; [c] suspended solids; [d] ammonium-nitrogen; [e] total phosphorus; [f] dissolved oxygen; [g] temperature; [h] faecal coliforms; [i] colony forming unit.

TABLE 7.8
Quality of the Effluent from a Representative Facultative Pond between July 2001 and June 2002

Variable	Units	Sample Number	Average	Standard Deviation
pH	–	24	8.3	0.18
Chemical oxygen demand	mg/L	24	289.8	2.3
Biochemical oxygen demand	mg/L	24	52.1	2.1
Suspended solids	mg/L	24	156.6	4.0
Ammonium-nitrogen	mg/L	24	22.3	1.9
Total phosphorus	mg/L	24	6.7	0.96
Dissolved oxygen	mg/L	24	1.4	0.1
Temperature	°C	24	24.3	2.9
Faecal coliforms	CFU[a]	24	578.2	106.2

[a] Colony forming unit/100 mL.

TABLE 7.9
Removal Efficiencies (%) Concerning the Effluent from a Representative Facultative Pond between July 2001 and June 2002

Date	COD[a]	BOD[b]	SS[c]	NH_4-N[d]	TP[e]	FC[f]
08/07/01	61	85	54	50	53	99
21/07/01	61	85	53	47	60	99
08/08/01	61	85	53	42	46	99
21/08/01	61	85	53	40	60	99
10/09/01	61	85	54	47	46	99
27/09/01	61	85	53	48	53	99
10/10/01	61	87	52	50	60	99
23/10/01	62	87	53	52	66	99
14/11/01	61	85	54	40	53	99
20/11/01	61	86	52	48	60	99
06/12/01	61	85	53	42	53	99
18/12/01	61	85	53	47	60	99
05/01/02	62	85	52	40	53	99
15/01/02	61	85	50	48	60	99
09/02/02	61	85	50	42	53	99
26/02/02	62	85	53	47	66	99
11/03/02	61	85	53	50	53	99
26/03/02	62	85	53	40	60	99
10/04/02	61	85	53	47	46	99
19/04/02	61	85	53	50	40	99
07/05/02	61	84	53	40	53	99
23/05/02	61	84	53	47	53	99
08/06/02	60	84	49	37	60	99
17/06/02	61	85	53	35	53	99

[a] Chemical oxygen demand; [b] biochemical oxygen demand; [c] suspended solids; [d] ammonium-nitrogen; [e] total phosphorus; [f] faecal coliforms

Table 7.9 highlights the treatment efficiency variations of the WSP for one year. The BOD removal efficiency was between 84% and 87%, while the COD removal efficiency was between 60% and 62%, and the SS removal efficiency was between 49% and 55%. Concerning nutrients, the NH_4-N treatment efficiency was between 35% and 52% and the TP removal efficacy was between 40% and 60%. The FC treatment proficiency was 99%.

Figures presented in Tables 7.7–7.9 indicate that the quality of the final effluent does not meet most of the recommended guidelines for effluent to be used for restricted irrigation of crops to be consumed either by humans or by cattle (Carr et al., 2004; Council of the European Communities, 1991; World Health Organization, 1989). It follows that the facultative ponds should be designed for higher BOD, COD and SS removal performances. This can be achieved by providing sufficient oxygen for aerobic bacteria, protozoa and other microorganisms involved in biodegradation of organic matter.

Photosynthesis of algae fuelled by sunlight is the key process governing the performance of facultative ponds. Algae utilize sunlight penetrating the pond surface as a source of energy and produce subsequently new cells of algae and oxygen available for further biodegradation processes (see above). The DO concentration within facultative ponds increases with increasing light penetration through the pond surface (Curtis et al., 1992). For a pilot-scale system, the maximum DO concentration due to photosynthesis may occur between 13:00 and 15:00, when the DO could be as high as 20 mg/L (Tadesse et al., 2004). Therefore, sunlight and solar radiation has a great implication on WSP treatment performance (Curtis et al., 1992; Mara, 1997; Ratchford and Fallowfield, 2003).

Most facultative ponds in Aden were pink in colour, and odour was released. The pink colour indicates that the facultative ponds are overloaded and mostly anaerobic (Mara, 2000). Table 7.10 shows that the amount of DO measured at different depths of the facultative ponds is too low (anaerobic conditions).

Red- and pink-coloured WSPs in Saana were researched by Veenstra et al. (1995). Insufficient organic matter metabolism prevailed in these facultative ponds. Organic overloading is the main cause of pond malfunctioning, and purple non-sulphide bacteria are good performance indicators (Calero et al., 2000; Mara, 2000). Their populations act as a biological filter preventing proper biodegradation of organic matter and causing odour nuisance (Almasi and Pescod, 1996).

7.2.4 Conclusions and Recommendations

The inflow to the representative example treatment plant is highly concentrated due to the arid climate and potable water shortage. This requires extra allowances for design loading recommendations of international guidelines.

The anaerobic pond fell short of its treatment targets due to the absence of a sludge removal strategy that complies with international guidelines. Sludge removal should take place every three years in Yemen and other arid countries. Moreover, an additional anaerobic pond (in parallel) would allow desludging of the existing anaerobic pond and future regular desludging of both anaerobic ponds.

The facultative ponds at the end of the treatment trains are overloaded and cause an odour problem. Therefore, it is recommended to expand the treatment train.

TABLE 7.10
pH (–) and Dissolved Oxygen, DO (mg/L), for Different Water Depths (m) of a Representative Facultative Pond

Depth	07/01		08/01		09/01		10/01		11/01		12/01	
	pH	DO	pH	DO	pH	DO	pH	DO	pH	DO	pH	DO
0.5	8.0	0.4	8.3	0.4	8.2	0.5	8.3	0.4	8.2	0.4	8.2	0.2
1.0	8.2	0.0	8.1	0.0	8.1	0.0	8.2	0.0	8.2	0.0	8.1	0.0

Anaerobic purple sulphide bacteria indicating overloading were also detected. However, key water quality variables were stable indicating a mature system able to cope with potential shock loads.

The positive effect of sunlight in generating oxygen for aerobic biodegradation processes should be taken into consideration in the design of facultative ponds, and the retention time of the facultative ponds should be between 15 and 20 days. Alternatively, a further facultative pond can be added to each series of the treatment train to allow for a better quality of the final effluent. This is essential to obtain good final quality effluent complying with WHO guidelines.

The FC counts comply with WHO guidelines. This allows for a restricted use of the final effluent for the irrigation of the Green Belt in Aden. Moreover, it is anticipated that the treated wastewater can be recycled for irrigation purposes after expansion of the treatment plant, assuming that social and religious barriers can be overcome.

7.3 IMPACT OF ANTHROPOGENIC RIVER REGULATION ON WATER AVAILABILITY IN WATERSHEDS

7.3.1 Introduction

The recognition of disparities among riparian states in terms of physiographic features, capacity and potential of human resources, infrastructure, socio-economic development, geopolitical supremacy, political orientation and short- to long-term interests and plans represent the challenges to sustainably manage and protect shared water resources at transboundary scale (Figure 7.16). These disparities have led to differences in perspectives, priorities and arrangements, observing networks,

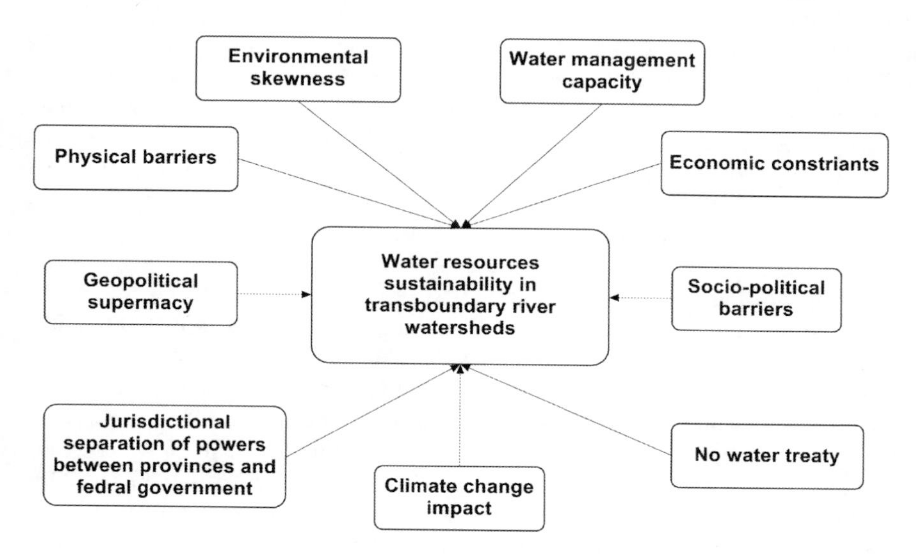

FIGURE 7.16 Driving forces affecting water resources sustainability in transboundary river watersheds.

measuring systems, data processing practices and management strategies. Climate change has become an additional concern, which is likely to exacerbate the differences in responses, policies and actions between upstream and downstream countries (Figure 7.16). However, the diverse range of interests, plans and arrangements as well as the necessity to sustainably use the shared water resources should open new opportunities for long-term transboundary collaboration. Despite the efforts of some co-watershed states to reach agreements on shared water apportionment, yet the current predominant trend is that the upstream country unilaterally endeavours in excessively utilized the shared water resources on its territory.

Governance of water resources in transboundary river watersheds is more complex and challenging when compared to water management at the national and subnational levels. The complexity and challenges are mainly due to (a) the excessive use of shared water resources in the upstream country at the expense of the downstream state, (b) the lack of collaboration and (c) the potential impacts of climate change.

The growing of upstream human-made disturbances such as damming, large-scale water abstraction schemes and inter-watershed water transfer network in transboundary watershed has unequivocally raised concerns about potential hydrological adverse impacts in the downstream states. Anthropogenic stressors coupled with droughts have considerably magnified the water management challenges. The impoundment of reservoirs and operation of hydraulic structures such as weirs, barrages and regulators have resulted in the sharp alteration of hydrologic regimes (Rosenberg et al., 2000; IUCN, 2000). Literature indicates that efforts have been undertaken to analyse hydrologic alterations (Olden and Poff, 2003; Shiau and Wu, 2004; Zolezzi et al., 2009) and characterize the impact of regulation on flow paradigms (Monk et al., 2007; Gao et al., 2009).

Montenegro and Ragab (2012) state that understanding the hydrology under current and future climate and land use changes is essential for the management and planning of water resources. The Intergovernmental Panel on Climate Change (IPCC, 2007) has defined climate change as “any change in climate over time, whether due to natural variability or as a result of human activity”. Concerning the climate alterations, the IPCC has reported that the average global temperature increased by approximately 0.6°C ± 0.20°C during the 20th century (IPCC, 2001). Recent investigations in the Diyala basin in Iraq (Waheed, 2013) have indicated that there was a rise in temperature between 0.2°C and 0.6°C over the water years 1960–2010. The increase in monthly evapotranspiration was between 4% and 13%. The average drop in runoff was as much as 4.7% in the last 20 years.

The existing data for the period 1980–2010 at the Derbandikhan meteorological station (located in the upper part of the Diyala basin in Iraq at 35.11°N and 45.69°E according to MA and WR-KRG (2013) show that the total annual precipitation ranged from 1,084 mm recorded in 1988 to 207 mm registered in 2008. Likewise, the mean annual temperature ranged from 18.8°C in 1982 to 23.1°C in 1999. The estimated total annual PET (using the Blaney-Criddle method) between 1980 and 2010 ranged from 1,467 to 1,632 mm. Moreover, the long-term mean annual precipitation, temperature and PET were estimated at 624 mm, 20.9°C and 1,547 mm, respectively.

It is worth noting that the reduction in the rate of runoff observed in Iraq in the last two decades cannot be isolated from the combined impact of climate alteration with respect to the entire watershed and the upstream anthropogenic river regulation. The impact of elevated temperature, evapotranspiration and reduction in precipitation rate is manifested by a series of successive dry years in the last two decades. The World Bank Climate Change Knowledge Portal (WB, 2014) has reported that the long-term mean annual precipitation in Iraq between 1960 and 1989 and between 1990 and 2009 were estimated at 185.3 and 172.9 mm, respectively. The long-term mean annual temperature over the period 1960–1989 was 21.25°C and between 1990 and 2009 was 22.1°C. The precipitation over the two examined time periods was reduced by as much as 6.7%, and the temperature was increased by 4%.

Miller and Yates (2006) have underlined that it is important to understand how environmental change may affect water resources upon which riparian countries depend on and compete with. The Third Assessment Report of the IPCC estimates that the global average temperature will rise by between 1.4°C and 5.8°C by the year 2100 (IPCC, 2001). This implies uncertainty in terms of the limit to which global mean temperature will rise. Most likely changes include increases in global average precipitation and alterations in runoff patterns.

The IPCC (2007, 2014) has repeatedly pointed out that many arid and semi-arid areas (e.g., the Mediterranean Watershed, western U.S., southern Africa, north-east Brazil and southern and eastern Australia) will suffer a decrease in water resources due to climate change. Annual average river runoff and water availability are projected to decrease for some dry regions at mid-latitudes and in the dry tropics, some of which are presently water-stressed areas.

Arnell and Chunzhen (2001) have highlighted that water will become either more scare in some parts of the world or over abundant in others, resulting in more extreme and frequent droughts and floods. Recent research has increasingly considered adaptation to climate change. Ragab and Prudhomme (2002) have pointed out some adaptation options for arid and semi-arid regions that include conventional solutions such as developing storage dams and irrigation schemes, inter-watershed transfers of water through networks of pipes and canals, and further development of groundwater resources.

According to IPCC (2007) and Arnell and Charlton (2009), four different types of limits or barriers to cope with environment change in terms of water quantity and quality were identified. Firstly, there may be physical barriers that constrain the performance of a particular adaptation option. Secondly, there could be economic constraints if some adaptations are considered to be too costly. There may be socio-political barriers to adaptation due to the attitudes of stakeholders to proposed options, and, finally, the capacity of water management institutions could limit the ability to promote or implement adaptation strategies.

With respect to developing adaptation strategies at transboundary development scale, the IPCC (2001) has reported that the current knowledge of adaptation and adaptive capacity to climate change is insufficient, and there are serious limitations in existing evaluations of adaptation options. The UNECE (2009) has underlined that the lack of experience is a challenge, and transboundary cooperation is almost non-existent in some regions. Therefore, the proposed transboundary development scale concept can help to address the challenge.

UNECE (2011) has highlighted the need for discussing water and climate change in the transboundary context. Very little has been done on any transboundary development scale to assess climate change impacts and to develop adaptation strategies to cope with the range and depths of anticipated impacts. The adaptation options have been schematized in this chapter (Figure 7.17) based on information drawn from various sources (Hiscock et al., 2002; Ragab and Prudhomme, 2002; Conway, 2005; IPCC, 2007; Kistin and Ashton, 2008; Tompkins et al., 2009; Arnell and Charlton, 2009; Bloetscher, 2012; Gain et al., 2012).

Section 7.3 aims to quantify the flow reduction rates in the downstream country Iraq associated with upstream river damming, large-scale irrigation projects, domestic and industrial water abstraction schemes, inter-watershed water transfer systems and landscape modifications. The key objectives are to assess the adverse impacts of upstream anthropogenic disruption on the natural monthly flow magnitudes, which were available to Iraq. Likewise, this chapter highlights the importance of developing strong communication channels and efficient networks between the upstream and downstream riparian countries, and among all stakeholders at national scale. Furthermore, Section 7.3 introduces a three-level system entitled the transboundary three-scalar framework, which helps to sustainably manage water resources.

The Diyala watershed shared between Iraq and Iran (Figure 7.18) has been chosen as one of the best options to achieve the chapter objectives. Water and land resources in the Diyala river basin are highly developed and committed. Irrigation is the

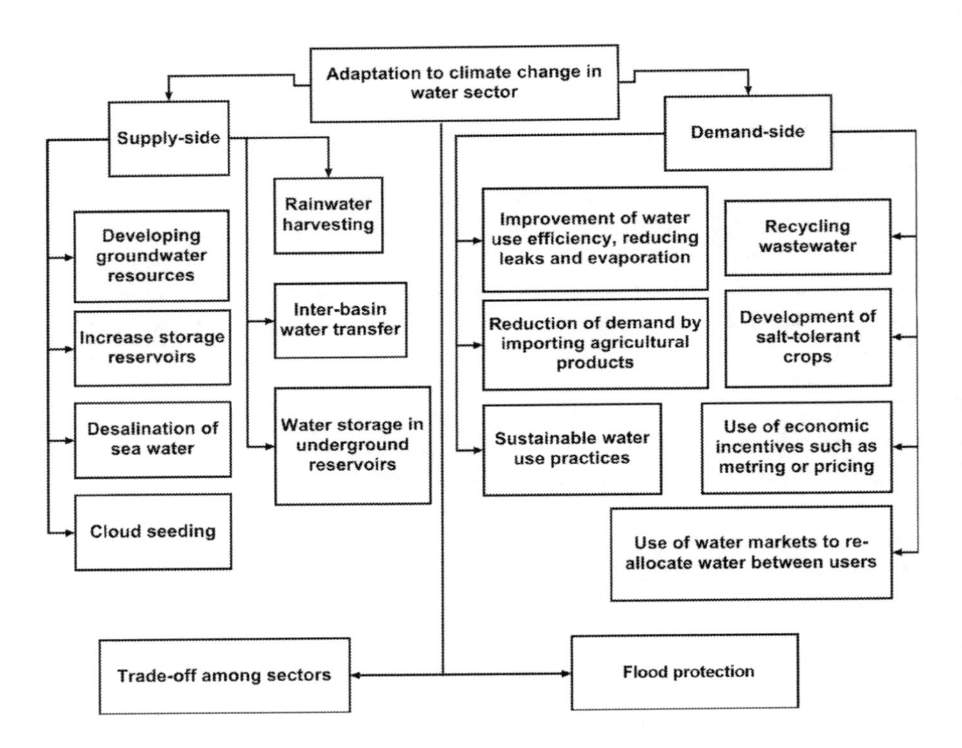

FIGURE 7.17 Adaptation options to climate change.

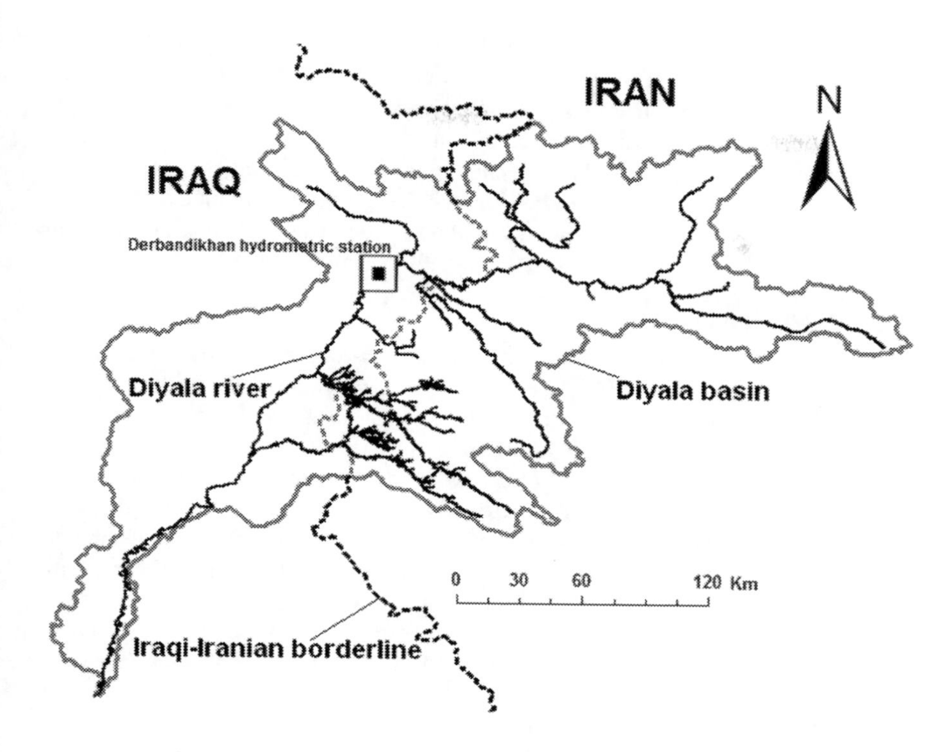

FIGURE 7.18 Diyala river watershed shared between Iraq and Iran.

dominant water use. Moreover, the basin is characterized by potential influences of upstream development measures on downstream demands. Note that Section 7.3 is based on an updated opinion piece published by Al-Faraj and Scholz (2015).

7.3.2 Materials and Methodology

7.3.2.1 Study Area, Conceptualized Framework and Data

Figure 7.18 shows the entire Diyala (Sīrvān) case study river watershed, which is one of the main river watersheds shared between Iraq and Iran. Its headwaters originate in the Zagros Mountains in western Iran. The watershed is situated between 33.216° N and 35.833° N and between 44.500° E and 46.833° E (Abdallah, 2010). The watershed drains a 32,600-km² area, of which about 43% and 57% lies in Iraq and Iran, respectively (Ministry of Water Resources, 1982).

Under natural conditions, the Diyala (Sīrvān) river is characterized by an irregular and perennial flow regime that strongly depends on rainfall. A period of no flow does not exist. The non-rainy period of low flow spans June to October. In contrast, high flows dominate from November to May. Section 7.3 limits its exploration to the upper part of the watershed, which is located just upstream of the Derbandikhan hydrometric station (DHS) at 35.110210°N and 45.703298°E. The upper part encompasses a drainage area of 17,900 km² in Iraq and Iran (Figure 7.18).

Section 7.3 introduces a new concept called the transboundary three-scalar framework to assess the impact of the interests of both riparian countries sharing the same example watershed. The conceptualized framework grouped the key driving forces, which are affecting the sustainable management of water resources in transboundary watersheds into three scales: (a) the upstream country national interest, (b) the transboundary scale (i.e. the entire shared watershed) where both riparian countries (i.e. Iraq and Iran in this case study) are involved and (c) the downstream country national interest. The key driving forces were schematized in Figure 7.16.

7.3.2.2 Methodology

Unaltered and artificially influenced daily flow datasets observed at the DHS in Iraq were analysed. A daily flow time series of 59 years (between 1955 and 2013) was examined. The datasets were obtained from various sources (Harza and Binnie, 1958, 1959, 1963; MA and WR-KRG, 2013).

The entire time series was first divided into two-time spans: (a) unregulated natural flow regime and (b) the regulated experimental case during which artificial alteration occurred. The year in which hydrologic impairment has started was used as a timeline to split the time series and compare between unaltered and altered flow conditions. The year 1983 was considered as a baseline. The pre-altered time series of 28 years comprises daily flow records between the water years 1955 and 1982, while the artificially influenced period encompasses 31 years spanning the hydrologic years 1983–2013. Furthermore, the artificially altered period was divided into two time frames depending on the scale of the anthropogenic regulation scheme and the volume of water abstraction across the main course of the river and its headwaters in the upstream country.

The natural flow regime was chosen as a reference condition (benchmark) to which the characteristics of the post-impacted time series were compared. The rate of flow alteration was determined using Equation 7.14. Data in Tables 7.11–7.13 and Figure 7.19 were linked to Equation 7.14 to estimate the watershed-averaged flow reduction rate with respect to the time-interval considered in the computational process. The flow reduction rate is an index reflecting the hydrologic changes due to both upstream artificial regulation and the type of the water year (i.e. dry, normal or wet).

$$\text{Rate of flow reduction} = \Big(\big(\text{Altered flow} - \text{Unaltered flow}\big)/\text{Unaltered flow}\Big) \times 100 \tag{7.14}$$

The analyses were undertaken using the HEC Data Storage System Visual Utility Engine (HEC-DSSVue) version 2.01 of the U.S. Army Corps of Engineers Hydrologic Engineering Centre. The MS Excel add-in associated with the HEC-DSSVue was used for retrieving and storing data directly from spreadsheets files.

7.3.2.3 Transboundary Three-Scalar Framework

The national interests of the upstream country Iran in terms of water resource management of the Diyala watershed have been reported by the Ministry of Energy (2003) and Shourian et al. (2008a, 2008b). Unilateral watershed development schemes in the

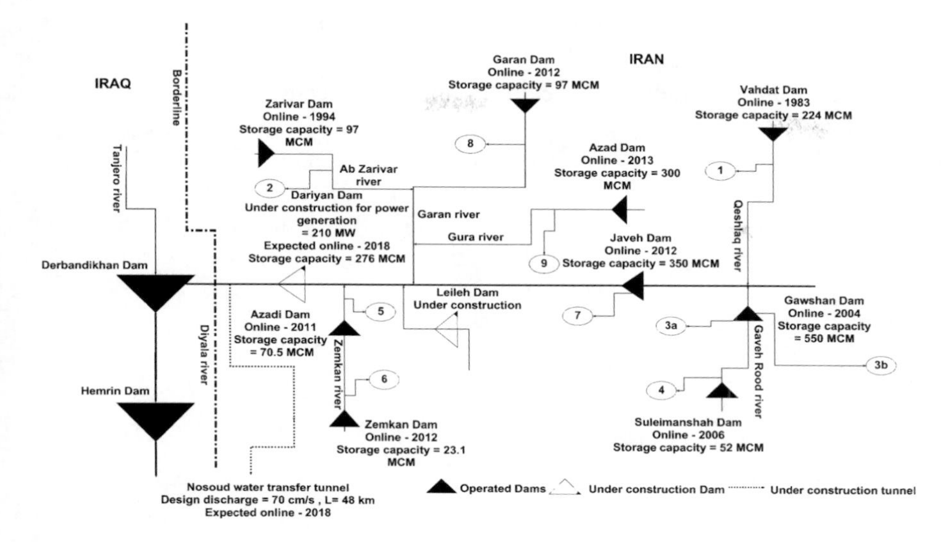

FIGURE 7.19 Development of dams and water abstraction in the upper Diyala (Sīrvān) river watershed in Iran. Note that all numbers are defined in Table 7.13.

TABLE 7.11
Development of Water Abstraction in the Upper Diyala Basin in Iran

Project Number	Irrigation Area (hectare)	Water Supply for Irrigation (million m³/year)	Fish Farm Area (hectare)	Domestic (million m³)	Industrial (million m³)
1	3,768	X	10	X	X
2	2,200	X	800	X	X
3a	31,000	189	X	X	X
3b	X	X	X	63	X
4	5,300	X	X	5 to 7	X
5	4,200	X	X	5.1	5
6	3,850	X	X	X	X
7	X	260	X	X	X
8	12,000	X	X	23 (domestic and industrial)	
9	19,100	190	X	X	X

Note that the design discharge for the Nosoud tunnel is 70 m³/s. X indicates either missing or not available information.

upstream country such as damming of the main watercourse and its tributaries, massive water abstraction benefiting large-scale irrigation schemes, fish farms as well as domestic and industrial demands cause a decline in flow magnitude, change in timing and deterioration of the quality of water entering the downstream nation (Figure 7.19 and Table 7.11). An indicator for the upstream riparian state (Iran) unilateral approach

TABLE 7.12
Long-Term Monthly Medians, Percentiles and Flow Recession Rates

Month	Condition	Median (m^3/s)	Flow Recession Rate (%)	25th Percentile (m^3/s)	75th Percentile (m^3/s)
October	Pre-developed	64.0	−48.4	37.5	91.5
	Developed	33.0		20.0	68.0
November	Pre-developed	72.3	−23.9	57.6	105.5
	Developed	55.0		33.8	88.5
December	Pre-developed	89.0	−31.5	65.3	112.5
	Developed	61.0		35.0	125.5
January	Pre-developed	111.5	−28.3	80.5	142.5
	Developed	80.0		55.0	156.0
February	Pre-developed	170.5	+5.3	133.6	239.5
	Developed	179.5		86.5	230.5
March	Pre-developed	298.5	−26.3	209.5	395.0
	Developed	220.0		124.5	372.0
April	Pre-developed	325.3	−11.2	249.9	463.0
	Developed	289.0		176.3	401.8
May	Pre-developed	190.0	−18.4	158.0	281.5
	Developed	155.0		112.5	212.5
June	Pre-developed	109.5	−37.9	74.9	166.5
	Developed	68.0		44.8	92.5
July	Pre-developed	85.0	−50.6	53.5	106.5
	Developed	42.0		28.0	63.0
August	Pre-developed	75.0	−53.3	42.0	99.5
	Developed	35.0		19.5	51.5
September	Pre-developed	71.0	−62.7	41.8	87.8
	Developed	26.5		16.5	54.3

to the shared river is the absence of a formal water-sharing arrangement with the downstream country Iraq (Figure 7.16). It is likely that an increase in hydrologic alteration will occur when more hydraulic structures are operated, irrigation schemes are expanded and the inter-watershed water transfer system is in operation (Figure 7.19 and Table 7.12).

The upper country Iran (Ministry of Energy, 2003) has reported that "The projects of transferring water from a watershed to another must be considered from the viewpoint of sustainable development while observing interested parties' rights and their technical, economic, social feasibility and explanation and national interests meeting various needs" (Figure 7.16). The achievement of national interests of the upstream country at the expense of the downstream state may cause adverse impacts to the environment of the Diyala watershed.

7.3.2.4 Transboundary Scale

The United Nations Department of Economic and Social Affairs (UNDESA, 2014) has reported that issues at transboundary scale predominantly include either the absence of a water treaty and the lack of an apportionment agreement among the riparian countries or significant weaknesses in most of the signed agreements and treaties. Underestimation of influences of excessive use of water is another important factor affecting the sustainability of water resources. Brels et al. (2008) have reported that despite the increase of formal agreements regarding shared river management, there are still numerous watersheds, which lack a sound legal framework for coordination and collaboration. Moreover, current arrangements are not always efficient. Some upper riparian countries such as Iran consider the use of water resources originating in their territory as their natural right; that is, to use rivers unilaterally regardless of downstream riparian concerns. The Ministry of Energy of the Islamic Republic of Iran (Ministry of Energy, 2003) has reported that "All the waters flowing out and join waters must be harnessed and consumed, and frontier rivers must be systematized observing economic and environmental standards" (Figure 7.16).

Climate change has become an important driving force, and its potential impacts on water allocation and management of water resources in shared river watersheds cannot be ignored (Figure 7.16). Eisenreich (2005) has stated that climate change impacts are evident from more frequently observed intense rainy seasons, longer dry periods, heavier storms and changes in rainfall and runoff patterns. More disastrous flood and drought events have been the most visible manifestation of this change. The United Nations Development Programme-Iraq (UNDP-Iraq, 2011) has reported that Iraq and Iran among other countries in the Middle East have experienced serious droughts since 1999. This can be assigned to the impact of climate change (downward trend of precipitation, and rising in temperature and PET rates), which still needs to be verified for the entire basin. Moreover, this may be a result of the downward trend of precipitation and rise in temperature and PET rates. It is worth mentioning that a transboundary early-warning system can reduce risks and effects caused by floods and drought on mainly the socio-economic and agriculture sectors.

Capacities and resilience of the existing structural measures or physical barriers such as dams, weirs, barrages and regulators are constraints to the viability of the system to deal with different water resources management options, and to accommodate the likely impacts of climate change. The present infrastructure of the Diyala basin comprises mainly a series of dams and hydraulic diversions (Figure 7.19). The reservoirs are operated to fulfil a variety of purposes such as flood control, domestic and industrial water supply, irrigation, fish farms demands and hydropower production. As with any set of multi-purpose reservoirs, these are managed to strike an appropriate balance among varying, and sometimes competing, usages. For this watershed, one main concern is to maximize water storage, while preserving space in the reservoirs for accommodating flood waters to prevent downstream flood losses. Attaining these multiple objectives is complicated by the fact that the reservoirs are managed in a multi-state, multi-jurisdictional and transboundary context (Figure 7.16).

The sharing of historical data and the installation of continuous hydro-meteorological monitoring networks at transboundary level are among the key challenges for efficiently managed shared water resources. Furthermore, the transparent exchange of data in a timely manner is of high importance in understanding how flow volumes and water quality are changing spatially and temporally in different parts of the shared watershed. At any rate, improved water resource management requires comprehensive monitoring networks and data sharing mechanisms to efficiently allocate water in time and space among various types of human-related water uses.

7.3.2.5 National Interests of the Downstream Country

Marcuello and Lallana (2003) as well as Goulden et al. (2009) have pointed out that the unilateral regulation measures on the main watercourse of the shared river and its tributaries including the water exploitation schemes of the upstream state are among the main drivers affecting the water management in the downstream states. The challenges facing the sustainable management of water resources are not limited to disputes among riparian states. Water conflicts can be extended to cover small political, spatial and geographical domains such as the local governments, local water authorities and tribal interests as well as between federal government and regional governorates (Figure 7.16). Various beneficiaries act differently. Therefore, there may be conflicts due to diverse values, goals and polices. Moreover, the main causatives of water stress and concern in the downstream state include the growing demand for water resources of good quality, the human resources and water management institutions capacities to handle and accommodate additional burdens and stress on the system under present operating rules, and the likely impacts of climate change.

7.3.2.6 Water Exploitation

Water resources in the Diyala river watershed have been highly regulated and are heavily committed for a variety of water demands. The watershed is exposed to unilateral water resources development actions by the upstream state (Iran). Massive river flow damming and significant utilization of corresponding water resources are currently being practiced; particularly due to large-scale irrigation projects such as the delivery of 260 million m^3/year from the Javeh dam to irrigate projects in the city of Sanandaj, and the Ghorveh and Dehgolan plains (both located to the East of Sanandaj) in the Kordestan province (Mahab Ghodss Consulting Engineering Company (2008,2013b). Further examples of large irrigation projects are the giant inter-watershed water transfer system; Nosoud tunnel (Mahab Ghodss Consulting Engineering Company (2008,2013b) and the Darian dam (Isdle, 2009).

A water treaty solution to the challenge of transboundary water management has not been pursued. Likewise, no international data sharing is currently being practiced. The increase in water exploitation associated with the main course of the river and its headwaters has created substantial needs to further understand the impacts of the upstream anthropogenic forces on the flow regime during the past three decades. The following paragraphs describe the river damming and regulation and water exploitation schemes in the upper reach of the Diyala watershed (Figure 7.20).

The Gheshlaq river, with a drainage area of 1,850 km^2 (Salim et al., 2009), is one of the main tributaries of the Diyala river in Iran. The Vahdat dam, inaugurated in 1983, is located 10 km to the North of Sanandaj on the river Gheshlaq. The objectives of this dam are flood protection, and water supply for irrigation to 4,000 ha of fertile land, domestic use, 10 ha of fish farms and 7 megawatt (MW) installed capacity of hydropower plant (Bozorgnia et al., 2012). The maximum storage capacity of

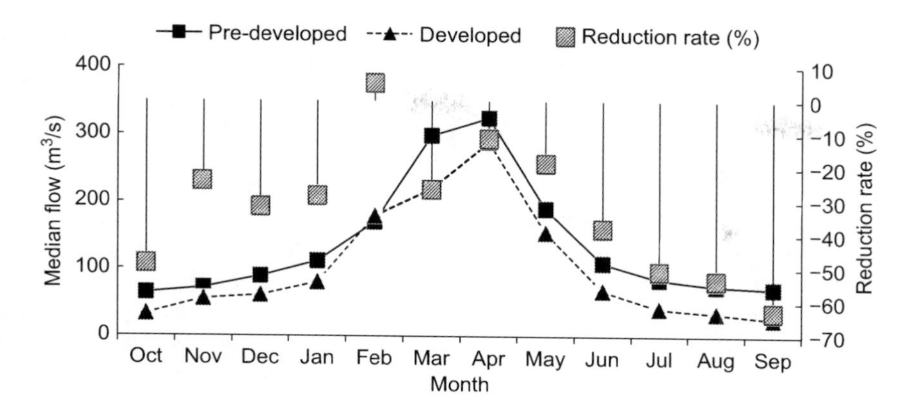

FIGURE 7.20 Long-term monthly median unaltered and altered flow conditions and deviations.

the reservoir is 224 million m^3 and the corresponding reservoir area is 9.34 km^2. Under normal pool elevation condition, the capacity of the reservoir is 190 million m^3 and the corresponding reservoir area is 8.3 km^2 (Faramarzi et al., 2009; Bozorgnia et al., 2012).

The rock-filled Garan Dam is located across the Gura (Garan) river in Marivan (Kurdistan province) and has a storage capacity of 97 million m^3 and a height of 57 m. The main objectives of this dam are to serve an area of 12,000 ha and supply 23 million m^3 to Marivan city to meet both domestic and industrial demands (Pooyab Consulting Engineers, 2013). The dam is in operation since 2012.

The Azadi dam is located 110 km north of Kermānshāh. It is a rock-filled dam with a clay core. The dam has a storage capacity of 70.5 million m^3 at normal pool elevation (Sad Afzar Engineering Service Company, 2013). The dam is located across the Zemkan river, Diyala's tributary, and was operated in 2012. The Zemkan watershed drains a 1,054 km^2 area up to the Azadi dam at 34.547144°N and 46.353938°E. The objectives of this dam are to serve agriculture land of 4,200 ha, and to supply 5.1 million m^3 for domestic and 5 million m^3 for industrial requirements. The dam height is 64 m (from foundation), and the crest has a length of 737 m and a corresponding width of 10 m.

The Darian dam is located on the main reach of the Diyala river (Sīrvān) (Isdle, 2009). The dam is rock-filled and comprises a clay core. The dam has 276 million m^3 of storage capacity and has 7.5 km^2 of associated reservoir storage area at normal pool elevation. The installed power plant capacity is 210 MW. The dam height is 160 m (from foundation), and the crest width is 15 m and the associated crest length is 330 m. The dam is currently the most downstream one before the river crosses the border between Iraq and Iran. The primary purpose of the dam is to supply up to 1,378 million m^3 of water annually to the 48-km-long Nosoud Water Conveyance Tunnel, which will irrigate areas of south-western Iran in the future.

Javeh earth fill dam site is located on the western part of the Iran and near the Sanandaj city (Aliasghari and Ahadi, 2007) on the main reach of the Sīrvān (Diyala) river and has a storage capacity of 350 million m^3. The dam was inaugurated in

2012. The dam height is 95 m and the crest length equals 334 m (Jyane Engineering Construction Company, 2013). The objectives of this dam and associated water conveyance tunnels are mainly to serve irrigation areas near Sanandaj city with 40 million m^3, and the Ghorveh and Dehgolan plains with 220 million m^3.

The Zemkan river, which is a tributary of the Sīrvān and located in Iran, has a drainage area of 2,400 km^2. The corresponding Zemkan dam has been commissioned in 2012. The dam is located 120 km west of Kermānshāh. The agriculture area served by this dam is 3,850 ha. The earth-filled dam has a clay core, a storage capacity of 23.1 million m^3, a corresponding reservoir area of 1.32 km^2 and a spillway discharge capacity 473 m^3/s. The dam height, crest width and crest length are 65 m, 10 m and 278 m, respectively. The dam crest level is 1,550 m above sea level (m.a.s.l.), the normal water level is 1,542 m.a.s.l. and the maximum water level is 1,549.7 m.a.s.l.

The Zarivar Lake is located in the West of the Zagros Mountain chains of western Iran (Asadi et al., 2012). The Zeribar Lake area comprises 2,090 ha and situated within the Zarivar watershed area of 15,827 ha. The lake has a maximum depth of 50 m and an average depth of 3 m. The Zarivar watershed has an annual average rainfall of 786 mm and a mean annual free surface evaporation of 1,900 mm. A water transmission line and hydraulic structures including a Zaribar Dam of 1.7-km length and 4.3-m height serve 2,200 ha of agriculture land and 800 ha of fish farms in the south of Marivan City.

Azad Dam is located on the Gura River (tributary of the Sīrvān), which was commissioned in 2013. The dam is situated at 40-km west of Sanandaj (Kurdestan province). The dam is associated with an electricity-generating capacity of 30 MW and an additional pumped storage power plant with a regenerating capacity of 500 MW (Tokmechi, 2011). Azad Dam has a storage capacity of 300 million m^3, a height of 125 m dam and a crest length and width of 600 and 14 m, respectively (Mahab Ghodss Consulting Engineering Company, 2008). The main objective of this dam project is to transfer 190 million m^3 of water annually to the agricultural plains of Ghorveh and Dehgolan via a conveyance system of 76-km length. The associated steel pipe has a diameter of 2 m, and the tunnel is 7-km long and 4 m in diameter. The design capacity of the transferring system is 7.5 m^3/s.

The Gaveh Rood watershed is a part of the Sīrvān drainage network in western Iran and is located between 47° E and 48° E and between 35° N and 45° N. Gaveh Rood is also known as Gaveh Rud, Gāveh Rūd and gawh rwd.

Gaveh Rood is located in the south of Kurdestan and the north of Kermānshāh. Topographically, it is a mountain watershed with a maximum altitude of 3,262 m.a.s.l. and a minimum altitude of 1,500 m.a.s.l. at the outlet of the watershed. The drainage area of this watershed is 2,092 km^2 and includes 23 sub-watersheds. Climatologically, the watershed is cold and semi-arid. Cold winters associated with snow and frost as well as moderate summers prevail. The rainfall regime is of Mediterranean nature, and a large part of the rainfall is recorded in the cold season. The mean annual rainfall is 450 mm, and most sub-watersheds have rainfall between 400 and 530 mm.

There are two dams currently located on the Gaveh rood river: Suleimanshah Dam and Gavoshan Dam. The Suleimanshah Dam was put into operation in 2006 and has a storage capacity of 52 million m^3. The main objectives of this dam are to supply water to 5,300 ha of agriculture area and to provide between 5 and 7 million m^3 of water for domestic use.

The Gavoshan Dam and associated conveyance tunnel are linked to the Gaveh Rood river. The dam is located 45 km south of Sanandaj city and 90 km north of Kermānshāh and has a storage capacity of 550 million m^3. The dam height is 123 m and the crest length is 620 m. The main purpose of the dam is to supply 63 million m^3 of water to Kermānshāh (domestic use) and 189 million m^3 via a 20-km-long conveyance tunnel to irrigate 31,000 ha of agricultural land in Kurdistan and Kermānshāh. Furthermore, the capacity of the power plant is 11 MW (Mahab Ghodss Consulting Engineering Company, 2013a). The coordinates of the dam site are 34.962° N and 46.994° E. The dam was inaugurated in 2005.

The Laileh Dam is located on the Laileh River at 34.90° N and 46.20° E. The river has a drainage area of 624 km^2. The average annual rainfall in the catchment is 703.7 mm and the average free surface evaporation is 1659.4 mm. The dam has been inaugurated in late 2013.

The Nosoud water tunnel is 48-km long and is in western Iran, northwest of Kermānshāh province (Mahab Ghodss Consulting Engineering Company, 2008, 2013b). The tunnel extends from the Sīrvān river to the Ezgeleh field. The design discharge is 70 m^3/s. The coordinates of the tunnel intake and outlet are 35.119° N and 46.237° E are 34.836° N and 45.839° E, respectively (Salimeh, 2012).

Figure 7.19 and Table 7.13 show the reservoirs, water diversions and inter-watershed water transfer systems. The first dam constructed on the Qeshlaq river (one of the river's tributaries) originating in Iran was inaugurated in 1983.

TABLE 7.13
Median Flow and Alteration (%) of Primary and High Alteration Periods

	Median Flow (m^3/s)				
Month	**Unaltered (1955–1982)**	**Primary Alteration (1983–2003)**	**Flow Recession Rate (%) (Primary Altered)**	**Median Flow (m^3/s) High Alteration (2004–2013)**	**Flow Recession Rate (%) (High Altered)**
October	64.0	58.0	−9.4	21.5	−66.4
November	72.3	67.0	−7.3	36.0	−50.2
December	89.0	117.0	31.5	39.5	−55.6
January	111.5	132.0	18.4	56.0	−49.8
February	170.5	183.0	7.3	130.3	−23.6
March	298.5	329.0	10.2	175.0	−41.4
April	325.3	372.0	14.4	176.3	−45.8
May	190.0	190.0	0.0	115.5	−39.2
June	109.5	87.0	−20.5	44.8	−59.1
July	85.0	56.0	−34.1	30.5	−64.1
August	75.0	47.0	−37.3	19.5	−74.0
September	71.0	41.0	−42.3	16.5	−76.8

7.3.3 Results and Discussions

The unimpaired daily flows observed between 1955 and 1982 at the DHS vary between 1 and 5,816 m^3/s. The long-term average and median daily flows were 173.10 and 108 m^3/s, respectively. The ratios between the maximum flow value and the long-term mean and median flows were 34:1 and 54:1, correspondingly. The flow variations during the natural flow regime can be mainly attributed to the deviations in annual precipitation. The range of hydrologic alterations is a function of the operated storage reservoirs, diversions, return flows, type of the water year, unregulated tributaries and annual watershed discharge.

Close examinations of Table 7.12 and Figure 7.20 reveal that the non-rainy months have significant reductions in median flow rates. The rates of reduction ranged from about –38% in June to nearly −63% in September. The median of the calculated reduction rates over this time window was estimated at –51%. This proportion represents the fraction of the long-term streamflow, which is no longer available to the downstream country during the dry season. Between November and May, the rates of anomaly varied between about −11.0% in April and –31.5% in December. An increase in median flow rate of about 4.3% was observed in February. Hence, this value was discarded while determining the alteration rate. The long-term annual median flows of the unaltered and artificially altered flow records were 99.25 and 64.50 m^3/s, respectively, indicating a reduction rate of about –35%. This is clear evidence of the impacts of the upstream human-induced river regulation on the water availability in the downstream nation.

Considering the impaired time series, two-time frames can be distinguished in terms of degree of flow alteration. The period of low level of river artificial regulation between 1983 and 2003, during which limited damming and water exploitation occurred. The total storage volume of reservoirs, and the total irrigation and fish farm areas that were put in place between 1983 and 1994 had capacities of 321 million m^3, and 5,968 ha and 810 ha, respectively. The currently available information for this case study (Figure 7.21 and Table 7.12) indicates that no considerable damming and water

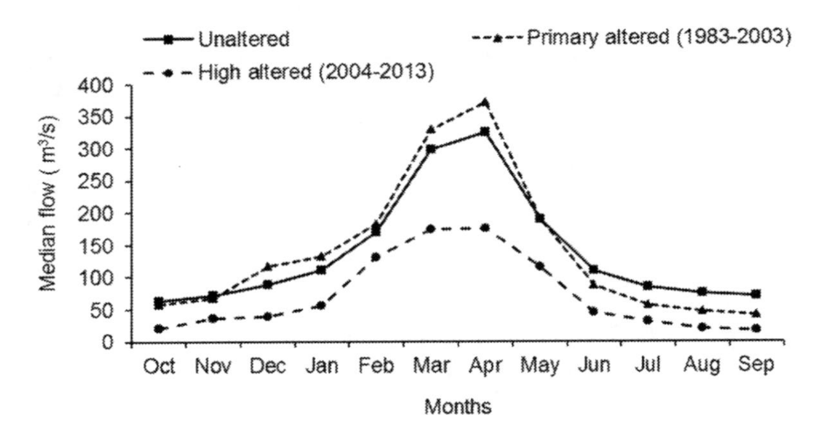

FIGURE 7.21 Long-term monthly median flow variation for three time frames.

abstraction were observed between 1995 and 2003. The water years between 2004 and 2013 represent the period of high level of man-made regulation during which substantial hydraulic works and severe water withdrawal and flow diversions were allocated. Between 2004 and 2013, storage reservoirs with capacities of nearly 1,500 million m^3 were operated and large-scale landscape modifications were implemented.

Findings demonstrating the three examined periods, natural (1955–1982), low level of man-made alteration (1983–2003) and high level of anthropogenic regulation (2004–2013), are illustrated in Table 7.13 and Figures 7.21 and 7.22. Results reveal that the impact of the human-made regulation system during the period of low level of artificial regulation was considerably limited to the dry season. The rate of flow deficiency ranged from –7.3% in November to −42.3% in September. For the non-rainy months June to September, the rates of flow anomaly varied between –20.5% and −42.3%. A growing trend in median flow was observed between December and April. The rates of increase ranged from +7.3% in February to +31.5% in December. No deviation was noticed in May.

The long-term monthly rates of flow reduction during the period of high level of human-induced alteration were far higher than those observed during the period of low level of regulation. The rates of flow reduction ranged from nearly –24% in February to about –77% in September. The median of the decline rates over the months June to October was estimated at −66.4%. The considerably high proportions of flow reduction can be assigned to upstream anthropogenic disturbances and dry climate conditions, which indicate a sharp decrease in the volume of water entering the downstream state.

Figure 7.23 shows the annual variations from the long-term mean annual unaltered flow. The anomalies in annual flow values over the period 1983–1998 fluctuated between −43.2% in 1984 and +79.9% in 1988. The annual flow values over the period 1999–2013 were lower than the long-term mean yearly unaltered flow. The anomalies ranged from –21% in 2003 to –79.1% in 2008. The mean annual anomaly over this period was estimated at –54.5%. Such fluctuation in anomaly values can be attributed to a combination of impacts associated with upstream anthropogenic river regulation and the characteristics of water years (dry, normal or wet). For instance, the historical flow records show that the year 1988 was a flood year, while the years 1999–2001 and 2008–2009 were registered as dry years. However, this does not weaken the prominent influence of upstream human-induced activities on the volume of water available to the downstream country.

7.3.4 Conclusions and Recommendations

The transboundary three-scalar framework has been successfully demonstrated by selecting the Diyala watershed as a representative case study. Upstream hydraulic structures and enormous water abstraction schemes have adversely impacted on the downstream multi-water uses and considerably modified the flow regime in the downstream state of Iraq. This is manifested by the significant rates of flow alteration observed for the downstream state.

Findings reveal that even with the low level of man-made regulation, the influence on water availability could be harmful to the need of Iraq, particularly during

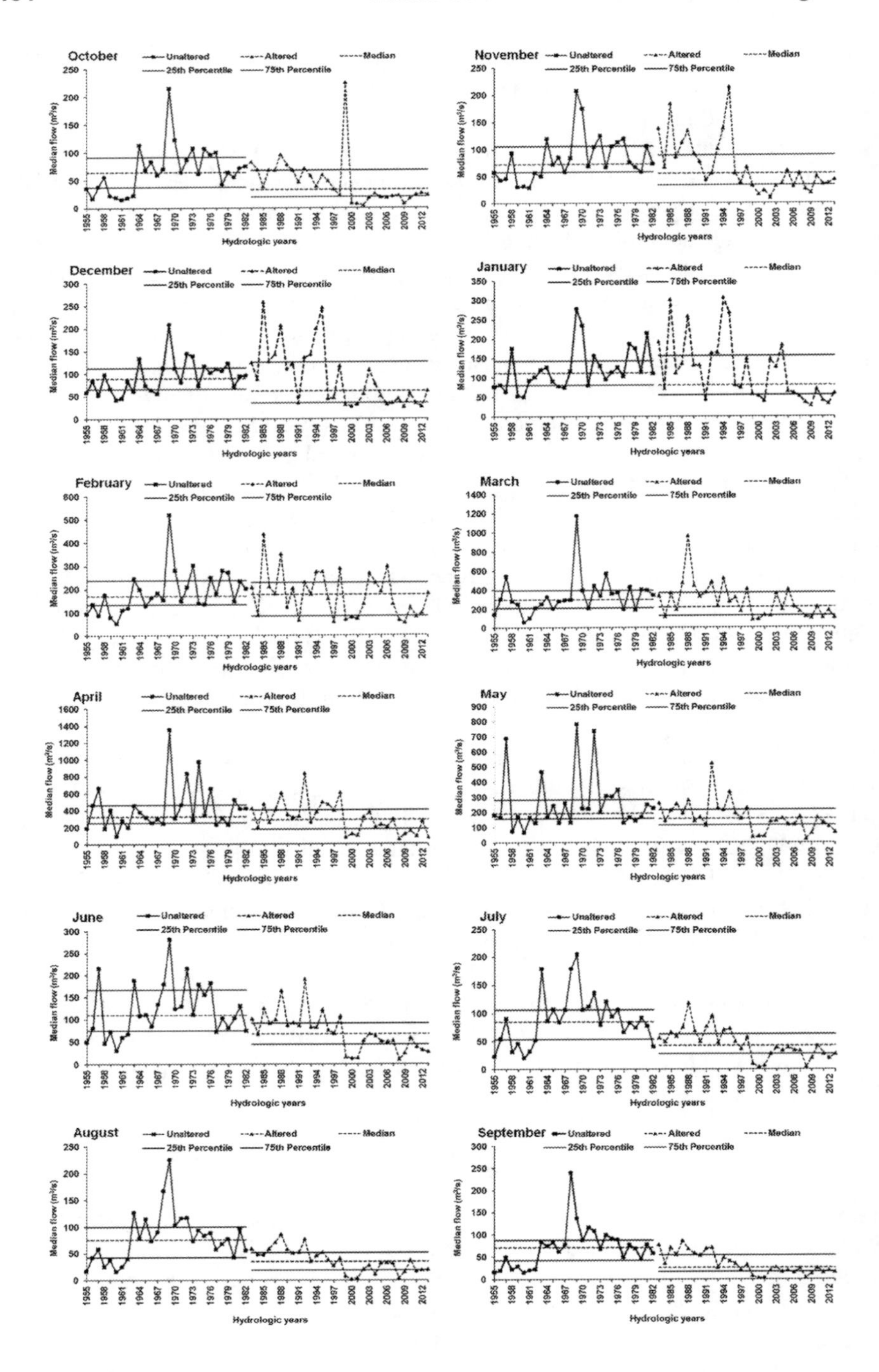

FIGURE 7.22 Long-term temporal hydrologic deviation (parts a to l) for October to September.

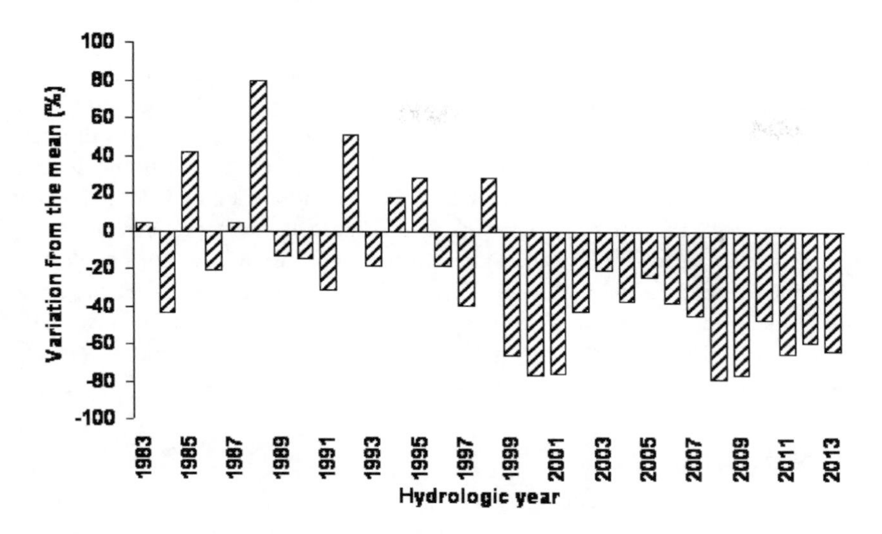

FIGURE 7.23 Annual variations from the long-term mean annual unaltered flow.

the non-rainy period. For the non-rainy months June to September, the rates of flow alteration ranged from −20.5 to −42.3%. The years between 2004 and 2013 were the most prominent dam building, water withdrawal and diversion period, which had profound adverse implications on the natural hydrologic regime observed in the downstream country. Correspondingly, the long-term monthly rates of reduction were notably higher than those during the period of low level of artificial regulation. The proportions of flow anomaly ranged from nearly −24% in February to about −77% in September. The median of the alteration rates over the months June to October was estimated at −66.4%. The mutual impact of upstream anthropogenic stressors and dry climate conditions was manifested by high rates of flow reduction.

The considerable rates of flow reduction require the downstream country (Iraq in this case) to improve its understanding with respect to the adverse impacts resulting from an upstream extensive water control and diversion infrastructure. Moreover, this country should identify and assess all previous failure experiences in the management of water within the watershed.

Minimum demands from both existing and potential future users need to be clearly identified and evaluated in relation to current and future flow reduction. It is noteworthy to highlight that the way the existing dams across the river in the downstream state are currently operating needs to be comprehensively reviewed. Reservoir operation guides should be updated to reduce adverse environmental impacts caused by changing the hydrological regime. A revision of the rule or guide curves does not necessarily affect the efficiency of existing reservoirs in terms of their main functions, namely irrigation, flood protection and hydropower generation. Improving the rule curves of the existing multi-purpose reservoirs may offer considerable scope for minimizing adverse impacts.

The author recommends further research on the impact of climate change at the transboundary scale to assess the possible changes in rainfall patterns and runoff as well as the frequency and intensity of extreme events. It is also recommended to improve the water governance, developing strong communication and efficient networks between the upstream and downstream riparian states, which may lead to the signing of international water treaties. It is also recommended to minimize water resources management conflicts between the federal government, provinces and other stakeholders at national level. This requires appropriate monitoring systems and transboundary coordination on methodologies for collecting and sharing basic data on climate and water resources such as river discharges at diversions, water quality and groundwater levels. A concrete water agreement also requires warning and alarming systems, which support the taking of right decisions.

REFERENCES

Abdallah, H.M., 2010. The problem of water in the province of Diyala and rationalization of water consumption. *Journal of Research Diyala Humanity*, **46**, 101–163 (in Arabic).

Abu-sharar, T.M., Hussenin, I.A., Al-Jayyousi, O.R., 2003. The use of treated sewage for irrigation in Jordan: Opportunities and constrains, *Journal of the Chartered Institute for Water and Environmental Management*, **17**(4), 232–238.

Agro Vision Holland, 1993. *Water Sanitation Performance of Waste Treatment Plants in Yemen*, Report to the Ministry of Electricity and Water (Sanaa: Agro Vision Holland).

Alcalde, L., Oron, G., Gillerman, P., Salgot, M. and Manor, Y., 2003. Removal of faecal coliforms, somatic coliphages and F-specific bacteriophages in a stabilization pond and reservoir system in arid regions, *Water, Science and Technology: Water Supply*, **3**(4), 177–184.

Al-Faraj, F.A.M. and Scholz, M., 2014. Incorporation of the flow duration curve method within digital filtering algorithms to estimate the baseflow contribution to total runoff. *Water Resources Management*, **28**(15), 5477–5489.

Al-Faraj, F.A.M. and Scholz, M., 2015. Impact of upstream anthropogenic river regulation on downstream water availability in transboundary river watersheds, *International Journal of Water Resources Development*, **31**(1), 28–49.

Al-Faraj, F.A.M., Scholz, M. and Tigkas, D., 2014. Sensitivity of surface runoff to drought and climate change: Application for Shared River Basins. *Water*, **6**, 3033–3048.

Ali, A., 1983. Reuse of wastewater from an Islamic Prospective, *Journal of Irrigation and Drainage Engineering - American Society of Civil Engineering*, **113**(2), 173–183.

Allen, R.G., Pereira, L.S., Raes, D. and Smith, M., 1998. Crop evapotranspiration: Guidelines for computing crop water requirements. Food and Agriculture Organization (FAO) Irrigation and Drainage Paper 56 (Rome: FAO).

Almas, A.A.M., 2001. *A Socio-Economic and Management Study of Water Resources in the Yemen*, MPhil Dissertation (Dundee: University of Abertay).

Almas, A.A.M., and Jefferies, C., 2001. Water resources and supply system in Yemen: Southern Yemen economic development and water as a physical basis (protocols from an excursion to Aden and Hadhramaut), *Würzburger Geographische Manuskripte*, **54**, 189–204.

Almas, A.A.M. and Scholz, M. 2006. Potential for wastewater reuse in irrigation: Case study from Aden (Yemen), *International Journal of Environmental Studies*, **63**(2), 131–142.

Almasi, A., and Pescod, M.B., 1996. Wastewater treatment mechanisms in anoxic stabilization ponds, *Water, Science and Technology*, **33**(7), 125–132.

Aliasghari, H. and Ahadi, B., 2007. Effect of anisotropy on rock mass deformation modulus in Javeh dam site. In: *11th ACUUS Conference on Underground Space: Expanding the Frontiers*, Athens, Greece.

American Public Health Association, 1995. *Standard Methods for the Examination of Water and Wastewater*, 19th edition (Washington DC: American Public Health Association, American Waterworks Association and Water and Environment Federation).

Arnell, N.W. and Charlton, M.B., 2009. Adapting to the effects of climate change on water supply reliability. In: W. N. Adger, I. Lorenzoni, and K. L. O'Brien (Eds.) *Adapting to Climate Change: Thresholds, Values, Governance* (Cambridge: Cambridge University Press) pp. 42–53.

Arnell, N.W. and Chunzhen, L., 2001. *Hydrology and Water Resources*. International Panel on Climate Change, Third Assessment Report-Working Group II: Impacts, Adaptation and Vulnerability. Available at https://www.ipcc.ch/ipccreports/tar/wg2 (accessed 12 April 2024).

Arnold, J.G. and Allen, P.M., 1999. Validation of automated methods for estimating baseflow and groundwater recharge from stream flow records. *Journal of the American Water Resources Association*, **35**(2), 411–424.

Arthur, J.P., 1983. *Note on the Design and Operation of Waste Stabilization Ponds in Warm Climates of Developing Countries* (Washington, DC: World Bank).

Asadi, M., Ghaderzadeh, H. and Seirafi, F., 2012. Economic valuation of the Zarivar Lake, Iran, *Lakes Reservoirs and Ponds*, **6**(2), 154–164.

Bahari, A., and Brssaud, F., 1996. Wastewater reuse in Tunisia: Assessing a national policy, *Water Science and Technology*, **33**(10–11), 87–94.

Bazza, M., 2003. Wastewater recycling and reuse in the Near East Region: experiences and issues, *Water Science and Technology: Water Supply*, **3**(4), 33–50.

Bloetscher, F., 2012. Protecting people, infrastructure, economies, and ecosystem assest; Water management in the face of climate change, *Water*, **4**(2), 367–388.

Bozorgnia, A., Youssefi, M.R., Barzegar, M., Hosseinifard, S.M. and Ebrahimpour, S., 2012. Biodiversity of parasites of fishes in Gheshlagh (Vahdat) Reservoir, Kurdistan Province, Iran, *World Journal of Fish and Marine Sciences*, **4**(3), 249–253.

Brels, S., Coates, D. and Loures, F., 2008. Transboundary water resources management: the role of international watercourse agreements in implementation of the CBD. CBD Technical Series no. 40 (Montreal: Secretariat of the Convention on Biological Diversity).

Brodie, R.S. and Hostetler, S., 2005. A review of techniques for analyzing base-flow from stream hydrographs. In: *Proceedings of the NZHSIAH-NZSSS 2005 Conference*, 28 November–2 December 2005, Auckland, New Zealand.

Calero, C.X, Mara, D.D. and Pena, M.R., 2000. Anoxic ponds in the sugar cane industry: a case study from Colombia, *Water, Science and Technology*, **42**(10–11), 67–74.

Carr, R.M., Blumenthal, U.J. and Mara, D.D., 2004. Guidelines for the safe use of wastewater in agriculture: Revisiting WHO guidelines, *Water Science and Technology*, **50**(2), 31–38.

Chapman, T., 1999. A comparison of algorithms for streamflow recession and baseflow separation, *Hydrological Processes*, **13**(5), 701–714.

Conway, D. 2005. From headwater tributaries to international river: observing and adapting to climate variability and change in the Nile watershed. *Global Environmental Change – Human and Policy Dimensions*, **15**(2), 99–114.

Council of the European Communities, 1991. *Official Journal of the European Communities*, No. L 135/40–52 (30 May 1991).

Curtis, T.P., Mara, D.D. and Sliva, S., 1992. Influence of pH, oxygen and humic substances on the ability of sunlight to damage faecal Coliforms in waste stabilization pond water, *Applied and Environmental Microbiology*, **58**(4), 1335–1343.

Eckhardt, K., 2005. How to construct recursive digital filters for baseflow separation, *Hydrological Processes*, **19**(2), 507–515.

Eisenreich, S.J., 2005. Climate change and the European water dimension. A Report to the European Water Directors. EU Report No. 21553 (Ispra: European Commission-Joint Research Centre).

Fadhil, M.A., 2011. Drought mapping using Geoinformation technology for some sites in the Iraqi Kurdistan region, *International Journal of Digital Earth*, **4**(3), 239–257.

Fan, Y., Chen, Y., Liu, Y. and Li, W., 2013. Variation of baseflows in the headstreams of the Tarim River Basin during 1960–2007, *Journal of Hydrology*, **487**, 98–108.

FAO, 2012. Food and Agriculture Organization of the UN. Adaptation to climate change in semi-arid environments. Experience and lessons from Mozambique. Environment and Natural Resources Management Series. Rome. pp. 1–83. https://www.fao.org/docrep/015/i2581e/i2581e00.pdf. (accessed 15 April 2024).

Faramarzi, M., Abbaspour, K.C., Schulin, R. and Yang, H., 2009. Modelling blue and green water resources availability in Iran, *Hydrological Processes*, **23**(3), 486–501.

Farooq, S., and Ansari, Z., 1983. Wastewater reuse in Muslim countries: An Islamic prospective, *Journal of Environmental Management*, **7**(2), 119–123.

Foehn, A., García Hernández, J., Roquier, B. and Paredes Arquiola, J., 2016. *RS MINERVE – User's manual v2.6*. RS MINERVE Group, Switzerland.

Foppen, J.W.A., 2003. Impact of high-strength infiltration on groundwater quality and drinking water supply: A case study of Sanaa in Yemen, *Journal of Hydrology*, **203**(2), 198–216.

GADM, Global Administrative Areas Database, 2012. Boundaries without limits. https://www.gadm.org (accessed 11 April 2024).

Gain, A.K., Giupponi, C. and Renaud, F.G., 2012. Climate change adaptation and vulnerability assessment of water resources systems in developing countries: A generalized framework and a feasibility study in Bangladesh, *Water*, **4**(2), 345–366.

Gao, Y., Vogel, R.M., Kroll, C.N., Poff, N.L. and Olden, J.D., 2009. Development of representative indicators of hydrologic alteration, *Journal of Hydrology*, **374**(1–2), 136–147.

Gitec Consultants and Dorch Consultants, 1997. *Aden Sewerage Project: Final Draft Design* (Düsseldorf: Gitec Consultants and Dorch Consultants).

GLCF, Global and Land Cover Facility, 2015. Earth science data interface. https://www.sciencebase.gov/catalog/item/4f4e4ae0e4b07f02db687f5d

Gordon, N.C., McMahon, T.A., Finlayson, B.L., Gippel, C.J. and Nathan, R.J., 2004. *Stream Hydrology: An Introduction for Ecologists*, 2nd edition (Chichester: John Wiley & Sons).

Goulden, M., Conway, D. and Persechino, A., 2009. Adaptation to climate change in international River watersheds in Africa: A review. *Hydrological Sciences Journal*, **54**(5), 805–828.

Harza and Binnie, 1958. *Hydrological Survey of Iraq: Discharge Records at Selected Gauging Stations*, report.

Harza and Binnie, 1959. *Hydrological Survey of Iraq: Discharge Records at Selected Gauging Stations in Iraq*, report.

Harza and Binnie, 1963. *Hydrological Survey of Iraq: Main Report.*

He, S., Li, S., Xie, R. and Lu, J., 2016. Baseflow separation based on a meteorology-corrected nonlinear reservoir algorithm in a typical rainy agricultural watershed, *Journal of Hydrology*, **535**, 418–428.

Hiscock, K.M., Rivett, M.O. and Davison, R.M., 2002. Sustainable groundwater development, *Geological Society, London, Special Publications*, **193**, 1–14.

Hussien, G. and Al-Saati, B. 1999. Wastewater quality and its reuse in agriculture in Saudi Arabia, *Desalination*, **123**(3), 241–251.

HydroOffice, 2015. https://hydrooffice.org/Downloads?Items=Software (accessed 11 April 2024).

Information Technology Services (ITS), 2016. IBM SPSS statistics 23 Part 3: Regression analysis. Winter 2016, Version 1.

IPCC, 2001. *Climate Change 2001: Synthesis Report.* https://www.ipcc.ch/report/ar3/syr/

IPCC, 2007. *Climate Change 2007: Impacts, Adaptation and Vulnerability. Contribution of Working Group II to the Fourth Assessment Report of the Intergovernmental Panel on Climate Change,* M. L. Parry, O. F. Canziani, J. P. Palutikof, P. J. van der Linden and C. E. Hanson (Eds.) (Cambridge and New York: Cambridge University Press) pp. 1–976.

IPCC, 2014. *Climate Change 2014: Impacts, Adaptation, and Vulnerability.* https://www.ipcc.ch/report/ar5/wg2 (accessed 5 April 2024).

Isdle, 2009. *Darian Dam Construction of the River Sirwan*, news item (in Arabic).

IUCN, 2000. *Vision for Water and Nature: A World Strategy for Conservation and Sustainable Management of Water Resources in the 21st Century* (Gland and Cambridge: International Union for Conservation of Nature (IUCN)).

JICA, 1990, *Study on the Improvement of the Maalla and Tawahi Sewerage System in Aden*, Final Report, Vol. 3, (Sanaa: JICA).

Jyane Engineering Construction Company, 2013. Javeh RCC dam.

Kaleris, V., and Langousis, A., 2017. Comparison of two rainfall-runoff models: Effects of conceptualization, model calibration and parameter variability, *Hydrological Sciences Journal*, **62**(5), 729–748.

Kistin, E.J. and Ashton, P.J., 2008. Adapting to change in transboundary rivers: An analysis of treaty flexibility on the Orange-Senqu River watershed, *International Journal of Water Resources Development*, **24**(3), 385–400.

Lim, K.J., Engel, B.A., Tang, Z., Choi, J., Kim, K.S., Muthukrishnan, S. and Tripathy, D., 2005. Automated web GIS based hydrograph analysis tool, WHAT, *Journal of the American Water Resources Association*, **41**(6), 1407–1416.

Linsley, R.K., Kohler, M.A. and Paulhus, J.L.H., 1988. *Hydrology for Engineers* (London: McGraw-Hill).

Lott, D.A. and Stewart, M.T., 2016. Baseflow separation: A comparison of analytical and mass balance Methods, *Journal of Hydrology*, **535**, 525–533.

Lu, S., Wu, B., Wei, Y., Yan, N. and Wang, H., 2015. Quantifying impacts of climate variability and human activities on the hydrological system of the Haihe River Basin, China, *Hydrology and Earth System Sciences*, **73**, 1491–1503.

MA and WR-KRG, 2013. General Directorate of Dams and Reservoirs and Directorate for Operation of Derbandikhan Dam, Ministry of Agriculture (MA) and Water Resources-Kurdistan Reginal Government (WR-KRG) (Erbil: MA and WR-KRG).

Mahab Ghodss Consulting Engineering Company, 2008. *Integrated Studies of Water Resources and Consumption in Garmsiri Watershed in Kermānshāh and Ilam Provinces Accompanied by Impact of Transferred Water from Sirvan Watershed.*

Mahab Ghodss Consulting Engineering Company, 2013a. *Gavoshan Dam and Conveyance Tunnel.*

Mahab Ghodss Consulting Engineering Company, 2013b. *Nosoud Water Tunnel* (in Arabic).

Mara, D.D., 1997. *Design Manual for Waste Stabilization Ponds in India* (Leeds: TPHE Publications).

Mara, D.D., 2000. The production of microbiologically safe effluent for wastewater reuse in the Middle East and North Africa, *Journal of Water, Air and Soil Pollution*, **123**(4), 595–603.

Mara, D.D., 2004. Natural sewerage treatment in the UK: Selection guidelines, *Water and Environment Journal (formerly Journal of the Chartered Institution of Water and Environmental Management)*, **18**(4), 41–48.

Marcuello, C. and Lallana, C., 2003. *Indicator Fact Sheet — (WQ01c) Water Exploitation Index*, European Environment Agency. https://www.eea.europa.eu/data-and-maps/indicators/water-exploitation-index/water-exploitation-index (accessed 16 April 2024).

Mbwele, L., Rubindamayugi, M., Kivaisi, A. and Dalhammar, G., 2004. Performance of a small wastewater stabilization pond system in a tropical climate; Dar Es Salam, Tanzania, *Water, Science and Technology*, **48**(11–12), 187–191.

Mei, Y. and Anagnostou, E.N., 2015. A hydrograph separation method based on information from rainfall and runoff records. *Journal of Hydrology*, **523**, 636–649.

Metwali, R.M., 2003. Water quality of some wells in Taiz City (Yemen) and its surroundings, *Folia Microbiologica*, **48**(1), 90–94.

Miller, K. and Yates, D., 2006. *Climate Change and Water Resources: A Primer for Municipal Water Providers*, report (Denver: Amearican Water Works Association Research Foundation) pp. 1–83.

Miller, M.P., Buto, S.G., Susong, D.D. and Rumsey, C.A., 2016. The importance of baseflow in sustaining surface water flow in the Upper Colorado River Basin, *Water Resources Research*, **52**(5), 3547–3562.

Ministry of Energy, 2003. *Long-term Development Strategies for Iran's Water Resources* (Theran: Iran Water Resources Management Company Islamic Republic of Iran).

Ministry of Water Resources, 1982. *General Scheme of Water Resources and Land Development in Iraq – Main Report*. Stage II: Main Report (Baghdad: Ministry of Water Resources).

Ministry of Planning, 2003. *Statistical Year Book for the Year 2004* (Sanaa: Ministry of Planning).

Ministry of Water and Irrigation, 1998. Wastewater Management Policy, Policy Paper No 2, The Hashemite Kingdom of Jordan (Amman: Ministry of Water and Irrigation).

Mohammed, R. and Scholz, M., 2016. Impact of climate variability and streamflow alteration on groundwater contribution to the baseflow of the Lower Zab River (Iran and Iraq), *Environmental Earth Sciences*, **75**(1392), 1–11.

Mohammed, R. and Scholz, M., 2017a. The reconnaissance drought index: a method for detecting regional arid climatic variability and potential drought risk, *Journal of Arid Environment*, **144**, 181–191.

Mohammed, R. and Scholz, M., 2017b. Adaptation strategy to mitigate the impact of climate change on water resources in arid and semi-arid regions: A case study, *Water Resources Management*, **31**(11), 3557–3573.

Mohammed, R. and Scholz, M., 2017c. Impact of evapotranspiration formulations at various elevations on the reconnaissance drought index, *Water Resources Management*, **31**, 531–538.

Mohammed, R. and Scholz, M., 2018a. Flow duration curve integration into digital filtering algorithms simulating climate variability on river baseflow, *Hydrological Sciences Journal*, **63**(10), 1558–1573.

Mohammed, R. and Scholz, M., 2018b. Climate change and anthropogenic intervention impact on the hydrologic anomalies in a semi-arid area: Lower Zab River Basin, Iraq, *Environmental Earth Sciences*, **77**(10), 357.

Mohammed, R., Scholz, M., Nanekely, M.A. and Mokhtari, Y., 2018. Assessment of models predicting anthropogenic interventions and climate variability on surface runoff of the Lower Zab River, *Stochastic Environmental Research and Risk Assessment*, **32**(1), 223–240.

Mohammed, R., Scholz, M. and Zounemat-Kermani, M., 2017. Temporal hydrologic alterations coupled with climate variability and drought for transboundary river basins, *Water Resources Management*, **31**(10), 1489–1502.

Monk, W.A., Wood, P.J., Hanneh, D.M. and Wislon, D.A., 2007. Selection of river flow indices for the assessment of hydroeclogical change, *River Research and Applications*, **23**(1), 113–122.

Montenegro, S. and Ragab, R., 2012. Impact of possible climate and land use changes in the semi arid regions: A case study from North Eastern Brazil, *Journal of Hydrology*, **434–435**, 55–68.

Mulder, G., Olsthoorn, T.N., Al-Manmi, D.A.A., Schrama, E.J.O. and Smidt, E.H. 2015. Identifying water mass depletion in northern Iraq observed by GRACE, *Hydrology and Earth System Sciences*, **19**(3), 1487–1500.

Nathan, R.J. and McMahon, T.A., 1990. Evaluation of automated techniques for baseflow and recession analysis, *Water Resources Research*, **26**(7), 1465–1473.

National Water Supply and Sanitation Authority, 2002. *Statistical Year Book for the Year 2002* (Aden: National Water Supply and Sanitation Authority).

National Water Supply and Sanitation Authority, 2003. *Statistical Year Book for the Year 2003* (Aden: National Water Supply and Sanitation Authority).

National Water Supply and Sanitation Authority, 2004. *Statistical Year Book for the Year 2004* (Aden: National Water Supply and Sanitation Authority).

Nelson, L.K., Cisneros, B.J., Tchobanoglous, G. and Jeannie, L., 2004. Sludge accumulation, characteristics and pathogen inactivation in four primary stabilization ponds in Central Mexico, *Water Research*, **38**(1), 111–127.

Olden, J.D. and Poff, N.L., 2003. Redundancy and the choice of hydrologic indices for characterizing streamflow regimes, *River Research and Applications*, **19**(2), 101–121.

Partington, D., Brunner, P., Simmons, C.T., Werner, A.D., Therrien, R., Maier, H.R. and Dandy, G.C., 2012. Evaluation of outputs from automated baseflow separation methods against simulated baseflow from a physically based, surface water-groundwater flow model, *Journal of Hydrology*, **458–459**, 28–39.

Pearson, H. and Mara, D.D., 1998. *Design Manual for Waste Stabilization Ponds in Mediterranean Countries* (Leeds: TPHE Publication).

Piggott, A.R., Moin, S. and Southam, C., 2005. A revised approach to the UKIH method for the calculation of baseflow/Une approche améliorée de la méthode de l'UKIH pour le calcul de l'écoulement de base. *Hydrological Sciences Journal*, **50**(5), 911–920.

Pooyab Consulting Engineers, 2013. *Garan Dam and Water Conveyance System*.

Price, K., 2011. Effects of watershed topography, soils, land use, and climate on baseflow hydrology in humid regions: A review, *Progress in Physical Geography*, **35**(4), 465–492.

Ragab, R. and Prudhomme, C., 2002. Climate change and water resources managment in arid and semi-arid regions: Prospective and challenges for the 21st Century, *Biosystem Engineering*, **81**(1), 3–34.

Ratchford, I.A.L. and Fallowfield, H.J., 2003. The effect of light and dark cycles of medium frequency on photosynthesis by *Chlorella vulgaris* and the implications for waste stabilization pond design and performance, *Water Science and Technology*, **48**(2), 69–74.

Reis, J., Culver, T.B., Block, P.J. and McCartney, M.P., 2016. Evaluating the impact and uncertainty of reservoir operation for malaria control as the climate changes in Ethiopia, *Climatic Change*, **136**, 601–614.

Robaa, S.M. and AL-Barazanji, Z.J., 2013. Trends of annual mean surface air temperature over Iraq, *Nature and Science*, **11**(12), 138–145.

Rosenberg, D.M., McCully, P. and Pringle, C.M., 2000. Global-scale environmental effects of hydrological alterations: Introduction, *BioScience*, **50**(9), 746–751.

RS MINERVE 2.5 Software, 2016. https://www.crealp.ch/down/rsm/install2/archives.html (accessed 11 April 2024).

Rumsey, C.A., Millera, M.P., Susonga, D.D., Tillman, F.D. and Anning, D.W., 2015. Regional scale estimates of baseflow and factors influencing base flow in the Upper Colorado River Basin, *Journal of Hydrology: Regional Studies*, **4**, 91–107.

Sad Afzar Engineering Service Company, 2013. news item.

Salim, J.B., Bidhendi, N.G., Salemi, A., Taheryioun, M. and Ardestani, M., 2009. Water quality assessment of Gheshlagh river using water quality indices, *Environmental Sciences*, **6**(4), 19–28.

Salimeh, R., 2012. Nasoud Tunnel-West Kermānshāh. *Iranian Journal of Mining Engineering*, **7**(15), 1–14.

Scholz, M., Morgan, R. and Picher, A., 2005. Stormwater resources development and management in Glasgow: Two case studies, *International Journal of Environmental Studies*, **62**(3), 263–282.

Shiau, J.T. and Wu, F.C., 2004. Assessment of hydrologic alterations caused by Chi-Chi diversion weir in Chou-Shui Creek, Taiwan, *River Research and Applications*, **20**(4), 401–412.

Shobha Shetty, 2005. Treated Wastewater Use in Tunisia: Lessons Learned and Road head. https://doi.org/10.1079/9780851998237.0163

Shourian, M., Mousavi, S.J., Menhaj, M.B. and Jabbari, E., 2008a. Neural-network-based simulation-optimization model for water allocation planning at basin scale, *Journal of Hydroinformatics,* **10**(4), 331–343.

Shourian, M., Mousavi, S.J. and Tahershamsi, A., 2008b. Basin-wide water Resources planning by integrating PSO Algorithm and MODSIM, *Water Resources Management*, **22**(10), 1347–1366.

Smakhtin, V.U., 2001. Low flow hydrology: A review, *Journal of Hydrology*, **240**(3–4), 147–186.

Smith, P.G. and Scott, J.S., 2002. *Dictionary of Waste and Waste Management* (London: International Water Association, and Butter, Worth and Heinemann Books).

Stewart, M.K., 2015. Promising new baseflow separation and recession analysis methods applied to streamflow at Glendhu Catchment, New Zealand, *Hydrology and Earth System Sciences*, **19**(6), 2587–2603.

Soundharajan, B.S., Adeloye, A.J. and Remesan, R., 2016. Evaluating the variability in surface water reservoir planning characteristics during climate change impacts assessment, *Journal of Hydrology*, **538**, 625–639.

Tabari, H. and Taalaee, P.H., 2011. Analysis of trend in temperature data in arid and semi-arid regions of Iran, *Global and Planetary Change*, **79**(1–2), 1–10.

Tadesse, I., Green, F.B., and Puhakka, J.A., 2004. Seasonal and diurnal variation of temperature, pH and dissolved oxygen in advanced integrated wastewater pond systems treating tannery effluent, *Water Research*, **38**(3), 645–654.

Tallaksen, L.M. and van Lanen, H.A., 2004. *Hydrological Drought – Processes and Estimation Methods for Streamflow and Groundwater*, Developments in Water Sciences 48 (Amsterdam: Elsevier).

The Nature Conservancy, 2009. Indicators of hydrologic alteration version 7.1 user's manual. The Nature Conservancy, June, 76.

Tigkas, D., 2008. Drought characterisation and monitoring in regions of Greece, *European Water*, **23**, 29–39.

Tigkas, D., Vangelis, H. and Tsakiris, G., 2012. Drought and climatic change impact on streamflow in small watersheds, *Science of the Total Environment*, **440**, 33–41.

Tokmechi, Z., 2011. The probability of Khuzestan environmental pollution due to seismic response of Azad Dam, *Research Journal of Fisheries and Hydrobiology*, **6**(3), 291–305.

Tompkins, E.L., Boyd, E., Nicholson-Cole, S.A., Weatherhead, K., Arnell, N.W. and Adger, W.N., 2009. *An Inventory of Adaptation to Climate Change in the UK: Challenges and Findings*. Working Paper 135, Tyndall Centre for Climate Change Research (Norwich: University of East Anglia).

Tsakiris, G. and Vangelis, H., 2005. Establishing a drought index incorporation evapotranspiration, *European Water*, **9**, 3–11.

Turtun P, 1999. Isalmic approach to demand management.

UNDESA, 2014. *International Decade for Action "Water for Life 2005–2015"*. The United Nations Department of Economic and Social Affairs. https://www.un.org/waterforlifedecade (accessed 15 March 2024).

UNDP-Iraq, 2011. Drought Impact Assessment, Recovery and Mitigation Framework and Regional Project Design in Kurdistan Region. https://reliefweb.int/report/iraq/iraq-drought-impact-assessment-recovery-and-mitigation-framework-and-regional-project

UNECE, 2009. *Guidence on Water and Adaptation to Climate Change*. United Nations Economic Commission for Europe (UNECE), Convention on the Protection and Use of Transboundary Watercourses and International Lakes. https://www.unece.org/fileadmin/DAM/env/water/publications/documents/Guidance_water_climate.pdf (accessed 13 April 2024).

UNECE, 2011. *Second Workshop on Adaptation to Climate Change in Transboundary Watersheds: Challenges, Progress and Lessons* (Beirut: United Nations Economic Commission for Europe (UNECE)).

UNESCO, 2014. *United Nations Educational, Scientific and Cultural Organization. Integrated Drought Risk Management-DRM National Framework for Iraq, An Analysis Report* (Beirut: United Nations Economic and Social Commission for Western Asia; Bundesanstalt für Geowissenschaften und Rohstoffe, Inventory of Shared Water Resources in Western Asia). https://unesdoc.unesco.org/images/0022/002283/228343E.pdfUN-ESCWA and BGR

Uzomah, V.C. and Scholz, M., 2002. Water availability assessment and corresponding public health implications for a rural area in Nigeria, *Journal of the Chartered Institution of Water and Environmental Management*, **16**(4), 296–299.

Vangelis, H., Tigkas, D. and Tsakiris, G., 2013. The effect of PET method on reconnaissance drought index (RDI) calculation, *Journal of Arid Environment*, **88**, 130–140.

Veenstra, S., Al Nozaily, F.A. and Alaerts, G.J., 1995. Purple non-sulphur bacteria and their influence on waste stabilization pond performance in the Republic of Yemen, *Water Science Technology*, **31**(12), 141–149.

Waheed, S.Q., 2013. Effects of climate change on water availability in Diyala watershed in Iraq. In: *Proceedings of the 2nd International Conference on Hydrology and Groundwater Expo*, August 26–27, 2013, USA. Available at https://omicsonline.org/2157-7587/2157-7587-S1.010-023.pdf. (accessed 7 April 2024).

Wan, L., Xia, J., Hong, S., Bu, H., Ning, L. and Chen, J., 2015. Decadal climate variability and vulnerability of water resources in arid regions of Northwest China, *Environmental Earth Sciences*, **73**, 6539–6552.

Welderufael, W. and Woyessa, Y., 2010. Stream flow analysis and comparison of methods for baseflow separation: Case study of the Modder River basin in central South Africa, *European Water*, **8**(2), 107–119.

World Bank, 2014. *The World Bank Climate Change Knowledge Portal*. https://sdwebx.worldbank.org/climateportal/index.cfm?page=country_historical_climateandThisRegion=AsiaandThisCCode=IRQ (accessed 7 April 2024).

World Health Organization, 1987. *Waste Treatment Stabilization Ponds, Principals of Planning and Practice*, WHO/EMRO Technical Publication No. 10 (Geneva: World Health Organization).

World Health Organization, 1989. *Health Guidelines for the Use of Wastewater in Agriculture and Aquaculture*, Technical Report Series No. 778 (Geneva: World Health Organization).

World Health Organization, 1993. *Guidelines for Water Quality* (Geneva: World Health Organization).

World Meteorological Organization, 2009. Manual of low-flow estimation and prediction, Operational Hydrology Report Number 50, World Meteorological Organization (WMO), Report Number 1029 (Geneva: WMO).

Yang, T., Zhang, Q., Xu, C.Y. and Chen, X., 2008. A spatial assessment of hydrologic alteration caused by dam construction in the middle and lower Yellow River, China, *Hydrological Processes*, **22**(18), 3829–3843.

Yoo, C., 2006. Long term analysis of wet and dry years in Seoul, Korea, *Journal of Hydrology*, **318**(1–4), 24–36.

Zolezzi, G., Bellin, A., Bruno, M. C., Maiolini, B. and Siviglia, A., 2009. Assessing hydrological alterations at multiple temporal scales: Adige River, Italy, *Water Resources Research*, **45**(12), 412–421.

Index

Printed in the United States
by Baker & Taylor Publisher Services